MILITARY AIRCRAFT MARKINGS 2015

MILITARY AIRCRAFT MARKINGS 2015

Howard J Curtis

Crécy Publishing Ltd

This 36th edition published by Crécy Publishing Ltd 2015

ISBN 9781857 803693

Printed in the UK by Martins The Printers Ltd

Crecy Publishing Ltd
1a Ringway Trading Est
Shadowmoss Rd
Manchester
M22 5LH
Tel +44 (0)161 499 0024
www.crecy.co.uk

Front cover: The RAF's first Airbus
A400M Atlas C1, ZM400, landing at
RAF Brize Norton, December 2014.
*Photograph Simon Gregory,
AirTeamImages*

CONTENTS

This 36th annual edition of *abc Military Aircraft Markings*, follows the pattern of previous years and lists in alphabetical and numerical order the aircraft that carry a United Kingdom military registration and which are normally based, or might be seen, in the UK. It also includes airworthy and current RAF/RN/Army aircraft that are based permanently or temporarily overseas. The term 'aircraft' used here covers powered, manned aeroplanes, helicopters, airships and gliders as well as target drones. Included are all the current Royal Air Force, Royal Navy, Army Air Corps, Ministry of Defence, QinetiQ - operated, manufacturers' test aircraft and civilian-owned aircraft with military markings or operated for the Ministry of Defence.

Aircraft withdrawn from operational use but which are retained in the UK for ground training purposes or otherwise preserved by the Services and in the numerous museums or collections are listed. The registrations of some incomplete aircraft have been included, such as the cockpit sections of machines displayed by the RAF, aircraft used by airfield fire sections and for service battle damage repair training (BDRT), together with significant parts of aircraft held by preservation groups and societies. Where only part of the aircraft fuselage remains, the abbreviation <ff> for front fuselage/cockpit section or <rf> for rear fuselage, is shown after the type. Many of these aircraft are allocated, and sometimes wear, a secondary identity, such as an RAF 'M' maintenance number; these numbers are listed against those aircraft to which they have been allocated.

A registration 'missing' from a sequence is either because it was never issued as it formed part of a 'black-out block' or because the aircraft has been written off, scrapped, sold abroad or allocated an alternative marking. Aircraft used as targets on MoD ranges to which access is restricted, and UK military aircraft that have been permanently grounded overseas and unlikely to return to Britain have generally been omitted. With the appearance of some military-registered UAVs, drones and small target aircraft at public events, these have now been included if they are likely to be seen.

In the main, the registrations listed are those markings presently displayed on the aircraft. Where an aircraft carries a false registration it is quoted in *italic type*. Very often these registrations are carried by replicas, that are denoted by <R> after the type. The manufacturer and aircraft type are given, together with recent alternative, previous, secondary or civil identity shown in round brackets. Complete records of multiple previous identities are only included where space permits this. The operating unit and its based location, along with any known unit and code markings, in square brackets, are given as accurately as possible. Where aircraft carry special or commemorative markings, a $ indicates this. Unit markings are normally carried boldly on the sides of the fuselage or on the aircraft's tail. In the case of RAF, RN and AAC machines currently in service, they are usually one or two letters or numbers, while the RN also continues to use a well-established system of three-figure codes between 000 and 999 often with a two-letter deck code on the tail, denoting the aircraft's operational base. RN squadrons, units and bases are allocated blocks of numbers from which individual aircraft codes are issued. To help identification of RN bases and landing platforms on ships, a list of tail-letter codes with their appropriate name, helicopter code number, ship pennant number and type of vessel, is included; as is a helicopter code number/ship's tail-letter code grid cross-reference.

Code changes, for example when aircraft move between units and therefore the markings currently painted on a particular aircraft, might not be those shown in this edition because of subsequent events. Aircraft still under manufacture or not yet delivered to the Service, such as Eurofighter Typhoons and Airbus A400Ms are listed under their allocated registration numbers. Likewise there are a number of newly built aircraft for overseas air arms that carry British registrations for their UK test and delivery flights. The airframes which will not appear in the next edition because of sale, accident, etc., have their fates, where known, shown in italic type in the *locations* column.

The Irish Army Air Corps fleet is listed, together with the registrations of other overseas air arms whose aircraft might be seen visiting the UK from time to time. The registration numbers are as usually presented on the individual machine or as they are normally identified. Where possible, the aircraft's base and operating unit have been shown.

USAF, US Army and US Navy aircraft based in the UK and in Western Europe, and types that regularly visit the UK from the USA, are each listed in separate sections by aircraft type. The serial number actually displayed on the aircraft is shown in full, with additional Fiscal Year (FY) or full serial information also provided. Where appropriate, details of the operating wing, squadron allocation and base are added. The USAF is, like the RAF, in a continuing period of change, resulting in the adoption of new unit titles, squadron and equipment changes and the closure of bases. Only details that concern changes effected by January 2015 are shown.

Veteran and vintage aircraft which carry overseas military markings but which are based in the UK or regularly visit from mainland Europe, have been separately listed, showing their principal means of identification. The growing list of aircraft in government or military service, often under contract to private operating companies, which carry civil registrations has again been included at the end of the respective country.

With the use of the Internet now very well established as a rich source of information, the section listing a selection of military aviation 'world wide web' sites, has again been updated this year. Although only a few of these sites provide details of aircraft registrations and markings, they do give interesting insights into air arms and their operating units, aircraft, museums and a broad range of associated topics.

'Military Aircraft Markings' has its own web site, www.militaryaircraftmarkings.co.uk, from which monthly updates and up to date code lists can be downloaded and you can even find 'MAM' on Facebook at www.facebook.com/MilitaryAircraftMarkings.

Information shown is believed to be correct at 31 January 2015.

ACKNOWLEDGEMENTS

The compiler wishes to thank the many people who have taken the trouble to send comments, additions, deletions and other useful information since the publication of the previous edition of Military Aircraft Markings. In particular the following individuals: Phil Adkin, Neil Brant, Ian Carroll, Glyn Coney, Patrick Dirksen, Ben Dunnell, Graham Gaff, Dylan Grant, Matt Hallam, Matt Hampton, Derek Hoddinott, Norman Hibberd, Peter R March, Andy Marden, Tony McCarthy, Mark Pople, Mark Ray, Norman Roberson, Mark Rourke, Mike Screech, Tim Senior, Vernon Simmonds-Dunne, Roger Syratt, Dave Taskis, David Thompson, Ian Thompson and Dave Walker.

The 2015 edition has also relied upon the printed publications and/or associated internet web-sites as follows: Aerodata Quantum+, 'Aeroplane' magazine, Airfields Yahoo! Group, 'Air Forces Monthly' magazine, BAEG Yahoo! Group, Bones' Aviation Page, CAA G-INFO Web Site, Delta Reflex, Fighter Control, 'FlyPast' magazine, Joe Baugher's Home Page, Brian Pickering/'Military Aviation Review', Mil Spotters' Forum, NAMAR Yahoo! Group, 'Pilot' magazine, Planebase, RAF Leeming Yahoo! Group, RAF Shawbury Yahoo! Group, 'Scramble' magazine, Souairport e-mail group, Tom McGhee/UK Serials Resource Centre, USMil Googlegroup, Mick Boulanger/Wolverhampton Aviation Group and 'Wrecks & Relics' .

HJC
January 2015

$	Aircraft in special markings
AAC	Army Air Corps
AACS	Airborne Air Control Squadron
AC	Air Cadets
ACCGS	Air Cadets Central Gliding School
ACCS	Airborne Command & Control Squadron
ACS	Air Control Squadron
ACW	Air Control Wing
AEF	Air Experience Flight
AESS	Air Engineering & Survival School
AEW	Airborne Early Warning
AFB	Air Force Base
AFD	Air Fleet Department
AFRC	Air Force Reserve Command
AG	Airlift Group
AGA	Academia General del Aire (General Air Academy)
AMD-BA	Avions Marcel Dassault-Breguet Aviation
AMG	Aircraft Maintenance Group
AM&SU	Aircraft Maintenance & Storage Unit
AMW	Air Mobility Wing
ANG	Air National Guard
ARS	Air Refueling Squadron
ARW	Air Refueling Wing
AS	Airlift Squadron/Air Squadron
ATC	Air Training Corps
ATCC	Air Traffic Control Centre
AVDEF	Aviation Defence Service
Avn	Aviation
AW	AgustaWestland/Airlift Wing/Armstrong Whitworth Aircraft
AWC	Air Warfare Centre
BAC	British Aircraft Corporation
BAe	British Aerospace
BAPC	British Aviation Preservation Council
BATUS	British Army Training Unit Suffield
BBMF	Battle of Britain Memorial Flight
BDRF	Battle Damage Repair Flight
BDRT	Battle Damage Repair Training
Be	Beech
Bf	Bayerische Flugzeugwerke
BG	Bomber Group
BGA	British Gliding & Soaring Association
bk	black (squadron colours and markings)
bl	blue (squadron colours and markings)
BP	Boulton & Paul/Boulton Paul
br	brown (squadron colours and markings)
BS	Bomber Squadron
B-V	Boeing-Vertol
BW	Bomber Wing
CAARP	Co-operative des Ateliers Air de la Région Parisienne
CAC	Commonwealth Aircraft Corporation
CASA	Construcciones Aeronautics SA
CC	County Council
CCF	Combined Cadet Force/Canadian Car & Foundry Company
CEAM	Centre d'Expérimentation Aériennes Militaires (Military Air Experimental Centre)
CEMAST	Centre of Excellence in Engineering & Manufacturing Advanced Skills Training
CEPA	Centre d'Expérimentation Pratique de l'Aéronautique Navale
CFIA	Centre de Formation Interarmées NH.90
CFS	Central Flying School
CGMF	Central Glider Maintenance Flight
CIEH	Centre d'Instruction des Equipages d'Hélicoptères
CHFMU	Commando Helicopter Force Maintenance Unit
CinC	Commander in Chief
CinCLANT	Commander in Chief Atlantic
CLV	Centrum Leteckeho Vycviku (Air Training Centre)
CMU	Central Maintenance Unit

Co	Company
COMALAT	Commandement de l'Aviation Légère de l'Armée de Terre
Comp	Composite with
CV	Chance-Vought
D-BA	Daimler-Benz Aerospace
D-BD	Dassault-Breguet Dornier
D&G	Dumfries and Galloway
DE&S	Defence Equipment and Support
DEFTS	Defence Elementary Flying Training School
DEODS	Defence Explosives Ordnance Disposal School
Det	Detachment
DFTDC	Defence Fire Training and Development Centre
DGA	Délégation Générale de l'Armement
DH	de Havilland
DHAHC	de Havilland Aircraft Heritage Centre
DHC	de Havilland Canada
DHFS	Defence Helicopter Flying School
DMS	Defence Movements School
DS&TL	Defence Science & Technology Laboratory
DSAE	Defence School of Aeronautical Engineering
DSEME	Defence School of Electro-Mechanical Engineering
DSG	Defence Support Group
DSMarE	Defence School of Marine Engineering
EA	Escadron Aérien (Air Squadron)
EAAT	Escadrille Avions de l'Armée de Terre
EAC	Ecole de l'Aviation de Chasse (Fighter Aviation School)
EALAT	Ecole de l'Aviation Légère de l'Armée de Terre
EAP	European Aircraft Project
EAT	Ecole de l'Aviation de Transport (Transport Aviation School)
EC	Escadre de Chasse (Fighter Wing)
ECG	Electronic Combat Group
ECM	Electronic Counter Measures
ECS	Electronic Countermeasures Squadron
EdC	Escadron de Convoyage
EDCA	Escadron de Détection et de Control Aéroportée (Airborne Detection & Control Sqn)
EE	English Electric/Escadrille Electronique
EEA	Escadron Electronique Aéroporté
EFA	Ecole Franco Allemande
EFTS	Elementary Flying Training School
EH	Escadron d'Helicoptères (Helicopter Flight)
EHADT	Escadrille Helicoptères de l'Armée de Terre
EHI	European Helicopter Industries
EKW	Eidgenössiches Konstruktionswerkstätte
el	Eskadra Lotnicza (Air Sqn)
EL	Escadre de Liaison (Liaison Wing)
elt	Eskadra Lotnictwa Taktycznego (Tactical Air Squadron)
eltr	Eskadra Lotnictwa Transportowego (Air Transport Squadron)
EMA	East Midlands Airport
EMB	Ecoles Militaires de Bourges
EMVO	Elementaire Militaire Vlieg Opleiding (Elementary Flying Training)
EoN	Elliot's of Newbury
EPAA	Ecole de Pilotage Elementaire de l'Armée de l'Air (Air Force Elementary Flying School)
EPNER	Ecole du Personnel Navigant d'Essais et de Reception
ER	Escadre de Reconnaissance (Reconnaissance Wing)
ES	Escadrille de Servitude
ESAM	Ecole Supérieur d'Application du Matériel
Esc	Escuadron (Squadron)
Esk	Eskadrille (Squadron)
Esq	Esquadra (Squadron)
ET	Escadre de Transport (Transport Squadron)
ETED	Escadron de Transformation des Equipages Mirage 2000D
ETO	Escadron de Transition Operationnelle
ETPS	Empire Test Pilots' School
ETR	Escadron de Tranformation Rafale
ETS	Engineering Training School

EVAA	Ecole de Voltige de l'Armée de l'Air (French Air Force Aerobatics School)
FAA	Fleet Air Arm/Federal Aviation Administration
FBS	Flugbereitschaftstaffel (Flight Readiness Squadron)
FBW	Fly-by-wire
FC	Forskokcentralen (Flight Centre)
FE	Further Education
FETC	Fire and Emergency Training Centre
ff	Front fuselage
FG	Fighter Group
FH	Fairchild-Hiller
FI	Falkland Islands
FJWOEU	Fast Jet & Weapons Operational Evaluation Unit
FLO	Forsvarets Logistikk Organisasjon (Defence Logistics Organisation)
FlSt	Flieger Staffel (Flight Squadron)
Flt	Flight
FMA	Fabrica Militar de Aviones
FMV	Forsvarets Materielwerk
FS	Fighter Squadron
FTS	Flying Training School
FTW	Flying Training Wing
Fw	Focke Wulf
FW	Fighter Wing/FlugWerk/Foster Wickner
FY	Fiscal Year
GAF	Government Aircraft Factory
GAFFTC	German Air Force Flight Training Centre
GAL	General Aircraft Ltd
GAM	Groupe Aerien Mixte (Composite Air Group)
GAM/STAT	Groupement Aéromobile/Section Technique de l'Armée de Terre
gd	gold (squadron colours and markings)
GD	General Dynamics
GI	Ground Instruction/Groupement d'Instruction (Instructional Group)
gn	green (squadron colours and markings)
GRD	Gruppe fur Rustunggdienste (Group for Service Preparation)
GRV	Groupe de Ravitaillement en Vol (Air Refuelling Group)
GT	Grupo de Transporte (Transport Wing)
GTT	Grupo de Transporte de Tropos (Troop Carrier Wing)
gy	grey (squadron colours and markings)
HAF	Historic Aircraft Flight
HAPS	High Altitude Pseudo-Satellite
HC	Helicopter Combat Support Squadron
HekoP	Helikopteripataljoona (Helicopter Battalion)
HF	Historic Flying Ltd
HFAZT	Heeresfliegerausbildungszentrum
HFWS	Heeresflieger Waffenschule (Army Air Weapons School)
HK	Helikopterikomppannia (Helicopter Company)
Hkp.Bat	Helikopter Bataljon (Helicopter Battalion)
HMA	Helicopter Maritime Attack
HMS	Her Majesty's Ship
HP	Handley-Page
HQ	Headquarters
HS	Hawker Siddeley
HSG	Hubschraubergeschwader (Helicopter Squadron)
IAF	Israeli Air Force
IAP	International Airport
IOW	Isle Of Wight
IWM	Imperial War Museum
JARTS	Joint Aircraft Recovery & Transportation Sqn
JATE	Joint Air Transport Establishment
JFACTSU	Joint Forward Air Control Training & Standards Unit
JHC	Joint Helicopter Command
KHR	Kampfhubschrauberregiment (Combat Helicopter Regiment)
Kridlo	Wing
lbvr	letka Bitevnich Vrtulníkú (Attack Helicopter Squadron)
Letka	Squadron
LTG	Lufttransportgeschwader (Air Transport Wing)

LTO	Letalska Transportni Oddelek (Air Transport Department)
LTV	Ling-Temco-Vought
LVG	Luftwaffen Versorgungs Geschwader (Air Force Maintenance Wing)/Luft Verkehrs Gesellschaft
LZO	Letecky Zku ební Odbor (Aviation Test Department)
m	multi-coloured (squadron colours and markings)
MAPK	Mira Anachestisis Pantos Kerou (All Weather Interception Sqn)
MBB	Messerschmitt Bolkow-Blohm
MCAS	Marine Corps Air Station
McD	McDonnell Douglas
MDMF	Merlin Depth Maintenance Facility
Med	Medical
MFG	Marine Flieger Geschwader (Naval Air Wing)
MFSU	Merlin Depth Forward Support Unit
MH	Max Holste
MI	Maritime Interdiction
MIB	Military Intelligence Battalion
MiG	Mikoyan-Gurevich
Mod	Modified
MoD	Ministry of Defence
MPSU	Multi-Platform Support Unit
MR	Maritime Reconnaissance
MRH	Multi-role Helikopters
M&RU	Marketing & Recruitment Unit
MS	Morane-Saulnier
MTHR	Mittlerer Transporthubschrauber Regiment (Medium Transport Helicopter Regiment)
MTM	Mira Taktikis Metaforon (Tactical Transport Sqn)
NA	North American
NACDS	Naval Air Command Driving School
NAEW&CF	NATO Airborne Early Warning & Control Force
NAF	Naval Air Facility
NAS	Naval Air Squadron (UK)/Naval Air Station (US)
NATO	North Atlantic Treaty Organisation
NAWC	Naval Air Warfare Center
NAWC-AD	Naval Air Warfare Center Aircraft Division
NBC	Nuclear, Biological and Chemical
NE	North-East
NI	Northern Ireland
NYARC	North Yorks Aircraft Restoration Centre
OEU	Operational Evaluation Unit
OFMC	Old Flying Machine Company
OGMA	Oficinas Gerais de Material Aeronautico
or	orange (squadron colours and markings)
OSAC	Operational Support Airlift Command
OSBL	Oddelek Sholskih Bojni Letal (Training & Combat School)
P2MF	Puma 2 Maintenance Flight
PAT	Priority Air Transport Detachment
PBN	Pilatus Britten-Norman
PDSH	Puma Depth Support Hub
PLM	Pulk Lotnictwa Mysliwskiego (Fighter Regiment)
pr	purple (squadron colours and markings)
r	red (squadron colours and markings)
R	Replica
RAeS	Royal Aeronautical Society
RAF	Royal Aircraft Factory/Royal Air Force
RAFC	Royal Air Force College
RAFM	Royal Air Force Museum
RAFGSA	Royal Air Force Gliding and Soaring Association
RE	Royal Engineers
Regt	Regiment
REME	Royal Electrical & Mechanical Engineers
rf	Rear fuselage
RFA	Royal Fleet Auxiliary
RHC	Régiment d'Helicoptères de Combat (Combat Helicopter Regiment)
RHFS	Régiment d'Helicoptères des Forces Spéciales (Special Forces Helicopter Regiment)

RJAF	Royal Jordanian Air Force
RM	Royal Marines
RMB	Royal Marines Base
RN	Royal Navy
RNAS	Royal Naval Air Station
RNGSA	Royal Navy Gliding and Soaring Association
RPAS	Remotely Piloted Air System
RQS	Rescue Squadron
R-R	Rolls-Royce
RS	Reid & Sigrist/Reconnaissance Squadron
RSV	Reparto Sperimentale Volo (Experimental Flight School)
RTP	Reduction To Produce
RW	Reconnaissance Wing
SA	Scottish Aviation
SAAB	Svenska Aeroplan Aktieboleg
SAL	Scottish Aviation Limited
SAR	Search and Rescue
Saro	Saunders-Roe
SARTU	Search and Rescue Training Unit
SCW	Strategic Communications Wing
SEAE	School of Electrical & Aeronautical Engineering
SEPECAT	Société Européenne de Production de l'avion Ecole de Combat et d'Appui Tactique
SFDO	School of Flight Deck Operations
SHAPE	Supreme Headquarters Allied Powers Europe
si	silver (squadron colours and markings)
SKAMG	Sea King Aircraft Maintenance Group
Skv	Skvadron (Squadron)
SLK	Stíhacie Letecké Kridlo (Fighter Air Wing)
slt	stíhací letka (Fighter Squadron)
SLV	School Licht Vliegwezen (Flying School)
Sm	Smaldeel (Squadron)
SNCAN	Société Nationale de Constructions Aéronautiques du Nord
SOG	Special Operations Group
SOS	Special Operations Squadron
SoTT	School of Technical Training
SOW	Special Operations Wing
SPAD	Société Pour les Appareils Deperdussin
SPP	Strojirny Prvni Petilesky
Sqn	Squadron
Sz.D.REB	'Szentgyörgyi Deszö' Harcászati Repülö Bázis
TA	Territorial Army
TAP	Transporten Avio Polk (Air Transport Regiment)
TAZLS	Technische Ausbildungs Zentrum Luftwaffe Süd (TAubZLwSüd, Technical Training Centre South)
TEF	Tornado Engineering Flight
TFC	The Fighter Collection
TGp	Test Groep
TIARA	Tornado Integrated Avionics Research Aircraft
tl	taktická tetka (Tactical Squadron)
TLG	Taktisches Luftwaffegeschwader (Air Force Tactical Squadron)
TMU	Typhoon Maintenance Unit
TS	Test Squadron
ts	transportnia speciálni letka (special transport squadron)
TsAGI	Tsentral'ny Aerogidrodinamicheski Instut (Central Aero & Hydrodynamics Institute)
TsLw	Technische Schule der Luftwaffe (Luftwaffe Technical School)
TW	Test Wing
UAS	University Air Squadron
UAV	Unmanned Air Vehicle
UK	United Kingdom
US	United States
USAF	United States Air Force
USAFE	United States Air Forces in Europe
USAREUR	US Army Europe
USCGS	US Coast Guard Station
USEUCOM	United States European Command

USMC	United States Marine Corps
USN	United States Navy
NWTSPM	United States Navy Test Pilots School
VAAC	Vectored thrust Advanced Aircraft flight Control
VFW	Vereinigte Flugtechnische Werke
VGS	Volunteer Gliding Squadron
vlt	výcviková letka (training squadron)
VMGR	Marine Aerial Refuelling/Transport Squadron
VMGRT	Marine Aerial Refuelling/Transport Training Squadron
VQ	Fleet Air Reconnaissance Squadron
VR	Fleet Logistic Support Squadron
vrl	vrtulníková letka (helicopter squadron)
VS	Vickers-Supermarine
w	white (squadron colours and markings)
Wg	Wing
WHL	Westland Helicopters Ltd
WLT	Weapons Loading Training
WRS	Weather Reconnaissance Squadron
WS	Westland
WSK	Wytwornia Sprzetu Kominikacyjnego
WTD	Wehrtechnische Dienstelle (Technical Support Unit)
WW2	World War II
y	yellow (squadron colours and markings)
zDL	základna Dopravního Letectva (Air Transport Base)
zL	základna Letectva (Air Force Base)
zSL	základna Speciálního Letectva (Training Air Base)
zTL	základna Taktického Letectva (Tactical Air Base)
zVrL	základna Vrtulníkového Letectva (Helicopter Air Base)

This section is to assist the reader to locate the places in the United Kingdom where operational military aircraft (including helicopters and gliders) are based.

The alphabetical order listing gives each location in relation to its county and to its nearest classified road(s) (*by* means adjoining; *of* means close to), together with its approximate direction and mileage from the centre of a nearby major town or city. Some civil airports are included where active military units are also based, but **excluded** are MoD sites with non-operational aircraft (e.g. *gate guardians*), the bases of privately-owned civil aircraft that wear military markings and museums. For GPS users, Latitude and Longitude are also listed.

User	Base name	County/Region	Location [Lat./long.]	Distance/direction from (town)
Army	Abingdon	Oxfordshire	W by B4017, W of A34 [N51°41'16" W001°18'58"]	5m SSW of Oxford
Army	Aldergrove/Belfast Airport	Co Antrim	W by A26 [N54°39'27" W006°12'578"]	13m W of Belfast
RM	Arbroath	Angus	E of A933 [N56°34'51" W002°36'55"]	2m NW of Arbroath
RAF	Barkston Heath	Lincolnshire	W by B6404, S of A153 [N52°57'46" W000°33'38"]	5m NNE of Grantham
RAF	Benson	Oxfordshire	E by A423 [N51°36'55" W001°05'45"]	1m NE of Wallingford
QinetiQ/ RAF	Boscombe Down	Wiltshire	S by A303, W of A338 [N51°09'12" W001°45'04"]	6m N of Salisbury
RAF	Boulmer	Northumberland	E of B1339 [N55°24'43" W001°35'37"]	4m E of Alnwick
RAF	Brize Norton	Oxfordshire	W of A4095 [N51°45'00" W001°35'01"]	5m SW of Witney
Marshall	Cambridge Airport/ Teversham	Cambridgeshire	S by A1303 [N52°12'18" E000°10'30"]	2m E of Cambridge
RM/RAF	Chivenor	Devon	S of A361 [N51°05'14" W004°09'01"]	4m WNW of Barnstaple
Army	Colerne	Wiltshire	S of A420, E of Fosse Way [N51°26'30" W002°16'46"]	5m NE of Bath
RAF	Coningsby	Lincolnshire	S of A153, W by B1192 [N53°05'35" W000°10'00"]	10m NW of Boston
DCAE	Cosford	Shropshire	W of A41, N of A464 [N52°38'25" W002°18'20"]	9m WNW of Wolverhampton
RAF	Cranwell	Lincolnshire	N by A17, S by B1429 [N53°01'49" W000°29'00"]	5m WNW of Sleaford
RN	Culdrose	Cornwall	E by A3083 [N50°05'08" W005°15'17"]	1m SE of Helston
Army	Dishforth	North Yorkshire	E by A1 [N54°08'14" W001°25'13"]	4m E of Ripon
USAF	Fairford	Gloucestershire	S of A417 [N51°41'01" W001°47'24"]	9m ESE of Cirencester
Vector Aerospace	Fleetlands	Hampshire	E by A32 [N50°50'06" W001°10'07"]	2m SE of Fareham
RAF	Halton	Buckinghamshire	N of A4011, S of B4544 [N51°47'28" W000°44'11"]	4m ESE of Aylesbury
RAF	Henlow	Bedfordshire	E of A600, W of A6001 [N52°01'10" W000°18'06"]	1m SW of Henlow
Army	Hullavington	Wiltshire	W of A429 [N51°31'47" W002°08'14"]	1m N of M4 jn 17
RAF	Kenley	Greater London	W of A22 [N51°18'20" W000°05'37"]	1m W of Warlingham
RAF	Kirknewton	Lothian	E by B7031, N by A70 [N55°52'32" W003°24'04"]	8m SW of Edinburgh
USAF	Lakenheath	Suffolk	W by A1065 [N52°24'33" E000°33'40"]	8m W of Thetford
RAF/Army	Leconfield (Normandy Barracks)	East Riding of Yorkshire	E by A164 [N53°52'39" W000°26'08"]	2m N of Beverley
RAF	Leeming	North Yorkshire	E by A1 [N54°17'33" W001°32'07"]	5m SW of Northallerton
RAF	Leuchars	Fife	E of A919 [N56°22'28" W002°51'50"]	7m SE of Dundee

User	Base name	County/Region	Location [Lat./long.]	Distance/direction from (town)
RAF	Lossiemouth	Grampian	W of B9135, S of B9040 [N57°42'22" W003°20'20"]	4m N of Elgin
RAF	Marham	Norfolk	N by A1122 [N52°38'54" E000°33'02"]	6m W of Swaffham
Army	Middle Wallop	Hampshire	S by A343 [N51°08'35" W001°34'14"]	6m SW of Andover
USAF	Mildenhall	Suffolk	S by A1101 [N52°21'42" E000°29'12"]	9m NNE of Newmarket
RAF	Northolt	Greater London	N by A40 [N51°33'11" W000°25'06"]	3m E of M40 jn 1
RAF	Odiham	Hampshire	E of A32 [N51°14'03" W000°56'34"]	2m S of M3 jn 5
RN	Predannack	Cornwall	W by A3083 [N50°00'07" W005°13'54"]	7m S of Helston
RAF	Scampton	Lincolnshire	W by A15 [N53°18'28" W000°33'04"]	6m N of Lincoln
RAF	Shawbury	Shropshire	W of B5063 [N52°47'53" W002°40'05"]	7m NNE of Shrewsbury
RAF	Syerston	Nottinghamshire	W by A46 [N53°01'22" W000°54'47"]	5m SW of Newark
RAF	Ternhill	Shropshire	SW by A41 [N52°52'24" W002°31'54"]	3m SW of Market Drayton
RAF/Army	Topcliffe	North Yorkshire	E of A167, W of A168 [N54°12'20" W001°22'55"]	3m SW of Thirsk
RAF	Valley	Gwynedd	S of A5 on Anglesey [N53°14'53" W004°32'06"]	5m SE of Holyhead
RAF	Waddington	Lincolnshire	E by A607, W by A15 [N53°09'58" W000°31'26"]	5m S of Lincoln
Army	Wattisham	Suffolk	N of B1078 [N52°07'38" E000°57'21"]	5m SSW of Stowmarket
RAF	Woodvale	Merseyside	W by A565 [N53°34'56" W003°03'24"]	5m SSW of Southport
RN	Yeovilton	Somerset	S by B3151, S of A303 [N51°00'30" W002°38'43"]	5m N of Yeovil

The Committee of Imperial Defence through its Air Committee introduced a standardised system of numbering aircraft in November 1912. The Air Department of the Admiralty was allocated the first batch 1-200 and used these to cover aircraft already in use and those on order. The Army was issued with the next block from 201-800, which included the number 304 which was given to the Cody Biplane now preserved in the Science Museum. By the outbreak of World War I, the Royal Navy was on its second batch of registrations 801-1600 and this system continued with alternating allocations between the Army and Navy until 1916 when number 10000, a Royal Flying Corps BE2C, was reached.

It was decided not to continue with five digit numbers but instead to start again from 1, prefixing RFC aircraft with the letter A and RNAS aircraft with the prefix N. The RFC allocations commenced with A1 an FE2D and before the end of the year had reached A9999 an Armstrong Whitworth FK8. The next group commenced with B1 and continued in logical sequence through the C, D, E and F prefixes. G was used on a limited basis to identify captured German aircraft, while H was the last block of wartime-ordered aircraft. To avoid confusion I was not used, so the new post-war machines were allocated registrations in the J range. A further minor change was made in the numbering system in August 1929 when it was decided to maintain four numerals after the prefix letter, thus omitting numbers 1 to 999. The new K series therefore commenced at K1000, which was allocated to an AW Atlas.

The Naval N prefix was not used in such a logical way. Blocks of numbers were allocated for specific types of aircraft such as seaplanes or flying-boats. By the late 1920s the sequence had largely been used up and a new series using the prefix S was commenced. In 1930 separate naval allocations were stopped and subsequent registrations were issued in the 'military' range which had by this time reached the K series. A further change in the pattern of allocations came in the L range. Commencing with L7272 numbers were issued in blocks with smaller blocks of registrations between not used. These were known as 'black-out blocks'. As M had already been used as a suffix for Maintenance Command instructional airframes it was not used as a prefix. Although N had previously been used for naval aircraft it was used again for registrations allocated from 1937.

With the build-up to World War II, the rate of allocations quickly accelerated and the prefix R was being used when war was declared. The letters O and Q were not allotted, nor was S, which had been used up to S1865 for naval aircraft before integration into the RAF series. By 1940 the registration Z9999 had been reached, as part of a black-out block, with the letters U and Y not used to avoid confusion.

The option to recommence registration allocation at A1000 was not taken up; instead it was decided to use an alphabetical two-letter prefix with three numerals running from 100 to 999. Thus, AA100 was allocated to a Blenheim IV and this two-letter, three-numeral registration system which started in 1940 continues today. The letters C, I, O, Q, U and Y were, with the exception of NC, not used. For various reasons the following letter combinations were not issued: DA, DB, DH, EA, GA to GZ, HA, HT, JE, JH, JJ, KR to KT, MR, NW, NZ, SA to SK, SV, TN, TR and VE. The first post-war registrations issued were in the VP range while the end of the WZs had been reached by the Korean War.

In January 1952 a civil servant at the then Air Ministry penned a memo to his superiors alerting them to the fact that a new military aircraft registration system would soon have to be devised. With allocations accelerating to accommodate a NATO response to the Korean War and a perceived Soviet threat building, he estimated that the end of the ZZs would quickly be reached. However, more than five decades later the allocations are only in the ZKs and at the present rate are unlikely to reach ZZ999 until the end of this century!

Military aircraft registrations are allocated by MoD Head Office and Corporate Services, where the Military Aircraft Register is maintained by the Military Aviation Authority. It should be pointed out that strictly the register places a space before the last three digits of the registration and, contrary to popular opinion, refers to them as 'registrations', not serials.

A change in policy in 2003 resulted in the use of the first 99 digits in the ZK sequence (ZK001 to ZK099), following on from ZJ999. The first of these, ZK001 to ZK004, were allocated to AgustaWestland Merlins. There is also a growing trend for 'out-of-sequence' registration numbers to be issued. At first this was to a manufacturer's prototype or development aircraft. However, following the Boeing C-17 Globemasters leased and subsequently purchased from Boeing (ZZ171-ZZ178), more allocations have been noted.

Since 2002 there has also been a new official policy concerning the use of military registration numbers on some types of UAV. 'Where a UAV is of modular construction the nationality and registration mark shall be applied to the fuselage of the vehicle or on the assembly forming the main part of the fuselage. To prevent the high usage of numbers for target drones which are eventually destroyed, a single registration mark (prefix) should be issued relating to the UAV type. The agency or service operating the target drone will be responsible for the identification of each individual UAV covered by that registration mark by adding a suffix.' This has resulted in the use of the same registration on a number of UAVs (or RPASs as they are now referred to) and drones with numbers following it - hence the appearance of ZZ420/001 etc. on Banshee drones. Aircraft using this system are denoted in the text by an asterisk (*).

Note: The compiler will be pleased to receive comments, corrections and further information for inclusion in subsequent editions of *Military Aircraft Markings* and the monthly up-date of additions and amendments. Please send your information to Military Aircraft Markings, Crécy Publishing Ltd, 1a Ringway Trading Estate, Shadowmoss Road, Manchester M22 5LH or by e-mail to admin@aviation-links.co.uk.

A serial in *italics* denotes that it is not the genuine marking for that airframe.

Serial	Type (code/other identity)	Owner/operator, location or fate	Notes
46	VS361 Spitfire LF IX <R> (*MH486*/BAPC 206) [FT-E]	RAF Museum, Hendon	
168	Sopwith Tabloid Scout <R> (G-BFDE)	RAF Museum, Hendon	
304	Cody Biplane (BAPC 62)	Science Museum, South Kensington	
347	RAF BE2c <R> (G-AWYI)	*Repainted as 471, July 2014*	
471	RAF BE2c <R> (G-AWYI)	Privately owned, Sywell	
687	RAF BE2b <R> (BAPC 181)	RAF Museum, Hendon	
2345	Vickers FB5 Gunbus <R> (G-ATVP)	RAF Museum, Hendon	
2699	RAF BE2c	Imperial War Museum, Duxford	
2783	RAF BE2b <R>	Boscombe Down Aviation Collection, Old Sarum	
3066	Caudron GIII (G-AETA/9203M)	RAF Museum, Hendon	
5964	DH2 <R> (BAPC 112)	Privately owned, Stretton on Dunsmore	
5964	DH2 <R> (G-BFVH)	Privately owned, Wickenby	
6232	RAF BE2c <R> (BAPC 41)	Yorkshire Air Museum, stored Elvington	
8359	Short 184 <ff>	FAA Museum, RNAS Yeovilton	
9917	Sopwith Pup (G-EBKY/N5180)	The Shuttleworth Collection, Old Warden	
A301	Morane BB (frame)	RAF Museum Reserve Collection, Stafford	
A653	Sopwith Pup <R> (*A7317*/BAPC 179)	Privately owned, Stow Maries, Essex	
A1452	Vickers FB5 Gunbus <R>	Spitfire Spares, Taunton, Somerset	
A1742	Bristol Scout D <R> (BAPC 38)	Privately owned, Old Warden	
A2767	RAF BE2e-1 <R> (ZK-KOZ)	WW1 Aviation Heritage Trust, Bicester	
A2943	RAF BE2e-1 <R> (ZK-TFZ)	WW1 Aviation Heritage Trust, Bicester	
A3930	RAF RE8 <R> (ZK-TVC)	RAF Museum, Hendon	
A6526	RAF FE2b <R>	RAF Museum, Hendon	
A7288	Bristol F2b Fighter <R>	Bristol Aero Collection, Filton	
A8226	Sopwith 1½ Strutter <R> (G-BIDW)	RAF Museum, Cosford	
B595	RAF SE5a <R> (G-BUOD) [W]	Privately owned, Defford	
B619	Sopwith 1½ Strutter <R>	RAF Manston History Museum	
B2458	Sopwith 1F.1 Camel <R> (G-BPOB/*F542*) [R]	Privately owned, Booker	
B5539	Sopwith 1F.1 Camel <R>	Privately owned, Booker	
B5577	Sopwith 1F.1 Camel <R> (*D3419*/BAPC 59) [W]	Montrose Air Station Heritage Centre	
B6401	Sopwith 1F.1 Camel <R> (G-AWYY/C1701)	FAA Museum, stored Cobham Hall, RNAS Yeovilton	
B7270	Sopwith 1F.1 Camel <R> (G-BFCZ)	Brooklands Museum, Weybridge	
C1096	Replica Plans SE5a <R> (G-ERFC) [W]	Privately owned, Egginton	
C1904	RAF SE5a <R> (G-PFAP) [Z]	Privately owned, Castle Bytham, Lincs	
C3009	Currie Wot (G-BFWD) [B]	Privately owned, Dunkeswell	
C3011	Phoenix Currie Super Wot (G-SWOT) [S]	Privately owned, Otherton, Staffs	
C3988	Sopwith 5F.1 Dolphin (comp D5329)	RAF Museum, Hendon	
C4451	Avro 504J <R> (BAPC 210)	Solent Sky, Southampton	
C4918	Bristol M1C <R> (G-BWJM)	The Shuttleworth Collection, Old Warden	
C4994	Bristol M1C <R> (G-BLWM)	RAF Museum, Cosford	
C5430	RAF SE5a <R> (G-CCXG) [V]	Privately owned, Wrexham	
C6468	RAF SE5a <R> (G-CEKL) [A]	Privately owned, RAF Halton	
C8996	RAF SE5a (G-ECAE/A2-25)	Privately owned, Milden	
C9533	RAF SE5a <R> (G-BUWE) [M]	Privately owned, Boscombe Down	
D276	RAF SE5a <R> (BAPC 208) [A]	Prince's Mead Shopping Centre, Farnborough	
D5649	Airco DH9	Imperial War Museum, Duxford	
D7560	Avro 504K	Science Museum, South Kensington	
D8096	Bristol F2b Fighter (G-AEPH) [D]	The Shuttleworth Collection, Old Warden	
E449	Avro 504K (G-EBJE/9205M)	RAF Museum, Hendon	
E2466	Bristol F2b Fighter (BAPC 165) [I]	RAF Museum, Hendon	
E2581	Bristol F2b Fighter [13]	Imperial War Museum, Duxford	
E2977	Avro 504K (G-EBHB)	Privately owned, Leighton Buzzard	

Notes	Serial	Type (code/other identity)	Owner/operator, location or fate
	E3273	Avro 504K (H5199/BK892/ G-ACNB/G-ADEV/3118M)	The Shuttleworth Collection, Old Warden
	E6655	Sopwith 7F.1 Snipe <R> [B]	RAF Museum, Hendon
	E8894	Airco DH9 (G-CDLI)	Aero Vintage, Westfield, Sussex
	F141	RAF SE5a <R> (G-SEVA) [G]	Privately owned, Boscombe Down
	F235	RAF SE5a <R> (G-BMDB) [B]	Privately owned, Stow Maries, Essex
	F904	RAF SE5a (G-EBIA)	The Shuttleworth Collection, Old Warden
	F904	RAF SE5a <R>	South Yorkshire Aircraft Museum, Doncaster
	F938	RAF SE5a (G-EBIC/9208M)	RAF Museum, Hendon
	F943	RAF SE5a <R> (G-BIHF) [S]	Privately owned, White Waltham
	F943	RAF SE5a <R> (G-BKDT)	Yorkshire Air Museum, Elvington
	F1010	Airco DH9A [C]	RAF Museum, Hendon
	F3556	RAF RE8	Imperial War Museum, Duxford
	F5447	RAF SE5a <R> (G-BKER) [N]	Privately owned, Bridge of Weir
	F5459	RAF SE5a <R> (G-INNY) [Y]	Privately owned, Temple Bruer, Lincs
	F5475	RAF SE5a <R> (BAPC 250)	Brooklands Museum, Weybridge
	F6314	Sopwith 1F.1 Camel (9206M) [B]	RAF Museum, Hendon
	F8010	RAF SE5a <R> (G-BDWJ) [Z]	Privately owned, Langport, Somerset
	F8614	Vickers FB27A Vimy IV <R> (G-AWAU)	RAF Museum, Hendon
	H1968	Avro 504K <R> (BAPC 42)	Yorkshire Air Museum, stored Elvington
	J7326	DH53 Humming Bird (G-EBQP)	DHAHC, London Colney
	J7904	Gloster Gamecock <R>	Jet Age Museum, Gloucester
	J8067	Westland Pterodactyl 1a	Science Museum, South Kensington
	J9941	Hawker Hart 2 (G-ABMR)	RAF Museum, Hendon
	K1786	Hawker Tomtit (G-AFTA)	The Shuttleworth Collection, Old Warden
	K2046	Isaacs Fury II (G-AYJY)	Privately owned, Little Rissington
	K2048	Isaacs Fury II (G-BZNW)	Privately owned, Linton-on-Ouse
	K2050	Isaacs Fury II (G-ASCM)	Privately owned, Enstone
	K2059	Isaacs Fury II (G-PFAR)	Privately owned, Netherthorpe
	K2060	Isaacs Fury II (G-BKZM)	Privately owned stored, Limetree, Ireland
	K2075	Isaacs Fury II (G-BEER)	Privately owned, Combrook, Warks
	K2227	Bristol 105 Bulldog IIA (G-ABBB)	RAF Museum, Hendon
	K2567	DH82A Tiger Moth (DE306/G-MOTH/7035M)	Privately owned, Tadlow
	K2572	DH82A Tiger Moth (NM129/G-AOZH)	Privately owned, Wanborough, Wilts
	K2585	DH82A Tiger Moth II (T6818/G-ANKT)	The Shuttleworth Collection, Old Warden
	K2587	DH82A Tiger Moth <R> (G-BJAP)	Privately owned, Shobdon
	K3241	Avro 621 Tutor (K3215/G-AHSA)	The Shuttleworth Collection, Old Warden
	K3661	Hawker Nimrod II (G-BURZ) [562]	Aero Vintage, Duxford
	K3731	Isaacs Fury <R> (G-RODI)	Privately owned, Hailsham
	K4232	Avro 671 Rota I (SE-AZB)	RAF Museum, Hendon
	K4259	DH82A Tiger Moth (G-ANMO) [71]	Privately owned, Sywell
	K4972	Hawker Hart Trainer IIA (1764M)	RAF Museum, Hendon
	K5054	Supermarine Spitfire <R> (EN398/BAPC 190)	Privately owned, Hawkinge
	K5054	Supermarine Spitfire <R> (BAPC 214)	Tangmere Military Aviation Museum
	K5054	Supermarine Spitfire <R> (G-BRDV)	Privately owned, Kent
	K5054	Supermarine Spitfire <R> (BAPC 297)	Kent Battle of Britain Museum, Hawkinge
	K5054	Supermarine Spitfire <R>	Southampton Airport, on display
	K5409	Hawker Hind	Currently not known
	K5414	Hawker Hind (G-AENP/BAPC 78) [XV]	The Shuttleworth Collection, Old Warden
	K5462	Hawker Hind	Currently not known
	K5554	Hawker Hind	Currently not known
	K5600	Hawker Audax I (G-BVVI/2015M)	Aero Vintage, Westfield, Sussex
	K5673	Isaacs Fury II (G-BZAS)	Privately owned, Morpeth
	K5673	Hawker Fury I <R> (BAPC 249)	Brooklands Museum, Weybridge
	K5674	Hawker Fury I (G-CBZP)	Historic Aircraft Collection Ltd, Goodwood
	K5682	Hawker Nimrod I <R> (S1579/G-BBVO) [6]	Privately owned, Felthorpe
	K6035	Westland Wallace II (2361M)	RAF Museum, Hendon
	K6618	Hawker Hind	Currently not known

Serial	Type (code/other identity)	Owner/operator, location or fate	Notes
K6833	Hawker Hind	*Currently not known*	
K7271	Hawker Fury II <R> (BAPC 148)	Shropshire Wartime Aircraft Recovery Grp Mus, Sleap	
K7271	Isaacs Fury II <R> (G-CCKV)	Privately owned, Temple Bruer, Lincs	
K7985	Gloster Gladiator I (L8032/G-AMRK)	The Shuttleworth Collection, Old Warden	
K8042	Gloster Gladiator II (8372M)	RAF Museum, Hendon	
K8203	Hawker Demon I (G-BTVE/2292M)	Demon Displays, Old Warden	
K8303	Isaacs Fury II (G-BWWN) [D]	Privately owned, RAF Henlow	
K9926	VS300 Spitfire I <R> (BAPC 217) [JH-C]	RAF Bentley Priory, on display	
K9942	VS300 Spitfire I (8383M) [SD-D]	RAF Museum, Cosford	
K9998	VS300 Spitfire I <R> [QJ-K]	RAF Biggin Hill, on display	
L1018	VS300 Spitfire I <R> [LO-S]	Privately owned, Edinburgh	
L1067	VS300 Spitfire I <R> (BAPC 227) [XT-D]	Edinburgh Airport, on display	
L1592	Hawker Hurricane I [KW-Z]	Science Museum, South Kensington	
L1639	Hawker Hurricane I	Cambridge Fighter & Bomber Society, Little Gransden	
L1679	Hawker Hurricane I <R> (BAPC 241) [JX-G]	Tangmere Military Aviation Museum	
L1684	Hawker Hurricane I <R> (BAPC 219)	RAF Northolt, on display	
L2301	VS Walrus I (G-AIZG)	FAA Museum, RNAS Yeovilton	
L2940	Blackburn Skua I	FAA Museum, RNAS Yeovilton	
L5343	Fairey Battle I	RAF Museum, Hendon	
L6739	Bristol 149 Bolingbroke IVT (G-BPIV/*R3821*) [YP-Q]	Blenheim(Duxford) Ltd, Duxford	
L6906	Miles M14A Magister I (G-AKKY/T9841/BAPC 44)	Museum of Berkshire Aviation, Woodley	
L7005	Boulton Paul P82 Defiant I <R> [PS-B]	Kent Battle of Britain Museum, Hawkinge	
L7181	Hawker Hind (G-CBLK)	Aero Vintage, Duxford	
L7191	Hawker Hind	*Currently not known*	
L7775	Vickers Wellington B IC <ff>	Lincolnshire Avn Heritage Centre, E Kirkby	
L8756	Bristol 149 Bolingbroke IVT (RCAF 10001) [XD-E]	RAF Museum, Hendon	
N248	Supermarine S6A (*S1596*)	Solent Sky, Southampton	
N500	Sopwith LC-1T Triplane <R> (G-PENY/G-BWRA)	Privately owned, Yarcombe, Devon/RNAS Yeovilton	
N540	Port Victoria PV8 Eastchurch Kitten <R>	Yorkshire Air Museum, Elvington	
N546	Wright Quadruplane 1 <R> (BAPC 164)	Solent Sky, Southampton	
N1671	Boulton Paul P82 Defiant I (8370M) [EW-D]	RAF Museum, Hendon	
N1854	Fairey Fulmar II (G-AIBE)	FAA Museum, RNAS Yeovilton	
N2078	Sopwith Baby (8214/8215)	FAA Museum, stored Cobham Hall, RNAS Yeovilton	
N2532	Hawker Hurricane I <R> (BAPC 272) [GZ-H]	Kent Battle of Britain Museum, Hawkinge	
N2980	Vickers Wellington IA [R]	Brooklands Museum, Weybridge	
N3200	VS300 Spitfire IA (G-CFGJ) [QV]	The Aircraft Restoration Company, Duxford	
N3289	VS300 Spitfire I <R> (BAPC 65) [DW-K]	Kent Battle of Britain Museum, Hawkinge	
N3290	VS300 Spitfire I <R> [AI-H]	Privately owned, St Mawgan	
N3310	VS361 Spitfire IX [A] <R>	Privately owned, Abingdon	
N3313	VS300 Spitfire I <R> (*MH314*/BAPC 69) [KL-B]	Kent Battle of Britain Museum, Hawkinge	
N3378	Boulton Paul P82 Defiant I (wreck)	RAF Museum, Cosford	
N3549	DH82A Tiger moth II (PG645/N3549)	Privately owned, Netherthorpe	
N3788	Miles M14A Magister I (V1075/G-AKPF)	Privately owned, Old Warden	
N4389	Fairey Albacore (N4172) [4M]	FAA Museum, stored Cobham Hall, RNAS Yeovilton	
N4877	Avro 652A Anson I (G-AMDA) [MK-V]	Imperial War Museum, Duxford	
N5137	DH82A Tiger Moth (N6638/G-BNDW)	Caernarfon Air World	
N5177	Sopwith 1½ Strutter <R>	Privately owned, Sedgensworth, Hants	
N5182	Sopwith Pup <R> (G-APUP/9213M)	RAF Museum, Cosford	
N5195	Sopwith Pup (G-ABOX)	Museum of Army Flying, Middle Wallop	
N5199	Sopwith Pup <R> (G-BZND)	Privately owned, Yarcombe, Devon	
N5459	Sopwith Triplane <R> (BAPC 111)	FAA Museum, stored Cobham Hall, RNAS Yeovilton	
N5518	Gloster Sea Gladiator	FAA Museum, RNAS Yeovilton	
N5628	Gloster Gladiator II	RAF Museum, Hendon	
N5719	Gloster Gladiator II (G-CBHO)	Privately owned, Dursley, Glos	
N5903	Gloster Gladiator II (*N2276*/G-GLAD)	The Fighter Collection, Duxford	
N5912	Sopwith Triplane (8385M)	RAF Museum, Hendon	
N5914	Gloster Gladiator II	Jet Age Museum, Gloucester	
N6290	Sopwith Triplane <R> (G-BOCK)	The Shuttleworth Collection, Old Warden (on repair)	
N6452	Sopwith Pup <R> (G-BIAU)	FAA Museum, RNAS Yeovilton	
N6466	DH82A Tiger Moth (G-ANKZ)	Privately owned, Compton Abbas	

Notes	Serial	Type (code/other identity)	Owner/operator, location or fate
	N6473	DH82A Tiger Moth (F-GTBO)	Privately owned, Orbigny, France
	N6537	DH82A Tiger Moth (G-AOHY)	Privately owned, Wickenby
	N6635	DH82A Tiger Moth (comp G-APAO & G-APAP) [25]	Imperial War Museum, Duxford
	N6720	DH82A Tiger Moth (G-BYTN/7014M) [VX]	Privately owned, Wickenby
	N6797	DH82A Tiger Moth (G-ANEH)	Privately owned, Swyncombe
	N6812	Sopwith 2F.1 Camel	Imperial War Museum, Duxford
	N6847	DH82A Tiger Moth (G-APAL)	Privately owned, Braceborough, Lincs
	N6965	DH82A Tiger Moth (G-AJTW) [FL-J]	Privately owned, Tibenham
	N7033	Noorduyn AT-16 Harvard IIB (FX442)	Kent Battle of Britain Museum, Hawkinge
	N9191	DH82A Tiger Moth (G-ALND)	Privately owned, Pontypool
	N9192	DH82A Tiger Moth (G-DHZF) [RCO-N]	Privately owned, Sywell
	N9328	DH82A Tiger Moth (G-ALWS) [69]	Privately owned, Henstridge
	N9389	DH82A Tiger Moth (G-ANJA)	Privately owned, Thruxton
	N9899	Supermarine Southampton I (fuselage)	RAF Museum, Hendon
	P1344	HP52 Hampden I (9175M) [PL-K]	Michael Beetham Conservation Centre, RAFM Cosford
	P2617	Hawker Hurricane I (8373M) [AF-F]	RAF Museum, Hendon
	P2725	Hawker Hurricane I (wreck)	Imperial War Museum, Lambeth
	P2725	Hawker Hurricane I <R> (BAPC 68) [TM-B]	Privately owned, Delabole, Cornwall
	P2793	Hawker Hurricane I <R> (BAPC 236) [SD-M]	Eden Camp Theme Park, Malton, North Yorkshire
	P2902	Hawker Hurricane I (G-ROBT) [DX-R]	Privately owned, Milden
	P2921	Hawker Hurricane I <R> (BAPC 273) [GZ-L]	Kent Battle of Britain Museum, Hawkinge
	P2921	Hawker Hurricane I <R> [GZ-L]	RAF Biggin Hill, on display
	P2921	Hawker Sea Hurricane X (AE977/G-CHTK) [GZ-L]	Privately owned, Biggin Hill
	P2954	Hawker Hurricane I <R> (BAPC 267) [WX-E]	Imperial War Museum, Duxford
	P2970	Hawker Hurricane I <R> (BAPC 291) [US-X]	Battle of Britain Memorial, Capel le Ferne, Kent
	P3059	Hawker Hurricane I <R> (BAPC 64) [SD-N]	Kent Battle of Britain Museum, Hawkinge
	P3175	Hawker Hurricane I (wreck)	RAF Museum, Hendon
	P3179	Hawker Hurricane I <ff>	Tangmere Military Aviation Museum
	P3208	Hawker Hurricane I <R> (BAPC 63/L1592) [SD-T]	Kent Battle of Britain Museum, Hawkinge
	P3351	Hawker Hurricane IIA (DR393/F-AZXR) [K]	Privately owned, Cannes, France
	P3386	Hawker Hurricane I <R> (BAPC 218) [FT-A]	RAF Bentley Priory, on display
	P3395	Hawker Hurricane IV (KX829)[JX-B]	Thinktank, Birmingham
	P3398	Supermarine Aircraft Spitfire 26 (G-CEPL)	Privately owned, Thurrock
	P3554	Hawker Hurricane I (composite)	The Air Defence Collection, Salisbury
	P3679	Hawker Hurricane I <R> (BAPC 278) [GZ-K]	Kent Battle of Britain Museum, Hawkinge
	P3708	Hawker Hurricane I	Norfolk & Suffolk Avn Museum, Flixton (on rebuild)
	P3717	Hawker Hurricane I (composite) (DR348/G-HITT) [SW-P]	Privately owned, Milden
	P3873	Hawker Hurricane I <R> (BAPC 265) [YO-H]	Yorkshire Air Museum, Elvington
	P3886	Hawker Sea Hurricane X (AE977/G-CHTK) [UF-K]	Repainted as P2921, October 2014
	P3901	Hawker Hurricane I <R> [RF-E]	Battle of Britain Bunker, Uxbridge
	P4139	Fairey Swordfish II (HS618) [5H]	FAA Museum, RNAS Yeovilton
	P6382	Miles M14A Hawk Trainer 3 (G-AJRS) [C]	The Shuttleworth Collection, Old Warden
	P7308	VS300 Spitfire IA (AR213/G-AIST) [XR-D]	Privately owned, Duxford
	P7350	VS329 Spitfire IIA (G-AWIJ) [EB-G]	RAF BBMF, Coningsby
	P7370	VS329 Spitfire II <R> [ZP-A]	Battle of Britain Experience, Canterbury
	P7540	VS329 Spitfire IIA [DU-W]	Dumfries & Galloway Avn Mus, Dumfries
	P7666	VS329 Spitfire II <R> [EB-Z]	RAF High Wycombe, on display
	P7819	VS329 Spitfire IIA (G-TCHZ)	Privately owned, Exeter
	P7966	VS329 Spitfire II <R> [D-B]	Manx Aviation & Military Museum, Ronaldsway
	P7895	VS329 Spitfire IIA <R> [RN-N]	Ulster Aviation Society, Long Kesh
	P8088	VS329 Spitfire IIA (G-CGRM) [NK-K]	Privately owned, Oxon
	P8140	VS329 Spitfire II <R> (P9390/BAPC 71) [ZF-K]	Norfolk & Suffolk Avn Museum, Flixton
	P8208	VS329 Spitfire IIB (G-RRFF)	Privately owned, Oxon
	P8448	VS329 Spitfire II <R> (BAPC 225) [UM-D]	RAF Cranwell, on display
	P9373	VS300 Spitfire IA (G-CFGN) (wreck)	Privately owned, Duxford
	P9374	VS300 Spitfire IA (G-MKIA) [J]	Privately owned, Duxford
	P9398	Supermarine Aircraft Spitfire 26 (G-CEPL) [KL-B]	Privately owned, Audley End
	P9444	VS300 Spitfire IA [RN-D]	Science Museum, South Kensington
	P9637	Supermarine Aircraft Spitfire 26 (G-RORB) [GR-B]	Privately owned, Perth

Serial	Type (code/other identity)	Owner/operator, location or fate	Notes
R1914	Miles M14A Magister (G-AHUJ)	Privately owned, Cheltenham	
R4118	Hawker Hurricane I (G-HUPW) [UP-W]	Privately owned, Didcot, Oxon	
R4229	Hawker Hurricane I <R> [GN-J]	Alexandra Park, Windsor	
R4922	DH82A Tiger Moth II (G-APAO)	Privately owned, Henlow	
R4959	DH82A Tiger Moth II (G-ARAZ) [59]	Privately owned, Temple Bruer, Lincs	
R5136	DH82A Tiger Moth II (G-APAP)	Privately owned, Henlow	
R5172	DH82A Tiger Moth II (G-AOIS) [FIJE]	Privately owned, Sywell	
R5246	DH82A Tiger Moth II (G-AMIV) [40]	Privately owned, Germany	
R5868	Avro 683 Lancaster I (7325M) [PO-S]	RAF Museum, Hendon	
R6690	VS300 Spitfire I <R> (BAPC 254) [PR-A]	Yorkshire Air Museum, Elvington	
R6775	VS300 Spitfire I <R> (BAPC 299) [YT-J]	Battle of Britain Memorial, Capel le Ferne, Kent	
R6904	VS300 Spitfire I <R> [BT-K]	Privately owned, Cornwall	
R6915	VS300 Spitfire I	Imperial War Museum, Lambeth	
R9125	Westland Lysander III (8377M)[LX-L]	RAF Museum, Hendon	
S1287	Fairey Flycatcher <R> (G-BEYB)	FAA Museum, RNAS Yeovilton	
S1581	Hawker Nimrod I (G-BWWK) [573]	The Fighter Collection, Duxford	
S1595	Supermarine S6B [1]	Science Museum, South Kensington	
S1615	Isaacs Fury II (G-BMEU)	Privately owned, stored Netherthorpe	
T5298	Bristol 156 Beaufighter I (4552M) <ff>	Midland Air Museum, Coventry	
T5424	DH82A Tiger Moth II (G-AJOA)	Privately owned, Swindon	
T5854	DH82A Tiger Moth II (G-ANKK)	Privately owned, Baxterley	
T5879	DH82A Tiger Moth II (G-AXBW) [RUC-W]	Privately owned, Frensham	
T6296	DH82A Tiger Moth II (8387M)	RAF Museum, Hendon	
T6562	DH82A Tiger Moth II (G-ANTE)	Privately owned, Sywell	
T6953	DH82A Tiger Moth II (G-ANNI)	Privately owned, Tisted, Hants	
T6991	DH82A Tiger Moth II (DE694/HB-UPY)	Privately owned, Lausanne, Switzerland	
T7109	DH82A Tiger Moth II (G-AOIM)	Privately owned, Turweston	
T7230	DH82A Tiger Moth II (G-AFVE)	Privately owned, Mazowiecke, Poland	
T7281	DH82A Tiger Moth II (G-ARTL)	Privately owned, Egton, nr Whitby	
T7290	DH82A Tiger Moth II (G-ANNK) [14]	Privately owned, Sywell	
T7793	DH82A Tiger Moth II (G-ANKV)	Privately owned, Wickenby	
T7842	DH82A Tiger Moth II (G-AMTF)	Privately owned, Westfield, Surrey	
T7909	DH82A Tiger Moth II (G-ANON)	Privately owned, Sherburn-in-Elmet	
T7997	DH82A Tiger Moth II (NL750/G-AHUF)	Privately owned, Barton	
T8191	DH82A Tiger Moth II (G-BWMK)	Privately owned, stored Sleap	
T9707	Miles M14A Magister I (G-AKKR/8378M/T9708)	Museum of Army Flying, Middle Wallop	
T9738	Miles M14A Magister I (G-AKAT)	Privately owned, Breighton	
V3388	Airspeed AS10 Oxford I (G-AHTW)	Imperial War Museum, Duxford	
V6028	Bristol 149 Bolingbroke IVT (G-MKIV) [GB-D] <rf>	The Aircraft Restoration Co, stored Duxford	
V6555	Hawker Hurricane I (P3144) <R> [DT-A]	Battle of Britain Experience, Canterbury	
V6799	Hawker Hurricane I <R> (BAPC 72/V7767) [SD-X]	Jet Age Museum, Gloucester	
V7313	Hawker Hurricane I <R> [US-F]	Privately owned, North Weald, on display	
V7350	Hawker Hurricane I (fuselage)	Romney Marsh Wartime Collection	
V7467	Hawker Hurricane I <R> [LE-D]	RAF High Wycombe, on display	
V7467	Hawker Hurricane I <R> (BAPC 223) [LE-D]	Gateguards UK, Newquay	
V7467	Hawker Hurricane I <R> (BAPC 288) [LE-D]	Wonderland Pleasure Park, Farnsfield, Notts	
V7497	Hawker Hurricane I (G-HRLI)	Hawker Restorations, Milden	
V9367	Westland Lysander IIIA (G-AZWT) [MA-B]	The Shuttleworth Collection, Old Warden	
V9673	Westland Lysander IIIA (V9300/G-LIZY) [MA-J]	Imperial War Museum, Duxford	
V9723	Westland Lysander IIIA (V9546/OO-SOT) [MA-D]	SABENA Old Timers, Brussels, Belgium	
V9312	Westland Lysander IIIA (G-CCOM)	The Aircraft Restoration Co, Duxford	
W1048	HP59 Halifax II (8465M) [TL-S]	RAF Museum, Hendon	
W2068	Avro 652A Anson I (9261M/VH-ASM) [68]	RAF Museum, Hendon	
W2718	VS Walrus I (G-RNLI)	Privately owned, Audley End	
W3257	Supermarine Aircraft Spitfire 26 (G-CENI) [FY-E]	Privately owned, Sibson	
W3632	VS509 Spitfire T9 (PV202/H-98/G-CCCA)	Historic Flying Ltd, Duxford	
W3644	VS349 Spitfire V <R> [QV-J]	Privately owned, Lake Fairhaven, Lancs	

Notes	Serial	Type (code/other identity)	Owner/operator, location or fate
	W3850	VS349 Spitfire V <R> [PR-A]	Privately owned, Cheshire
	W4041	Gloster E28/39	Science Museum, South Kensington
	W4041	Gloster E28/39 <R>	Jet Age Museum, Gloucester
	W4050	DH98 Mosquito	Mosquito Aircraft Museum, London Colney
	W5856	Fairey Swordfish II (G-BMGC) [4A]	RN Historic Flight, Yeovilton
	W9385	DH87B Hornet Moth (G-ADND) [YG-L,3]	Privately owned, Hullavington
	X4178	VS300 Spitfire I <R> [EB-K]	Imperial War Museum, Duxford
	X4276	VS300 Spitfire I (G-CDGU)	Privately owned, Sandown
	X4590	VS300 Spitfire I (8384M) [PR-F]	RAF Museum, Hendon
	X4650	VS300 Spitfire I (G-CGUK) [KL-A]	Privately owned, Biggin Hill
	X4683	Jurca MJ10 Spitfire (G-MUTS) [EB-N]	Privately owned, Fishburn
	X7688	Bristol 156 Beaufighter I (3858M/G-DINT)	Privately owned, Hatch
	Z1206	Vickers Wellington IV (fuselage)	Privately owned, Kenilworth
	Z2033	Fairey Firefly I (G-ASTL) [275/N]	FAA Museum, RNAS Yeovilton
	Z2315	Hawker Hurricane IIA [JU-E]	Imperial War Museum, Duxford
	Z2389	Hawker Hurricane IIA [XR-T]	Brooklands Museum, Weybridge
	Z3427	Hawker Hurricane IIC <R> (BAPC 205) [AV-R]	RAF Museum, Hendon
	Z5140	Hawker Hurricane XIIA (Z7381/G-HURI) [HA-C]	Historic Aircraft Collection, Duxford
	Z5207	Hawker Hurricane IIB (G-BYDL)	Privately owned, Thruxton
	Z5252	Hawker Hurricane IIB (G-BWHA/Z5053) [GO-B]	Privately owned, Milden
	Z7015	Hawker Sea Hurricane IB (G-BKTH) [7-L]	The Shuttleworth Collection, Old Warden
	Z7197	Percival P30 Proctor III (G-AKZN/8380M)	RAF Museum Reserve Collection, Stafford
	Z7258	DH89A Dragon Rapide (NR786/G-AHGD)	Privately owned, Membury (wreck)
	AB196	Supermarine Aircraft Spitfire 26 (G-CCGH)	Privately owned, Enstone
	AB550	VS349 Spitfire VB <R> (BAPC 230/AA908) [GE-P]	Eden Camp Theme Park, Malton, North Yorkshire
	AB910	VS349 Spitfire VB (G-AISU) [SH-F]	RAF BBMF, Coningsby
	AD189	VS349 Spitfire VB (G-CHVJ)	Privately owned, Raglan, Monmouthshire
	AD540	VS349 Spitfire VB (wreck)	Kennet Aviation, Tollerton (on rebuild)
	AE436	HP52 Hampden I [PL-J] (parts)	Lincolnshire Avn Heritage Centre, E Kirkby
	AJ841	CCF T-6J Texan (G-BJST/KF729)	Privately owned, Duxford
	AL246	Grumman Martlet I	FAA Museum, RNAS Yeovilton
	AP506	Cierva C30A (G-ACWM) (wreck)	The Helicopter Museum, Weston-super-Mare
	AP507	Cierva C30A (G-ACWP) [KX-P]	Science Museum, South Kensington
	AR501	VS349 Spitfire LF VC (G-AWII/AR4474)	The Shuttleworth Collection, Old Warden
	AV511	EHI-101 Merlin <R> [511]	SFDO, RNAS Culdrose
	BB803	DH82A Tiger Moth (G-ADWJ) [75]	Privately owned, Henstridge
	BB807	DH82A Tiger Moth (G-ADWO)	Solent Sky, Southampton
	BD713	Hawker Hurricane IIB	Privately owned, Taunton
	BE505	Hawker Hurricane IIB (RCAF 5403/G-HHII) [XP-L]	Hangar 11 Collection, North Weald
	BH238	Hawker Hurricane IIB	Airframe Assemblies, stored Sandown
	BL614	VS349 Spitfire VB (4354M) [ZD-F]	RAF Museum, Hendon
	BL655	VS349 Spitfire VB (wreck)	Lincolnshire Avn Heritage Centre, East Kirkby
	BL735	Supermarine Aircraft Spitfire 26 (G-HABT) [BT-A]	Privately owned, North Coates
	BL924	VS349 Spitfire VB <R> (BAPC 242) [AZ-G]	Castle Motors, Liskeard, Cornwall
	BL927	Supermarine Aircraft Spitfire 26 (G-CGWI) [YH-I]	Privately owned, Perth
	BM361	VS349 Spitfire VB <R> (BAPC 269) [XR-C]	RAF Lakenheath, on display
	BM481	VS349 Spitfire VB <R> (BAPC 301) [YO-T]	Thornaby Aerodrome Memorial
		(also wears PK651/RAO-B)	
	BM539	VS349 Spitfire LF VB (G-SSVB)	Privately owned, Hastings
	BM597	VS349 Spitfire LF VB (G-MKVB/5718M) [JH-C]	Historic Aircraft Collection, Duxford
	BN230	Hawker Hurricane IIC (LF751/5466M) [FT-A]	RAF Manston, Memorial Pavilion
	BP926	VS353 Spitfire PR IV (G-PRIV)	Privately owned, Newport Pagnell
	BR954	VS353 Spitfire PR IV <R> (N3194/BAPC 220) [JP-A]	Aircraft Restoration Group, stored Dalton, N Yorks
	BS239	VS361 Spitfire IX <R> (BAPC 222) [5R-E]	Battle of Britain Bunker, Uxbridge
	BS410	VS361 Spitfire IXC (G-TCHI)	Airframe Assemblies Ltd, Sandown
	BS435	VS361 Spitfire IX <R> [FY-F]	Privately owned, Lytham St Annes
	BW853	Hawker Hurricane XIIA (G-BRKE) (fuselage)	Privately owned, Cotswold Airport

Serial	Type (code/other identity)	Owner/operator, location or fate	Notes
DD931	Bristol 152 Beaufort VIII (9131M) [L]	RAF Museum, Hendon	
DE208	DH82A Tiger Moth II (G-AGYU)	Privately owned, Treswell, Notts	
DE470	DH82A Tiger Moth II (G-ANMY) [16]	Privately owned, Garford, Oxon	
DE623	DH82A Tiger Moth II (G-ANFI)	Privately owned, Cardiff	
DE673	DH82A Tiger Moth II (G-ADNZ/6948M)	Privately owned, Tibenham	
DE971	DH82A Tiger Moth II (G-OOSY)	Privately owned, Wickenby	
DE974	DH82A Tiger Moth II (G-ANZZ)	Privately owned, Clacton	
DE992	DH82A Tiger Moth II (G-AXXV)	Privately owned, Membury, Berks	
DF112	DH82A Tiger Moth II (G-ANRM)	Privately owned, Clacton/Duxford	
DF128	DH82A Tiger Moth II (G-AOJJ) [RCO-U]	Privately owned, White Waltham	
DF198	DH82A Tiger Moth II (G-BBRB)	Privately owned, stored West Wickham, Kent	
DG202	Gloster F9/40 (5758M)	RAF Museum, Hendon	
DG590	Miles M2H Hawk Major (G-ADMW/8379M)	RAF Museum Reserve Collection, Stafford	
DP872	Fairey Barracuda II <ff>	FAA Museum, Kiltech Vehicle Protection, Newcastle	
DV372	Avro 683 Lancaster I <ff>	Imperial War Museum, Lambeth	
DZ313	DH98 Mosquito B IV <R>	Privately owned, Little Rissington	
EB518	Airspeed AS10 Oxford V	Privately owned, Kenilworth	
EE416	Gloster Meteor F3 <ff>	Martin Baker Aircraft, Chalgrove, fire section	
EE425	Gloster Meteor F3 <ff>	Jet Age Museum, stored Gloucester	
EE531	Gloster Meteor F4 (7090M)	Midland Air Museum, Coventry	
EE549	Gloster Meteor F4 (7008M) [A]	Tangmere Military Aviation Museum	
EE602	VS349 Spitfire LF VC (G-IBSY) [DV-V]	Privately owned, Biggin Hill	
EF545	VS349 Spitfire LF VC (G-CDGY)	Aero Vintage, Rye	
EJ693	Hawker Tempest V (N7027E) [SA-J]	Privately owned, Booker	
EJ922	Hawker Typhoon IB <ff>	Privately owned, Booker	
EM720	DH82A Tiger Moth II (G-AXAN)	Privately owned, Doncaster	
EM840	DH82A Tiger Moth II (G-ANBY)	Privately owned, Middle Wallop	
EN179	VS361 Spitfire F IX (G-TCHO)	Privately owned, Exeter	
EN224	VS366 Spitfire F XII (G-FXII)	Privately owned, Newport Pagnell	
EN398	VS361 Spitfire F IX <R> [JE-J]	AMSS, Pyle, Bridgend	
EN398	VS361 Spitfire F IX <R> [JE-J]	RAF Coningsby, on display	
EN398	VS361 Spitfire F IX <R> (BAPC 184) [JE-J]	Shropshire Wartime Aircraft Recovery Grp Mus, Sleap	
EN398	VS361 Spitfire F IX <R> [JE-J]	Spitfire Spares, Taunton, Somerset	
EN961	Isaacs Spitfire <R> (G-CGIK) [SD-X]	Privately owned,	
EP120	VS349 Spitfire LF VB (G-LFVB/5377M/8070M) [AE-A]	The Fighter Collection, Duxford	
EP121	VS349 Spitfire LF VB <R> [LO-D]	Montrose Air Station Heritage Centre	
EP122	VS349 Spitfire F VB	Privately owned, Biggin Hill	
EV771	Fairchild UC-61 Argus <R> (BAPC.294)	Thorpe Camp Preservation Group, Lincs	
EX976	NA AT-6D Harvard III (FAP 1657)	FAA Museum, RNAS Yeovilton	
FE695	Noorduyn AT-16 Harvard IIB (G-BTXI) [94]	The Fighter Collection, Duxford	
FE788	CCF Harvard IV (MM54137/G-CTKL)	Privately owned, Biggin Hill	
FE905	Noorduyn AT-16 Harvard IIB (LN-BNM)	RAF Museum, Hendon	
FJ992	Boeing-Stearman PT-17 Kaydet (OO-JEH) [44]	Privately owned, Wevelgem, Belgium	
FK338	Fairchild 24W-41 Argus I (G-AJOZ)	Yorkshire Air Museum, Elvington	
FL586	Douglas C-47B Dakota (OO-SMA) [AI-N] (fuselage)	WWII Remembrance Museum, Handcross, W Sussex	
FR886	Piper L-4J Cub (G-BDMS)	Privately owned, Old Sarum	
FS628	Fairchild Argus 2 (43-14601/G-AIZE)	RAF Museum, Cosford	
FS728	Noorduyn AT-16 Harvard IIB (D-FRCP)	Privately owned, Gelnhausen, Germany	
FT118	Noorduyn AT-16 Harvard IIB (G-BZHL)	Privately owned, Wickenby	
FT323	NA AT-6D Harvard III (FAP 1513/G-CCOY)	Privately owned, Bruntingthorpe	
FT391	Noorduyn AT-16 Harvard IIB (G-AZBN)	Privately owned, Goodwood	
FX322	Noorduyn AT-16 Harvard IIB <ff>	Privately owned, Doncaster	
FX760	Curtiss P-40N Kittyhawk IV (A29-556/9150M) [GA-?]	RAF Museum, Hendon	
FZ626	Douglas Dakota III (KN566/G-AMPO) [YS-DH]	RAF Brize Norton, for display	
HB751	Fairchild Argus III (G-BCBL)	Privately owned, Woolsery, Devon	
HG691	DH89A Dragon Rapide (G-AIYR)	Privately owned, Clacton/Duxford	
HH268	GAL48 Hotspur II (HH379/BAPC 261) [H]	Museum of Army Flying, Middle Wallop	
HJ711	DH98 Mosquito NF II [VI-C]	Night-Fighter Preservation Tm, Elvington	
HM580	Cierva C-30A (G-ACUU) [KX-K]	Imperial War Museum, Duxford	

Notes	Serial	Type (code/other identity)	Owner/operator, location or fate
	HS503	Fairey Swordfish IV (BAPC 108)	RAF Museum Reserve Collection, Stafford
	IR206	Eurofighter Typhoon F2 <R> [IR]	RAF M&RU, Bottesford
	IR808	B-V Chinook HC2 <R>	RAF M&RU, Bottesford
	JG668	VS359 Spitfire T8 (A58-441/G-CFGA)	Privately owned, Haverfordwest
	JN768	Hawker Tempest V (4887M/G-TMPV)	Privately owned, Bentwaters
	JP843	Hawker Typhoon IB [Y]	Privately owned, Shrewsbury
	JR505	Hawker Typhoon IB <ff>	Midland Air Museum, Coventry
	JV482	Grumman Wildcat V	Ulster Aviation Society, Long Kesh
	JV579	Grumman FM-2 Wildcat (N4845V/G-RUMW) [F]	The Fighter Collection, Duxford
	JV928	Consolidated PBY-5A Catalina (N423RS) [Y]	Sold to the USA, January 2015
	KB889	Avro 683 Lancaster B X (G-LANC) [NA-I]	Imperial War Museum, Duxford
	KB976	Avro 683 Lancaster B X <ff>	RAF Scampton Heritage Centre
	KB976	Avro 683 Lancaster B X (G-BCOH) <rf>	South Yorkshire Aircraft Museum, Doncaster
	KB994	Avro 683 Lancaster B X (G-BVBP) <ff>	Currently not known
	KD345	Goodyear FG-1D Corsair (88297/G-FGID) [130-A]	The Fighter Collection, Duxford
	KD431	CV Corsair IV [E2-M]	FAA Museum, RNAS Yeovilton
	KE209	Grumman Hellcat II	FAA Museum, RNAS Yeovilton
	KE418	Hawker Tempest <rf>	Currently not known
	KF183	Noorduyn AT-16 Harvard IIB [3]	MoD/AFD/QinetiQ, Boscombe Down
	KF388	Noorduyn AT-16 Harvard IIB (composite)	Bournemouth Aviation Museum
	KF435	Noorduyn AT-16 Harvard IIB	Privately owned, Maidenhead
	KF532	Noorduyn AT-16 Harvard IIB <ff>	Privately owned, Bruntingthorpe
	KF584	CCF T-6J Texan (G-RAIX) [RAI-X]	Privately owned, Lee-on-Solent
	KF650	Noorduyn AT-16 Harvard IIB <ff>	Sywell Aviation Museum
	KF729	CCF T-6J Texan (G-BJST)	Repainted as AJ841, 2014
	KF741	Noorduyn AT-16 Harvard IIB <ff>	Privately owned, Kenilworth
	KG374	Douglas Dakota IV (KP208) [YS-DM]	Merville Barracks, Colchester, on display
	KG651	Douglas Dakota III (G-AMHJ)	RAF Transport Command Memorial, North Weald
	KH774	NA P-51D Mustang IV (44-73877/N167F) [GA-S]	Privately owned, Shoreham/Bournemouth
	KJ351	Airspeed AS58 Horsa II (TL659/BAPC 80) [23]	Museum of Army Flying, Middle Wallop
	KJ994	Douglas Dakota III (F-AZTE)	Dakota et Cie, La Ferté Alais, France
	KK116	Douglas Dakota IV (G-AMPY)	Classic Air Force, Coventry
	KK995	Sikorsky Hoverfly I [E]	RAF Museum, Hendon
	KL216	Republic P-47D Thunderbolt (45-49295/9212M) [RS-L]	RAF Museum, Hendon
	KN353	Douglas Dakota IV (G-AMYJ)	Yorkshire Air Museum, Elvington
	KN645	Douglas Dakota IV (KG374/8355M)	RAF Museum, Cosford
	KN751	Consolidated Liberator C VI (IAF HE807) [F]	RAF Museum, Hendon
	KP110	Beech C-45 Expeditor II (RCAF 2324/G-BKGM/HB275)	Privately owned, Dunkeswell
	KZ191	Hawker Hurricane IV (frame only)	Privately owned, East Garston, Bucks
	LA198	VS356 Spitfire F21 (7118M) [RAI-G]	Kelvingrove Art Gallery & Museum, Glasgow
	LA226	VS356 Spitfire F21 (7119M)	RAF Museum Reserve Collection, Stafford
	LA255	VS356 Spitfire F21 (6490M)	RAF, Lossiemouth
	LA543	VS474 Seafire F46 <ff>	The Air Defence Collection, Salisbury
	LA546	VS474 Seafire F46 (G-CFZJ)	Privately owned, Colchester
	LA564	VS474 Seafire F46 (G-FRSX)	Kennet Aviation, North Weald
	LB264	Taylorcraft Plus D (G-AIXA)	RAF Museum, Hendon
	LB294	Taylorcraft Plus D (G-AHWJ)	Saywell Heritage Centre, Worthing
	LB312	Taylorcraft Plus D (HH982/G-AHXE)	Privately owned, Netheravon
	LB323	Taylorcraft Plus D (G-AHSD)	Privately owned, Tibenham (rebuild)
	LB367	Taylorcraft Plus D (G-AHGZ)	Privately owned, Melksham
	LB375	Taylorcraft Plus D (G-AHGW)	Privately owned, Coventry
	LF363	Hawker Hurricane IIC [JX-B]	RAF BBMF, Coningsby
	LF738	Hawker Hurricane IIC (5405M) [UH-A]	RAF Museum, Cosford
	LF789	DH82 Queen Bee (K3584/BAPC 186) [R2-K]	DHAHC, London Colney
	LF858	DH82 Queen Bee (G-BLUZ)	Privately owned, Henlow
	LH291	Airspeed AS51 Horsa I <R> (BAPC 279)	RAF Museum, stored Cosford
	LS326	Fairey Swordfish II (G-AJVH) [L2]	RN Historic Flight, Yeovilton
	LV907	HP59 Halifax III (HR792) [NP-F] (marked NP763 [H7-N] on port side)	Yorkshire Air Museum, Elvington

Serial	Type (code/other identity)	Owner/operator, location or fate	Notes
LZ551	DH100 Vampire	FAA Museum, RNAS Yeovilton	
LZ766	Percival P34 Proctor III (G-ALCK)	Imperial War Museum, Duxford	
LZ842	VS361 Spitfire F IX (G-CGZU) [EF-D]	Privately owned, Biggin Hill	
LZ844	VS349 Spitfire F VC [UP-X]	Vintage Flyers, stored Nailsworth, Glos	
MA764	VS361 Spitfire F IX (G-MCDB)	Privately owned, Haverfordwest	
MB293	VS357 Seafire IIC (G-CFGI) (wreck)	Privately owned, Duxford	
MD338	VS359 Spitfire LF VIII	Privately owned, Sandown	
MF628	Vickers Wellington T10 (9210M)	Michael Beetham Conservation Centre, RAFM Cosford	
MH314	VS361 Spitfire IX <R> (*EN526*/BAPC 221) [SZ-G]	RAF Northolt, on display	
MH415	VS361 Spitfire IX <R> (*MJ751*/BAPC 209) [DU-V]	The Aircraft Restoration Co, Duxford	
MH434	VS361 Spitfire LF IXB (G-ASJV) [ZD-B]	The Old Flying Machine Company, Duxford	
MJ627	VS509 Spitfire T9 (G-BMSB) [9G-P]	Privately owned, Biggin Hill	
MJ832	VS361 Spitfire IX <R> (*L1096*/BAPC 229) [DN-Y]	RAF Digby, on display	
MK356	VS361 Spitfire LF IXC (5690M) [5J-K]	RAF BBMF, Coningsby	
MK356	VS361 Spitfire LF IXC <R> [2I-V]	Kent Battle of Britain Museum, Hawkinge	
MK356	VS361 Spitfire LF IXC (BAPC 298) <R>	RAF Cosford, on display	
MK805	VS361 Spitfire IX <R> [SH-B]	Simply Spitfire, Oulton Broad, Suffolk	
MK912	VS361 Spitfire LF IXC (G-BRRA) [SH-L]	Privately owned, Biggin Hill	
ML295	VS 361 Spitfire LF IXB (NH341/G-CICK)	Privately owned, Duxford	
ML407	VS509 Spitfire T9 (G-LFIX) [OU-V]	Privately owned, Bentwaters	
ML411	VS361 Spitfire LF IXE (G-CBNU)	Privately owned, Ashford, Kent	
ML427	VS361 Spitfire IX (6457M) [HK-A]	Thinktank, Birmingham	
ML796	Short S25 Sunderland V [NS-F]	Imperial War Museum, Duxford	
ML824	Short S25 Sunderland V [NS-Z]	RAF Museum, Hendon	
MN235	Hawker Typhoon IB	*To Canada, February 2014*	
MP425	Airspeed AS10 Oxford I (G-AITB) [G]	RAF Museum, Hendon	
MS902	Miles M25 Martinet TT1 (TF-SHC)	Museum of Berkshire Aviation, Woodley	
MT182	Auster J/1 Autocrat (G-AJDY)	Privately owned, Spanhoe	
MT197	Auster IV (G-ANHS)	Privately owned, Spanhoe	
MT438	Auster III (G-AREI)	Privately owned, Eggesford	
MT818	VS502 Spitfire T8 (G-AIDN)	Privately owned, Booker	
MT847	VS379 Spitfire FR XIVE (6960M) [AX-H]	RAF Museum, Hendon	
MT928	VS359 Spitfire HF VIIIC (D-FEUR/MV154/*AR654*)[ZX-M]	Privately owned, Bremgarten, Germany	
MV268	VS379 Spitfire FR XIVE (MV293/G-SPIT) [JE-J]	The Fighter Collection, Duxford	
MW401	Hawker Tempest II (IAF HA604/G-PEST)	Privately owned, Wickenby	
MW763	Hawker Tempest II (IAF HA586/G-TEMT) [HF-A]	Privately owned, North Weald	
MW810	Hawker Tempest II (IAF HA591) <ff>	Privately owned, Bentwaters	
NF370	Fairey Swordfish III [NH-L]	Imperial War Museum, Duxford	
NF389	Fairey Swordfish III [D]	RN Historic Flight, Yeovilton	
NH238	VS361 Spitfire LF IXE (G-MKIX) [D-A]	Privately owned, stored Greenham Common	
NJ633	Auster 5D (G-AKXP)	Privately owned, Old Sarum	
NJ673	Auster 5D (G-AOCR)	Privately owned, Shenington, Oxon	
NJ689	Auster AOP5 (G-ALXZ)	Privately owned, Breighton	
NJ695	Auster AOP5 (G-AJXV)	Privately owned, Newark	
NJ719	Auster AOP5 (TW385/G-ANFU)	North-East Aircraft Museum, Usworth	
NJ728	Auster AOP5 (G-AIKE)	Privately owned, Wickenby	
NJ889	Auster AOP3 (G-AHLK)	Privately owned, Leicester East	
NL750	DH82A Tiger Moth II (T7997/G-AOBH)	Privately owned, Eaglescott	
NL985	DH82A Tiger Moth I (G-BWIK/7015M)	Privately owned, Sywell	
NM138	DH82A Tiger Moth I (G-ANEW) [41]	Privately owned, Henstridge	
NM181	DH82A Tiger Moth I (G-AZGZ)	Privately owned, Rush Green	
NP294	Percival P31 Proctor IV [TB-M]	Lincolnshire Avn Heritage Centre, E Kirkby	
NV778	Hawker Tempest TT5 (8386M)	RAF Museum, Hendon	
NX534	Auster III (G-BUDL)	Privately owned, Spanhoe	
NX611	Avro 683 Lancaster B VII (G-ASXX/8375M) [DX-F,LE-H]	Lincolnshire Avn Heritage Centre, E Kirkby	
PA474	Avro 683 Lancaster B I [KC-A]	RAF BBMF, Coningsby	
PD685	Slingsby T7 Cadet TX1	Tettenhall Transport Heritage Centre	

Notes	Serial	Type (code/other identity)	Owner/operator, location or fate
	PF179	HS Gnat T1 (XR541/8602M)	Privately owned, North Weald
	PG712	DH82A Tiger Moth II (PH-CSL) [2]	Privately owned, Hilversum, The Netherlands
	PK519	VS356 Spitfire F22 (G-SPXX)	Privately owned, Newport Pagnell
	PK624	VS356 Spitfire F22 (8072M)	The Fighter Collection, Duxford
	PK664	VS356 Spitfire F22 (7759M) [V6-B]	Kennet Aviation, North Weald
	PK683	VS356 Spitfire F24 (7150M)	Solent Sky, Southampton
	PK724	VS356 Spitfire F24 (7288M)	RAF Museum, Hendon
	PL256	VS361 Spitfire IX <R> [TM-L]	Privately owned, Leicester
	PL279	VS361 Spitfire IX <R> (N3317/BAPC 268) [ZF-Z]	Privately owned, St Mawgan
	PL788	Supermarine Aircraft Spitfire 26 (G-CIEN)	Privately owned, Perth
	PL904	VS365 Spitfire PR XI <R> (EN343/BAPC 226)	RAF Benson, on display
	PL965	VS365 Spitfire PR XI (G-MKXI) [R]	Hangar 11 Collection, North Weald
	PL983	VS365 Spitfire PR XI (G-PRXI)	Privately owned, Duxford (on rebuild)
	PM631	VS390 Spitfire PR XIX	RAF BBMF, Coningsby
	PM651	VS390 Spitfire PR XIX (7758M) [X]	To Kuwait, February 2013
	PN323	HP Halifax VII <ff>	Imperial War Museum, Duxford
	PP566	Fairey Firefly I <rf>	Privately owned, Newton Abbott, Devon
	PP972	VS358 Seafire LF IIIC (G-BUAR)	Privately owned, Bentwaters
	PR536	Hawker Tempest II (IAF HA457)[OQ-H]	RAF Museum, Hendon
	PS001	Airbus Zephyr 7 HAPS	Airbus Defence & Space/QinetiQ, UK
	PS853	VS390 Spitfire PR XIX (G-RRGN) [C]	Rolls-Royce, East Midlands
	PS890	VS390 Spitfire PR XIX (F-AZJS) [UM-E]	Privately owned, Dijon, France
	PS915	VS390 Spitfire PR XIX (7548M/7711M)	RAF BBMF, Coningsby
	PT462	VS509 Spitfire T9 (G-CTIX/N462JC) [SW-A]	Privately owned, Caernarfon/Duxford
	PT462	VS361 Spitfire IX <R> [SW-A]	Privately owned, Moffat, Dumfries & Galloway
	PV303	Supermarine Aircraft Spitfire 26 (G-CCJL) [ON-B]	Privately owned, Barton
	PZ460	BBC Mosquito (F-PMOZ) [NE-K]	Privately owned, Fontenay-le-Comte, France
	PZ865	Hawker Hurricane IIC (G-AMAU) [EG-S]	RAF BBMF, Coningsby
	QQ100	Agusta A109E Power Elite (G-CFVB)	MoD/AFD/QinetiQ, Boscombe Down
	QQ101	BAe RJ.100 (G-BZAY)	MoD/ETPS, Boscombe Down
	QQ102	BAe RJ.70ER (G-BVRJ)	MoD/AFD/QinetiQ, Boscombe Down
	RA848	Slingsby T7 Cadet TX1	Privately owned, Leeds
	RA854	Slingsby T7 Cadet TX1	Yorkshire Air Museum, Elvington
	RA897	Slingsby T7 Cadet TX1	Newark Air Museum, Winthorpe
	RA905	Slingsby T7 Cadet TX1 (BGA1143)	Trenchard Museum, RAF Halton
	RB142	Supermarine Aircraft Spitfire 26 (G-CEFC) [DW-B]	Privately owned, Basingstoke, Hants
	RB159	VS379 Spitfire F XIV <R> [DW-D]	Privately owned, Delabole, Cornwall
	RD220	Bristol 156 Beaufighter TF X	Royal Scottish Mus'm of Flight, stored E Fortune
	RD253	Bristol 156 Beaufighter TF X (7931M)	RAF Museum, Hendon
	RF398	Avro 694 Lincoln B II (8376M)	RAF Museum, Cosford
	RG333	Miles M38 Messenger IIA (G-AIEK)	Privately owned, Felton, Bristol
	RG904	VS Spitfire <R> [BT-K]	RAF Museum, Cosford
	RH746	Bristol 164 Brigand TF1 (fuselage)	RAF Museum, stored Cosford
	RK855	Supermarine Aircraft Spitfire 26 (G-PIXY) [FT-C]	Privately owned, Henstridge
	RL962	DH89A Dominie II (G-AHED)	RAF Museum Reserve Collection, Stafford
	RM169	Percival P31 Proctor IV (SE-CEA) [4-47]	Privately owned, Great Oakley, Essex
	RM221	Percival P31 Proctor IV (G-ANXR)	Privately owned, Headcorn
	RM689	VS379 Spitfire F XIV (G-ALGT)	Rolls-Royce, stored East Midlands Airport
	RM694	VS379 Spitfire F XIV (G-DBKL/6640M)	Privately owned, Booker
	RM927	VS379 Spitfire F XIV (G-JNMA)	Privately owned, Sandown
	RN218	Isaacs Spitfire <R> (G-BBJI) [N]	Privately owned, Builth Wells
	RR232	VS361 Spitfire HF IXC (G-BRSF)	Privately owned, Colerne
	RT486	Auster 5 (G-AJGJ)	Privately owned, Lee-on-Solent
	RT520	Auster 5 (G-ALYB)	South Yorkshire Aircraft Museum, Doncaster
	RT610	Auster 5A-160 (G-AKWS)	Privately owned, Shobdon
	RW382	VS361 Spitfire LF XVIE (G-PBIX/7245M/8075M) [3W-P]	Privately owned, Biggin Hill
	RW386	VS361 Spitfire LF XVIE (SE-BIR/6944M) [NG-D]	Privately owned, Angelholm, Sweden
	RW388	VS361 Spitfire LF XVIE (6946M) [U4-U]	Stoke-on-Trent City Museum, Hanley
	RX168	VS358 Seafire L IIIC (IAC 157/G-BWEM)	Privately owned, Exeter

Serial	Type (code/other identity)	Owner/operator, location or fate	Notes
SE001*	Boeing Scan Eagle RM1 RPAS	RN No 700X NAS, Culdrose	
SL611	VS361 Spitfire LF XVIE	Supermarine Aero Engineering, Stoke-on-Trent	
SL674	VS361 Spitfire LF IX (8392M) [RAS-H]	RAF Museum Reserve Collection, Stafford	
SM520	VS509 Spitfire T9 (H-99/G-ILDA) [KJ-I]	Privately owned, Kidlington	
SM845	VS394 Spitfire FR XVIII (G-BUOS) [R]	Spitfire Ltd, Humberside	
SN280	Hawker Tempest V <ff>	South Yorkshire Aircraft Museum, Doncaster	
SR462	VS377 Seafire F XV (G-TGVP)	Privately owned, North Weald	
SX137	VS384 Seafire F XVII	FAA Museum, RNAS Yeovilton	
SX300	VS384 Seafire F XVII (G-RIPH)	Kennet Aviation, North Weald	
SX336	VS384 Seafire F XVII (G-KASX) [105/VL]	Kennet Aviation, North Weald	
TA122	DH98 Mosquito FB VI [UP-G]	Mosquito Aircraft Museum, London Colney	
TA634	DH98 Mosquito TT35 (G-AWJV) [8K-K]	Mosquito Aircraft Museum, London Colney	
TA639	DH98 Mosquito TT35 (7806M) [AZ-E]	RAF Museum, Cosford	
TA719	DH98 Mosquito TT35 (G-ASKC)	Imperial War Museum, Duxford	
TA805	VS361 Spitfire HF IX (G-PMNF) [FX-M]	Privately owned, Biggin Hill	
TB382	VS361 Spitfire LF XVIE (X4277/MK673)	RAF BBMF, stored Coningsby	
TB675	VS361 Spitfire LF XVIE (RW393/7293M) [4D-V]	RAF Museum Reserve Collection, Stafford	
TB752	VS361 Spitfire LF XVIE (8086M) [KH-Z]	RAF Manston, Memorial Pavilion	
TD248	VS361 Spitfire LF XVIE (G-OXVI/7246M) [CR-S]	Spitfire Ltd, Humberside	
TD248	VS361 Spitfire LF XVIE [8Q-T] (fuselage)	Norfolk & Suffolk Avn Mus'm, Flixton	
TD314	VS361 Spitfire LF IX (G-CGYJ) [FX-P]	Privately owned, Biggin Hill	
TE184	VS361 Spitfire LF XVIE (G-MXVI/6850M) [DU-N]	Privately owned, Biggin Hill	
TE311	VS361 Spitfire LF XVIE (MK178/7241M) [4D-V]	RAF BBMF, Coningsby	
TE462	VS361 Spitfire LF XVIE (7243M)	Royal Scottish Mus'm of Flight, E Fortune	
TE517	VS361 Spitfire LF IXE (G-JGCA) [HL-K]	Privately owned, Biggin Hill	
TE566	VS509 Spitfire T9 (VH-IXT)	Vintage Flyers, Cotswold Airport (rebuild)	
TG263	Saro SR A1 (G-12-1)	Solent Sky, Southampton	
TG511	HP67 Hastings T5 (8554M) [511]	RAF Museum, Cosford	
TG517	HP67 Hastings T5 [517]	Newark Air Museum, Winthorpe	
TG528	HP67 Hastings C1A [528,T]	Imperial War Museum, Duxford	
TJ118	DH98 Mosquito TT35 <ff>	Mosquito Aircraft Museum, stored London Colney	
TJ138	DH98 Mosquito B35 (7607M) [VO-L]	RAF Museum, Hendon	
TJ207	Auster AOP5 (NJ703/G-AKPI) [P]	Privately owned, Spanhoe	
TJ343	Auster AOP5 (G-AJXC)	Privately owned, Henstridge	
TJ518	Auster J/1 Autocrat (G-AJIH)	Privately owned, Bidford	
TJ534	Auster AOP5 (G-AKSY)	Privately owned, Dunsfold	
TJ569	Auster AOP5 (G-AKOW)	Museum of Army Flying, Middle Wallop	
TJ652	Auster AOP5 (TJ565/G-AMVD)	Privately owned, Hardwick, Norfolk	
TJ672	Auster 5D (G-ANIJ) [TS-D]	Privately owned, Netheravon	
TK718	GAL59 Hamilcar I (fuselage)	The Tank Museum, Bovington	
TK777	GAL59 Hamilcar I (fuselage)	Museum of Army Flying, Middle Wallop	
TP298	VS394 Spitfire FR XVIII [UM-T]	Privately owned, Sandown	
TS291	Slingsby T7 Cadet TX1 (BGA852)	Royal Scottish Mus'm of Flight, stored Granton	
TS798	Avro 685 York C1 (G-AGNV/MW100)	RAF Museum, Cosford	
TW439	Auster AOP5 (G-ANRP)	Privately owned, Strubby	
TW467	Auster AOP5 (G-ANIE)	Privately owned, Elmsett	
TW477	Auster AOP5 (OY-EFI)	Privately owned, Ringsted, Denmark	
TW501	Auster AOP5 (G-ALBJ)	Privately owned, Dunkeswell	
TW511	Auster AOP5 (G-APAF)	Privately owned, Chiseldon, Wilts	
TW536	Auster AOP6 (G-BNGE/7704M) [TS-V]	Privately owned, Netheravon	
TW591	Auster 6A (G-ARIH) [6]	Privately owned, Eggesford	
TW641	Beagle A61 Terrier 2 (G-ATDN)	Privately owned, Biggin Hill	
TX213	Avro 652A Anson C19 (G-AWRS)	North-East Aircraft Museum, Usworth	
TX214	Avro 652A Anson C19 (7817M)	RAF Museum, Cosford	
TX226	Avro 652A Anson C19 (7865M)	Classic Air Force, stored Compton Verney	
TX235	Avro 652A Anson C19	Classic Air Force, stored Compton Verney	
TX310	DH89A Dragon Rapide 6 (G-AIDL)	Classic Air Force, Newquay	
VF301	DH100 Vampire F1 (7060M) [RAL-G]	Midland Air Museum, Coventry	
VF512	Auster 6A (G-ARRX) [PF-M]	Privately owned, Popham	
VF516	Beagle A61 Terrier 2 (G-ASMZ)	Privately owned, Eggesford	

Notes	Serial	Type (code/other identity)	Owner/operator, location or fate
	VF519	Auster AOP6 (G-ASYN)	Privately owned, Doncaster
	VF526	Auster 6A (G-ARXU) [T]	Privately owned, Netheravon
	VF557	Auster 6A (G-ARHM) [H]	Privately owned, Spanhoe
	VF560	Auster 6A (frame)	South Yorkshire Aircraft Museum, Doncaster
	VF581	Beagle A61 Terrier 1 (G-ARSL) [G]	Privately owned, East Fortune
	VF631	Auster AOP6 (G-ASDK)	Privately owned, Hibaldstow
	VH127	Fairey Firefly TT4 [200/R]	FAA Museum, stored Cobham Hall, RNAS Yeovilton
	VL348	Avro 652A Anson C19 (G-AVVO)	Newark Air Museum, Winthorpe
	VL349	Avro 652A Anson C19 (G-AWSA) [V7-Q]	Norfolk & Suffolk Avn Museum, Flixton
	VM325	Avro 652A Anson C19	Privately owned, Carew Cheriton, Pembrokeshire
	VM360	Avro 652A Anson C19 (G-APHV)	Royal Scottish Mus'm of Flight, E Fortune
	VM687	Slingsby T8 Tutor (BGA794)	Privately owned, Lee-on-Solent
	VM791	Slingsby Cadet TX3 (XA312/8876M)	RAF Manston History Museum
	VN485	VS356 Spitfire F24 (7326M)	Imperial War Museum, Duxford
	VN799	EE Canberra T4 (WJ874/G-CDSX)	Classic Air Force, Newquay
	VP293	Avro 696 Shackleton T4 [X] <ff>	Shackleton Preservation Trust, Coventry
	VP519	Avro 652A Anson C19 (G-AVVR) <ff>	South Yorkshire Aircraft Museum, Doncaster
	VP952	DH104 Devon C2 (8820M)	RAF Museum, Cosford
	VP955	DH104 Devon C2 (G-DVON)	Privately owned, Cricklade, Wilts
	VP957	DH104 Devon C2 (8822M) <ff>	No 1137 Sqn ATC, Long Kesh
	VP967	DH104 Devon C2 (G-KOOL)	Yorkshire Air Museum, Elvington
	VP975	DH104 Devon C2 [M]	Science Museum, Wroughton
	VP981	DH104 Devon C2 (G-DHDV)	Classic Air Force, Coventry
	VR137	Westland Wyvern TF1	FAA Museum, RNAS Yeovilton
	VR192	Percival P40 Prentice T1 (G-APIT)	Privately owned, Cambs
	VR249	Percival P40 Prentice T1 (G-APIY) [FA-EL]	Newark Air Museum, Winthorpe
	VR259	Percival P40 Prentice T1 (G-APJB) [M]	Classic Air Force, Newquay
	VR930	Hawker Sea Fury FB11 (8382M) [110/Q]	RN Historic Flight, Yeovilton
	VS356	Percival P40 Prentice T1 (G-AOLU)	Privately owned, Montrose
	VS610	Percival P40 Prentice T1 (G-AOKL)[K-L]	The Shuttleworth Collection, Old Warden
	VS618	Percival P40 Prentice T1 (G-AOLK)	RAF Museum, Hendon
	VS623	Percival P40 Prentice T1 (G-AOKZ)[KQ-F]	Midland Air Museum, Coventry
	VT812	DH100 Vampire F3 (7200M) [N]	RAF Museum, Hendon
	VT935	Boulton Paul P111A (VT769)	Midland Air Museum, Coventry
	VT987	Auster AOP6 (G-BKXP)	Privately owned, Thruxton
	VV106	Supermarine 510 (7175M)	FAA Museum, stored Cobham Hall, RNAS Yeovilton
	VV217	DH100 Vampire FB5 (7323M)	Mosquito Aircraft Museum, stored London Colney
	VV400	EoN Olympia 2 (BGA1697)	Privately owned, Rivar Hill, Wilts
	VV401	EoN Olympia 2 (BGA1125) [99]	Privately owned, Ringmer, E Sussex
	VV901	Avro 652A Anson T21	Yorkshire Air Museum, Elvington
	VW453	Gloster Meteor T7 (8703M) [Z]	Jet Age Museum, Gloucester
	VW957	DH103 Sea Hornet NF21 <rf>	Privately owned, Chelmsford
	VW993	Beagle A61 Terrier 2 (G-ASCD)	Yorkshire Air Museum, Elvington
	VX113	Auster AOP6 (G-ARNO) [36]	Privately owned, Stow Maries, Essex
	VX147	Alon A2 Aircoupe (G-AVIL)	Privately owned, Stow Maries
	VX185	EE Canberra B(I)8 (7631M) <ff>	Royal Scottish Mus'm of Flight, E Fortune
	VX250	DH103 Sea Hornet NF21 [48] <rf>	DHAHC, London Colney
	VX272	Hawker P.1052 (7174M)	FAA Museum, stored RNAS Yeovilton
	VX275	Slingsby T21B Sedbergh TX1 (BGA572/8884M)	RAF Museum Reserve Collection, Stafford
	VX281	Hawker Sea Fury T20S (G-RNHF) [120/VL]	Naval Aviation Ltd, North Weald (on repair)
	VX573	Vickers Valetta C2 (8389M)	RAF Museum, stored Cosford
	VX580	Vickers Valetta C2 [580]	Norfolk & Suffolk Avn Museum, Flixton
	VX595	WS51 Dragonfly HR1	FAA Museum, RNAS Yeovilton
	VX665	Hawker Sea Fury FB11 <rf>	RN Historic Flight, at BAE Systems Brough
	VX926	Auster T7 (G-ASKJ)	Privately owned, Gamlingay, Cambs
	VX927	Auster T7 (G-ASYG)	Privately owned, Wickenby
	VZ193	DH100 Vampire FB5 <ff>	Privately owned, Hooton Park
	VZ345	Hawker Sea Fury T20S	The Fighter Collection, Duxford
	VZ440	Gloster Meteor F8 (WA984) [X]	Tangmere Military Aviation Museum
	VZ477	Gloster Meteor F8 (7741M) <ff>	Midland Air Museum, Coventry
	VZ608	Gloster Meteor FR9	Newark Air Museum, Winthorpe
	VZ634	Gloster Meteor T7 (8657M)	Newark Air Museum, Winthorpe

Serial	Type (code/other identity)	Owner/operator, location or fate	Notes
VZ638	Gloster Meteor T7 (G-JETM) [HF]	Gatwick Aviation Museum, Charlwood, Surrey	
VZ728	RS4 Desford Trainer (G-AGOS)	Privately owned, stored Spanhoe Lodge	
WA346	DH100 Vampire FB5	RAF Museum, stored Cosford	
WA473	VS Attacker F1 [102/J]	FAA Museum, RNAS Yeovilton	
WA576	Bristol 171 Sycamore 3 (G-ALSS/7900M)	Dumfries & Galloway Avn Mus, Dumfries	
WA577	Bristol 171 Sycamore 3 (G-ALST/7718M)	North-East Aircraft Museum, Usworth	
WA591	Gloster Meteor T7 (G-BWMF/7917M) [FMK-Q]	Classic Air Force, Newquay	
WA630	Gloster Meteor T7 [69] <ff>	Robertsbridge Aviation Society, Newhaven	
WA634	Gloster Meteor T7/8	RAF Museum, stored Cosford	
WA638	Gloster Meteor T7(mod)	Martin Baker Aircraft, Chalgrove	
WA662	Gloster Meteor T7	South Yorkshire Aircraft Museum, Doncaster	
WB188	Hawker Hunter F3 (7154M)	Tangmere Military Aviation Museum	
WB188	Hawker Hunter GA11 (WV256/G-BZPB)	Classic Air Force, Newquay	
WB188	Hawker Hunter GA11 (XF300/G-BZPC)	Privately owned, Melksham, Wilts	
WB440	Fairey Firefly AS6 <ff>	Privately owned, Newton Abbott, Devon	
WB491	Avro 706 Ashton 2 (TS897/G-AJJW) <ff>	Newark Air Museum, Winthorpe	
WB555	DHC1 Chipmunk T10 <ff>	Privately owned, Ellerton	
WB556	DHC1 Chipmunk T10 (fuselage) [O-C,P]	*Sold to Australia, November 2013*	
WB560	DHC1 Chipmunk T10 (comp WG403)	Privately owned, South Molton, Devon	
WB565	DHC1 Chipmunk T10 (G-PVET) [X]	Privately owned, Rendcomb	
WB569	DHC1 Chipmunk T10 (G-BYSJ) [R]	Privately owned, Cotswold Airport	
WB571	DHC1 Chipmunk T10 (D-EOSR) [34]	Privately owned, Porta Westfalica, Germany	
WB584	DHC1 Chipmunk T10 (comp WG303/7706M)	Solwauy Aviation Museum, Carlisle	
WB585	DHC1 Chipmunk T10 (G-AOSY) [M]	Privately owned, Audley End	
WB588	DHC1 Chipmunk T10 (G-AOTD) [D]	Privately owned, Old Sarum	
WB615	DHC1 Chipmunk T10 (G-BXIA) [E]	Privately owned, Blackpool	
WB624	DHC1 Chipmunk T10	Newark Air Museum, Winthorpe	
WB626	DHC1 Chipmunk T10 <ff>	Trenchard Museum, RAF Halton	
WB627	DHC1 Chipmunk T10 (9248M) (fuselage) [N]	Dulwich College CCF	
WB654	DHC1 Chipmunk T10 (G-BXGO) [U]	Privately owned, Finmere	
WB657	DHC1 Chipmunk T10 [908]	RN Historic Flight, Yeovilton	
WB670	DHC1 Chipmunk T10 (comp WG303/8361M)	Privately owned, Carlisle	
WB671	DHC1 Chipmunk T10 (G-BWTG) [910]	Privately owned, Teuge, The Netherlands	
WB685	DHC1 Chipmunk T10 (comp WP969/G-ATHC)	Mosquito Aircraft Museum, London Colney	
WB685	DHC1 Chipmunk T10 <rf>	North-East Aircraft Museum, stored Usworth	
WB697	DHC1 Chipmunk T10 (G-BXCT) [95]	Privately owned, Wickenby	
WB702	DHC1 Chipmunk T10 (G-AOFE)	Privately owned, Goodwood	
WB703	DHC1 Chipmunk T10 (G-ARMC)	Privately owned, Compton Abbas	
WB711	DHC1 Chipmunk T10 (G-APPM)	Privately owned, Sywell	
WB726	DHC1 Chipmunk T10 (G-AOSK) [E]	Privately owned, Turweston	
WB733	DHC1 Chipmunk T10 (comp WG422)	South Yorkshire Aircraft Museum, Doncaster	
WB763	DHC1 Chipmunk T10 (G-BBMR) [14]	*Repainted as G-BBMR*	
WB922	Slingsby T21B Sedbergh TX1 (BGA4366)	Privately owned, Hullavington	
WB924	Slingsby T21B Sedbergh TX1 (BGA3901)	Privately owned, Dunstable	
WB944	Slingsby T21B Sedbergh TX1 (BGA3160)	Privately owned, Bicester	
WB945	Slingsby T21B Sedbergh TX1 (BGA1254)	Privately owned, Lasham	
WB971	Slingsby T21B Sedbergh TX1 (BGA3324)	Privately owned, Felthorpe	
WB975	Slingsby T21B Sedbergh TX1 (BGA3288) [FJB]	Privately owned, Shipdham	
WB980	Slingsby T21B Sedbergh TX1 (BGA3290)	Privately owned, Husbands Bosworth	
WB981	Slingsby T21B Sedbergh TX1 (BGA3238)	Privately owned, Keevil	
WD286	DHC1 Chipmunk T10 (G-BBND)	Privately owned, Little Gransden	
WD292	DHC1 Chipmunk T10 (G-BCRX)	Privately owned, White Waltham	
WD293	DHC1 Chipmunk T10 (7645M) <ff>	Privately owned, St Athan	
WD310	DHC1 Chipmunk T10 (G-BWUN) [B]	Privately owned, Deanland	
WD319	DHC1 Chipmunk T10 (OY-ATF)	Privately owned, Stauning, Denmark	
WD321	DHC1 Chipmunk T10 (G-BDCC)	Boscombe Down Aviation Collection, Old Sarum	
WD325	DHC1 Chipmunk T10 [N]	RN Historic Flight, Yeovilton	
WD331	DHC1 Chipmunk T10 (G-BXDH)	Privately owned, Farnborough	
WD355	DHC1 Chipmunk T10 (WD335/G-CBAJ)	Privately owned, Eastleigh	

Notes	Serial	Type (code/other identity)	Owner/operator, location or fate
	WD363	DHC1 Chipmunk T10 (G-BCIH) [5]	Privately owned, Netheravon
	WD370	DHC1 Chipmunk T10 <ff>	No 225 Sqn ATC, Brighton
	WD373	DHC1 Chipmunk T10 (G-BXDI) [12]	Privately owned, Turweston
	WD377	DHC1 Chipmunk T10 <ff>	Privately owned, Lancs
	WD386	DHC1 Chipmunk T10 (comp WD377)	Ulster Aviation Soc, stored Upper Ballinderry, NI
	WD388	DHC1 Chipmunk T10 (D-EPAK) [68]	Quax Flieger, Hamm, Germany
	WD390	DHC1 Chipmunk T10 (G-BWNK) [68]	Privately owned, Wickenby
	WD413	Avro 652A Anson T21 (7881M/G-VROE)	Classic Air Force, Newquay
WD615		Gloster Meteor TT20 (WD646/8189M) [R]	RAF Manston History Museum
	WD686	Gloster Meteor NF11 [S]	Muckleburgh Collection, Weybourne
	WD790	Gloster Meteor NF11 (8743M)<ff>	North-East Aircraft Museum, Usworth
	WD889	Fairey Firefly AS5 (comp VT809)	Privately owned, Newton Abbot, Devon
	WD931	EE Canberra B2 <ff>	RAF Museum, stored Cosford
	WD935	EE Canberra B2 (8440M) <ff>	South Yorkshire Aircraft Museum, Doncaster
	WD954	EE Canberra B2 <ff>	Privately owned, St Mawgan
	WE113	EE Canberra B2 <ff>	Privately owned, Tangmere
	WE122	EE Canberra TT18 [845] <ff>	Blyth Valley Aviation Collection, Walpole, Suffolk
	WE139	EE Canberra PR3 (8369M)	RAF Museum, Hendon
	WE168	EE Canberra PR3 (8049M) <ff>	Norfolk & Suffolk Avn Museum, Flixton
	WE173	EE Canberra PR3 (8740M) <ff>	Robertsbridge Aviation Society, Mayfield
	WE188	EE Canberra T4	Solway Aviation Society, Carlisle
	WE192	EE Canberra T4 <ff>	Blyth Valley Aviation Collection, Walpole, Suffolk
WE275		DH112 Venom FB50 (J-1601/G-VIDI)	BAE Systems Hawarden, Fire Section
	WE558	Auster T7 (frame)	East Midlands Airport Aeropark
	WE569	Auster T7 (G-ASAJ)	Classic Air Force, Newquay
	WE570	Auster T7 (G-ASBU)	Privately owned, Stonehaven
	WE591	Auster T7 (F-AZTJ)	Privately owned, Toussus-le-Noble, France
	WE600	Auster T7 Antarctic (7602M)	RAF Museum, Cosford
WE724		Hawker Sea Fury FB11 (VX653/G-BUCM) [062]	The Fighter Collection, Duxford
	WE982	Slingsby T30B Prefect TX1 (8781M)	RAF Museum, stored Cosford
	WE987	Slingsby T30B Prefect TX1 (BGA2517)	South Yorkshire Aircraft Museum, Doncaster
	WE990	Slingsby T30B Prefect TX1 (BGA2523)	Privately owned, Tibenham
	WE992	Slingsby T30B Prefect TX1 (BGA2692)	Privately owned, Hullavington
	WF118	Percival P57 Sea Prince T1 (G-DACA) [569/CU]	Gatwick Aviation Museum, Charlwood, Surrey
	WF122	Percival P57 Sea Prince T1 [575/CU]	South Yorkshire Aircraft Museum, Doncaster
	WF128	Percival P57 Sea Prince T1 (8611M)	Norfolk & Suffolk Avn Museum, Flixton
	WF137	Percival P57 Sea Prince C1	*Scrapped at St Athan, September 2014*
	WF145	Hawker Sea Hawk F1 <ff>	Privately owned, Newton Abbot, Devon
	WF219	Hawker Sea Hawk F1 <rf>	FAA Museum, stored Cobham Hall, RNAS Yeovilton
	WF225	Hawker Sea Hawk F1 [CU]	RNAS Culdrose, at main gate
	WF259	Hawker Sea Hawk F2 [171/A]	Royal Scottish Mus'm of Flight, E Fortune
	WF369	Vickers Varsity T1 [F]	Newark Air Museum, Winthorpe
	WF372	Vickers Varsity T1 [A]	Brooklands Museum, Weybridge
	WF408	Vickers Varsity T1 (8395M) <ff>	Privately owned, Ashford, Kent
	WF643	Gloster Meteor F8 [F]	Norfolk & Suffolk Avn Museum, Flixton
	WF784	Gloster Meteor T7 (7895M)	Jet Age Museum, stored Gloucester
	WF825	Gloster Meteor T7 (8359M) [A]	Montrose Air Station Heritage Centre
	WF877	Gloster Meteor T7 (G-BPOA)	Privately owned, stored Duxford (dismantled)
	WF911	EE Canberra B2 [CO] <ff>	Ulster Aviation Society, Long Kesh
	WF922	EE Canberra PR3	Midland Air Museum, Coventry
	WG303	DHC1 Chipmunk T10 (8208M) <ff>	Privately owned, Partridge Green, W Sussex
	WG308	DHC1 Chipmunk T10 (G-BYHL) [8]	Privately owned, Averham, Notts
	WG316	DHC1 Chipmunk T10 (G-BCAH)	Privately owned, Linton-on-Ouse
	WG319	DHC1 Chipmunk T10 <ff>	Privately owned, Blandford Forum, Dorset
	WG321	DHC1 Chipmunk T10 (G-DHCC)	Privately owned, Wevelgem, Belgium
	WG348	DHC1 Chipmunk T10 (G-BBMV)	Privately owned, Biggin Hill
	WG350	DHC1 Chipmunk T10 (G-BPAL)	Privately owned, Cascais, Portugal
	WG362	DHC1 Chipmunk T10 (8437M/8630M/*WX643*) <ff>	No 1094 Sqn ATC, Ely
	WG407	DHC1 Chipmunk T10 (G-BWMX) [67]	Privately owned, Croydon, Cambs

Serial	Type (code/other identity)	Owner/operator, location or fate	Notes
WG418	DHC1 Chipmunk T10 (8209M/G-ATDY) <ff>	No 1940 Sqn ATC, Levenshulme, Gr Manchester	
WG419	DHC1 Chipmunk T10 (8206M) <ff>	Sywell Aviation Museum	
WG422	DHC1 Chipmunk T10 (8394M/G-BFAX) [16]	Privately owned, Eggesford	
WG432	DHC1 Chipmunk T10 [L]	Museum of Army Flying, Middle Wallop	
WG458	DHC1 Chipmunk T10 (N458BG) [2]	Privately owned, Breighton	
WG465	DHC1 Chipmunk T10 (G-BCEY)	Privately owned, White Waltham	
WG469	DHC1 Chipmunk T10 (G-BWJY) [72]	Privately owned, Sligo, Eire	
WG471	DHC1 Chipmunk T10 (8210M) <ff>	Thameside Aviation Museum, East Tilbury	
WG472	DHC1 Chipmunk T10 (G-AOTY)	Privately owned, Bryngwyn Bach, Clwyd	
WG477	DHC1 Chipmunk T10 (8362M/G-ATDP) <ff>	No 281 Sqn ATC, Birkdale, Merseyside	
WG486	DHC1 Chipmunk T10 [G]	RAF BBMF, Coningsby	
WG498	Slingsby T21B Sedbergh TX1 (BGA3245)	Privately owned, Aston Down	
WG511	Avro 696 Shackleton T4 (fuselage)	Flambards Village Theme Park, Helston	
WG655	Hawker Sea Fury T20 (G-CHFP) [910/GN]	The Fighter Collection, Duxford	
WG719	WS51 Dragonfly HR5 (G-BRMA)	The Helicopter Museum, Weston-super-Mare	
WG724	WS51 Dragonfly HR5 [932]	North-East Aircraft Museum, Usworth	
WG751	WS51 Dragonfly HR5 [710/GJ]	World Naval Base, Chatham	
WG760	EE P1A (7755M)	RAF Museum, Cosford	
WG763	EE P1A (7816M)	Museum of Science & Industry, Manchester	
WG768	Short SB5 (8005M)	RAF Museum, Cosford	
WG774	BAC 221	Science Museum, at FAA Museum, RNAS Yeovilton	
WG777	Fairey FD2 (7986M)	RAF Museum, Cosford	
WG789	EE Canberra B2/6 <ff>	Norfolk & Suffolk Avn Museum, Flixton	
WH132	Gloster Meteor T7 (7906M) [J]	RAF Leconfield, for display	
WH166	Gloster Meteor T7 (8052M) [A]	Privately owned, Birlingham, Worcs	
WH291	Gloster Meteor F8	Privately owned, Liverpool Airport	
WH301	Gloster Meteor F8 (7930M) [T]	RAF Museum, Hendon	
WH364	Gloster Meteor F8 (8169M)	Jet Age Museum, Gloucester	
WH453	Gloster Meteor F8	Bentwaters Cold War Air Museum	
WH646	EE Canberra T17A <ff>	Midland Air Museum, Coventry	
WH657	EE Canberra B2 <ff>	Romney Marsh Wartime Collection	
WH725	EE Canberra B2	Imperial War Museum, Duxford	
WH734	EE Canberra B2(mod) <ff>	Privately owned, Pershore	
WH739	EE Canberra B2 <ff>	No 2475 Sqn ATC, Ammanford, Dyfed	
WH740	EE Canberra T17 (8762M) [K]	East Midlands Airport Aeropark	
WH773	EE Canberra PR7 (8696M)	Gatwick Aviation Museum, Charlwood, Surrey	
WH775	EE Canberra PR7 (8128M/8868M) <ff>	Privately owned, Welshpool	
WH779	EE Canberra PR7 <ff>	South Yorkshire Aircraft Museum, Doncaster	
WH779	EE Canberra PR7 [BP] <rf>	RAF AM&SU, stored Shawbury	
WH792	EE Canberra PR7 (WH791/8165M/8176M/8187M)	Newark Air Museum, Winthorpe	
WH798	EE Canberra PR7 (8130M) <ff>	Privately owned, Kesgrave, Suffolk	
WH840	EE Canberra T4 (8350M)	Privately owned, Flixton	
WH846	EE Canberra T4	Yorkshire Air Museum, Elvington	
WH850	EE Canberra T4 <ff>	Privately owned, Narborough	
WH863	EE Canberra T17 (8693M) <ff>	Newark Air Museum, Winthorpe	
WH876	EE Canberra B2(mod) <ff>	Boscombe Down Aviation Collection, Old Sarum	
WH887	EE Canberra TT18 [847] <ff>	Sywell Aviation Museum	
WH903	EE Canberra B2 <ff>	Yorkshire Air Museum, Elvington	
WH904	EE Canberra T19	Newark Air Museum, Winthorpe	
WH953	EE Canberra B6(mod) <ff>	Blyth Valley Aviation Collection, Walpole, Suffolk	
WH957	EE Canberra E15 (8869M) <ff>	Lincolnshire Avn Heritage Centre, East Kirkby	
WH960	EE Canberra B15 (8344M) <ff>	Rolls-Royce Heritage Trust, Derby	
WH964	EE Canberra E15 (8870M) <ff>	Privately owned, Lewes	
WH984	EE Canberra B15 (8101M) <ff>	City of Norwich Aviation Museum	
WH991	WS51 Dragonfly HR3	Yorkshire Helicopter Preservation Group, Elvington	
WJ231	Hawker Sea Fury FB11 (*WE726*) [115/O]	FAA Museum, RNAS Yeovilton	
WJ306	Slingsby T21B Sedbergh TX1 (BGA3240)	Privately owned, Weston-on-the-Green	
WJ306	Slingsby T21B Sedbergh TX1 (WB957/BGA2720)	Privately owned, Parham Park, Sussex	
WJ358	Auster AOP6 (G-ARYD)	Museum of Army Flying, Middle Wallop	
WJ368	Auster AOP6 (G-ASZX)	Privately owned, Eggesford	

Notes	Serial	Type (code/other identity)	Owner/operator, location or fate
	WJ404	Auster AOP6 (G-ASOI)	Privately owned, Bidford-on-Avon, Warks
	WJ476	Vickers Valetta T3 <ff>	South Yorkshire Aircraft Museum, Doncaster
	WJ565	EE Canberra T17 (8871M) <ff>	South Yorkshire Aircraft Museum, Doncaster
	WJ567	EE Canberra B2 <ff>	Privately owned, Houghton, Cambs
	WJ576	EE Canberra T17 <ff>	Privately owned, stored Baxterley
	WJ633	EE Canberra T17 <ff>	City of Norwich Aviation Museum
	WJ639	EE Canberra TT18 [39]	North-East Aircraft Museum, Usworth
	WJ677	EE Canberra B2 <ff>	Privately owned, Redruth
	WJ717	EE Canberra TT18 (9052M) <ff>	Privately owned, Crewe, Cheshire
	WJ721	EE Canberra TT18 [21] <ff>	No 2405 Det Flt ATC, Gairloch
	WJ731	EE Canberra B2T [BK] <ff>	Privately owned, Golders Green
	WJ775	EE Canberra B6 (8581M) <ff>	Privately owned, Upwood, Cambs
	WJ865	EE Canberra T4 <ff>	Boscombe Down Aviation Collection, Old Sarum
	WJ880	EE Canberra T4 (8491M) <ff>	Dumfries & Galloway Avn Mus, Dumfries
	WJ903	Vickers Varsity T1 <ff>	South Yorkshire Aircraft Museum, Doncaster
	WJ945	Vickers Varsity T1 (G-BEDV) [21]	Classic Air Force, Newquay
	WJ975	EE Canberra T19 <ff>	South Yorkshire Aircraft Museum, Doncaster
	WJ992	EE Canberra T4	Bournemouth Airport Fire Section
	WK001	Thales Watchkeeper 450 RPAS (4X-USC)	Army 43 Battery Royal Artillery, Boscombe Down
	WK002	Thales Watchkeeper 450 RPAS (4X-USD)	Army
	WK003	Thales Watchkeeper 450 RPAS	Army
	WK004	Thales Watchkeeper 450 RPAS	Army
	WK005	Thales Watchkeeper 450 RPAS	Army 43 Battery Royal Artillery, Boscombe Down
	WK006	Thales Watchkeeper 450 RPAS	Army 43 Battery Royal Artillery, Boscombe Down
	WK007	Thales Watchkeeper 450 RPAS	Army
	WK008	Thales Watchkeeper 450 RPAS	Army 47 Regt Royal Artillery, Boscombe Down
	WK009	Thales Watchkeeper 450 RPAS	Army
	WK010	Thales Watchkeeper 450 RPAS	Army 43 Battery Royal Artillery, Boscombe Down
	WK011	Thales Watchkeeper 450 RPAS	Army
	WK012	Thales Watchkeeper 450 RPAS	Army
	WK013	Thales Watchkeeper 450 RPAS	Army
	WK014	Thales Watchkeeper 450 RPAS	Army
	WK015	Thales Watchkeeper 450 RPAS	Army
	WK016	Thales Watchkeeper 450 RPAS	Army
	WK017	Thales Watchkeeper 450 RPAS	Army
	WK018	Thales Watchkeeper 450 RPAS	Army
	WK019	Thales Watchkeeper 450 RPAS	Army
	WK020	Thales Watchkeeper 450 RPAS	Army
	WK021	Thales Watchkeeper 450 RPAS	Army
	WK022	Thales Watchkeeper 450 RPAS	Army
	WK023	Thales Watchkeeper 450 RPAS	Army
	WK024	Thales Watchkeeper 450 RPAS	Army
	WK025	Thales Watchkeeper 450 RPAS	Army
	WK026	Thales Watchkeeper 450 RPAS	Army
	WK027	Thales Watchkeeper 450 RPAS	Army
	WK028	Thales Watchkeeper 450 RPAS	Army
	WK029	Thales Watchkeeper 450 RPAS	Army 43 Battery Royal Artillery, Boscombe Down
	WK030	Thales Watchkeeper 450 RPAS	Army 43 Battery Royal Artillery, Boscombe Down
	WK031	Thales Watchkeeper 450 RPAS	*Crashed 16 October 2014, Aberporth*
	WK032	Thales Watchkeeper 450 RPAS	Thales, for Army
	WK033	Thales Watchkeeper 450 RPAS	Thales, for Army
	WK034	Thales Watchkeeper 450 RPAS	Thales, for Army
	WK035	Thales Watchkeeper 450 RPAS	Thales, for Army
	WK036	Thales Watchkeeper 450 RPAS	Thales, for Army
	WK037	Thales Watchkeeper 450 RPAS	Thales, for Army
	WK038	Thales Watchkeeper 450 RPAS	Thales, for Army
	WK039	Thales Watchkeeper 450 RPAS	Thales, for Army
	WK040	Thales Watchkeeper 450 RPAS	Thales, for Army
	WK041	Thales Watchkeeper 450 RPAS	Thales, for Army
	WK042	Thales Watchkeeper 450 RPAS	Thales, for Army
	WK043	Thales Watchkeeper 450 RPAS	Thales, for Army

Serial	Type (code/other identity)	Owner/operator, location or fate	Notes
WK044	Thales Watchkeeper 450 RPAS	Thales, for Army	
WK045	Thales Watchkeeper 450 RPAS	Thales, for Army	
WK046	Thales Watchkeeper 450 RPAS	Thales, for Army	
WK047	Thales Watchkeeper 450 RPAS	Thales, for Army	
WK048	Thales Watchkeeper 450 RPAS	Thales, for Army	
WK049	Thales Watchkeeper 450 RPAS	Thales, for Army	
WK050	Thales Watchkeeper 450 RPAS	Thales, for Army	
WK051	Thales Watchkeeper 450 RPAS	Thales, for Army	
WK052	Thales Watchkeeper 450 RPAS	Thales, for Army	
WK053	Thales Watchkeeper 450 RPAS	Thales, for Army	
WK054	Thales Watchkeeper 450 RPAS	Thales, for Army	
WK060	Thales Watchkeeper 450 RPAS	MoD/Thales, Aberporth	
WK102	EE Canberra T17 (8780M) <ff>	Privately owned, Welshpool	
WK118	EE Canberra TT18 [CQ] <ff>	Privately owned, Holt Heath, Worcs	
WK122	EE Canberra TT18 <ff>	Privately owned, Wesham, Lancs	
WK124	EE Canberra TT18 (9093M) [CR]	MoD DFTDC, Manston	
WK126	EE Canberra TT18 (N2138J) [843]	Jet Age Museum, stored Gloucester	
WK127	EE Canberra TT18 (8985M) <ff>	Privately owned, Peterborough	
WK146	EE Canberra B2 <ff>	Gatwick Aviation Museum, Charlwood, Surrey	
WK163	EE Canberra B2/6 (G-BVWC)	Privately owned, Coventry	
WK198	VS Swift F4 (7428M) (fuselage)	Brooklands Museum, Weybridge	
WK275	VS Swift F4	Privately owned, Thorpe Wood, N Yorks	
WK277	VS Swift FR5 (7719M) [N]	Newark Air Museum, Winthorpe	
WK281	VS Swift FR5 (7712M) [S]	Tangmere Military Aviation Museum	
WK393	DH112 Venom FB1 <ff>	South Yorkshire Aircraft Museum, stored Doncaster	
WK436	DH112 Venom FB50 (J-1614/G-VENM)	Classic Air Force, Newquay	
WK512	DHC1 Chipmunk T10 (G-BXIM) [A]	Privately owned, South Cerney	
WK514	DHC1 Chipmunk T10 (G-BBMO)	Privately owned, Wellesbourne Mountford	
WK517	DHC1 Chipmunk T10 (G-ULAS)	Privately owned, Turweston	
WK518	DHC1 Chipmunk T10 [C]	RAF BBMF, Coningsby	
WK518	DHC1 Chipmunk T10 (WP772)	RAF Manston History Museum	
WK522	DHC1 Chipmunk T10 (G-BCOU)	Privately owned, Duxford	
WK549	DHC1 Chipmunk T10 (G-BTWF)	Privately owned, Breighton	
WK562	DHC1 Chipmunk T10 (F-AZUR) [91]	Privately owned, La Baule, France	
WK570	DHC1 Chipmunk T10 (8211M) <ff>	No 424 Sqn ATC, Solent Sky, Southampton	
WK576	DHC1 Chipmunk T10 (8357M) <ff>	Tettenhall Transport Heritage Centre	
WK577	DHC1 Chipmunk T10 (G-BCYM)	Privately owned, Oaksey Park	
WK584	DHC1 Chipmunk T10 (7556M) <ff>	No 511 Sqn ATC, Ramsey, Cambs	
WK585	DHC1 Chipmunk T10 (9265M/G-BZGA)	Privately owned, Duxford	
WK586	DHC1 Chipmunk T10 (G-BXGX) [V]	Privately owned, Shoreham	
WK590	DHC1 Chipmunk T10 (G-BWVZ) [69]	Privately owned, Grimbergen, Belgium	
WK608	DHC1 Chipmunk T10 [906]	RN Historic Flight, Yeovilton	
WK609	DHC1 Chipmunk T10 (G-BXDN) [93]	Privately owned, Booker	
WK611	DHC1 Chipmunk T10 (G-ARWB)	Privately owned, Thruxton	
WK620	DHC1 Chipmunk T10 [T] (fuselage)	Privately owned, Enstone	
WK622	DHC1 Chipmunk T10 (G-BCZH)	Privately owned, Horsford	
WK624	DHC1 Chipmunk T10 (G-BWHI)	Privately owned, Blackpool	
WK626	DHC1 Chipmunk T10 (8213M) <ff>	South Yorkshire Aircraft Museum, stored Doncaster	
WK628	DHC1 Chipmunk T10 (G-BBMW)	Privately owned, Goodwood	
WK630	DHC1 Chipmunk T10 (G-BXDG)	Privately owned, Felthorpe	
WK633	DHC1 Chipmunk T10 (G-BXEC) [A]	Privately owned, Duxford	
WK635	DHC1 Chipmunk T10 (G-HFRH)	Privately owned, Hawarden	
WK638	DHC1 Chipmunk T10 (G-BWJZ) (fuselage)	Privately owned, South Marston, Swindon	
WK640	DHC1 Chipmunk T10 (G-BWUV) [C]	Privately owned, Aylesbury	
WK640	OGMA/DHC1 Chipmunk T20 (G-CERD)	Privately owned, Spanhoe	
WK642	DHC1 Chipmunk T10 (G-BXDP) [94]	Privately owned, Kilrush, Eire	
WK654	Gloster Meteor F8 (8092M)	City of Norwich Aviation Museum	
WK800	Gloster Meteor D16 [Z]	Boscombe Down Aviation Collection, Old Sarum	
WK864	Gloster Meteor F8 (WL168/7750M) [C]	Yorkshire Air Museum, Elvington	
WK935	Gloster Meteor Prone Pilot (7869M)	RAF Museum, Cosford	
WK991	Gloster Meteor F8 (7825M)	Imperial War Museum, Duxford	

Notes	Serial	Type (code/other identity)	Owner/operator, location or fate
	WL131	Gloster Meteor F8 (7751M) <ff>	South Yorkshire Aircraft Museum, Doncaster
	WL181	Gloster Meteor F8 [X]	North-East Aircraft Museum, Usworth
	WL332	Gloster Meteor T7 [888]	Privately owned, Long Marston
	WL345	Gloster Meteor T7 (comp WL360)	Privately owned, Booker
	WL349	Gloster Meteor T7	Gloucestershire Airport, Staverton, on display
	WL375	Gloster Meteor T7(mod)	Dumfries & Galloway Avn Mus, Dumfries
	WL405	Gloster Meteor T7 <ff>	Privately owned, Parbold, Lancs
	WL419	Gloster Meteor T7(mod)	Martin Baker Aircraft, Chalgrove
	WL505	DH100 Vampire FB9 (7705M/G-FBIX)	*Currently not known*
	WL626	Vickers Varsity T1 (G-BHDD) [P]	East Midlands Airport Aeropark
	WL627	Vickers Varsity T1 (8488M) [D] <ff>	Privately owned, Preston, E Yorkshire
	WL679	Vickers Varsity T1 (9155M)	RAF Museum, Cosford
	WL732	BP P108 Sea Balliol T21	RAF Museum, Cosford
	WL756	Avro 696 Shackleton AEW2 (9101M) <ff>	Privately owned,
	WL795	Avro 696 Shackleton MR2C (8753M) [T]	Newquay Airport, on display
	WL798	Avro 696 Shackleton MR2C (8114M) <ff>	Privately owned, Elgin
	WM145	AW Meteor NF11 <ff>	Privately owned, Over Dinsdale, N Yorks
	WM167	AW Meteor NF11 (G-LOSM)	Classic Air Force, Newquay
	WM224	AW Meteor TT20 (*WM311*/8177M) [X]	East Midlands Airport Aeropark
	WM267	AW Meteor NF11 <ff>	City of Norwich Aviation Museum
	WM292	AW Meteor TT20 [841]	FAA Museum, stored Cobham Hall, RNAS Yeovilton
	WM366	AW Meteor NF13 (4X-FNA) (comp VZ462)	Jet Age Museum, Gloucester
	WM367	AW Meteor NF13 <ff>	East Midlands Airport Aeropark
	WM571	DH112 Sea Venom FAW21 [VL]	Solent Sky, stored Romsey
	WM729	DH113 Vampire NF10 <ff>	Mosquito Aircraft Museum, stored London Colney
	WM913	Hawker Sea Hawk FB5 (8162M) [456/J]	Newark Air Museum, Winthorpe
	WM961	Hawker Sea Hawk FB5 [J]	Caernarfon Air World
	WM969	Hawker Sea Hawk FB5 [10/Z]	Imperial War Museum, Duxford
	WN105	Hawker Sea Hawk FB3 (WF299/8164M)	Privately owned, Birlingham, Worcs
	WN108	Hawker Sea Hawk FB5 [033]	Ulster Aviation Society, Long Kesh
	WN149	BP P108 Balliol T2 [AT]	RAF Museum, Cosford
	WN411	Fairey Gannet AS1 (fuselage)	Privately owned, Sholing, Hants
	WN493	WS51 Dragonfly HR5	FAA Museum, RNAS Yeovilton
	WN499	WS51 Dragonfly HR5	South Yorkshire Aircraft Museum, Doncaster
	WN516	BP P108 Balliol T2 <ff>	Privately owned, stored Otherton, Staffs
	WN534	BP P108 Balliol T2 <ff>	Tettenhall Transport Heritage Centre
	WN890	Hawker Hunter F2 <ff>	Boscombe Down Aviation Collection, Old Sarum
	WN904	Hawker Hunter F2 (7544M)	Sywell Aviation Museum
	WN907	Hawker Hunter F2 (7416M) <ff>	Robertsbridge Aviation Society, Newhaven
	WN957	Hawker Hunter F5 <ff>	Privately owned, Stockport
	WP185	Hawker Hunter F5 (7583M)	Privately owned, Great Dunmow, Essex
	WP190	Hawker Hunter F5 (7582M/8473M/*WP180*) [K]	Tangmere Military Aviation Museum
	WP255	DH113 Vampire NF10 <ff>	South Yorkshire Aircraft Museum, stored Doncaster
	WP269	EoN Eton TX1 (BGA3214)	Privately owned, stored Keevil
	WP270	EoN Eton TX1 (8598M)	RAF Museum Reserve Collection, Stafford
	WP308	Percival P57 Sea Prince T1 (G-GACA) [572/CU]	Gatwick Aviation Museum, Charlwood, Surrey
	WP313	Percival P57 Sea Prince T1 [568/CU]	FAA Museum, stored Cobham Hall, RNAS Yeovilton
	WP314	Percival P57 Sea Prince T1 (8634M) [573/CU]	Privately owned, Carlisle Airport
	WP321	Percival P57 Sea Prince T1 (G-BRFC) [750/CU]	Privately owned, St Athan
	WP772	DHC1 Chipmunk T10 [Q] (wreck)	*Repainted as WK518*
	WP784	DHC1 Chipmunk T10 (comp WZ876) [RCY-E]	East Midlands Airport Aeropark
	WP788	DHC1 Chipmunk T10 (G-BCHL)	Privately owned, Sleap
	WP790	DHC1 Chipmunk T10 (G-BBNC) [T]	DHAHC, London Colney
	WP795	DHC1 Chipmunk T10 (G-BVZZ) [901]	Privately owned, Lee-on-Solent
	WP800	DHC1 Chipmunk T10 (G-BCXN) [2]	Privately owned, Halton
	WP803	DHC1 Chipmunk T10 (G-HAPY) [G]	Privately owned, Booker
	WP805	DHC1 Chipmunk T10 (G-MAJR) [D]	Privately owned, Lee-on-Solent
	WP809	DHC1 Chipmunk T10 (G-BVTX) [78]	Privately owned, Husbands Bosworth
	WP835	DHC1 Chipmunk T10 (D-ERTY)	Privately owned, Teuge, The Netherlands

Serial	Type (code/other identity)	Owner/operator, location or fate	Notes
WP840	DHC1 Chipmunk T10 (F-AZQM) [9]	Privately owned, Reims, France	
WP844	DHC1 Chipmunk T10 (G-BWOX) [85]	Privately owned, Shobdon	
WP848	DHC1 Chipmunk T10 (8342M/G-BFAW)	Privately owned, Old Buckenham	
WP859	DHC1 Chipmunk T10 (G-BXCP) [E]	Privately owned, Fishburn	
WP860	DHC1 Chipmunk T10 (G-BXDA) [6]	Privately owned, Kirknewton	
WP863	DHC1 Chipmunk T10 (8360M/G-ATJI) <ff>	No 1011 Sqn ATC, Boscombe Down	
WP869	DHC1 Chipmunk T10 (8215M) <ff>	Mosquito Aircraft Museum, London Colney	
WP870	DHC1 Chipmunk T10 (G-BCOI) [12]	Privately owned, Rayne Hall Farm, Essex	
WP896	DHC1 Chipmunk T10 (G-BWVY)	Privately owned, RAF Halton	
WP901	DHC1 Chipmunk T10 (G-BWNT) [B]	Privately owned, Tollerton	
WP903	DHC1 Chipmunk T10 (G-BCGC)	Privately owned, Henlow	
WP912	DHC1 Chipmunk T10 (8467M)	RAF Museum, Cosford	
WP921	DHC1 Chipmunk T10 (G-ATJJ) <ff>	Privately owned, Brooklands	
WP925	DHC1 Chipmunk T10 (G-BXHA) [C]	Privately owned, Seppe, The Netherlands	
WP927	DHC1 Chipmunk T10 (8216M/G-ATJK) <ff>	Privately owned, St Neots, Cambs	
WP928	DHC1 Chipmunk T10 (G-BXGM) [D]	Privately owned, Shoreham	
WP929	DHC1 Chipmunk T10 (G-BXCV) [F]	Privately owned, Duxford	
WP930	DHC1 Chipmunk T10 (G-BXHF) [J]	Privately owned, Duxford	
WP962	DHC1 Chipmunk T10 (9287M) [C]	RAF Museum, Hendon	
WP971	DHC1 Chipmunk T10 (G-ATHD)	Privately owned, Denham	
WP977	DHC1 Chipmunk T10 (G-BHRD) <ff>	Privately owned, Yateley, Hants	
WP983	DHC1 Chipmunk T10 (G-BXNN) [B]	Privately owned, Eggesford	
WP984	DHC1 Chipmunk T10 (G-BWTO) [H]	Privately owned, Little Gransden	
WR410	DH112 Venom FB54 (J-1790/G-BLKA) [N]	DHAHC, London Colney	
WR470	DH112 Venom FB50 (J-1542/G-DHVM)	Classic Air Force, Newquay	
WR539	DH112 Venom FB4 (8399M) <ff>	Privately owned, Cantley, Norfolk	
WR960	Avro 696 Shackleton AEW2 (8772M)	Museum of Science & Industry, Manchester	
WR963	Avro 696 Shackleton AEW2 (G-SKTN) [B-M]	Privately owned, Coventry	
WR971	Avro 696 Shackleton MR3 (8119M) [Q] (fuselage)	Fenland & W Norfolk Aviation Museum, Wisbech	
WR974	Avro 696 Shackleton MR3 (8117M) [K]	Privately owned, Bruntingthorpe	
WR977	Avro 696 Shackleton MR3 (8186M) [B]	Newark Air Museum, Winthorpe	
WR982	Avro 696 Shackleton MR3 (8106M) [J]	Gatwick Aviation Museum, Charlwood, Surrey	
WR985	Avro 696 Shackleton MR3 (8103M) [H]	Privately owned, Long Marston	
WS103	Gloster Meteor T7 [709]	FAA Museum, stored Cobham Hall, RNAS Yeovilton	
WS692	Gloster Meteor NF12 (7605M) [C]	Newark Air Museum, Winthorpe	
WS726	Gloster Meteor NF14 (7960M) [H]	No 1855 Sqn ATC, Royton, Gr Manchester	
WS739	Gloster Meteor NF14 (7961M)	Newark Air Museum, Winthorpe	
WS760	Gloster Meteor NF14 (7964M)	East Midlands Airport Aeropark, stored	
WS776	Gloster Meteor NF14 (7716M) [K]	Bournemouth Aviation Museum	
WS788	Gloster Meteor NF14 (7967M) [Z]	Yorkshire Air Museum, Elvington	
WS792	Gloster Meteor NF14 (7965M) [K]	Brighouse Bay Caravan Park, Borgue, D&G	
WS807	Gloster Meteor NF14 (7973M) [N]	Jet Age Museum, stored Gloucester	
WS832	Gloster Meteor NF14	Solway Aviation Society, Carlisle	
WS838	Gloster Meteor NF14 [D]	Midland Air Museum, Coventry	
WS840	Gloster Meteor NF14 (7969M) [N] <rf>	Privately owned, Upper Ballinderry, NI	
WS843	Gloster Meteor NF14 (7937M) [J]	RAF Museum, Cosford	
WT121	Douglas Skyraider AEW1 [415/CU]	FAA Museum, stored Cobham Hall, RNAS Yeovilton	
WT205	EE Canberra B15 <ff>	RAF Manston History Museum	
WT308	EE Canberra B(I)6	RN, Predannack Fire School	
WT309	EE Canberra B(I)6	Farnborough Air Sciences Trust, Farnborough	
WT319	EE Canberra B(I)6 <ff>	Privately owned, Lavendon, Bucks	
WT333	EE Canberra B6(mod) (G-BVXC)	Privately owned, Bruntingthorpe	
WT339	EE Canberra B(I)8 (8198M)	RAF Barkston Heath Fire Section	
WT482	EE Canberra T4 <ff>	Privately owned, Marske by the Sea, Durham	
WT486	EE Canberra T4 (8102M) <ff>	Privately owned, Newtownards	
WT507	EE Canberra PR7 (8131M/8548M) [44] <ff>	No 384 Sqn ATC, Mansfield	
WT520	EE Canberra PR7 (8094M/8184M) <ff>	No 967 Sqn ATC, Warton	
WT525	EE Canberra T22 [855] <ff>	Privately owned, Camborne	
WT532	EE Canberra PR7 (8728M/8890M) <ff>	Bournemouth Aviation Museum	

Notes	Serial	Type (code/other identity)	Owner/operator, location or fate
	WT534	EE Canberra PR7 (8549M) [43] <ff>	Privately owned, Upwood
	WT536	EE Canberra PR7 (8063M) <ff>	Privately owned, Shirrell Heath, Hants
	WT555	Hawker Hunter F1 (7499M)	Vanguard Haulage, Greenford, London
	WT569	Hawker Hunter F1 (7491M)	No 2117 Sqn ATC, Kenfig Hill, Mid-Glamorgan
	WT612	Hawker Hunter F1 (7496M)	RAF Henlow, on display
	WT619	Hawker Hunter F1 (7525M)	RAF Museum Reserve Collection, Stafford
	WT648	Hawker Hunter F1 (7530M) <ff>	Boscombe Down Aviation Collection, Old Sarum
	WT651	Hawker Hunter F1 (7532M) [C]	Newark Air Museum, Winthorpe
	WT660	Hawker Hunter F1 (7421M) [C]	Highland Aviation Museum, Inverness
	WT680	Hawker Hunter F1 (7533M) [J]	Privately owned, Holbeach, Lincs
	WT684	Hawker Hunter F1 (7422M) <ff>	Privately owned, Lavendon, Bucks
	WT694	Hawker Hunter F1 (7510M)	Caernarfon Air World
	WT711	Hawker Hunter GA11 [833/DD]	Lakes Lightnings, Spark Bridge, Cumbria
	WT720	Hawker Hunter F51 (RDAF E-408/8565M) [B]	Privately owned, North Scarle, Lincs
	WT722	Hawker Hunter T8C (G-BWGN) [878/VL]	Classic Air Force, Newquay
	WT723	Hawker Hunter PR11 (XG194/G-PRII) [692/LM]	Hunter Flight Academy, St Athan
	WT741	Hawker Hunter GA11 [791] <ff>	Privately owned, Doncaster
	WT744	Hawker Hunter GA11 [868/VL]	Privately owned, Ilfracombe
	WT799	Hawker Hunter T8C [879]	Blue Lagoon Diving Centre, Womersley, N Yorks
	WT804	Hawker Hunter GA11 [831/DD]	FETC, Moreton-in-Marsh, Glos
	WT806	Hawker Hunter GA11	Privately owned, Bruntingthorpe
	WT859	Supermarine 544 <ff>	Boscombe Down Aviation Collection, Old Sarum
	WT867	Slingsby T31B Cadet TX3	Privately owned, Eaglescott
	WT874	Slingsby T31B Cadet TX3 (BGA1255)	Privately owned, Lasham
	WT877	Slingsby T31B Cadet TX3	Staffs Aircraft Restoration Team, Baxterley
	WT900	Slingsby T31B Cadet TX3 (BGA3272)	Currently not known
	WT905	Slingsby T31B Cadet TX3	Privately owned, Keevil
	WT908	Slingsby T31B Cadet TX3 (BGA3487)	Privately owned, Dunstable
	WT910	Slingsby T31B Cadet TX3 (BGA3953)	Privately owned, Llandegla, Denbighshire
	WT914	Slingsby T31B Cadet TX3 (BGA3194) (fuselage)	East Midlands Airport Aeropark
	WT933	Bristol 171 Sycamore 3 (G-ALSW/7709M)	Newark Air Museum, Winthorpe
	WV106	Douglas Skyraider AEW1 [427/C]	FAA Museum, stored Cobham Hall, RNAS Yeovilton
	WV198	Sikorsky S55 Whirlwind HAR21 (G-BJWY) [K]	Solway Aviation Society, Carlisle
	WV318	Hawker Hunter T7A (9236M/G-FFOX)	Hunter Flight Academy, Cranfield
	WV322	Hawker Hunter T8C (G-BZSE/9096M) [Y]	Hunter Flying Ltd, North Weald
	WV332	Hawker Hunter F4 (7673M) <ff>	Tangmere Military Aircraft Museum
	WV372	Hawker Hunter T7 (G-BXFI) [R]	Privately owned, North Weald
	WV381	Hawker Hunter GA11 [732] <ff>	Privately owned, Chiltern Park, Wallingford
	WV382	Hawker Hunter GA11 [830/VL]	East Midlands Airport Aeropark
	WV383	Hawker Hunter T7	Farnborough Air Sciences Trust, Farnborough
	WV396	Hawker Hunter T8C (9249M) [91]	RAF Valley, at main gate
	WV493	Percival P56 Provost T1 (G-BDYG/7696M) [29]	Royal Scottish Mus'm of Flight, stored E Fortune
	WV499	Percival P56 Provost T1 (G-BZRF/7698M) [P-G]	Privately owned, Westonzoyland, Somerset
	WV514	Percival P56 Provost T51 (G-BLIW) [N-C]	Privately owned, Shoreham
	WV562	Percival P56 Provost T1 (7606M) [P-C]	Repainted in Omani markings as XF688, June 2014
	WV605	Percival P56 Provost T1 [T-B]	Norfolk & Suffolk Avn Museum, Flixton
	WV606	Percival P56 Provost T1 (7622M)[P-B]	Newark Air Museum, Winthorpe
	WV679	Percival P56 Provost T1 (7615M) [O-J]	Wellesbourne Wartime Museum
	WV705	Percival P66 Pembroke C1 <ff>	Privately owned, Awbridge, Hants
	WV740	Percival P66 Pembroke C1 (G-BNPH)	Privately owned, St Athan
	WV746	Percival P66 Pembroke C1 (8938M)	RAF Museum, Cosford
	WV781	Bristol 171 Sycamore HR12 (G-ALTD/7839M) <ff>	Caernarfon Air World
	WV783	Bristol 171 Sycamore HR12 (G-ALSP/7841M)	RAF Museum, Hendon
	WV787	EE Canberra B2/8 (8799M)	Newark Air Museum, Winthorpe
	WV795	Hawker Sea Hawk FGA6 (8151M)	Privately owned, Dunsfold
	WV797	Hawker Sea Hawk FGA6 (8155M) [491/J]	Midland Air Museum, Coventry
	WV798	Hawker Sea Hawk FGA6 [026/CU]	Classic Air Force, Newquay
	WV838	Hawker Sea Hawk FGA4 [182] <ff>	Norfolk & Suffolk Avn Museum, Flixton
	WV856	Hawker Sea Hawk FGA4 [163]	FAA Museum, RNAS Yeovilton
	WV903	Hawker Sea Hawk FGA4 (8153M) [128] <ff>	Privately owned, Hooton Park
	WV908	Hawker Sea Hawk FGA6 (8154M) [188/A]	RN Historic Flight, Yeovilton

Serial	Type (code/other identity)	Owner/operator, location or fate	Notes
WV910	Hawker Sea Hawk FGA6 <ff>	Boscombe Down Aviation Collection, Old Sarum	
WV911	Hawker Sea Hawk FGA4 [115/C]	RNAS Yeovilton, Fire Section	
WW138	DH112 Sea Venom FAW22 [227/Z]	FAA Museum, stored Cobham Hall, RNAS Yeovilton	
WW145	DH112 Sea Venom FAW22 [680/LM]	Royal Scottish Mus'm of Flight, E Fortune	
WW217	DH112 Sea Venom FAW22 [351]	Newark Air Museum, Winthorpe	
WW388	Percival P56 Provost T1 (7616M) [O-F]	Privately owned, stored Hinstock, Shrops	
WW421	Percival P56 Provost T1 (WW450/G-BZRE/7689M) [P-B]	Bournemouth Aviation Museum	
WW442	Percival P56 Provost T1 (7618M) [N]	East Midlands Airport Aeropark	
WW444	Percival P56 Provost T1 [D]	Privately owned, Brownhills, Staffs	
WW447	Percival P56 Provost T1 [F]	Privately owned, Shoreham	
WW453	Percival P56 Provost T1 (G-TMKI) [W-S]	Privately owned, Westonzoyland, Somerset	
WW654	Hawker Hunter GA11 [834/DD]	Privately owned, Ford, W Sussex	
WW664	Hawker Hunter F4 <ff>	Privately owned, Norfolk	
WX788	DH112 Venom NF3 <ff>	South Yorkshire Aircraft Museum, Doncaster	
WX853	DH112 Venom NF3 (7443M)	Mosquito Aircraft Museum, stored London Colney	
WX905	DH112 Venom NF3 (7458M)	Newark Air Museum, Winthorpe	
WZ425	DH115 Vampire T11	Privately owned, Birlingham, Worcs	
WZ450	DH115 Vampire T11	Privately owned, Corscombe, Dorset	
WZ507	DH115 Vampire T11 (G-VTII) [74]	Privately owned, North Weald	
WZ515	DH115 Vampire T11 [60]	Solway Aviation Society, Carlisle	
WZ518	DH115 Vampire T11 [B]	North-East Aircraft Museum, Usworth	
WZ549	DH115 Vampire T11 (8118M) [F]	Ulster Aviation Society, Long Kesh	
WZ553	DH115 Vampire T11 (G-DHYY) <ff>	Privately owned, Stockton, Warks	
WZ557	DH115 Vampire T11	Privately owned, Over Dinsdale, N Yorks	
WZ572	DH115 Vampire T11 (8124M) [65] <ff>	Privately owned, Sholing, Hants	
WZ581	DH115 Vampire T11 <ff>	The Vampire Collection, Hemel Hempstead	
WZ584	DH115 Vampire T11 (G-BZRC) [K]	Privately owned, Binbrook	
WZ589	DH115 Vampire T11 [19]	Privately owned, Wigmore, Kent	
WZ589	DH115 Vampire T55 (U-1230/LN-DHZ)	Privately owned, Norway	
WZ590	DH115 Vampire T11 [49]	Imperial War Museum, Duxford	
WZ662	Auster AOP9 (G-BKVK)	Privately owned, Eggesford	
WZ706	Auster AOP9 (7851M/G-BURR)	Privately owned, Eggesford	
WZ711	Auster AOP9/Beagle E3 (G-AVHT)	Privately owned, Spanhoe	
WZ721	Auster AOP9	Museum of Army Flying, Middle Wallop	
WZ724	Auster AOP9 (7432M)	AAC Middle Wallop, at main gate	
WZ729	Auster AOP9 (G-BXON)	Privately owned, Newark-on-Trent	
WZ736	Avro 707A (7868M)	Museum of Science & Industry, Manchester	
WZ744	Avro 707C (7932M)	RAF Museum, stored Cosford	
WZ753	Slingsby T38 Grasshopper TX1	Solent Sky, Southampton	
WZ755	Slingsby T38 Grasshopper TX1 (BGA3481)	Privately owned, stored Baxterley	
WZ757	Slingsby T38 Grasshopper TX1 (comp XK820)	Privately owned, Kirton-in-Lindsey, Lincs	
WZ767	Slingsby T38 Grasshopper TX1	North-East Aircraft Museum, stored Usworth	
WZ772	Slingsby T38 Grasshopper TX1	Trenchard Museum, RAF Halton	
WZ773	Slingsby T38 Grasshopper TX1	Edinburgh Academy	
WZ784	Slingsby T38 Grasshopper TX1	Privately owned, stored Southend	
WZ791	Slingsby T38 Grasshopper TX1 (8944M)	RAF Museum, Hendon	
WZ793	Slingsby T38 Grasshopper TX1	Privately owned, Keevil	
WZ796	Slingsby T38 Grasshopper TX1	Privately owned, stored Aston Down	
WZ798	Slingsby T38 Grasshopper TX1	Privately owned, stored Hullavington	
WZ816	Slingsby T38 Grasshopper TX1 (BGA3979)	Privately owned, Redhill	
WZ818	Slingsby T38 Grasshopper TX1 (BGA4361)	Privately owned, Nympsfield	
WZ819	Slingsby T38 Grasshopper TX1 (BGA3498)	Privately owned, Halton	
WZ820	Slingsby T38 Grasshopper TX1	Sywell Aviation Museum	
WZ822	Slingsby T38 Grasshopper TX1	South Yorkshire Aircraft Museum, stored Doncaster	
WZ824	Slingsby T38 Grasshopper TX1	Solway Aviation Society, Carlisle	
WZ826	Vickers Valiant B(K)1 (XD826/7872M) <ff>	Privately owned, Rayleigh, Essex	
WZ828	Slingsby T38 Grasshopper TX1 (BGA4421)	Privately owned, Hullavington	
WZ831	Slingsby T38 Grasshopper TX1	Privately owned, stored Nympsfield, Glos	

Notes	Serial	Type (code/other identity)	Owner/operator, location or fate
	WZ846	DHC1 Chipmunk T10 (G-BCSC/8439M)	No 2427 Sqn ATC, Biggin Hill
	WZ847	DHC1 Chipmunk T10 (G-CPMK) [F]	Privately owned, Sleap
	WZ868	DHC1 Chipmunk T10 (WG322/G-ARMF)	Vintage Flyers, stored Nailsworth, Glos
	WZ869	DHC1 Chipmunk T10 (8019M) <ff>	Privately owned, Leicester
	WZ872	DHC1 Chipmunk T10 (G-BZGB) [E]	Privately owned, Blackpool
	WZ876	DHC1 Chipmunk T10 (G-BBWN) <ff>	Tangmere Military Aviation Museum
	WZ879	DHC1 Chipmunk T10 (G-BWUT) [X]	Privately owned, Duxford
	WZ882	DHC1 Chipmunk T10 (G-BXGP) [K]	Privately owned, Eaglescott
	XA109	DH115 Sea Vampire T22	Montrose Air Station Heritage Centre
	XA127	DH115 Sea Vampire T22 <ff>	FAA Museum, RNAS Yeovilton
	XA129	DH115 Sea Vampire T22	FAA Museum, stored Cobham Hall, RNAS Yeovilton
	XA225	Slingsby T38 Grasshopper TX1	Privately owned, RAF Odiham
	XA226	Slingsby T38 Grasshopper TX1	Norfolk & Suffolk Avn Museum, Flixton
	XA228	Slingsby T38 Grasshopper TX1	Royal Scottish Mus'm of Flight, stored Granton
	XA230	Slingsby T38 Grasshopper TX1 (BGA4098)	Privately owned, Henlow
	XA231	Slingsby T38 Grasshopper TX1 (8888M)	RAF Manston History Museum
	XA240	Slingsby T38 Grasshopper TX1 (BGA4556)	Privately owned, Portmoak, Perth & Kinross
	XA241	Slingsby T38 Grasshopper TX1	Shuttleworth Collection, Old Warden
	XA243	Slingsby T38 Grasshopper TX1 (8886M)	Privately owned, Gransden Lodge, Cambs
	XA244	Slingsby T38 Grasshopper TX1	Privately owned, Brent Tor, Devon
	XA282	Slingsby T31B Cadet TX3	Caernarfon Air World
	XA289	Slingsby T31B Cadet TX3	Privately owned, Eaglescott
	XA290	Slingsby T31B Cadet TX3	Privately owned, Portmoak, Perth & Kinross
	XA293	Slingsby T31B Cadet TX3 <ff>	Privately owned, Breighton
	XA295	Slingsby T31B Cadet TX3 (BGA3336)	Privately owned, Eaglescott
	XA302	Slingsby T31B Cadet TX3 (BGA3786)	RAF Museum, Hendon
	XA310	Slingsby T31B Cadet TX3 (BGA4963)	Privately owned, Hullavington
	XA459	Fairey Gannet ECM6 [E]	Privately owned, White Waltham
	XA460	Fairey Gannet ECM6 [768/BY]	Ulster Aviation Society, Long Kesh
	XA466	Fairey Gannet COD4 [777/LM]	FAA Museum, stored Cobham Hall, RNAS Yeovilton
	XA508	Fairey Gannet T2 [627/GN]	FAA Museum, at Midland Air Museum, Coventry
	XA564	Gloster Javelin FAW1 (7464M)	RAF Museum, Cosford
	XA634	Gloster Javelin FAW4 (7641M)	Jet Age Museum, Gloucester
	XA699	Gloster Javelin FAW5 (7809M)	Midland Air Museum, Coventry
	XA847	EE P1B (8371M)	Privately owned, Stowmarket, Suffolk
	XA862	WS55 Whirlwind HAR1 (G-AMJT) <ff>	South Yorkshire Aircraft Museum, Doncaster
	XA864	WS55 Whirlwind HAR1	FAA Museum, stored Cobham Hall, RNAS Yeovilton
	XA870	WS55 Whirlwind HAR1 [911]	South Yorkshire Aircraft Museum, Doncaster
	XA880	DH104 Devon C2 (G-BVXR)	Privately owned, Little Rissington
	XA893	Avro 698 Vulcan B1 (8591M) <ff>	RAF Museum, Cosford
	XA903	Avro 698 Vulcan B1 <ff>	Privately owned, Stoneykirk, D&G
	XA917	HP80 Victor B1 (7827M) <ff>	Privately owned, Cupar, Fife
	XB259	Blackburn B101 Beverley C1 (G-AOAI)	Fort Paull Armoury
	XB261	Blackburn B101 Beverley C1 <ff>	Newark Air Museum, Winthorpe
	XB446	Grumman TBM-3 Avenger ECM6B	FAA Museum, Yeovilton
	XB480	Hiller HT1 [537]	FAA Museum, stored Cobham Hall, RNAS Yeovilton
	XB812	Canadair CL-13 Sabre F4 (9227M) [U]	RAF Museum, Cosford
	XD145	Saro SR53	RAF Museum, Cosford
	XD163	WS55 Whirlwind HAR10 (8645M) [X]	The Helicopter Museum, Weston-super-Mare
	XD165	WS55 Whirlwind HAR10 (8673M)	Caernarfon Airfield Fire Section
	XD215	VS Scimitar F1 <ff>	Privately owned, Cheltenham
	XD235	VS Scimitar F1 <ff>	Privately owned, Olney, Bucks
	XD317	VS Scimitar F1 [112/R]	FAA Museum, RNAS Yeovilton
	XD332	VS Scimitar F1 [194/C]	Solent Sky, stored Romsey
	XD375	DH115 Vampire T11 (7887M)	Privately owned, Elland, W Yorks
	XD377	DH115 Vampire T11 (8203M) <ff>	South Yorkshire Aircraft Museum, stored Doncaster
	XD425	DH115 Vampire T11 <ff>	Morayvia, Kinloss
	XD434	DH115 Vampire T11 [25]	Fenland & W Norfolk Aviation Museum, Wisbech
	XD445	DH115 Vampire T11 [51]	Privately owned, stored Baxterley

Serial	Type (code/other identity)	Owner/operator, location or fate	Notes
XD447	DH115 Vampire T11 [50]	East Midlands Airport Aeropark	
XD452	DH115 Vampire T11 (7990M) [66] <ff>	Privately owned, Dursley, Glos	
XD459	DH115 Vampire T11 [63] <ff>	East Midlands Airport Aeropark, stored	
XD506	DH115 Vampire T11 (7983M)	Jet Age Museum, stored Gloucester	
XD515	DH115 Vampire T11 (7998M/XM515)	RAF Museum, stored Cosford	
XD525	DH115 Vampire T11 (7882M) <ff>	Privately owned, Templepatrick, NI	
XD534	DH115 Vampire T11 [41]	East Midlands Airport Aeropark	
XD542	DH115 Vampire T11 (7604M) [N]	Montrose Air Station Heritage Centre	
XD547	DH115 Vampire T11 [Z] (composite)	Privately owned, Cantley, Norfolk	
XD593	DH115 Vampire T11	Newark Air Museum, Winthorpe	
XD595	DH115 Vampire T11 <ff>	Privately owned, Glentham, Lincs	
XD596	DH115 Vampire T11 (7939M)	Solent Sky, stored Timsbury, Hants	
XD599	DH115 Vampire T11 [A] <ff>	Sywell Aviation Museum	
XD616	DH115 Vampire T11 [56]	Mosquito Aircraft Museum, stored Gloucester	
XD622	DH115 Vampire T11 (8160M)	No 2214 Sqn ATC, Usworth	
XD624	DH115 Vampire T11	Privately owned, Hooton Park	
XD626	DH115 Vampire T11 [Q]	Midland Air Museum, stored Coventry	
XD674	Hunting Jet Provost T1 (7570M)	RAF Museum, Cosford	
XD693	Hunting Jet Provost T1 (XM129/G-AOBU) [Z-Q]	Kennet Aviation, North Weald	
XD816	Vickers Valiant B(K)1 <ff>	Brooklands Museum, Weybridge	
XD818	Vickers Valiant B(K)1 (7894M)	RAF Museum, Cosford	
XD857	Vickers Valiant B(K)1 <ff>	Norfolk & Suffolk Aviation Museum, Flixton	
XD875	Vickers Valiant B(K)1 <ff>	Highland Aviation Museum, Inverness	
XE317	Bristol 171 Sycamore HR14 (G-AMWO) [S-N]	South Yorkshire Aircraft Museum, stored Doncaster	
XE339	Hawker Sea Hawk FGA6 (8156M) [149] <ff>	Privately owned, Glos	
XE339	Hawker Sea Hawk FGA6 (8156M) [E] <rf>	Privately owned, Booker	
XE340	Hawker Sea Hawk FGA6 [131/Z]	FAA Museum, stored Cobham Hall, RNAS Yeovilton	
XE364	Hawker Sea Hawk FGA6 (G-JETH) (comp WM983) [485/J]	Gatwick Aviation Museum, Charlwood, Surrey	
XE368	Hawker Sea Hawk FGA6 [200/J]	Privately owned, Barrow-in-Furness	
XE521	Fairey Rotodyne Y (parts)	The Helicopter Museum, Weston-super-Mare	
XE584	Hawker Hunter FGA9 <ff>	Privately owned, Hooton Park	
XE597	Hawker Hunter FGA9 (8874M) <ff>	Privately owned, Bromsgrove	
XE620	Hawker Hunter F6A (XE606/8841M) [B]	RAF Waddington, on display	
XE624	Hawker Hunter FGA9 (8875M) [G]	Privately owned, Wickenby	
XE627	Hawker Hunter F6A [T]	Imperial War Museum, Duxford	
XE643	Hawker Hunter FGA9 (8586M) <ff>	Ulster Aviation Society, Long Kesh	
XE650	Hawker Hunter FGA9 (G-9-449) <ff>	Farnborough Air Sciences Trust, Farnborough	
XE664	Hawker Hunter F4 <ff>	Jet Age Museum, stored Gloucester	
XE665	Hawker Hunter T8C (G-BWGM)	Privately owned, Cotswold Airport	
XE668	Hawker Hunter GA11 [832/DD]	Hamburger Hill Paintball, Marksbury, Somerset	
XE670	Hawker Hunter F4 (7762M/8585M) <ff>	RAF Museum, Cosford	
XE683	Hawker Hunter F51 (RDAF E-409) [G]	City of Norwich Aviation Museum	
XE685	Hawker Hunter GA11 (G-GAII) [861/VL]	Hawker Hunter Aviation, Scampton	
XE689	Hawker Hunter GA11 (G-BWGK) <ff>	Privately owned, Cotswold Airport	
XE707	Hawker Hunter GA11 (N707XE) [865]	Bentwaters Cold War Museum	
XE786	Slingsby T31B Cadet TX3 (BGA4033)	Privately owned, Arbroath	
XE793	Slingsby T31B Cadet TX3 (8666M)	Privately owned, Tamworth	
XE799	Slingsby T31B Cadet TX3 (8943M) [R]	Privately owned, stored Lasham	
XE802	Slingsby T31B Cadet TX3 (BGA5283)	Privately owned, Shipdham	
XE852	DH115 Vampire T11 [H]	No 2247 Sqn ATC, Hawarden	
XE855	DH115 Vampire T11	Midland Air Museum, stored Coventry	
XE856	DH115 Vampire T11 (G-DUSK) [V]	Bournemouth Aviation Museum	
XE864	DH115 Vampire T11(comp XD435) <ff>	Privately owned, Ingatstone, Essex	
XE872	DH115 Vampire T11 [62]	Midland Air Museum, Coventry	
XE874	DH115 Vampire T11 (8582M)	Paintball Commando, Birkin, W Yorks	
XE897	DH115 Vampire T11 (XD403)	Privately owned, Errol, Tayside	
XE921	DH115 Vampire T11 [64] <ff>	Privately owned, St Mawgan	
XE935	DH115 Vampire T11	South Yorkshire Aircraft Museum, Doncaster	
XE946	DH115 Vampire T11 (7473M) <ff>	RAF Cranwell Aviation Heritage Centre	
XE956	DH115 Vampire T11 (G-OBLN)	De Havilland Aviation, stored Rochester	

Notes	Serial	Type (code/other identity)	Owner/operator, location or fate
	XE979	DH115 Vampire T11 [54]	Privately owned, Cantley, Norfolk
	XE982	DH115 Vampire T11 (7564M) (fuselage)	Privately owned, Saggart, Eire
	XE985	DH115 Vampire T11 (WZ476)	Privately owned, New Inn, Torfaen
	XF113	VS Swift F7 [19] <ff>	Boscombe Down Aviation Collection, Old Sarum
	XF114	VS Swift F7 (G-SWIF)	Solent Sky, stored Romsey
	XF314	Hawker Hunter F51 (RDAF E-412) [N]	Brooklands Museum, Weybridge
	XF321	Hawker Hunter T7 <ff>	Privately owned, Welshpool
	XF321	Hawker Hunter T7 <rf>	Phoenix Aviation, Bruntingthorpe
	XF375	Hawker Hunter F6A (8736M/G-BUEZ) [6]	Boscombe Down Aviation Collection, Old Sarum
	XF382	Hawker Hunter F6A [15]	Midland Air Museum, Coventry
	XF383	Hawker Hunter F6 (8706M) <ff>	Gloster Aviation Club, Gloucester
	XF418	Hawker Hunter F51 (RDAF E-430)	Gatwick Aviation Museum, Charlwood, Surrey
	XF506	Hawker Hunter F4 (WT746/7770M) [A]	Dumfries & Galloway Avn Mus, Dumfries
	XF509	Hawker Hunter F6 (8708M)	Fort Paull Armoury
	XF522	Hawker Hunter F6 <ff>	Herts & Bucks ATC Wing, RAF Halton
	XF526	Hawker Hunter F6 (8679M) [78/E]	Privately owned, Birlingham, Worcs
	XF527	Hawker Hunter F6 (8680M)	RAF Halton, on display
	XF545	Percival P56 Provost T1 (7957M) [O-K]	Privately owned, Bucklebury, Berks
	XF597	Percival P56 Provost T1 (G-BKFW) [AH]	Privately owned, Brimpton, Berks
	XF603	Percival P56 Provost T1 (G-KAPW)	Shuttleworth Collection, Old Warden
	XF690	Percival P56 Provost T1 (8041M/G-MOOS)	Kennet Aviation, Yeovilton
	XF708	Avro 716 Shackleton MR3 [C]	Imperial War Museum, Duxford
	XF785	Bristol 173 (7648M/G-ALBN)	Bristol Aero Collection, stored Filton
	XF836	Percival P56 Provost T1 (8043M/G-AWRY) [JG]	Privately owned stored, Newbury
	XF840	Percival P56 Provost T1 <ff>	Tangmere Military Aviation Museum
	XF926	Bristol 188 (8368M)	RAF Museum, Cosford
	XF940	Hawker Hunter F4 <ff>	Privately owned, Kew Stoke, Somerset
	XF994	Hawker Hunter T8C (G-CGHU) [873/VL]	Hawker Hunter Aviation, Scampton
	XF995	Hawker Hunter T8B (G-BZSF/9237M) [K]	Hawker Hunter Aviation, Scampton
	XG154	Hawker Hunter FGA9 (8863M)	RAF Museum, Hendon
	XG160	Hawker Hunter F6A (8831M/G-BWAF) [U]	Bournemouth Aviation Museum
	XG164	Hawker Hunter F6 (8681M)	Davidstow Airfield & Cornwall At War Museum
	XG168	Hawker Hunter F6A (XG172/8832M) [10]	City of Norwich Aviation Museum
	XG190	Hawker Hunter F51 (RDAF E-425) [C]	Solway Aviation Society, Carlisle
	XG193	Hawker Hunter FGA9 (XG297)	South Yorkshire Aircraft Museum, Doncaster
		(comp with WT741) <ff>	
	XG194	Hawker Hunter FGA9 (8839M)	Wattisham Airfield Museum
	XG194	Hawker Hunter PR11 (WT723/G-PRII) [N]	Repainted as WT723, March 2014
	XG195	Hawker Hunter FGA9 <ff>	Privately owned, Lewes
	XG196	Hawker Hunter F6A (8702M) [31]	Privately owned, Bentwaters
	XG209	Hawker Hunter F6 (8709M) <ff>	Privately owned, Kingston-on-Thames
	XG210	Hawker Hunter F6	Privately owned, Beck Row, Suffolk
	XG225	Hawker Hunter F6A (8713M)	DSAE Cosford, at main gate
	XG226	Hawker Hunter F6A (8800M) <ff>	RAF Manston History Museum
	XG254	Hawker Hunter FGA9 (8881M) [A]	Norfolk & Suffolk Avn Museum, Flixton
	XG274	Hawker Hunter F6 (8710M) [71]	Privately owned, Newmarket
	XG290	Hawker Hunter F6 (8711M) <ff>	Boscombe Down Aviation Collection, Old Sarum
	XG290	Hawker Hunter T7 (comp XL578 & XL586)	Privately owned, Binbrook
	XG297	Hawker Hunter FGA9 <ff>	South Yorkshire Aircraft Museum, Doncaster
	XG325	EE Lightning F1 <ff>	Privately owned, Norfolk
	XG329	EE Lightning F1 (8050M)	Privately owned, Flixton
	XG331	EE Lightning F1 <ff>	Privately owned, Glos
	XG337	EE Lightning F1 (8056M) [M]	RAF Museum, Cosford
	XG452	Bristol 192 Belvedere HC1 (7997M/G-BRMB)	The Helicopter Museum, Weston-super-Mare
	XG454	Bristol 192 Belvedere HC1 (8366M)	Museum of Science & Industry, Manchester
	XG462	Bristol 192 Belvedere HC1 <ff>	The Helicopter Museum, stored Weston-super-Mare
	XG474	Bristol 192 Belvedere HC1 (8367M) [O]	RAF Museum, Hendon
	XG502	Bristol 171 Sycamore HR14	Museum of Army Flying, Middle Wallop
	XG518	Bristol 171 Sycamore HR14 (8009M) [S-E]	Norfolk & Suffolk Avn Museum, Flixton
	XG523	Bristol 171 Sycamore HR14 <ff> [V]	Norfolk & Suffolk Avn Museum, Flixton

Serial	Type (code/other identity)	Owner/operator, location or fate	Notes
XG574	WS55 Whirlwind HAR3 [752/PO]	FAA Museum, RNAS Yeovilton	
XG588	WS55 Whirlwind HAR3 (G-BAMH/VR-BEP)	East Midlands Airport Aeropark	
XG592	WS55 Whirlwind HAS7 [54]	*Task Force* Adventure Park, Cowbridge, S Glam	
XG594	WS55 Whirlwind HAS7 [517]	FAA Museum, stored Cobham Hall, RNAS Yeovilton	
XG596	WS55 Whirlwind HAS7 [66]	The Helicopter Museum, stored Weston-super-Mare	
XG629	DH112 Sea Venom FAW22	Privately owned, Stone, Staffs	
XG680	DH112 Sea Venom FAW22 [438]	North-East Aircraft Museum, Usworth	
XG692	DH112 Sea Venom FAW22 [668/LM]	Privately owned, Stockport	
XG730	DH112 Sea Venom FAW22	DHAHC, London Colney	
XG736	DH112 Sea Venom FAW22	Privately owned, East Midlands	
XG737	DH112 Sea Venom FAW22 [220/Z]	East Midlands Airport Aeropark	
XG743	DH115 Sea Vampire T22 [597/LM]	Privately owned, Pickhill, N Yorks	
XG797	Fairey Gannet ECM6 [277]	Imperial War Museum, Duxford	
XG831	Fairey Gannet ECM6 [396]	Davidstow Airfield & Cornwall At War Museum	
XG882	Fairey Gannet T5 (8754M) [771/LM]	Privately owned, Errol, Tayside	
XG883	Fairey Gannet T5 [773/BY]	FAA Museum, at Museum of Berkshire Aviation, Woodley	
XG900	Short SC1	Science Museum, South Kensington	
XG905	Short SC1	Ulster Folk & Transport Mus, Holywood, Co Down	
XH131	EE Canberra PR9	Ulster Aviation Society, Long Kesh	
XH134	EE Canberra PR9 (G-OMHD)	Midair Squadron, Cotswold Airport	
XH135	EE Canberra PR9	Privately owned, Cotswold Airport	
XH136	EE Canberra PR9 (8782M) [W] <ff>	Romney Marsh Wartime Collection	
XH165	EE Canberra PR9 <ff>	Blyth Valley Aviation Collection, Walpole	
XH168	EE Canberra PR9	RAF Marham Fire Section	
XH169	EE Canberra PR9	RAF Marham, on display	
XH170	EE Canberra PR9 (8739M)	RAF Wyton, on display	
XH171	EE Canberra PR9 (8746M) [U]	RAF Museum, Cosford	
XH174	EE Canberra PR9 <ff>	Privately owned, Leicester	
XH175	EE Canberra PR9 <ff>	Privately owned, Bewdley, Worcs	
XH177	EE Canberra PR9 <ff>	Newark Air Museum, Winthorpe	
XH278	DH115 Vampire T11 (8595M/7866M) [42]	Yorkshire Air Museum, Elvington	
XH313	DH115 Vampire T11 (G-BZRD) [E]	Tangmere Military Aviation Museum	
XH318	DH115 Vampire T11 (7761M) [64]	Privately owned, Sholing, Hants	
XH328	DH115 Vampire T11 <ff>	Privately owned, Cantley, Norfolk	
XH330	DH115 Vampire T11 [73]	Privately owned, Milton Keynes	
XH537	Avro 698 Vulcan B2MRR (8749M) <ff>	Bournemouth Aviation Museum	
XH558	Avro 698 Vulcan B2 (G-VLCN)	Vulcan To The Sky Trust, Doncaster Sheffield Airport	
XH560	Avro 698 Vulcan K2 <ff>	Privately owned, Foulness	
XH563	Avro 698 Vulcan B2MRR <ff>	Privately owned, Over Dinsdale, N Yorks	
XH584	EE Canberra T4 (G-27-374) <ff>	South Yorkshire Aircraft Museum, Doncaster	
XH592	HP80 Victor B1A (8429M) <ff>	Phoenix Aviation, Bruntingthorpe	
XH648	HP80 Victor K1A	Imperial War Museum, Duxford	
XH669	HP80 Victor K2 (9092M) <ff>	Privately owned, Foulness	
XH670	HP80 Victor SR2 <ff>	Privately owned, Foulness	
XH672	HP80 Victor K2 (9242M)	RAF Museum, Cosford	
XH673	HP80 Victor K2 (8911M)	RAF Marham, on display	
XH767	Gloster Javelin FAW9 (7955M) [L]	Yorkshire Air Museum, Elvington	
XH783	Gloster Javelin FAW7 (7798M) <ff>	Privately owned, Catford	
XH837	Gloster Javelin FAW7 (8032M) <ff>	Caernarfon Air World	
XH892	Gloster Javelin FAW9R (7982M) [J]	Norfolk & Suffolk Avn Museum, Flixton	
XH897	Gloster Javelin FAW9	Imperial War Museum, Duxford	
XH903	Gloster Javelin FAW9 (7938M)	Jet Age Museum, Gloucester	
XH992	Gloster Javelin FAW8 (7829M) [P]	Newark Air Museum, Winthorpe	
XJ314	RR Thrust Measuring Rig	Science Museum, South Kensington	
XJ380	Bristol 171 Sycamore HR14 (8628M)	Boscombe Down Aviation Collection, Old Sarum	
XJ389	Fairey Jet Gyrodyne (XD759/G-AJJP)	Museum of Berkshire Aviation, Woodley	
XJ398	WS55 Whirlwind HAR10 (XD768/G-BDBZ)	South Yorkshire Aircraft Museum, Doncaster	
XJ407	WS55 Whirlwind HAR10 (N7013H)	Privately owned, Tattershall Thorpe	
XJ435	WS55 Whirlwind HAR10 (XD804/8671M) [V]	RAF Manston History Museum, spares use	
XJ476	DH110 Sea Vixen FAW1 <ff>	Boscombe Down Aviation Collection, Old Sarum	

Notes	Serial	Type (code/other identity)	Owner/operator, location or fate
	XJ481	DH110 Sea Vixen FAW1 [VL]	FAA Museum, stored Cobham Hall, RNAS Yeovilton
	XJ482	DH110 Sea Vixen FAW1 [713/VL]	Norfolk & Suffolk Avn Museum, Flixton
	XJ488	DH110 Sea Vixen FAW1 <ff>	Robertsbridge Aviation Society, Mayfield
	XJ494	DH110 Sea Vixen FAW2 [121/E]	Privately owned, Bruntingthorpe
	XJ560	DH110 Sea Vixen FAW2 (8142M) [243/H]	Newark Air Museum, Winthorpe
	XJ565	DH110 Sea Vixen FAW2 [127/E]	DHAHC, London Colney
	XJ571	DH110 Sea Vixen FAW2 (8140M) [242/R]	Solent Sky, Southampton
	XJ575	DH110 Sea Vixen FAW2 <ff> [SAH-13]	Wellesbourne Wartime Museum
	XJ579	DH110 Sea Vixen FAW2 <ff>	Midland Air Museum, Coventry
	XJ580	DH110 Sea Vixen FAW2 [131/E]	Tangmere Military Aviation Museum
	XJ714	Hawker Hunter FR10 (comp XG226)	East Midlands Airport Aeropark
	XJ723	WS55 Whirlwind HAR10	Privately owned, Newcastle upon Tyne
	XJ726	WS55 Whirlwind HAR10	Caernarfon Air World
	XJ727	WS55 Whirlwind HAR10 (8661M) [L]	Privately owned, Ramsgate
	XJ729	WS55 Whirlwind HAR10 (8732M/G-BVGE)	Privately owned, Crewkerne, Somerset
	XJ758	WS55 Whirlwind HAR10 (8464M) <ff>	Privately owned, Welshpool
	XJ771	DH115 Vampire T55 (U-1215/G-HELV)	Classic Air Force, Newquay
	XJ772	DH115 Vampire T11 [H]	DHAHC, London Colney
	XJ823	Avro 698 Vulcan B2A	Solway Aviation Society, Carlisle
	XJ824	Avro 698 Vulcan B2A	Imperial War Museum, Duxford
	XJ917	Bristol 171 Sycamore HR14 [H-S]	Bristol Sycamore Group, stored Filton
	XJ918	Bristol 171 Sycamore HR14 (8190M)	RAF Museum, Cosford
	XK416	Auster AOP9 (7855M/G-AYUA)	Privately owned, Widmerpool
	XK417	Auster AOP9 (G-AVXY)	Privately owned, Messingham, Lincs
	XK418	Auster AOP9 (7976M)	93rd Bomb Group Museum, Hardwick, Norfolk
	XK421	Auster AOP9 (8365M) (frame)	Privately owned, South Molton, Devon
	XK488	Blackburn NA39 Buccaneer S1	FAA Museum, stored Cobham Hall, RNAS Yeovilton
	XK526	Blackburn NA39 Buccaneer S2 (8648M)	RAF Honington, at main gate
	XK527	Blackburn NA39 Buccaneer S2D (8818M) <ff>	Privately owned, North Wales
	XK532	Blackburn NA39 Buccaneer S1 (8867M) [632/LM]	Highland Aviation Museum, Inverness
	XK533	Blackburn NA39 Buccaneer S1 <ff>	Royal Scottish Mus'm of Flight, stored Granton
	XK590	DH115 Vampire T11 [V]	Wellesbourne Wartime Museum
	XK623	DH115 Vampire T11 (*G-VAMP*) [56]	Caernarfon Air World
	XK624	DH115 Vampire T11 [32]	Norfolk & Suffolk Avn Museum, Flixton
	XK625	DH115 Vampire T11 [14]	Romney Marsh Wartime Collection
	XK627	DH115 Vampire T11 <ff>	Davidstow Airfield & Cornwall At War Museum
	XK632	DH115 Vampire T11 <ff>	Privately owned, Greenford, London
	XK637	DH115 Vampire T11 [56]	Top Gun Flight Simulation Centre, Stalybridge
	XK695	DH106 Comet C2(RC) (G-AMXH/9164M) <ff>	DHAHC, London Colney
	XK699	DH106 Comet C2 (7971M) <ff>	Boscombe Down Aviation Collection, Old Sarum
	XK724	Folland Gnat F1 (7715M)	RAF Museum, Cosford
	XK740	Folland Gnat F1 (8396M)	Solent Sky, Southampton
	XK776	ML Utility 1	Museum of Army Flying, Middle Wallop
	XK789	Slingsby T38 Grasshopper TX1	Midland Air Museum, stored Coventry
	XK790	Slingsby T38 Grasshopper TX1	Privately owned, stored Husbands Bosworth
	XK819	Slingsby T38 Grasshopper TX1	Privately owned, stored Rufforth
	XK822	Slingsby T38 Grasshopper TX1	Privately owned, Partridge Green, W Sussex
	XK885	Percival P66 Pembroke C1 (8452M/N46EA)	Privately owned, St Athan
	XK895	DH104 Sea Devon C20 (G-SDEV) [19/CU]	Classic Air Force, Newquay
	XK907	WS55 Whirlwind HAS7	Midland Air Museum, stored Coventry
	XK911	WS55 Whirlwind HAS7 [519/PO]	Privately owned, stored Dagenham
	XK936	WS55 Whirlwind HAS7 [62]	Imperial War Museum, Duxford
	XK940	WS55 Whirlwind HAS7 (G-AYXT) [911]	The Helicopter Museum, Weston-super-Mare
	XK970	WS55 Whirlwind HAR10 (8789M)	Army, Bramley, Hants (for disposal)
	XL149	Blackburn B101 Beverley C1 (7988M) <ff>	South Yorkshire Aircraft Museum, Doncaster
	XL160	HP80 Victor K2 (8910M) <ff>	Norfolk & Suffolk Avn Museum, Flixton
	XL164	HP80 Victor K2 (9215M) <ff>	Bournemouth Aviation Museum
	XL190	HP80 Victor K2 (9216M) <ff>	RAF Manston History Museum
	XL231	HP80 Victor K2	Yorkshire Air Museum, Elvington
	XL318	Avro 698 Vulcan B2 (8733M)	RAF Museum, Hendon

Serial	Type (code/other identity)	Owner/operator, location or fate	Notes
XL319	Avro 698 Vulcan B2	North-East Aircraft Museum, Usworth	
XL360	Avro 698 Vulcan B2A	Midland Air Museum, Coventry	
XL388	Avro 698 Vulcan B2 <ff>	South Yorkshire Aircraft Museum, Doncaster	
XL426	Avro 698 Vulcan B2 (G-VJET)	Vulcan Restoration Trust, Southend	
XL445	Avro 698 Vulcan K2 (8811M) <ff>	Norfolk & Suffolk Avn Museum, Flixton	
XL449	Fairey Gannet AEW3 <ff>	Privately owned, Camberley, Surrey	
XL472	Fairey Gannet AEW3 [044/R]	Hunter Flying Ltd, St Athan (spares use)	
XL497	Fairey Gannet AEW3 [041/R]	Dumfries & Galloway Avn Mus, Dumfries	
XL500	Fairey Gannet AEW3 (G-KAEW) [CU]	Hunter Flying Ltd, St Athan	
XL502	Fairey Gannet AEW3 (8610G/G-BMYP)	Yorkshire Air Museum, Elvington	
XL503	Fairey Gannet AEW3 [070/E]	FAA Museum, RNAS Yeovilton	
XL563	Hawker Hunter T7 (9218M)	Farnborough Air Sciences Trust, stored Farnborough	
XL564	Hawker Hunter T7 <ff>	Privately owned, Tilehurst, Berks	
XL565	Hawker Hunter T7 (parts of WT745) [Y]	Privately owned, Bruntingthorpe	
XL568	Hawker Hunter T7A (9224M) [X]	RAF Museum, Cosford	
XL569	Hawker Hunter T7 (8833M)	East Midlands Airport Aeropark	
XL571	Hawker Hunter T7 (XL572/8834M/G-HNTR) [V]	Yorkshire Air Museum, Elvington	
XL573	Hawker Hunter T7 (G-BVGH)	Hunter Flying Ltd, St Athan	
XL577	Hawker Hunter T7 (8676M/G-XMHD)	Midair Squadron, Cotswold Airport	
XL580	Hawker Hunter T8M [723]	FAA Museum, RNAS Yeovilton	
XL586	Hawker Hunter T7 (comp XL578)	Action Park, Wickford, Essex	
XL587	Hawker Hunter T7 (8807M/G-HPUX) [Z]	Hawker Hunter Aviation, stored Scampton	
XL591	Hawker Hunter T7	Gatwick Aviation Museum, Charlwood, Surrey	
XL592	Hawker Hunter T7 (8836M) [Y]	Privately owned, Maidenhead, Berks	
XL600	Hawker Hunter T7 (G-RAXA)	Midair Squadron, Cotswold Airport	
XL601	Hawker Hunter T7 (G-BZSR) [874/VL]	Classic Fighters, Brustem, Belgium	
XL602	Hawker Hunter T8M (G-BWFT)	Hunter Flying Ltd, St Athan	
XL609	Hawker Hunter T7 <ff>	Privately owned, South Molton, Devon	
XL612	Hawker Hunter T7 [2]	Privately owned, Swansea	
XL618	Hawker Hunter T7 (8892M)	Newark Air Museum, Winthorpe	
XL621	Hawker Hunter T7 (G-BNCX)	Privately owned, Dunsfold	
XL623	Hawker Hunter T7 (8770M)	The Planets Leisure Centre, Woking	
XL629	EE Lightning T4	MoD/QinetiQ Boscombe Down, at main gate	
XL703	SAL Pioneer CC1 (8034M) [Z]	*Repainted in Omani markings as XL554, June 2014*	
XL714	DH82A Tiger Moth II (T6099/G-AOGR)	Privately owned, Boughton, Lincs	
XL736	Saro Skeeter AOP12	The Helicopter Museum stored, Weston-super-Mare	
XL738	Saro Skeeter AOP12 (XM565/7861M)	Privately owned, Storwood, Yorkshire	
XL739	Saro Skeeter AOP12	Privately owned, Forncett St Peter, Norfolk	
XL762	Saro Skeeter AOP12 (8017M)	Royal Scottish Mus'm of Flight, E Fortune	
XL763	Saro Skeeter AOP12	Privately owned, Storwood, Yorkshire	
XL764	Saro Skeeter AOP12 (7940M) [J]	Newark Air Museum, Winthorpe	
XL765	Saro Skeeter AOP12	Privately owned, Melksham, Wilts	
XL767	Saro Skeeter AOP12 <ff>	The Helicopter Museum, stored Weston-super-Mare	
XL770	Saro Skeeter AOP12 (8046M)	Solent Sky, Southampton	
XL809	Saro Skeeter AOP12 (G-BLIX)	Privately owned, Wilden, Beds	
XL811	Saro Skeeter AOP12	The Helicopter Museum, Weston-super-Mare	
XL812	Saro Skeeter AOP12 (G-SARO)	AAC Historic Aircraft Flight, stored Middle Wallop	
XL813	Saro Skeeter AOP12	Museum of Army Flying, Middle Wallop	
XL814	Saro Skeeter AOP12	AAC Historic Aircraft Flight, Middle Wallop	
XL824	Bristol 171 Sycamore HR14 (8021M)	RAF Museum Reserve Collection, Stafford	
XL829	Bristol 171 Sycamore HR14	The Helicopter Museum, Weston-super-Mare	
XL840	WS55 Whirlwind HAS7	Privately owned, Bawtry	
XL853	WS55 Whirlwind HAS7 [PO]	FAA Museum, stored Cobham Hall, RNAS Yeovilton	
XL875	WS55 Whirlwind HAR9	Perth Technical College	
XL929	Percival P66 Pembroke C1 (G-BNPU)	Classic Air Force, stored Compton Verney	
XL954	Percival P66 Pembroke C1 (9042M/N4234C/G-BXES)	Classic Air Force, Coventry	
XL993	SAL Twin Pioneer CC1 (8388M)	RAF Museum, Cosford	
XM135	BAC Lightning F1 [B]	Imperial War Museum, Duxford	
XM169	BAC Lightning F1A (8422M) <ff>	Highland Aviation Museum, Inverness	
XM172	BAC Lightning F1A (8427M)	Lakes Lightnings, Spark Bridge, Cumbria	

Notes	Serial	Type (code/other identity)	Owner/operator, location or fate
	XM173	BAC Lightning F1A (8414M) [A]	Privately owned, Cotswold Airport
	XM191	BAC Lightning F1A (7854M/8590M) <ff>	Privately owned, Thorpe Wood, N Yorks
	XM192	BAC Lightning F1A (8413M) [K]	Thorpe Camp Preservation Group, Lincs
	XM223	DH104 Devon C2 (G-BWWC) [J]	Classic Air Force, stored Compton Verney
	XM279	EE Canberra B(I)8 <ff>	Privately owned, Flixton
	XM300	WS58 Wessex HAS1	Privately owned, Nantgarw, Rhondda
	XM328	WS58 Wessex HAS3 [653/PO]	The Helicopter Museum, Weston-super-Mare
	XM330	WS58 Wessex HAS1	The Helicopter Museum, Weston-super-Mare
	XM350	Hunting Jet Provost T3A (9036M) [89]	South Yorkshire Aircraft Museum, Doncaster
	XM351	Hunting Jet Provost T3 (8078M) [Y]	RAF Museum, Cosford
	XM355	Hunting Jet Provost T3 (8229M)	Newcastle Aviation Academy
	XM358	Hunting Jet Provost T3A (8987M) [53]	Privately owned, Newbridge, Powys
	XM362	Hunting Jet Provost T3 (8230M)	DSAE, No 1 SoTT, Cosford
	XM365	Hunting Jet Provost T3 (G-BXBH)	Privately owned, Bruntingthorpe
	XM373	Hunting Jet Provost T3 (7726M) [2] <ff>	Yorkshire Air Museum, Elvington
	XM383	Hunting Jet Provost T3 [90]	Newark Air Museum, Winthorpe
	XM402	Hunting Jet Provost T3 (8055AM) [18]	Fenland & W Norfolk Aviation Museum, Wisbech
	XM404	Hunting Jet Provost T3 (8055BM) <ff>	Privately owned, Bruntingthorpe
	XM409	Hunting Jet Provost T3 (8082M) <ff>	Air Scouts, Guernsey Airport
	XM410	Hunting Jet Provost T3 (8054AM) [B]	Privately owned, Gillingham, Kent
	XM411	Hunting Jet Provost T3 (8434M) <ff>	South Yorkshire Aircraft Museum, Doncaster
	XM412	Hunting Jet Provost T3A (9011M) [41]	Privately owned, Balado Bridge, Scotland
	XM414	Hunting Jet Provost T3A (8996M)	Ulster Aviation Society, Long Kesh
	XM417	Hunting Jet Provost T3 (8054BM) [D] <ff>	Privately owned, Cannock
	XM419	Hunting Jet Provost T3A (8990M) [102]	Newcastle Aviation Academy
	XM424	Hunting Jet Provost T3 (G-BWDS)	Classic Air Force, Newquay
	XM425	Hunting Jet Provost T3A (8995M) [88]	Privately owned, Longton, Staffs
	XM463	Hunting Jet Provost T3A [38] (fuselage)	RAF Museum, Hendon
	XM468	Hunting Jet Provost T3 (8081M) <ff>	Privately owned, Terrington St Clement, Norfolk
	XM473	Hunting Jet Provost T3A (8974M/G-TINY)	Privately owned, Wethersfield
	XM474	Hunting Jet Provost T3 (8121M) <ff>	Privately owned, South Reddish, Stockport
	XM479	Hunting Jet Provost T3A (G-BVEZ) [U]	Privately owned, Newcastle
	XM480	Hunting Jet Provost T3 (8080M)	4x4 Car Centre, Chesterfield
	XM496	Bristol 253 Britannia C1 (EL-WXA) [496]	Britannia Preservation Society, Cotswold Airport
	XM497	Bristol 175 Britannia 312F (G-AOVF) [497]	RAF Museum, Cosford
	XM529	Saro Skeeter AOP12 (7979M/G-BDNS)	Privately owned, Handforth
	XM553	Saro Skeeter AOP12 (G-AWSV)	Yorkshire Air Museum, Elvington
	XM555	Saro Skeeter AOP12 (8027M)	RAF Museum Reserve Collection, Stafford
	XM557	Saro Skeeter AOP12 <ff>	The Helicopter Museum, stored Weston-super-Mare
	XM564	Saro Skeeter AOP12	The Tank Museum, Bovington
	XM569	Avro 698 Vulcan B2 <ff>	Jet Age Museum, stored Gloucester
	XM575	Avro 698 Vulcan B2A (G-BLMC)	East Midlands Airport Aeropark
	XM594	Avro 698 Vulcan B2	Newark Air Museum, Winthorpe
	XM597	Avro 698 Vulcan B2	Royal Scottish Mus'm of Flight, E Fortune
	XM598	Avro 698 Vulcan B2 (8778M)	RAF Museum, Cosford
	XM602	Avro 698 Vulcan B2 (8771M) <ff>	Manchester Museum of Science & Industry, stored
	XM603	Avro 698 Vulcan B2	Avro Aircraft Heritage Society, Woodford
	XM607	Avro 698 Vulcan B2 (8779M)	RAF Waddington, on display
	XM612	Avro 698 Vulcan B2	City of Norwich Aviation Museum
	XM651	Saro Skeeter AOP12 (XM561/7980M)	South Yorkshire Aircraft Museum, Doncaster
	XM652	Avro 698 Vulcan B2 <ff>	Privately owned, Welshpool
	XM655	Avro 698 Vulcan B2 (G-VULC)	Privately owned, Wellesbourne Mountford
	XM685	WS55 Whirlwind HAS7 (G-AYZJ) [513/PO]	Newark Air Museum, Winthorpe
	XM692	HS Gnat T1 <ff>	Privately owned, Welshpool
	XM693	HS Gnat T1 (7891M)	BAE Systems Hamble, on display
	XM697	HS Gnat T1 (G-NAAT)	Reynard Garden Centre, Carluke, S Lanarkshire
	XM708	HS Gnat T1 (8573M)	Privately owned, Lytham St Annes
	XM715	HP80 Victor K2	Cold War Jets Collection, Bruntingthorpe
	XM717	HP80 Victor K2 <ff>	RAF Museum, Hendon
	XM819	Lancashire EP9 Prospector (G-APXW)	Museum of Army Flying, Middle Wallop
	XM833	WS58 Wessex HAS3	North-East Aircraft Museum, Usworth

Serial	Type (code/other identity)	Owner/operator, location or fate	Notes
XN126	WS55 Whirlwind HAR10 (8655M) [S]	Pinewood Studios, Elstree	
XN137	Hunting Jet Provost T3 <ff>	Privately owned, Little Addington, Notts	
XN156	Slingsby T21B Sedbergh TX1 (BGA3250)	Privately owned, Portmoak, Perth & Kinross	
XN157	Slingsby T21B Sedbergh TX1 (BGA3255)	Privately owned, Long Mynd	
XN185	Slingsby T21B Sedbergh TX1 (8942M/BGA4077)	RAF Scampton Heritage Centre	
XN186	Slingsby T21B Sedbergh TX1 (BGA3905) [HFG]	Privately owned, Wethersfield	
XN187	Slingsby T21B Sedbergh TX1 (BGA3903)	Privately owned, Halton	
XN198	Slingsby T31B Cadet TX3	Privately owned, Bodmin	
XN238	Slingsby T31B Cadet TX3 <ff>	South Yorkshire Aircraft Museum, Doncaster	
XN239	Slingsby T31B Cadet TX3 (8889M) [G]	Imperial War Museum, Duxford	
XN246	Slingsby T31B Cadet TX3	Solent Sky, Southampton	
XN258	WS55 Whirlwind HAR9 [589/CU]	North-East Aircraft Museum, Usworth	
XN297	WS55 Whirlwind HAR9 (XN311) [12]	Privately owned, Hull	
XN298	WS55 Whirlwind HAR9 [810/LS]	Privately owned	
XN299	WS55 Whirlwind HAS7 [758]	Tangmere Military Aviation Museum	
XN304	WS55 Whirlwind HAS7 [WW/B]	Norfolk & Suffolk Avn Museum, Flixton	
XN332	Saro P531 (G-APNV) [759]	FAA Museum, stored Cobham Hall, RNAS Yeovilton	
XN334	Saro P531	FAA Museum, stored Cobham Hall, RNAS Yeovilton	
XN341	Saro Skeeter AOP12 (8022M)	Stondon Transport Mus & Garden Centre, Beds	
XN344	Saro Skeeter AOP12 (8018M)	Science Museum, South Kensington	
XN345	Saro Skeeter AOP12 <ff>	The Helicopter Museum, stored Weston-super-Mare	
XN351	Saro Skeeter AOP12 (G-BKSC)	Privately owned, Clopton, Northants	
XN380	WS55 Whirlwind HAS7	RAF Manston History Museum	
XN385	WS55 Whirlwind HAS7	Battleground Paintball, Yarm, Cleveland	
XN386	WS55 Whirlwind HAR9 [435/ED]	South Yorkshire Aircraft Museum, Doncaster	
XN412	Auster AOP9	Auster 9 Group, Melton Mowbray	
XN437	Auster AOP9 (G-AXWA)	Privately owned, North Weald	
XN441	Auster AOP9 (G-BGKT)	Privately owned, Week St Mary, Cornwall	
XN458	Hunting Jet Provost T3 (8234M/XN594)	Privately owned, Northallerton	
XN459	Hunting Jet Provost T3A (G-BWOT)	Kennet Aviation, North Weald	
XN462	Hunting Jet Provost T3A [17]	FAA Museum, stored Cobham Hall, RNAS Yeovilton	
XN466	Hunting Jet Provost T3A [29] <ff>	No 184 Sqn ATC, Gr Manchester	
XN492	Hunting Jet Provost T3 (8079M) <ff>	No 2434 Sqn ATC, Church Fenton	
XN493	Hunting Jet Provost T3 (XN137) <ff>	Privately owned, Camberley	
XN494	Hunting Jet Provost T3A (9012M) [43]	Privately owned, Bruntingthorpe	
XN500	Hunting Jet Provost T3A	Norfolk & Suffolk Avn Museum, Flixton	
XN503	Hunting Jet Provost T3 <ff>	Boscombe Down Aviation Collection, Old Sarum	
XN508	Hunting Jet Provost T3A <ff>	Scrapped	
XN511	Hunting Jet Provost T3 [64] <ff>	South Yorkshire Aircraft Museum, Doncaster	
XN549	Hunting Jet Provost T3 (8235M) <ff>	Privately owned, Warrington	
XN551	Hunting Jet Provost T3A (8984M)	Privately owned, Felton Common, Bristol	
XN554	Hunting Jet Provost T3 (8436M) [K]	Gunsmoke Paintball, Hadleigh, Suffolk	
XN573	Hunting Jet Provost T3 [E] <ff>	Newark Air Museum, Winthorpe	
XN579	Hunting Jet Provost T3A (9137M) [14]	Gunsmoke Paintball, Hadleigh, Suffolk	
XN582	Hunting Jet Provost T3A (8957M) [95,H]	Privately owned, Bruntingthorpe	
XN584	Hunting Jet Provost T3A (9014M) [E]	Phoenix Aviation, Bruntingthorpe	
XN586	Hunting Jet Provost T3A (9039M) [91,S]	Brooklands Museum, Weybridge	
XN589	Hunting Jet Provost T3 (9143M) [46]	RAF Linton-on-Ouse, on display	
XN597	Hunting Jet Provost T3 (7984M) <ff>	Privately owned, Market Drayton, Shrops	
XN607	Hunting Jet Provost T3 <ff>	Highland Aviation Museum, Inverness	
XN623	Hunting Jet Provost T3 (XN632/8352M)	Privately owned, Birlingham, Worcs	
XN629	Hunting Jet Provost T3A (G-BVEG/G-KNOT) [49]	Bentwaters Cold War Museum	
XN634	Hunting Jet Provost T3A <ff>	Privately owned, Blackpool	
XN634	Hunting Jet Provost T3A [53] <rf>	BAE Systems Warton Fire Section	
XN637	Hunting Jet Provost T3 (G-BKOU) [03]	Privately owned, North Weald	
XN647	DH110 Sea Vixen FAW2 <ff>	Privately owned, Steventon, Oxon	
XN650	DH110 Sea Vixen FAW2 [456] <ff>	Privately owned, Norfolk	
XN651	DH110 Sea Vixen FAW2 <ff>	Privately owned, Lavendon, Bucks	
XN685	DH110 Sea Vixen FAW2 (8173M) [703/VL]	Midland Air Museum, Coventry	
XN696	DH110 Sea Vixen FAW2 [751] <ff>	North-East Aircraft Museum, Usworth	
XN714	Hunting H126	RAF Museum, Cosford	
XN726	EE Lightning F2A (8545M) <ff>	Boscombe Down Aviation Collection, Old Sarum	

Notes	Serial	Type (code/other identity)	Owner/operator, location or fate
	XN728	EE Lightning F2A (8546M) <ff>	Privately owned, Binbrook
	XN774	EE Lightning F2A (8551M) <ff>	Privately owned, Boston
	XN776	EE Lightning F2A (8535M) [C]	Royal Scottish Mus'm of Flight, E Fortune
	XN795	EE Lightning F2A <ff>	Privately owned, Rayleigh, Essex
	XN819	AW660 Argosy C1 (8205M) <ff>	Newark Air Museum, Winthorpe
	XN923	HS Buccaneer S1 [13]	Gatwick Aviation Museum, Charlwood, Surrey
	XN928	HS Buccaneer S1 (8179M) <ff>	Privately owned, Gravesend
	XN957	HS Buccaneer S1 [630/LM]	FAA Museum, RNAS Yeovilton
	XN964	HS Buccaneer S1 [118/V]	Newark Air Museum, Winthorpe
	XN967	HS Buccaneer S1 [233] <ff>	City of Norwich Aviation Museum
	XN972	HS Buccaneer S1 (8183M/XN962) <ff>	RAF Museum, Cosford
	XN974	HS Buccaneer S2A	Yorkshire Air Museum, Elvington
	XN981	HS Buccaneer S2B (fuselage)	Privately owned, Errol
	XN983	HS Buccaneer S2B <ff>	Fenland & W Norfolk Aviation Museum, Wisbech
	XP110	WS58 Wessex HAS3 (A2636)	DSMarE AESS, *HMS Sultan*, Gosport
	XP137	WS58 Wessex HAS3 [11/DD]	RN, Predannack Fire School
	XP142	WS58 Wessex HAS3	FAA Museum, stored Cobham Hall, RNAS Yeovilton
	XP150	WS58 Wessex HAS3 [LS]	FETC, Moreton-in-Marsh, Glos
	XP165	WS Scout AH1	The Helicopter Museum, Weston-super-Mare
	XP190	WS Scout AH1	South Yorkshire Aircraft Museum, Doncaster
	XP191	WS Scout AH1	Privately owned, Prenton, The Wirral
	XP226	Fairey Gannet AEW3	Newark Air Museum, Winthorpe
	XP241	Auster AOP9 (G-CEHR)	Privately owned, Eggesford
	XP242	Auster AOP9 (G-BUCI)	AAC Historic Aircraft Flight, Middle Wallop
	XP244	Auster AOP9 (7864M/M7922)	Privately owned, Stretton on Dunsmore
	XP248	Auster AOP9 (7863M/WZ679)	Privately owned, Coventry
	XP254	Auster AOP11 (G-ASCC)	Privately owned, Cambs
	XP279	Auster AOP9 (G-BWKK)	Privately owned, Popham
	XP280	Auster AOP9	Snibston Discovery Park, Coalville
	XP281	Auster AOP9	Imperial War Museum, Duxford
	XP286	Auster AOP9	Privately owned, South Molton, Devon
	XP299	WS55 Whirlwind HAR10 (8726M)	RAF Museum, Hendon
	XP328	WS55 Whirlwind HAR10 (G-BKHC)	Privately owned, Tattershall Thorpe (wreck)
	XP329	WS55 Whirlwind HAR10 (8791M) [V]	Privately owned, Tattershall Thorpe (wreck)
	XP330	WS55 Whirlwind HAR10	CAA Fire School, Durham/Tees Valley
	XP344	WS55 Whirlwind HAR10 (8764M) [H723]	RAF North Luffenham Training Area
	XP345	WS55 Whirlwind HAR10 (8792M) [N]	Yorkshire Helicopter Preservation Group, Doncaster
	XP346	WS55 Whirlwind HAR10 (8793M)	Privately owned, Honeybourne, Worcs
	XP350	WS55 Whirlwind HAR10	Privately owned, Bassetts Pole, Staffs
	XP351	WS55 Whirlwind HAR10 (8672M) [Z]	Gatwick Aviation Museum, Charlwood, Surrey
	XP354	WS55 Whirlwind HAR10 (8721M)	Privately owned, Mullingar, Eire
	XP355	WS55 Whirlwind HAR10 (8463M/G-BEBC)	City of Norwich Aviation Museum
	XP360	WS55 Whirlwind HAR10 [V]	Privately owned, Bicton, nr Leominster
	XP398	WS55 Whirlwind HAR10 (8794M)	Gatwick Aviation Museum, Charlwood, Surrey
	XP404	WS55 Whirlwind HAR10 (8682M)	The Helicopter Museum, stored Weston-super-Mare
	XP411	AW660 Argosy C1 (8442M) [C]	RAF Museum, Cosford
	XP454	Slingsby T38 Grasshopper TX1	Privately owned, Sywell
	XP459	Slingsby T38 Grasshopper TX1	Privately owned, stored Nayland, Suffolk
	XP463	Slingsby T38 Grasshopper TX1 (BGA4372)	Privately owned, Lasham
	XP488	Slingsby T38 Grasshopper TX1	Privately owned, Keevil
	XP490	Slingsby T38 Grasshopper TX1 (BGA4552)	Privately owned, stored Watton
	XP492	Slingsby T38 Grasshopper TX1 (BGA3480)	Privately owned, Gallows Hill, Dorset
	XP493	Slingsby T38 Grasshopper TX1	Privately owned, stored Aston Down
	XP494	Slingsby T38 Grasshopper TX1	Privately owned, stored Baxterley
	XP505	HS Gnat T1	Science Museum, Wroughton
	XP516	HS Gnat T1 (8580M) [16]	Farnborough Air Sciences Trust, Farnborough
	XP540	HS Gnat T1 (8608M) [62]	Privately owned, North Weald
	XP542	HS Gnat T1 (8575M)	Solent Sky, Hamble
	XP556	Hunting Jet Provost T4 (9027M) [B]	RAF Cranwell Aviation Heritage Centre
	XP557	Hunting Jet Provost T4 (8494M) [72]	Dumfries & Galloway Avn Mus, Dumfries
	XP558	Hunting Jet Provost T4 (8627M) <ff>	Privately owned, Stoneykirk, D&G

Serial	Type (code/other identity)	Owner/operator, location or fate	Notes
XP558	Hunting Jet Provost T4 (8627M)[20] <rf>	Privately owned, Sproughton	
XP563	Hunting Jet Provost T4 (9028M) [C]	Privately owned, Sproughton	
XP568	Hunting Jet Provost T4	East Midlands Airport Aeropark	
XP573	Hunting Jet Provost T4 (8236M) [19]	Jersey Airport Fire Section	
XP585	Hunting Jet Provost T4 (8407M) [24]	NE Wales Institute, Wrexham	
XP627	Hunting Jet Provost T4	North-East Aircraft Museum, stored Usworth	
XP629	Hunting Jet Provost T4 (9026M) [P]	Gunsmoke Paintball, Hadleigh, Suffolk	
XP640	Hunting Jet Provost T4 (8501M) [M]	Yorkshire Air Museum, Elvington	
XP642	Hunting Jet Provost T4 <ff>	Privately owned, Welshpool	
XP672	Hunting Jet Provost T4 (8458M/G-RAFI) [03]	Privately owned, Bruntingthorpe	
XP680	Hunting Jet Provost T4 (8460M)	FETC, Moreton-in-Marsh, Glos	
XP686	Hunting Jet Provost T4 (8401M/8502M) [G]	Gunsmoke Paintball, Hadleigh, Suffolk	
XP701	BAC Lightning F3 (8924M) <ff>	Robertsbridge Aviation Society, Mayfield	
XP703	BAC Lightning F3 <ff>	Lightning Preservation Group, Bruntingthorpe	
XP706	BAC Lightning F3 (8925M)	South Yorkshire Aircraft Museum, Doncaster	
XP743	BAC Lightning F3 <ff>	Wattisham Airfield Museum	
XP745	BAC Lightning F3 (8453M) <ff>	Vanguard Haulage, Greenford, London	
XP757	BAC Lightning F3 <ff>	Privately owned, Boston, Lincs	
XP765	BAC Lightning F6 (XS897) [A]	Lakes Lightnings, RAF Coningsby	
XP820	DHC2 Beaver AL1 (G-CICP)	Army Historic Aircraft Flight, Middle Wallop	
XP821	DHC2 Beaver AL1 [MCO]	Museum of Army Flying, Middle Wallop	
XP822	DHC2 Beaver AL1	Museum of Army Flying, Middle Wallop	
XP831	Hawker P.1127 (8406M)	Science Museum, South Kensington	
XP841	Handley-Page HP115	FAA Museum, RNAS Yeovilton	
XP847	WS Scout AH1	Museum of Army Flying, Middle Wallop	
XP848	WS Scout AH1	AAC MPSU, Middle Wallop (for disposal)	
XP849	WS Scout AH1 (XP895)	Privately owned, Woodley, Berks	
XP853	WS Scout AH1	Privately owned, Sutton, Surrey	
XP854	WS Scout AH1 (7898M/TAD 043)	Mayhem Paintball, Abridge, Essex	
XP855	WS Scout AH1	DSEME SEAE, Arborfield	
XP883	WS Scout AH1	Privately owned, Bruntingthorpe	
XP883	WS Scout AH1 (XW281/G-BYNZ) [T]	Privately owned, Dungannon, NI	
XP884	WS Scout AH1	AAC, stored Middle Wallop	
XP885	WS Scout AH1	AAC Wattisham, instructional use	
XP886	WS Scout AH1	The Helicopter Museum, stored Weston-super-Mare	
XP888	WS Scout AH1	Privately owned, Sproughton	
XP890	WS Scout AH1 [G] (fuselage)	Privately owned, Ipswich	
XP893	WS Scout AH1	AAC Middle Wallop, BDRT	
XP899	WS Scout AH1 [D]	AAC MPSU, Middle Wallop (for disposal)	
XP900	WS Scout AH1	AAC Wattisham, instructional use	
XP902	WS Scout AH1 <ff>	South Yorkshire Aircraft Museum, Doncaster	
XP905	WS Scout AH1	Privately owned, stored Sproughton	
XP907	WS Scout AH1 (G-SROE)	Privately owned, Wattisham	
XP910	WS Scout AH1	Museum of Army Flying, Middle Wallop	
XP924	DH110 Sea Vixen D3 (G-CVIX) [134/E]	Fly Navy Heritage Trust, Yeovilton	
XP925	DH110 Sea Vixen FAW2 [752] <ff>	Privately owned,	
XP980	Hawker P.1127	FAA Museum, RNAS Yeovilton	
XP984	Hawker P.1127	Brooklands Museum, Weybridge	
XR220	BAC TSR2 (7933M)	RAF Museum, Cosford	
XR222	BAC TSR2	Imperial War Museum, Duxford	
XR232	Sud Alouette AH2 (F-WEIP)	Museum of Army Flying, Middle Wallop	
XR239	Auster AOP9	Privately owned, Stretton on Dunsmore	
XR240	Auster AOP9 (G-BDFH)	Privately owned, Yeovilton	
XR241	Auster AOP9 (G-AXRR)	Privately owned, Eggesford	
XR244	Auster AOP9 (G-CICR)	Army Historic Aircraft Flight, Middle Wallop	
XR246	Auster AOP9 (7862M/G-AZBU)	Privately owned, Melton Mowbray	
XR267	Auster AOP9 (G-BJXR)	Privately owned, Hucknall	
XR271	Auster AOP9	Royal Artillery Experience, Woolwich	
XR346	Northrop Shelduck D1 (comp XW578)	Bournemouth Aviation Museum	
XR371	SC5 Belfast C1	RAF Museum, Cosford	
XR379	Sud Alouette AH2 (G-CICS)	Army Historic Aircraft Flight, Middle Wallop	

Notes	Serial	Type (code/other identity)	Owner/operator, location or fate
	XR453	WS55 Whirlwind HAR10 (8873M) [A]	RAF Odiham, on gate
	XR485	WS55 Whirlwind HAR10 [Q]	Norfolk & Suffolk Avn Museum, Flixton
	XR486	WS55 Whirlwind HCC12 (8727M/G-RWWW)	The Helicopter Museum, Weston-super-Mare
	XR498	WS58 Wessex HC2 (9342M) [X]	RAF St Mawgan, for display
	XR501	WS58 Wessex HC2	*Sold to France 2014*
	XR503	WS58 Wessex HC2	MoD DFTDC, Manston
	XR506	WS58 Wessex HC2 (9343M) [V]	Privately owned, Corley Moor, Warks
	XR516	WS58 Wessex HC2 (9319M) [V]	RAF Shawbury, on display
	XR517	WS58 Wessex HC2 [N]	Ulster Aviation Society, Long Kesh
	XR518	WS58 Wessex HC2	DSMarE AESS, *HMS Sultan*, Gosport
	XR523	WS58 Wessex HC2 [M]	RN *HMS Raleigh*, Torpoint, instructional use
	XR525	WS58 Wessex HC2 [G]	RAF Museum, Cosford
	XR526	WS58 Wessex HC2 (8147M)	The Helicopter Museum, stored Weston-super-Mare
	XR528	WS58 Wessex HC2	Privately owned, Little Rissington (GI use)
	XR529	WS58 Wessex HC2 (9268M) [E]	RAF Aldergrove, on display
	XR534	HS Gnat T1 (8578M) [65]	Newark Air Museum, Winthorpe
	XR537	HS Gnat T1 (8642M/G-NATY)	Privately owned, Bournemouth
	XR538	HS Gnat T1 (8621M/G-RORI) [01]	Heritage Aircraft Trust, North Weald
	XR540	HS Gnat T1 (XP502/8576M) [2]	Privately owned, Cotswold Airport
	XR571	HS Gnat T1 (8493M)	RAF *Red Arrows*, Scampton, on display
	XR574	HS Gnat T1 (8631M) [72]	Trenchard Museum, Halton
	XR595	WS Scout AH1 (G-BWHU) [M]	Privately owned, North Weald
	XR601	WS Scout AH1	Army Whittington Barracks, Lichfield, on display
	XR627	WS Scout AH1 [X]	Privately owned, Storwood, Yorkshire
	XR628	WS Scout AH1	Privately owned, Ipswich
	XR629	WS Scout AH1 (fuselage)	Privately owned, Ipswich
	XR635	WS Scout AH1	Midland Air Museum, Coventry
	XR650	Hunting Jet Provost T4 (8459M) [28]	Boscombe Down Aviation Collection, Old Sarum
	XR654	Hunting Jet Provost T4 <ff>	Privately owned, Chester
	XR658	Hunting Jet Provost T4 (8192M)	Deeside College, Connah's Quay, Clwyd
	XR662	Hunting Jet Provost T4 (8410M) [25]	Privately owned, stored Baxterley
	XR673	Hunting Jet Provost T4 (G-BXLO/9032M) [L]	Privately owned, Linton-on-Ouse
	XR681	Hunting Jet Provost T4 (8588M) <ff>	Robertsbridge Aviation Society, Mayfield
	XR700	Hunting Jet Provost T4 (8589M) <ff>	Ulster Aviation Society, Long Kesh
	XR713	BAC Lightning F3 (8935M) [C]	Lightning Preservation Grp, Bruntingthorpe
	XR718	BAC Lightning F6 (8932M) [DA]	Privately owned, Over Dinsdale, N Yorks
	XR724	BAC Lightning F6 (G-BTSY)	The Lightning Association, Binbrook
	XR725	BAC Lightning F6	Privately owned, Binbrook
	XR726	BAC Lightning F6 <ff>	Privately owned, Harrogate
	XR728	BAC Lightning F6 [JS]	Lightning Preservation Grp, Bruntingthorpe
	XR747	BAC Lightning F6 <ff>	No 20 Sqn ATC, Bideford, Devon
	XR749	BAC Lightning F3 (8934M) [DA]	Privately owned, Peterhead
	XR751	BAC Lightning F3 <ff>	Privately owned, Thorpe Wood, N Yorks
	XR753	BAC Lightning F6 (8969M) [XI]	RAF Coningsby on display
	XR753	BAC Lightning F53 (ZF578) [A]	Tangmere Military Aviation Museum
	XR754	BAC Lightning F6 (8972M) <ff>	Privately owned, Upwood, Cambs
	XR755	BAC Lightning F6	Privately owned, Callington, Cornwall
	XR757	BAC Lightning F6 <ff>	RAF Scampton Heritage Centre
	XR759	BAC Lightning F6 <ff>	Privately owned, Haxey, Lincs
	XR770	BAC Lightning F6 [AA]	RAF Waddington, for display
	XR771	BAC Lightning F6 [BF]	Midland Air Museum, Coventry
	XR806	BAC VC10 C1K (9285M) <ff>	RAF Brize Norton, BDRT
	XR808	BAC VC10 C1K [R] $	RAF Museum, Cosford (at Bruntingthorpe)
	XR810	BAC VC10 C1K <ff>	Privately owned, Crondall, Hants
	XR944	Wallis WA116 (G-ATTB)	Privately owned, Old Buckenham
	XR977	HS Gnat T1 (8640M) [3]	RAF Museum, Cosford
	XR992	HS Gnat T1 (8624M/XS102/G-MOUR)	Heritage Aircraft Trust, North Weald
	XR993	HS Gnat T1 (8620M/XP534/G-BVPP)	Privately owned, Bruntingthorpe
	XS100	HS Gnat T1 (8561M) <ff>	Privately owned, London SW3
	XS100	HS Gnat T1 (8561M) <rf>	Privately owned, North Weald
	XS104	HS Gnat T1 (8604M/G-FRCE)	Privately owned, North Weald

Serial	Type (code/other identity)	Owner/operator, location or fate	Notes
XS111	HS Gnat T1 (8618M/XP504/G-TIMM)	Heritage Aircraft Trust, North Weald	
XS149	WS58 Wessex HAS3 [661/GL]	The Helicopter Museum, Weston-super-Mare	
XS176	Hunting Jet Provost T4 (8514M) <ff>	Highland Aviation Museum, Dallachy, Moray	
XS177	Hunting Jet Provost T4 (9044M) [N]	Privately owned, Hexham, Northumberland	
XS179	Hunting Jet Provost T4 (8237M) [20]	Secret Nuclear Bunker, Hack Green, Cheshire	
XS180	Hunting Jet Provost T4 (*8238M/8338M*) [21]	MoD, Boscombe Down	
XS181	Hunting Jet Provost T4 (9033M) <ff>	Lakes Lightnings, Spark Bridge, Cumbria	
XS183	Hunting Jet Provost T4 <ff>	Privately owned, Plymouth	
XS186	Hunting Jet Provost T4 (8408M) [M]	Metheringham Airfield Visitors Centre	
XS209	Hunting Jet Provost T4 (8409M)	Solway Aviation Society, Carlisle	
XS216	Hunting Jet Provost T4 <ff>	South Yorkshire Aircraft Museum, Doncaster	
XS218	Hunting Jet Provost T4 (8508M) <ff>	No 447 Sqn ATC, Henley-on-Thames, Berks	
XS231	BAC Jet Provost T5 (G-ATAJ)	Boscombe Down Aviation Collection, Old Sarum	
XS235	DH106 Comet 4C (G-CPDA)	Cold War Jets Collection, Bruntingthorpe	
XS238	Auster AOP9 (TAD 200)	Newark Air Museum, stored Winthorpe	
XS416	BAC Lightning T5	Privately owned, New York, Lincs	
XS417	BAC Lightning T5 [DZ]	Newark Air Museum, Winthorpe	
XS420	BAC Lightning T5	Lakes Lightnings, FAST, Farnborough	
XS421	BAC Lightning T5 <ff>	Privately owned, Foulness	
XS456	BAC Lightning T5 [DX]	Skegness Water Leisure Park	
XS457	BAC Lightning T5 <ff>	Privately owned, Binbrook	
XS458	BAC Lightning T5 [T]	T5 Projects, Cranfield	
XS459	BAC Lightning T5 [AW]	Fenland & W Norfolk Aviation Museum, Wisbech	
XS481	WS58 Wessex HU5	South Yorkshire Aircraft Museum, Doncaster	
XS482	WS58 Wessex HU5	RAF Manston History Museum	
XS486	WS58 Wessex HU5 (9272M) [524/CU,F]	The Helicopter Museum, Weston-super-Mare	
XS488	WS58 Wessex HU5 (9056M) [F]	Privately owned, Tiptree, Essex	
XS489	WS58 Wessex HU5 [R]	Privately owned, Westerham, Kent	
XS493	WS58 Wessex HU5	Vector Aerospace, stored Fleetlands	
XS507	WS58 Wessex HU5	RAF Benson, for display	
XS508	WS58 Wessex HU5	FAA Museum, stored Cobham Hall, RNAS Yeovilton	
XS510	WS58 Wessex HU5 [626/PO]	No 1414 Sqn ATC, Crowborough, Sussex	
XS511	WS58 Wessex HU5 [M]	Tangmere Military Aircraft Museum	
XS513	WS58 Wessex HU5	RNAS Yeovilton Fire Section	
XS514	WS58 Wessex HU5 (A2740) [L/PO]	DSMarE AESS, *HMS Sultan*, Gosport	
XS515	WS58 Wessex HU5 [N]	Army, Keogh Barracks, Aldershot, instructional use	
XS516	WS58 Wessex HU5 [Q]	Privately owned, Redruth, Cornwall	
XS520	WS58 Wessex HU5 [F]	RN, Predannack Fire School	
XS522	WS58 Wessex HU5 [ZL]	Blackball Paintball, Truro, Cornwall	
XS527	WS Wasp HAS1	FAA Museum, stored Cobham Hall, RNAS Yeovilton	
XS529	WS Wasp HAS1	Privately owned, Redruth, Cornwall	
XS539	WS Wasp HAS1 [435]	Vector Aerospace Fleetlands Apprentice School	
XS567	WS Wasp HAS1 [434/E]	Imperial War Museum, Duxford	
XS568	WS Wasp HAS1 (A2715) [441]	DSMarE AESS, *HMS Sultan*, Gosport	
XS570	WS Wasp HAS1 [445/P]	MSS Holdings, Kirkham, Lancs	
XS574	Northrop Shelduck D1 <R>	FAA Museum, stored Cobham Hall, RNAS Yeovilton	
XS576	DH110 Sea Vixen FAW2 [125/E]	Imperial War Museum, Duxford	
XS587	DH110 Sea Vixen FAW(TT)2 (8828M/G-VIXN)	Gatwick Aviation Museum, Charlwood, Surrey	
XS590	DH110 Sea Vixen FAW2 [131/E]	FAA Museum, RNAS Yeovilton	
XS598	HS Andover C1 (fuselage)	FETC, Moreton-in-Marsh, Glos	
XS639	HS Andover E3A (9241M)	RAF Museum, Cosford	
XS641	HS Andover C1PR (9198M) (fuselage)	Privately owned, Sandbach, Cheshire	
XS643	HS Andover E3A (9278M) <ff>	Privately owned, Stock, Essex	
XS646	HS Andover C1(mod)	MoD, Boscombe Down (wfu)	
XS651	Slingsby T45 Swallow TX1 (BGA1211)	Privately owned, Keevil	
XS652	Slingsby T45 Swallow TX1 (BGA1107)	Privately owned, Chipping, Lancs	
XS674	WS58 Wessex HC2 [R]	Privately owned, Biggin Hill	
XS695	HS Kestrel FGA1 [5]	RAF Museum, Cosford	
XS709	HS125 Dominie T1 [M]	RAF Museum, Cosford	
XS710	HS125 Dominie T1 (9259M) [O]	RAF Cranwell Fire Section	
XS713	HS125 Dominie T1 [C]	RAF Shawbury Fire Section	
XS714	HS125 Dominie T1 (9246M) [P]	MoD DFTDC, Manston	

Notes	Serial	Type (code/other identity)	Owner/operator, location or fate
	XS726	HS125 Dominie T1 (9273M) [T]	Newark Air Museum, Winthorpe
	XS727	HS125 Dominie T1 [D]	RAF Cranwell, on display
	XS733	HS125 Dominie T1 (9276M) [Q]	Privately owned, Sproughton
	XS734	HS125 Dominie T1 (9260M) [N]	Privately owned, Sproughton
	XS735	HS125 Dominie T1 (9264M) [R]	Privately owned, Walcott, Lincs
	XS736	HS125 Dominie T1 [S]	MoD Winterbourne Gunner, Wilts
	XS738	HS125 Dominie T1 (9274M) [U]	RN, Predannack Fire School
	XS743	Beagle B206Z	QinetiQ, Boscombe Down, Apprentice School
	XS765	Beagle B206 Basset CC1 (G-BSET)	MoD, QinetiQ, Boscombe Down (spares use)
	XS770	Beagle B206 Basset CC1 (G-HRHI)	MoD, QinetiQ, Boscombe Down (spares use)
	XS790	HS748 Andover CC2 <ff>	Boscombe Down Aviation Collection, Old Sarum
	XS791	HS748 Andover CC2 (fuselage)	Privately owned, Stock, Essex
	XS863	WS58 Wessex HAS1 [304/R]	Imperial War Museum, Duxford
	XS865	WS58 Wessex HAS1 (A2694)	Privately owned, Ballygowan, NI
	XS885	WS58 Wessex HAS1 [512/DD]	SFDO, RNAS Culdrose
	XS886	WS58 Wessex HAS1 [527/CU]	Privately owned, Ditchling, E Sussex
	XS887	WS58 Wessex HAS1 [403/FI]	South Yorkshire Aircraft Museum, Doncaster
	XS888	WS58 Wessex HAS1 [521]	Guernsey Airport Fire Section
	XS898	BAC Lightning F6 <ff>	Privately owned, Lavendon, Bucks
	XS899	BAC Lightning F6 <ff>	Privately owned, Binbrook
	XS903	BAC Lightning F6 [BA]	Yorkshire Air Museum, Elvington
	XS904	BAC Lightning F6 [BQ]	Lightning Preservation Grp, Bruntingthorpe
	XS919	BAC Lightning F6	Privately owned, stored Henstridge
	XS922	BAC Lightning F6 (8973M) <ff>	Lakes Lightnings, Spark Bridge, Cumbria
	XS923	BAC Lightning F6 <ff>	Privately owned, Welshpool
	XS925	BAC Lightning F6 (8961M) [BA]	RAF Museum, Hendon
	XS928	BAC Lightning F6 [AD]	BAE Systems Warton, on display
	XS932	BAC Lightning F6 <ff>	Privately owned, Walcott, Lincs
	XS933	BAC Lightning F6 <ff>	Privately owned, Farnham
	XS933	BAC Lightning F53 (ZF594) [BF]	North-East Aircraft Museum, Usworth
	XS936	BAC Lightning F6	Castle Motors, Liskeard, Cornwall
	XT108	Agusta-Bell 47G-3 Sioux AH1 [U]	Museum of Army Flying, Middle Wallop
	XT123	WS Sioux AH1 (XT827) [D]	AAC Middle Wallop, at main gate
	XT131	Agusta-Bell 47G-3 Sioux AH1 (G-CICN) [B]	Army Historic Aircraft Flight, Middle Wallop
	XT140	Agusta-Bell 47G-3 Sioux AH1	Perth Technical College
	XT141	Agusta-Bell 47G-3 Sioux AH1	Privately owned, Newcastle
	XT150	Agusta-Bell 47G-3 Sioux AH1 (7883M) [R]	South Yorkshire Aircraft Museum, Doncaster
	XT151	WS Sioux AH1	Museum of Army Flying, stored Middle Wallop
	XT176	WS Sioux AH1 [U]	FAA Museum, stored Cobham Hall, RNAS Yeovilton
	XT190	WS Sioux AH1	The Helicopter Museum, Weston-super-Mare
	XT200	WS Sioux AH1 [F]	Newark Air Museum, Winthorpe
	XT208	WS Sioux AH1 (wreck)	Blessingbourne Museum, Fivemiletown, Co Tyrone, NI
	XT223	WS Sioux AH1 (G-XTUN)	Privately owned, Sherburn-in-Elmet
	XT236	WS Sioux AH1 (frame only)	South Yorkshire Aircraft Museum, Doncaster
	XT242	WS Sioux AH1 (composite) [12]	South Yorkshire Aircraft Museum, Doncaster
	XT257	WS58 Wessex HAS3 (8719M)	Bournemouth Aviation Museum
	XT277	HS Buccaneer S2A (8853M) <ff>	Privately owned, Welshpool
	XT280	HS Buccaneer S2A <ff>	Dumfries & Galloway Avn Mus, Dumfries
	XT284	HS Buccaneer S2A (8855M) <ff>	Privately owned, Felixstowe
	XT288	HS Buccaneer S2B (9134M)	Royal Scottish Museum of Flight, E Fortune
	XT420	WS Wasp HAS1 (G-CBUI) [606]	*Currently not known*
	XT427	WS Wasp HAS1 [606]	FAA Museum, stored RNAS Yeovilton
	XT431	WS Wasp HAS1 (comp XS463)	Bournemouth Aviation Museum
	XT434	WS Wasp HAS1 (G-CGGK) [455]	Privately owned, Breighton
	XT435	WS Wasp HAS1 (NZ3907/G-RIMM) [430]	Privately owned, Badwell Green, Suffolk
	XT437	WS Wasp HAS1 [423]	Boscombe Down Aviation Collection, Old Sarum
	XT439	WS Wasp HAS1 [605]	Privately owned, Hemel Hempstead
	XT443	WS Wasp HAS1 [422/AU]	The Helicopter Museum, Weston-super-Mare
	XT453	WS58 Wessex HU5 (A2756) [B/PO]	DSMarE AESS, HMS Sultan, Gosport
	XT455	WS58 Wessex HU5 (A2654) [U]	Privately owned, Hixon, Staffs
	XT456	WS58 Wessex HU5 (8941M) [XZ]	Belfast Airport Fire Section

Serial	Type (code/other identity)	Owner/operator, location or fate	Notes
XT466	WS58 Wessex HU5 (A2617/8921M) [XV]	Army Whittington Barracks, Lichfield, on display	
XT467	WS58 Wessex HU5 (8922M) [BF]	Gunsmoke Paintball, Hadleigh, Suffolk	
XT469	WS58 Wessex HU5 (8920M)	Privately owned, Weeton, Lancs	
XT472	WS58 Wessex HU5 [XC]	The Helicopter Museum, Weston-super-Mare	
XT480	WS58 Wessex HU5 [468/RG]	Rednal Paintball, Shropshire	
XT482	WS58 Wessex HU5 [ZM/VL]	FAA Museum, RNAS Yeovilton	
XT484	WS58 Wessex HU5 (A2742) [H]	DSMarE AESS, HMS Sultan, Gosport	
XT485	WS58 Wessex HU5 (A2680)	DSMarE, stored HMS Sultan, Gosport	
XT486	WS58 Wessex HU5 (8919M)	Dumfries & Galloway Avn Mus, Dumfries	
XT550	WS Sioux AH1 [D]	AAC, stored Middle Wallop	
XT575	Vickers Viscount 837 <ff>	Brooklands Museum, Weybridge	
XT581	Northrop Shelduck D1	Imperial War Museum, Duxford	
XT583	Northrop Shelduck D1	Royal Artillery Experience, Woolwich	
XT596	McD F-4K Phantom FG1	FAA Museum, RNAS Yeovilton	
XT597	McD F-4K Phantom FG1	Privately owned, Bentwaters	
XT601	WS58 Wessex HC2 (9277M) (composite)	RAF Odiham, BDRT	
XT604	WS58 Wessex HC2	East Midlands Airport Aeropark	
XT617	WS Scout AH1	Wattisham Airfield Museum	
XT621	WS Scout AH1	Defence Academy of the UK, Shrivenham	
XT623	WS Scout AH1	DSEME SEAE, Arborfield	
XT626	WS Scout AH1 (G-CIBW) [Q]	Army Historic Aircraft Flt, Middle Wallop	
XT630	WS Scout AH1 (G-BXRL) [X]	Privately owned, Bruntingthorpe	
XT631	WS Scout AH1 [D]	Privately owned, Ipswich	
XT633	WS Scout AH1	AAC MPSU, Middle Wallop (for disposal)	
XT634	WS Scout AH1 (G-BYRX) [T]	Privately owned, Tollerton	
XT638	WS Scout AH1 [N]	AAC Middle Wallop, at gate	
XT640	WS Scout AH1	Privately owned, Sproughton	
XT643	WS Scout AH1 [Z]	Army, Thorpe Camp, East Wretham	
XT653	Slingsby T45 Swallow TX1 (BGA3469)	Privately owned, Oakhill, Somerset	
XT672	WS58 Wessex HC2 [WE]	RAF Stafford, on display	
XT681	WS58 Wessex HC2 (9279M) [U] <ff>	Privately owned, Wallingford, Oxon	
XT761	WS58 Wessex HU5	RNAS Culdrose	
XT762	WS58 Wessex HU5	Hamburger Hill Paintball, Marksbury, Somerset	
XT765	WS58 Wessex HU5 [J]	FAA Museum, RNAS Yeovilton	
XT765	WS58 Wessex HU5 (XT458/A2768) [P/VL]	RNAS Yeovilton, for display	
XT769	WS58 Wessex HU5 [823]	FAA Museum, RNAS Yeovilton	
XT771	WS58 Wessex HU5 [620/PO]	RNAS Culdrose, stored	
XT773	WS58 Wessex HU5 (9123M)	RAF Shawbury Fire Section	
XT778	WS Wasp HAS1 [430]	FAA Museum, stored Cobham Hall, RNAS Yeovilton	
XT780	WS Wasp HAS1 [636]	CEMAST, Fareham College, Lee-on-Solent	
XT787	WS Wasp HAS1 (NZ3905/G-KAXT)	Privately owned, Middle Wallop	
XT788	WS Wasp HAS1 (G-BMIR) [474]	Privately owned, Storwood, Yorkshire	
XT793	WS Wasp HAS1 (G-BZPP) [456]	Privately owned, Babcary, Somerset (wreck)	
XT863	McD F-4K Phantom FG1 <ff>	Privately owned, Cowes, IOW	
XT864	McD F-4K Phantom FG1 (8998M/XT684) [BJ]	Ulster Aviation Society, Long Kesh	
XT891	McD F-4M Phantom FGR2 (9136M)	RAF Coningsby, at main gate	
XT903	McD F-4M Phantom FGR2 <ff>	Michael Beetham Conservation Centre, RAFM Cosford	
XT905	McD F-4M Phantom FGR2 (9286M) [P]	Privately owned, Bentwaters	
XT907	McD F-4M Phantom FGR2 (9151M) [W]	Privately owned, Bentwaters	
XT914	McD F-4M Phantom FGR2 (9269M) [Z]	Wattisham Station Heritage Museum	
XV104	BAC VC10 C1K [U]	*Scrapped at Bruntingthorpe, 2013*	
XV106	BAC VC10 C1K [W]	*Scrapped at Bruntingthorpe, 2013*	
XV108	BAC VC10 C1K <ff>	East Midlands Airport Aeropark	
XV109	BAC VC10 C1K <ff>	Privately owned, Bruntingthorpe	
XV118	WS Scout AH1 (9141M) <ff>	Kennet Aviation, North Weald	
XV122	WS Scout AH1 [D]	Defence Academy of the UK, Shrivenham	
XV123	WS Scout AH1	RAF Shawbury, on display	
XV124	WS Scout AH1 [W]	Privately owned, Godstone, Surrey	
XV127	WS Scout AH1	Museum of Army Flying, Middle Wallop	
XV130	WS Scout AH1 (G-BWJW) [R]	Privately owned, Babcary, Somerset	
XV131	WS Scout AH1 [Y]	AAC 70 Aircraft Workshops, Middle Wallop, BDRT	

Notes	Serial	Type (code/other identity)	Owner/operator, location or fate
	XV136	WS Scout AH1 [X]	Ulster Aviation Society, Long Kesh
	XV137	WS Scout AH1 (G-CRUM)	Privately owned, Chiseldon, Wilts
	XV137	WS Scout AH1 (XV139)	South Yorkshire Aircraft Museum, stored Doncaster
	XV138	WS Scout AH1 (G-SASM)	Privately owned, Babcary, Somerset
	XV141	WS Scout AH1	REME Museum, Arborfield
	XV148	HS Nimrod MR1(mod) <ff>	Privately owned, Malmesbury
	XV161	HS Buccaneer S2B (9117M) <ff>	Dundonald Aviation Centre
	XV165	HS Buccaneer S2B <ff>	Privately owned, Ashford, Kent
	XV168	HS Buccaneer S2B [AF]	Yorkshire Air Museum, Elvington
	XV177	Lockheed C-130K Hercules C3A [177]	RAF, St Athan (wfu)
	XV188	Lockheed C-130K Hercules C3A [188]	RAF, St Athan (wfu)
	XV196	Lockheed C-130K Hercules C1 [196]	RAF, St Athan (wfu)
	XV197	Lockheed C-130K Hercules C3 [197]	Scrapped at Hixon, 2014
	XV200	Lockheed C-130K Hercules C1 [200]	RAF, St Athan (wfu)
	XV201	Lockheed C-130K Hercules C1K <ff>	Marshalls, Cambridge
	XV202	Lockheed C-130K Hercules C3 [202]	RAF Museum, Cosford
	XV208	Lockheed C-130K Hercules W2	Marshalls, Cambridge (wfu)
	XV209	Lockheed C-130K Hercules C3A [209]	RAF, St Athan (wfu)
	XV214	Lockheed C-130K Hercules C3A [214]	RAF, St Athan (wfu)
	XV217	Lockheed C-130K Hercules C3	Scrapped at Hixon, 2014
	XV220	Lockheed C-130K Hercules C3	Scrapped at Hixon, 2014
	XV221	Lockheed C-130K Hercules C3 [221]	Scrapped at Hixon, 2014
	XV226	HS Nimrod MR2 $	Cold War Jets Collection, Bruntingthorpe
	XV229	HS Nimrod MR2 [29]	MoD DFTDC, Manston
	XV231	HS Nimrod MR2 [31]	Aviation Viewing Park, Manchester
	XV232	HS Nimrod MR2 [32]	AIRBASE, Coventry
	XV235	HS Nimrod MR2 [35] <ff>	Privately owned, RAF Scampton
	XV240	HS Nimrod MR2 [40] <ff>	Spey Bay Salvage, Dallachy, Moray
	XV241	HS Nimrod MR2 [41] <ff>	Royal Scottish Mus'm of Flight, E Fortune
	XV244	HS Nimrod MR2 [44] <ff>	Morayavia, RAF Kinloss
	XV249	HS Nimrod R1 $	RAF Museum, Cosford
	XV250	HS Nimrod MR2 [50]	Yorkshire Air Museum, Elvington
	XV252	HS Nimrod MR2 [52] <ff>	Privately owned, Cullen, Moray
	XV254	HS Nimrod MR2 [54] <ff>	Highland Aviation Museum, Inverness
	XV255	HS Nimrod MR2 [55]	City of Norwich Aviation Museum
	XV259	BAe Nimrod AEW3 <ff>	Privately owned, Wales
	XV263	BAe Nimrod AEW3P (8967M) <ff>	BAE Systems, Brough
	XV263	BAe Nimrod AEW3P (8967M) <rf>	MoD/BAE Systems, Woodford
	XV268	DHC2 Beaver AL1 (G-BVER)	Privately owned, Cumbernauld
	XV277	HS P.1127(RAF)	Royal Scottish Mus'm of Flight, E Fortune
	XV279	HS P.1127(RAF) (8566M)	Harrier Heritage Centre, RAF Wittering
	XV280	HS P.1127(RAF) <ff>	RNAS Yeovilton, GI use
	XV295	Lockheed C-130K Hercules C1 [295]	RAF, St Athan (wfu)
	XV301	Lockheed C-130K Hercules C3 [301]	Scrapped at Hixon, 2014
	XV302	Lockheed C-130K Hercules C3 [302]	Marshalls, Cambridge, fatigue test airframe
	XV303	Lockheed C-130K Hercules C3A [303]	RAF, St Athan (wfu)
	XV304	Lockheed C-130K Hercules C3A	RAF Brize Norton, instructional use
	XV305	Lockheed C-130K Hercules C3 $	Scrapped at Hixon, 2014
	XV328	BAC Lightning T5 <ff>	Lightning Preservation Group, Bruntingthorpe
	XV333	HS Buccaneer S2B [234/H]	FAA Museum, RNAS Yeovilton
	XV344	HS Buccaneer S2C	QinetiQ Farnborough, on display
	XV350	HS Buccaneer S2B	East Midlands Airport Aeropark
	XV352	HS Buccaneer S2B <ff>	RAF Manston History Museum
	XV359	HS Buccaneer S2B [035/R]	Privately owned, Topsham, Devon
	XV361	HS Buccaneer S2B	Ulster Aviation Society, Long Kesh
	XV370	Sikorsky SH-3D (A2682) [260]	DSMarE AESS, HMS Sultan, Gosport
	XV371	WS61 Sea King HAS1(DB) [61/DD]	SFDO, RNAS Culdrose
	XV372	WS61 Sea King HAS1	RAF, St Mawgan, instructional use
	XV383	Northrop MQM-57A/3 (fuselage)	Privately owned, Wimborne, Dorset
	XV401	McD F-4M Phantom FGR2 [I]	Privately owned, Bentwaters
	XV402	McD F-4M Phantom FGR2 <ff>	Privately owned, Kent
	XV406	McD F-4M Phantom FGR2 (9098M) [CK]	Solway Aviation Society, Carlisle

Serial	Type (code/other identity)	Owner/operator, location or fate	Notes
XV408	McD F-4M Phantom FGR2 (9165M) [Z]	Tangmere Military Aviation Museum	
XV411	McD F-4M Phantom FGR2 (9103M) [L]	MoD DFTDC, Manston	
XV415	McD F-4M Phantom FGR2 (9163M) [E]	RAF Boulmer, on display	
XV424	McD F-4M Phantom FGR2 (9152M) [I]	RAF Museum, Hendon	
XV426	McD F-4M Phantom FGR2 <ff>	City of Norwich Aviation Museum	
XV426	McD F-4M Phantom FGR2 [P] <rf>	RAF Coningsby, BDRT	
XV460	McD F-4M Phantom FGR2 <ff>	Privately owned, Bentwaters	
XV474	McD F-4M Phantom FGR2 [T]	The Old Flying Machine Company, Duxford	
XV490	McD F-4M Phantom FGR2 [R] <ff>	Newark Air Museum, Winthorpe	
XV497	McD F-4M Phantom FGR2 (9295M) [D]	Privately owned, Bentwaters	
XV499	McD F-4M Phantom FGR2 <ff>	Privately owned, Hixon, Staffs	
XV581	McD F-4K Phantom FG1 (9070M) <ff>	No 2481 Sqn ATC, Bridge of Don	
XV582	McD F-4K Phantom FG1 (9066M) [M]	RAF Leuchars	
XV586	McD F-4K Phantom FG1 (9067M) [010-R]	Privately owned, RNAS Yeovilton	
XV591	McD F-4K Phantom FG1 [013] <ff>	RAF Museum, Cosford	
XV625	WS Wasp HAS1 (A2649) [471]	Privately owned, Colsterworth, Leics	
XV631	WS Wasp HAS1 (fuselage)	Farnborough Air Sciences Trust, Farnborough	
XV642	WS61 Sea King HAS2A (A2614) [259]	DSMarE AESS, *HMS Sultan*, Gosport	
XV643	WS61 Sea King HAS6 [262]	DSAE, No 1 SoTT, Cosford	
XV647	WS61 Sea King HU5 [28]	RN Culdrose (wfu)	
XV648	WS61 Sea King HU5 [18/CU]	RN No 771 NAS, Culdrose	
XV649	WS61 Sea King ASaC7 [180]	RN No 849 NAS, Culdrose	
XV651	WS61 Sea King HU5	MoD/AFD/QinetiQ, Boscombe Down	
XV653	WS61 Sea King HAS6 (9326M) [63/CU]	DSAE, No 1 SoTT, Cosford	
XV654	WS61 Sea King HAS6 [705/DD] (wreck)	RN, Predannack Fire School	
XV655	WS61 Sea King HAS6 [270/N]	DSMarE AESS, *HMS Sultan*, Gosport	
XV656	WS61 Sea King ASaC7 [185]	RN No 849 NAS, Culdrose	
XV657	WS61 Sea King HAS5 (ZA135) [32/DD]	SFDO, RNAS Culdrose	
XV659	WS61 Sea King HAS6 (9324M) [62/CU]	DSAE, No 1 SoTT, Cosford	
XV660	WS61 Sea King HAS6 [69/N]	*Currently not known*	
XV661	WS61 Sea King HU5 [26]	RN No 771 NAS, Culdrose	
XV663	WS61 Sea King HAS6	DSMarE, stored *HMS Sultan*, Gosport	
XV664	WS61 Sea King ASaC7 [190]	RN No 849 NAS, Culdrose	
XV665	WS61 Sea King HAS6 [507/CU]	DSMarE AESS, *HMS Sultan*, Gosport	
XV666	WS61 Sea King HU5 [21]	RN No 771 NAS, Culdrose	
XV670	WS61 Sea King HU5 [17]	RN No 771 NAS, Culdrose	
XV671	WS61 Sea King ASaC7 [183]	RN No 849 NAS, Culdrose	
XV672	WS61 Sea King ASaC7 [187]	RN No 849 NAS, Culdrose	
XV673	WS61 Sea King HU5 [27/CU]	RN No 771 NAS, Culdrose	
XV674	WS61 Sea King HAS6	Privately owned, Horsham	
XV675	WS61 Sea King HAS6 [701/PW]	DSMarE AESS, *HMS Sultan*, Gosport	
XV676	WS61 Sea King HC6 [ZE]	DSMarE, stored *HMS Sultan*, Gosport	
XV677	WS61 Sea King HAS6 [269]	South Yorkshire Aircraft Museum, Doncaster	
XV696	WS61 Sea King HAS6 [267/L]	DSMarE AESS, *HMS Sultan*, Gosport	
XV697	WS61 Sea King ASaC7 [181]	RN No 849 NAS, Culdrose	
XV699	WS61 Sea King HU5 [823/PW]	DSMarE, stored *HMS Sultan*, Gosport	
XV700	WS61 Sea King HC6 [ZC]	DSMarE AESS, *HMS Sultan*, Gosport	
XV701	WS61 Sea King HAS6 [268/N,64]	DSAE, No 1 SoTT, Cosford	
XV703	WS61 Sea King HC6 [ZD]	DSMarE, stored *HMS Sultan*, Gosport	
XV705	WS61 Sea King HU5 [29]	RN No 771 NAS, Culdrose	
XV706	WS61 Sea King HAS6 (9344M) [017/L]	RN ETS, Culdrose	
XV707	WS61 Sea King ASaC7 [184]	RN No 849 NAS, Culdrose	
XV708	WS61 Sea King HAS6 [501/CU]	DSMarE AESS, *HMS Sultan*, Gosport	
XV709	WS61 Sea King HAS6 (9303M) [263]	RAF Valley, instructional use	
XV711	WS61 Sea King HAS6 [15/CW]	DSMarE AESS, *HMS Sultan*, Gosport	
XV712	WS61 Sea King HAS6 [66]	Imperial War Museum, Duxford	
XV713	WS61 Sea King HAS6 (A2646) [018/L]	DSMarE AESS, *HMS Sultan*, Gosport	
XV714	WS61 Sea King ASaC7 [188]	RN No 849 NAS, Culdrose	
XV720	WS58 Wessex HC2 (A2701)	Privately owned, Hixon, Staffs	
XV722	WS58 Wessex HC2 (8805M) [WH]	Privately owned, Badgers Mount, Kent	
XV724	WS58 Wessex HC2	DSMarE AESS, *HMS Sultan*, Gosport	
XV725	WS58 Wessex HC2 [C]	MoD DFTDC, Manston	

Notes	Serial	Type (code/other identity)	Owner/operator, location or fate
	XV726	WS58 Wessex HC2 [J]	Privately owned, Biggin Hill
	XV728	WS58 Wessex HC2 [A]	Newark Air Museum, Winthorpe
	XV731	WS58 Wessex HC2 [Y]	Privately owned, Badgers Mount, Kent
	XV732	WS58 Wessex HCC4	RAF Museum, Hendon
	XV733	WS58 Wessex HCC4	The Helicopter Museum, Weston-super-Mare
	XV741	HS Harrier GR3 (A2608) [41/DD]	Privately owned, Thorpe Wood, N Yorks
	XV744	HS Harrier GR3 (9167M) [3K]	Tangmere Military Aviation Museum
	XV748	HS Harrier GR3 [3D]	Yorkshire Air Museum, Elvington
	XV751	HS Harrier GR3 [AU]	Gatwick Aviation Museum, Charlwood, Surrey
	XV752	HS Harrier GR3 (9075M) [B]	South Yorkshire Aircraft Museum, Doncaster
	XV753	HS Harrier GR3 (9078M) [53/DD]	Classic Air Force, Newquay
	XV755	HS Harrier GR3 [M]	RNAS Yeovilton Fire Section
	XV759	HS Harrier GR3 [O] <ff>	Privately owned, Hitchin, Herts
	XV760	HS Harrier GR3 <ff>	Solent Sky, Southampton
	XV779	HS Harrier GR3 (8931M)	Harrier Heritage Centre, RAF Wittering
	XV783	HS Harrier GR3 [83/DD]	Privately owned, Corley Moor, Warks
	XV784	HS Harrier GR3 (8909M) <ff>	Boscombe Down Aviation Collection, Old Sarum
	XV786	HS Harrier GR3 <ff>	RNAS Culdrose Fire Section
	XV786	HS Harrier GR3 [S] <rf>	RN, Predannack Fire School
	XV798	HS Harrier GR1(mod)	The Helicopter Museum, Weston-super-Mare
	XV804	HS Harrier GR3 (9280M) [O]	*Sold to the USA, July 2014*
	XV806	HS Harrier GR3 <ff>	Privately owned, Worksop
	XV808	HS Harrier GR3 (9076M/A2687) [08/DD]	Privately owned, Chetton, Shrops
	XV810	HS Harrier GR3 (9038M) [K]	Privately owned, Walcott, Lincs
	XV814	DH106 Comet 4 (G-APDF) <ff>	Privately owned, Chipping Campden
	XV863	HS Buccaneer S2B (9115M/9139M/9145M) [S]	Privately owned, Weston, Eire
	XV864	HS Buccaneer S2B (9234M)	MoD DFTDC, Manston
	XV865	HS Buccaneer S2B (9226M)	Imperial War Museum, Duxford
	XV867	HS Buccaneer S2B <ff>	Highland Aviation Museum, Inverness
	XW175	HS Harrier T4(VAAC)	MoD, Boscombe Down (wfu)
	XW198	WS Puma HC1	RAF Benson (wfu)
	XW199	WS Puma HC2	MoD/Eurocopter, Brasov, Romania (conversion)
	XW200	WS Puma HC1 (wreck)	*Scrapped, 2014*
	XW201	WS Puma HC1	*Scrapped, 2014*
	XW202	WS Puma HC1	Army, Imphal Barracks, York, on display
	XW204	WS Puma HC2	RAF No 33 Sqn/No 230 Sqn, Benson
	XW206	WS Puma HC1	*Scrapped, 2014*
	XW207	WS Puma HC1	*Scrapped, 2014*
	XW208	WS Puma HC1	RAFC Cranwell, GI use
	XW209	WS Puma HC2	RAF No 33 Sqn/No 230 Sqn, Benson
	XW210	WS Puma HC1 (comp XW215)	RAF AM&SU, stored Shawbury
	XW212	WS Puma HC2	RAF No 33 Sqn/No 230 Sqn, Benson
	XW213	WS Puma HC2	RAF No 33 Sqn/No 230 Sqn, Benson
	XW214	WS Puma HC2	RAF No 33 Sqn/No 230 Sqn, Benson
	XW216	WS Puma HC2 (F-ZWDD)	RAF No 33 Sqn/No 230 Sqn, Benson
	XW217	WS Puma HC2	RAF No 33 Sqn/No 230 Sqn, Benson
	XW218	WS Puma HC1 (wreck)	*Scrapped, 2014*
	XW219	WS Puma HC2	RAF No 33 Sqn/No 230 Sqn, Benson
	XW220	WS Puma HC2	RAF No 33 Sqn/No 230Sqn, Benson
	XW222	WS Puma HC1	Ulster Aviation Society, Long Kesh
	XW223	WS Puma HC1	RAF AM&SU, stored Shawbury
	XW224	WS Puma HC2	MoD/AFD/QinetiQ, Boscombe Down
	XW226	WS Puma HC1	RAF AM&SU, stored Shawbury
	XW227	WS Puma HC1	Privately owned, Colsterworth, Leics
	XW229	WS Puma HC2	RAF No 33 Sqn/No 230 Sqn, Benson
	XW231	WS Puma HC2	RAF No 33 Sqn/No 230 Sqn, Benson
	XW232	WS Puma HC2 (F-ZWDE)	MoD/AFD/QinetiQ, Boscombe Down
	XW235	WS Puma HC2	RAF PDSH, Benson
	XW236	WS Puma HC1	RAF Defence Movements School, Brize Norton
	XW237	WS Puma HC2	RAF No 33 Sqn/No 230 Sqn, Benson
	XW241	Sud SA330E Puma	Farnborough Air Sciences Trust, Farnborough

Serial	Type (code/other identity)	Owner/operator, location or fate	Notes
XW264	HS Harrier T2 <ff>	Harrier Trust, Dorset	
XW265	HS Harrier T4A (9258M) <ff>	No 2345 Sqn ATC, RAF Leuchars	
XW267	HS Harrier T4 (9263M)	Privately owned, Bentwaters	
XW268	HS Harrier T4N	City of Norwich Aviation Museum	
XW269	HS Harrier T4 [TB]	Caernarfon Air World	
XW270	HS Harrier T4 [T]	Coventry University, instructional use	
XW271	HS Harrier T4 [71/DD]	Privately owned, Sproughton	
XW272	HS Harrier T4 (8783M) (fuselage) (comp XV281)	Privately owned, Welshpool	
XW276	Aérospatiale SA341 Gazelle (F-ZWRI)	Newark Air Museum, Winthorpe	
XW281	WS Scout AH1 (G-BYNZ) [T]	Now marked as XP883, 2014	
XW283	WS Scout AH1 (G-CIMX) [U]	Kennet Aviation, North Weald	
XW289	BAC Jet Provost T5A (G-BVXT/G-JPVA) [73]	Kennet Aviation, Yeovilton	
XW290	BAC Jet Provost T5A (9199M) [41,MA]	Privately owned, Bruntingthorpe	
XW293	BAC Jet Provost T5 (G-BWCS) [Z]	Privately owned, Bournemouth	
XW299	BAC Jet Provost T5A (9146M) [60,MB]	QinetiQ Boscombe Down, Apprentice School	
XW301	BAC Jet Provost T5A (9147M) [63,MC]	Sold to The Netherlands, January 2014	
XW303	BAC Jet Provost T5A (9119M) [127]	RAF Halton	
XW304	BAC Jet Provost T5 (9172M) [MD]	Privately owned, Eye, Suffolk	
XW309	BAC Jet Provost T5 (9179M) [V,ME]	Hartlepool College of Further Education	
XW311	BAC Jet Provost T5 (9180M) [W,MF]	Privately owned, North Weald	
XW315	BAC Jet Provost T5A <ff>	Privately owned, Preston	
XW318	BAC Jet Provost T5A (9190M) [78,MG]	Sold to The Netherlands, January 2014	
XW320	BAC Jet Provost T5A (9015M) [71]	DSAE, No 1 SoTT, Cosford	
XW321	BAC Jet Provost T5A (9154M) [62,MH]	Privately owned, Bentwaters	
XW323	BAC Jet Provost T5A (9166M) [86]	RAF Museum, Hendon	
XW324	BAC Jet Provost T5 (G-BWSG) [U]	Privately owned, East Midlands	
XW325	BAC Jet Provost T5B (G-BWGF) [E]	Privately owned, Carlisle	
XW327	BAC Jet Provost T5A (9130M) [62]	DSAE, No 1 SoTT, Cosford	
XW328	BAC Jet Provost T5A (9177M) [75,MI]	Sold to Belgium, April 2014	
XW330	BAC Jet Provost T5A (9195M) [82,MJ]	Privately owned,	
XW333	BAC Jet Provost T5A (G-BVTC)	Global Aviation, Humberside	
XW353	BAC Jet Provost T5A (9090M) [3]	RAF Cranwell, on display	
XW354	BAC Jet Provost T5A (XW355/G-JPTV)	Privately owned, Linton-on-Ouse	
XW358	BAC Jet Provost T5A (9181M) [59,MK]	CEMAST, Fareham College, Lee-on-Solent	
XW360	BAC Jet Provost T5A (9153M) [61,ML]	Privately owned, Thorpe Wood, N Yorks	
XW361	BAC Jet Provost T5A (9192M) [81,MM]	Sold to The Netherlands, January 2014	
XW363	BAC Jet Provost T5A [36]	Dumfries & Galloway Avn Mus, Dumfries	
XW364	BAC Jet Provost T5A (9188M) [35,MN]	RAF Halton	
XW367	BAC Jet Provost T5A (9193M) [64,MO]	Privately owned, Sproughton	
XW370	BAC Jet Provost T5A (9196M) [72,MP]	Privately owned, Sproughton	
XW375	BAC Jet Provost T5A (9149M) [52]	DSAE, No 1 SoTT, Cosford	
XW404	BAC Jet Provost T5A (9049M) [77]	Hartlepool FE College	
XW405	BAC Jet Provost T5A (9187M)	Hartlepool FE College, on display	
XW409	BAC Jet Provost T5A (9047M)	Privately owned, Hawarden	
XW410	BAC Jet Provost T5A (9125M) [80] <ff>	Privately owned, Norwich	
XW416	BAC Jet Provost T5A (9191M) [84,MS]	Privately owned, Bentwaters	
XW418	BAC Jet Provost T5A (9173M) [MT]	RAF Museum, Cosford	
XW419	BAC Jet Provost T5A (9120M) [125]	Privately owned, Bruntingthorpe	
XW420	BAC Jet Provost T5A (9194M) [83,MU]	RAF Woodvale, on display	
XW422	BAC Jet Provost T5A (G-BWEB) [3]	Privately owned, Cotswold Airport	
XW423	BAC Jet Provost T5A (G-BWUW) [14]	Deeside College, Connah's Quay, Clwyd	
XW425	BAC Jet Provost T5A (9200M) [H,MV]	Sold to The Netherlands, 2014	
XW430	BAC Jet Provost T5A (9176M) [77,MW]	DSAE, No 1 SoTT, Cosford	
XW432	BAC Jet Provost T5A (9127M) [76,MX]	Privately owned, Bentwaters	
XW433	BAC Jet Provost T5A (G-JPRO)	Classic Air Force, Newquay	
XW434	BAC Jet Provost T5A (9091M) [78,MY]	Halfpenny Green Airport, on display	
XW436	BAC Jet Provost T5A (9148M) [68]	DSAE, No 1 SoTT, Cosford	
XW530	HS Buccaneer S2B [530]	Buccaneer Service Station, Elgin	
XW541	HS Buccaneer S2B (8858M) <ff>	Privately owned, Lavendon, Bucks	
XW544	HS Buccaneer S2B (8857M) [O]	The Buccaneer Aviation Group, Bruntingthorpe	
XW547	HS Buccaneer S2B (9095M/9169M) [R]	RAF Museum, Hendon	
XW550	HS Buccaneer S2B <ff>	Privately owned, West Horndon, Essex	

Notes	Serial	Type (code/other identity)	Owner/operator, location or fate
	XW560	SEPECAT Jaguar S <ff>	Boscombe Down Aviation Collection, Old Sarum
	XW563	SEPECAT Jaguar S (XX822/8563M)	County Hall, Norwich, on display
	XW566	SEPECAT Jaguar B	Farnborough Air Sciences Trust, Farnborough
	XW612	WS Scout AH1 (G-KAXW)	Privately owned, North Weald
	XW613	WS Scout AH1 (G-BXRS)	Privately owned, North Weald
	XW616	WS Scout AH1	AAC Dishforth, instructional use
	XW630	HS Harrier GR3	RNAS Yeovilton, Fire Section
	XW635	Beagle D5/180 (G-AWSW)	Privately owned, Spanhoe
	XW664	HS Nimrod R1	East Midlands Airport Aeropark
	XW666	HS Nimrod R1 <ff>	South Yorkshire Aircraft Museum, Doncaster
	XW763	HS Harrier GR3 (9002M/9041M) <ff>	Privately owned, Wigston, Leics
	XW768	HS Harrier GR3 (9072M) [N]	MoD DFTDC, Manston
	XW784	Mitchell-Procter Kittiwake I (G-BBRN) [VL]	Privately owned, RNAS Yeovilton
	XW795	WS Scout AH1	Blessingbourne Museum, Fivemiletown, Co Tyrone, NI
	XW796	WS Scout AH1	Gunsmoke Paintball, Hadleigh, Suffolk
	XW838	WS Lynx (TAD 009)	DSEME SEAE, Arborfield
	XW839	WS Lynx	The Helicopter Museum, Weston-super-Mare
	XW844	WS Gazelle AH1	Vector Aerospace Fleetlands, preserved
	XW846	WS Gazelle AH1	AAC No 665 Sqn/5 Regt, Aldergrove
	XW847	WS Gazelle AH1	AAC MPSU, Middle Wallop
	XW848	WS Gazelle AH1 [D]	Privately owned, Stapleford Tawney
	XW849	WS Gazelle AH1 [G]	Privately owned, Hurstbourne Tarrant, Hants
	XW851	WS Gazelle AH1	Fitted with the boom of ZA726, 2014
	XW852	WS Gazelle HCC4 (9331M)	DSAE, No 1 SoTT, Cosford
	XW855	WS Gazelle HCC4	RAF Museum, Hendon
	XW858	WS Gazelle HT3 (G-ONNE) [C]	Privately owned, Steeple Bumstead, Cambs
	XW860	WS Gazelle HT2 (TAD 021)	DSEME SEAE, Arborfield
	XW862	WS Gazelle HT3 (G-CBKC) [D]	Privately owned, Fowlmere
	XW863	WS Gazelle HT2 (TAD 022)	Farnborough Air Sciences Trust, Farnborough
	XW864	WS Gazelle HT2 [54/CU]	FAA Museum, stored Cobham Hall, RNAS Yeovilton
	XW865	WS Gazelle AH1 [5C]	AAC No 29 Flt, BATUS, Suffield, Canada
	XW870	WS Gazelle HT3 (9299M) [F]	MoD DFTDC, Manston
	XW888	WS Gazelle AH1 (TAD 017)	DSEME SEAE, Arborfield
	XW889	WS Gazelle AH1 (TAD 018)	DSEME SEAE, Arborfield
	XW890	WS Gazelle HT2	RNAS Yeovilton, on display
	XW892	WS Gazelle AH1 (G-CGJX/9292M) [C]	Privately owned, Hurstbourne Tarrant, Hants
	XW897	WS Gazelle AH1	DSAE, No 1 SoTT, Cosford
	XW899	WS Gazelle AH1 [Z]	DSAE, No 1 SoTT, Cosford
	XW900	WS Gazelle AH1 (TAD 900)	Army, Bramley, Hants
	XW902	WS Gazelle HT3 (G-RBIL) [H]	Repainted as G-RBIL, October 2014
	XW904	WS Gazelle AH1 [H]	AAC MPSU, Middle Wallop
	XW906	WS Gazelle HT3 [J]	QinetiQ Boscombe Down, Apprentice School
	XW908	WS Gazelle AH1 [A]	QinetiQ, Boscombe Down (spares use)
	XW909	WS Gazelle AH1	Privately owned, Stapleford Tawney
	XW912	WS Gazelle AH1 (TAD 019)	DSEME SEAE, Arborfield
	XW913	WS Gazelle AH1	Privately owned, Stapleford Tawney
	XW917	HS Harrier GR3 (8975M)	NATS Air Traffic Control Centre, Swanwick
	XW922	HS Harrier GR3 (8885M)	MoD DFTDC, Manston
	XW923	HS Harrier GR3 (8724M) <ff>	Harrier Heritage Centre, RAF Wittering
	XW924	HS Harrier GR3 (9073M) [G]	RAF Coningsby, preserved
	XW927	HS Harrier T4 <ff>	Privately owned, South Molton, Devon
	XW934	HS Harrier T4 [Y]	Farnborough Air Sciences Trust, Farnborough
	XW994	Northrop Chukar D1	FAA Museum, stored Cobham Hall, RNAS Yeovilton
	XW999	Northrop Chukar D1	Davidstow Airfield & Cornwall At War Museum
	XX108	SEPECAT Jaguar GR1(mod)	Imperial War Museum, Duxford
	XX109	SEPECAT Jaguar GR1 (8918M) [GH]	City of Norwich Aviation Museum
	XX110	SEPECAT Jaguar GR1 (8955M) [EP]	DSAE, No 1 SoTT, Cosford
	XX110	SEPECAT Jaguar GR1 <R> (BAPC 169)	DSAE, No 1 SoTT, Cosford
	XX112	SEPECAT Jaguar GR3A [EA]	DSAE, No 1 SoTT, Cosford
	XX115	SEPECAT Jaguar GR1 (8821M) (fuselage)	DSAE, No 1 SoTT, Cosford
	XX116	SEPECAT Jaguar GR3A [EO]	MoD DFTDC, Manston

Serial	Type (code/other identity)	Owner/operator, location or fate	Notes
XX117	SEPECAT Jaguar GR3A [ES]	DSAE, No 1 SoTT, Cosford	
XX119	SEPECAT Jaguar GR3A (8898M) [AI]$	DSAE, No 1 SoTT, Cosford	
XX121	SEPECAT Jaguar GR1 [EQ]	Privately owned	
XX139	SEPECAT Jaguar T4 [PT]	Privately owned, Sproughton	
XX140	SEPECAT Jaguar T2 (9008M) <ff>	Privately owned, Chesterfield	
XX141	SEPECAT Jaguar T2A (9297M) [T]	DSAE, No 1 SoTT, Cosford	
XX144	SEPECAT Jaguar T2A [U]	Privately owned, Sproughton	
XX145	SEPECAT Jaguar T2A	Privately owned, Bruntingthorpe	
XX146	SEPECAT Jaguar T4 [GT]	Privately owned, Sproughton	
XX150	SEPECAT Jaguar T4 [FY]	Sold to the USA	
XX153	WS Lynx AH1 (9320M)	Museum of Army Flying, Middle Wallop	
XX154	HS Hawk T1	MoD/ETPS, Boscombe Down	
XX156	HS Hawk T1 [156]	RAF No 4 FTS/208(R) Sqn, Valley	
XX157	HS Hawk T1A $	RN No 736 NAS, Culdrose	
XX158	HS Hawk T1A [158]	RAF No 4 FTS/208(R) Sqn, Valley	
XX159	HS Hawk T1A $	RN No 736 NAS, Culdrose	
XX160	HS Hawk T1 [160]	RAF AM&SU, stored Shawbury	
XX161	HS Hawk T1W [161]	RAF, Valley	
XX162	HS Hawk T1	RAF Centre of Aviation Medicine, Boscombe Down	
XX165	HS Hawk T1 [165]	RAF AM&SU, stored Shawbury	
XX167	HS Hawk T1W [167]	RAF AM&SU, stored Shawbury	
XX168	HS Hawk T1 [168]	RAF AM&SU, stored Shawbury	
XX169	HS Hawk T1 [169]	RAF AM&SU, stored Shawbury	
XX170	HS Hawk T1 [170]	RN No 736 NAS, Culdrose	
XX171	HS Hawk T1 [171]	RAF AM&SU, stored Shawbury	
XX172	HS Hawk T1 [172]	RAF AM&SU, stored Shawbury	
XX173	HS Hawk T1	RAF AM&SU, stored Shawbury	
XX174	HS Hawk T1 [174]	RAF AM&SU, stored Shawbury	
XX175	HS Hawk T1 [175]	RAF HSF, Valley	
XX176	HS Hawk T1W [176]	RAF AM&SU, stored Shawbury	
XX177	HS Hawk T1 $	RAF Red Arrows, Scampton (on repair)	
XX178	HS Hawk T1W [178]	RAF AM&SU, stored Shawbury	
XX181	HS Hawk T1W [181]	RAF AM&SU, stored Shawbury	
XX184	HS Hawk T1 [CQ]	RAF No 100 Sqn, Leeming	
XX185	HS Hawk T1 [185]	RAF AM&SU, stored Shawbury	
XX187	HS Hawk T1A [187]	RN No 736 NAS, Culdrose	
XX188	HS Hawk T1A [188]	RAF No 4 FTS/208(R) Sqn, Valley	
XX189	HS Hawk T1A [CR]	RAF No 100 Sqn, Leeming	
XX190	HS Hawk T1A [CN]	RAF AM&SU, stored Shawbury	
XX191	HS Hawk T1A [191]	MoD/Vector Aerospace, Fleetlands	
XX194	HS Hawk T1A [CP]	RAF AM&SU, stored Shawbury	
XX195	HS Hawk T1W [195]	RAF AM&SU, stored Shawbury	
XX198	HS Hawk T1A [CG]	RAF No 100 Sqn, Leeming	
XX199	HS Hawk T1A [199]	RAF No 4 FTS/208(R) Sqn, Valley	
XX200	HS Hawk T1A [CO]	RAF No 100 Sqn, Leeming	
XX201	HS Hawk T1A	RAF No 4 FTS/208(R) Sqn, Valley	
XX202	HS Hawk T1A [CF]	RAF No 100 Sqn, Leeming	
XX203	HS Hawk T1A [CC]	RAF No 100 Sqn, Leeming	
XX204	HS Hawk T1A [204]	RAF No 4 FTS/208(R) Sqn, Valley	
XX205	HS Hawk T1A [846/CU]	RN No 736 NAS, Culdrose	
XX217	HS Hawk T1A [217]	RN No 736 NAS, Culdrose	
XX218	HS Hawk T1A [218]	AM&SU, stored Shawbury	
XX219	HS Hawk T1A $	RAF Red Arrows, Scampton	
XX220	HS Hawk T1A [220]	RAF AM&SU, stored Shawbury	
XX221	HS Hawk T1A [221]	RN No 736 NAS, Culdrose	
XX222	HS Hawk T1A [CI]	RAF AM&SU, stored Shawbury	
XX223	HS Hawk T1 <ff>	Privately owned, Charlwood, Surrey	
XX224	HS Hawk T1W [224]	RAF AM&SU, stored Shawbury	
XX225	HS Hawk T1 [322]	RAF AM&SU, stored Shawbury	
XX226	HS Hawk T1 [226]	RAF AM&SU, stored Shawbury	
XX227	HS Hawk T1 <R> (XX226/BAPC 152)	RAF M&RU, Bottesford	
XX227	HS Hawk T1A $	RAF Red Arrows, Scampton	

Notes	Serial	Type (code/other identity)	Owner/operator, location or fate
	XX228	HS Hawk T1A [CG]	RAF AM&SU, stored Shawbury
	XX230	HS Hawk T1A $	RAF No 4 FTS/208(R) Sqn, Valley
	XX231	HS Hawk T1W [213]	RAF AM&SU, stored Shawbury
	XX232	HS Hawk T1 [232]	RAF *Red Arrows*, Scampton
	XX234	HS Hawk T1 [234]	RAF AM&SU, stored Shawbury
	XX235	HS Hawk T1W [235]	RAF AM&SU, stored Shawbury
	XX236	HS Hawk T1W [236]	RAF No 4 FTS/208(R) Sqn, Valley
	XX237	HS Hawk T1	RAF AM&SU, stored Shawbury
	XX238	HS Hawk T1 [238]	RAF AM&SU, stored Shawbury
	XX239	HS Hawk T1W [239]	RAF No 4 FTS/208(R) Sqn, Valley
	XX240	HS Hawk T1 [840/CU]	RN No 736 NAS, Culdrose
	XX242	HS Hawk T1 $	RAF *Red Arrows*, Scampton
	XX244	HS Hawk T1 $	RAF *Red Arrows*, Scampton
	XX245	HS Hawk T1 $	RAF *Red Arrows*, Scampton
	XX246	HS Hawk T1A [95-Y] $	RAF No 100 Sqn, Leeming
	XX247	HS Hawk T1A [247]	RAF AM&SU, stored Shawbury
	XX248	HS Hawk T1A [CJ]	RAF AM&SU, stored Shawbury
	XX250	HS Hawk T1 [250]	RAF No 4 FTS/208(R) Sqn, Valley
	XX253	HS Hawk T1A	RAF Scampton, on display
	XX254	HS Hawk T1A <R>	Privately owned, Marlow, Bucks
	XX255	HS Hawk T1A [CL]	RAF No 100 Sqn, Leeming
	XX256	HS Hawk T1A [256]	RAF No 4 FTS/208(R) Sqn, Valley
	XX257	HS Hawk T1A (fuselage)	Privately owned, Charlwood, Surrey
	XX258	HS Hawk T1A [CE]	RAF No 100 Sqn, Leeming
	XX260	HS Hawk T1A	RAF AM&SU, stored Shawbury
	XX261	HS Hawk T1A $	RN No 736 NAS, Culdrose
	XX263	HS Hawk T1A	RAF AM&SU, stored Shawbury
	XX264	HS Hawk T1A $	RAF AM&SU, stored Shawbury
	XX265	HS Hawk T1A [265]	RAF AM&SU, stored Shawbury
	XX266	HS Hawk T1A	RAF AM&SU, stored Shawbury
	XX278	HS Hawk T1A $	RAF *Red Arrows*, Scampton
	XX280	HS Hawk T1A [CM]	RAF No 100 Sqn, Leeming
	XX281	HS Hawk T1A	RN No 736 NAS, Culdrose
	XX283	HS Hawk T1W [283]	RAF AM&SU, stored Shawbury
	XX284	HS Hawk T1A [CA]	RAF AM&SU, stored Shawbury
	XX285	HS Hawk T1A $	RAF No 100 Sqn, Leeming
	XX286	HS Hawk T1A [286]	RAF AM&SU, stored Shawbury
	XX287	HS Hawk T1A [287]	RAF No 4 FTS/208(R) Sqn, Valley
	XX289	HS Hawk T1A [CO]	RAF AM&SU, stored Shawbury
	XX290	HS Hawk T1W	RAF AM&SU, stored Shawbury
	XX292	HS Hawk T1	RAF AM&SU, stored Shawbury
	XX294	HS Hawk T1	RAF AM&SU, stored Shawbury
	XX295	HS Hawk T1W [295]	RAF AM&SU, stored Shawbury
	XX296	HS Hawk T1 [296]	RAF AM&SU, stored Shawbury
	XX299	HS Hawk T1W [299]	RAF AM&SU, stored Shawbury
	XX301	HS Hawk T1A $	RN No 736 NAS, Culdrose
	XX303	HS Hawk T1A	RAF No 4 FTS/208(R) Sqn, Valley
	XX304	HS Hawk T1A <rf>	Cardiff International Airport Fire Section
	XX306	HS Hawk T1A	RAF AM&SU, stored Shawbury
	XX307	HS Hawk T1 [307]	RAF AM&SU, stored Shawbury
	XX308	HS Hawk T1	RAF AM&SU, stored Shawbury
	XX308	HS Hawk T1 <R> (*XX263*/BAPC 171)	RAF M&RU, Bottesford
	XX309	HS Hawk T1	RAF AM&SU, stored Shawbury
	XX310	HS Hawk T1W $	RAF *Red Arrows*, Scampton
	XX311	HS Hawk T1 $	RAF *Red Arrows*, Scampton
	XX312	HS Hawk T1W	RAF AM&SU, stored Shawbury
	XX313	HS Hawk T1W [313]	RAF AM&SU, stored Shawbury
	XX314	HS Hawk T1W [314]	RAF AM&SU, stored Shawbury
	XX315	HS Hawk T1A [315]	RAF No 4 FTS/208(R) Sqn, Valley
	XX316	HS Hawk T1A [849/CU]	RN No 736 NAS, Culdrose
	XX317	HS Hawk T1A [317]	RAF No 4 FTS/208(R) Sqn, Valley
	XX318	HS Hawk T1A [95-Y] $	RAF No 100 Sqn, Leeming

Serial	Type (code/other identity)	Owner/operator, location or fate	Notes
XX319	HS Hawk T1A $	RAF *Red Arrows*, Scampton	
XX320	HS Hawk T1A <ff>	RAF Scampton Heritage Centre	
XX321	HS Hawk T1A [CI]	RAF No 100 Sqn, Leeming	
XX322	HS Hawk T1A $	RAF *Red Arrows*, Scampton	
XX323	HS Hawk T1A $	RAF *Red Arrows*, Scampton	
XX324	HS Hawk T1A	RAF No 4 FTS/208(R) Sqn, Valley	
XX325	HS Hawk T1 $	RAF *Red Arrows*, Scampton	
XX326	HS Hawk T1A <ff>	*Scrapped*	
XX326	HS Hawk T1A	MoD/BAE Systems, Brough (on rebuild)	
XX327	HS Hawk T1	RAF Centre of Aviation Medicine, Boscombe Down	
XX329	HS Hawk T1A [CJ]	RAF No 100 Sqn, Leeming	
XX330	HS Hawk T1A [330]	RAF Valley	
XX331	HS Hawk T1A [331]	RAF AM&SU, stored Shawbury	
XX332	HS Hawk T1A	RAF No 100 Sqn, Leeming	
XX335	HS Hawk T1A [335]	RAF AM&SU, stored Shawbury	
XX337	HS Hawk T1A	RN No 736 NAS, Culdrose	
XX338	HS Hawk T1	RAF AM&SU, stored Shawbury	
XX339	HS Hawk T1A [CK]	RAF No 100 Sqn, Leeming	
XX341	HS Hawk T1 ASTRA	MoD/ETPS, Boscombe Down	
XX342	HS Hawk T1 [2]	MoD/ETPS, Boscombe Down	
XX343	HS Hawk T1 [3] (fuselage)	Boscombe Down Aviation Collection, Old Sarum	
XX345	HS Hawk T1A [CE]	RAF AM&SU, stored Shawbury	
XX346	HS Hawk T1A [CH]	RAF No 100 Sqn, Leeming	
XX348	HS Hawk T1A [348]	RAF No 4 FTS/208(R) Sqn, Valley	
XX349	HS Hawk T1W <ff>	RAF Scampton, instructional use	
XX350	HS Hawk T1A [350]	RAF Valley	
XX351	HS Hawk T1A	RAF AM&SU, stored Shawbury	
XX371	WS Gazelle AH1 (G-CHLU)	*Repainted as G-CHLU*	
XX372	WS Gazelle AH1	QinetiQ, Boscombe Down, spares use	
XX375	WS Gazelle AH1	Privately owned, Shepherds Bush	
XX378	WS Gazelle AH1 [Q]	AAC MPSU, Middle Wallop	
XX379	WS Gazelle AH1 [Y]	*Currently not known*	
XX380	WS Gazelle AH1 [A]	AAC Wattisham, on display	
XX381	WS Gazelle AH1	Privately owned, Welbeck	
XX383	WS Gazelle AH1 [D]	Privately owned, Stapleford Tawney	
XX384	WS Gazelle AH1	AAC, stored Dishforth	
XX386	WS Gazelle AH1	Privately owned, Stapleford Tawney	
XX387	WS Gazelle AH1 (TAD 014)	Privately owned, stored Cranfield	
XX392	WS Gazelle AH1	Army, Middle Wallop, preserved	
XX394	WS Gazelle AH1 [X]	Privately owned, Stapleford Tawney	
XX396	WS Gazelle HT3 (8718M) [N]	Privately owned,	
XX398	WS Gazelle AH1	Privately owned, Stapleford Tawney	
XX399	WS Gazelle AH1 [B]	QinetiQ, Boscombe Down, spares use	
XX403	WS Gazelle AH1 [U]	AAC MPSU, Middle Wallop	
XX405	WS Gazelle AH1	AAC No 667 Sqn/7 Regt, Middle Wallop	
XX406	WS Gazelle HT3 (G-CBSH) [P]	Privately owned, Hurstbourne Tarrant, Hants	
XX409	WS Gazelle AH1 (G-CHYV)	Privately owned, Stapleford Tawney	
XX411	WS Gazelle AH1 [X]	South Yorkshire Aircraft Museum, Doncaster	
XX411	WS Gazelle AH1 <rf>	FAA Museum, RNAS Yeovilton	
XX412	WS Gazelle AH1 [B]	DSAE, No 1 SoTT, Cosford	
XX414	WS Gazelle AH1 [V]	Privately owned, Badgers Mount, Kent	
XX416	WS Gazelle AH1	Privately owned, Stapleford Tawney	
XX418	WS Gazelle AH1	Privately owned, Hurstbourne Tarrant, Hants	
XX419	WS Gazelle AH1	AAC MPSU, Middle Wallop	
XX431	WS Gazelle HT2 (9300M) [43/CU]	RAF Shawbury, for display	
XX433	WS Gazelle AH1 <ff>	Privately owned, Hurstbourne Tarrant, Hants	
XX435	WS Gazelle AH1 (fuselage)	QinetiQ, Boscombe Down (spares use)	
XX436	WS Gazelle AH1 (G-ZZLE)	Privately owned, Hurstbourne Tarrant, Hants	
XX437	WS Gazelle AH1	Privately owned, Stapleford Tawney	
XX438	WS Gazelle AH1 [F]	Privately owned, Stapleford Tawney	
XX439	WS Gazelle AH1 (G-CHLW)	*Repainted as G-CHLW*	
XX440	WS Gazelle AH1 (G-BCHN/G-CHBJ)	*Sold to Russia, 13 February 2014*	

Notes	Serial	Type (code/other identity)	Owner/operator, location or fate
	XX442	WS Gazelle AH1 [E]	AAC AM&SU, stored Shawbury
	XX443	WS Gazelle AH1 [Y]	AAC MPSU, Middle Wallop
	XX444	WS Gazelle AH1	Wattisham Airfield Museum
	XX445	WS Gazelle AH1 [T]	Privately owned, Stapleford Tawney
	XX447	WS Gazelle AH1 [D1]	AAC MPSU, Middle Wallop
	XX449	WS Gazelle AH1	MoD/QinetiQ, Boscombe Down
	XX453	WS Gazelle AH1	MoD/QinetiQ, Boscombe Down
	XX454	WS Gazelle AH1 (TAD 023) (fuselage)	DSEME SEAE, Arborfield
	XX455	WS Gazelle AH1	Privately owned, Stapleford Tawney
	XX456	WS Gazelle AH1	Privately owned, Stapleford Tawney
	XX457	WS Gazelle AH1 (TAD 001)	East Midlands Airport Aeropark
	XX460	WS Gazelle AH1	AAC AM&SU, stored Shawbury
	XX462	WS Gazelle AH1 [W]	Privately owned, Stapleford Tawney
	XX466	HS Hunter T66B/T7 [830] <ff>	Guernsey Airport Fire Section
	XX467	HS Hunter T66B/T7 (XL605/G-TVII) [86]	Privately owned, Bruntingthorpe
	XX477	HP137 Jetstream T1 (G-AXXS/8462M) <ff>	South Yorkshire Aircraft Museum, Doncaster
	XX478	HP137 Jetstream T2 (G-AXXT) [564/CU]	Privately owned, Sproughton
	XX479	HP137 Jetstream T2 (G-AXUR)	RN, Predannack Fire School
	XX481	HP137 Jetstream T2 (G-AXUP) [560/CU]	Privately owned, Sproughton
	XX482	SA Jetstream T1 [J]	Privately owned, Hixon, Staffs
	XX483	SA Jetstream T2 [562] <ff>	Dumfries & Galloway Avn Mus, Dumfries
	XX486	SA Jetstream T2 [567/CU]	Privately owned, Bentwaters
	XX487	SA Jetstream T2 [568/CU]	Barry Technical College, instructional use
	XX491	SA Jetstream T1 [K]	Northbrook College, Shoreham, instructional use
	XX492	SA Jetstream T1 [A]	Newark Air Museum, Winthorpe
	XX494	SA Jetstream T1 [B]	Privately owned, Bruntingthorpe
	XX495	SA Jetstream T1 [C]	Bedford College, instructional use
	XX496	SA Jetstream T1 [D]	RAF Museum, Cosford
	XX499	SA Jetstream T1 [G]	Brooklands Museum, Weybridge
	XX500	SA Jetstream T1 [H]	Privately owned, Pinewood Studios
	XX510	WS Lynx HAS2 [69/DD]	SFDO, RNAS Culdrose
	XX513	SA Bulldog T1 (G-KKKK) [10]	Privately owned, Rufforth
	XX515	SA Bulldog T1 (G-CBBC) [4]	Privately owned, Blackbushe
	XX518	SA Bulldog T1 (G-UDOG) [S]	Privately owned, Ursel, Belgium
	XX520	SA Bulldog T1 (9288M) [A]	No 172 Sqn ATC, Haywards Heath
	XX521	SA Bulldog T1 (G-CBEH) [H]	Privately owned, Charney Bassett, Oxon
	XX522	SA Bulldog T1 (G-DAWG) [06]	Privately owned, Blackpool
	XX524	SA Bulldog T1 (G-DDOG) [04]	Privately owned, Malaga, Spain
	XX528	SA Bulldog T1 (G-BZON) [D]	Privately owned, Earls Colne
	XX530	SA Bulldog T1 (XX637/9197M) [F]	No 2175 Sqn ATC, RAF Kinloss
	XX534	SA Bulldog T1 (G-EDAV) [B]	Privately owned, Tollerton
	XX537	SA Bulldog T1 (G-CBCB) [C]	Privately owned, RAF Halton
	XX538	SA Bulldog T1 (G-TDOG) [O]	Privately owned, Shobdon
	XX539	SA Bulldog T1 [L]	Privately owned, Derbyshire
	XX543	SA Bulldog T1 (G-CBAB) [U]	Privately owned, Duxford
	XX546	SA Bulldog T1 (G-WINI) [03]	Privately owned, Conington
	XX549	SA Bulldog T1 (G-CBID) [6]	Privately owned, White Waltham
	XX550	SA Bulldog T1 (G-CBBL) [Z]	Privately owned, Abbeyshrule, Eire
	XX551	SA Bulldog T1 (G-BZDP) [E]	Privately owned, Boscombe Down
	XX554	SA Bulldog T1 (G-BZMD) [09]	Privately owned, Wellesbourne Mountford
	XX557	SA Bulldog T1	Privately owned, stored Fort Paull, Yorks
	XX561	SA Bulldog T1 (G-BZEP) [7]	Privately owned, Eggesford
	XX611	SA Bulldog T1 (G-CBDK) [7]	Privately owned, Coventry
	XX612	SA Bulldog T1 (G-BZXC) [A,03]	Ayr College, instructional use
	XX614	SA Bulldog T1 (G-GGRR) [I]	Privately owned, Turweston
	XX619	SA Bulldog T1 (G-CBBW) [T]	Privately owned, Coventry
	XX621	SA Bulldog T1 (G-CBEF) [H]	Privately owned, Leicester
	XX622	SA Bulldog T1 (G-CBGX) [B]	Privately owned, Shoreham
	XX623	SA Bulldog T1 [M]	Privately owned, Hurstbourne Tarrant, Hants
	XX624	SA Bulldog T1 (G-KDOG) [E]	Privately owned, Ursel, Belgium
	XX626	SA Bulldog T1 (G-CDVV/9290M) [W,02]	Privately owned, Abbots Bromley, Staffs
	XX628	SA Bulldog T1 (G-CBFU) [9]	Privately owned, Faversham

Serial	Type (code/other identity)	Owner/operator, location or fate	Notes
XX629	SA Bulldog T1 (G-BZXZ)	Privately owned, Wellesbourne Mountford	
XX630	SA Bulldog T1 (G-SIJW) [5]	Privately owned, Audley End	
XX631	SA Bulldog T1 (G-BZXS) [W]	Privately owned, Sligo, Eire	
XX633	SA Bulldog T1 [X]	Privately owned, Diseworth, Leics	
XX634	SA Bulldog T1 [T]	Newark Air Museum, Winthorpe	
XX636	SA Bulldog T1 (G-CBFP) [Y]	Privately owned, Empingham, Rutland	
XX638	SA Bulldog T1 (G-DOGG)	Privately owned, Thruxton	
XX639	SA Bulldog T1 (F-AZTF) [D]	Privately owned, La Baule, France	
XX654	SA Bulldog T1 [3]	RAF Museum, Cosford	
XX655	SA Bulldog T1 (9294M) [V] <ff>	South Yorkshire Aircraft Museum, Doncaster	
XX656	SA Bulldog T1 [C]	Privately owned, Derbyshire	
XX658	SA Bulldog T1 (G-BZPS) [07]	Privately owned, Wellesbourne Mountford	
XX659	SA Bulldog T1 [E]	Privately owned, Derbyshire	
XX664	SA Bulldog T1 (F-AZTV) [04]	Privately owned, Pontoise, France	
XX665	SA Bulldog T1 (9289M)	No 2409 Sqn ATC, Halton	
XX667	SA Bulldog T1 (G-BZFN) [16]	Privately owned, Ronaldsway, IoM	
XX668	SA Bulldog T1 (G-CBAN) [1]	Privately owned, St Athan	
XX669	SA Bulldog T1 (8997M) [B]	South Yorkshire Aircraft Museum, Doncaster	
XX671	SA Bulldog T1 [D]	Privately owned, Diseworth, Leics	
XX687	SA Bulldog T1 [F]	Barry Technical College, Cardiff Airport	
XX690	SA Bulldog T1 [A]	Privately owned, Strathaven, S Lanarks	
XX692	SA Bulldog T1 (G-BZMH) [A]	Privately owned, Wellesbourne Mountford	
XX693	SA Bulldog T1 (G-BZML) [07]	Privately owned, Elmsett	
XX694	SA Bulldog T1 (G-CBBS) [E]	Privately owned, Newcastle	
XX695	SA Bulldog T1 (G-CBBT)	Privately owned, Perth	
XX698	SA Bulldog T1 (G-BZME) [9]	Privately owned, Sleap	
XX700	SA Bulldog T1 (G-CBEK) [17]	Privately owned, Blackbushe	
XX702	SA Bulldog T1 (G-CBCR) [π]	Privately owned, Egginton	
XX704	SA122 Bulldog (G-BCUV/G-112)	Privately owned, Bournemouth	
XX705	SA Bulldog T1 [5]	QinetiQ Boscombe Down, Apprentice School	
XX707	SA Bulldog T1 (G-CBDS) [4]	Privately owned, Caernarfon	
XX711	SA Bulldog T1 (G-CBBU) [X]	Privately owned, Perth	
XX720	SEPECAT Jaguar GR3A [FL]	Privately owned, Sproughton	
XX722	SEPECAT Jaguar GR1 (9252M) <ff>	*Currently not known*	
XX723	SEPECAT Jaguar GR3A [EU]	DSAE, No 1 SoTT, Cosford	
XX724	SEPECAT Jaguar GR3A [EC]	DSAE, No 1 SoTT, Cosford	
XX725	SEPECAT Jaguar GR3A [T]	DSAE, No 1 SoTT, Cosford	
XX726	SEPECAT Jaguar GR1 (8947M) [EB]	DSAE, No 1 SoTT, Cosford	
XX727	SEPECAT Jaguar GR1 (8951M) [ER]	DSAE, No 1 SoTT, Cosford	
XX729	SEPECAT Jaguar GR3A [EL]	DSAE, No 1 SoTT, Cosford	
XX734	SEPECAT Jaguar GR1 (8816M)	Boscombe Down Aviation Collection, Old Sarum	
XX736	SEPECAT Jaguar GR1 (9110M) <ff>	South Yorkshire Aircraft Museum, Doncaster	
XX738	SEPECAT Jaguar GR3A [ED]	DSAE, No 1 SoTT, Cosford	
XX739	SEPECAT Jaguar GR1 (8902M) [I]	Privately owned, Bentwaters	
XX741	SEPECAT Jaguar GR1A [EJ]	Bentwaters Cold War Museum	
XX743	SEPECAT Jaguar GR1 (8949M) [EG]	DSAE, No 1 SoTT, Cosford	
XX744	SEPECAT Jaguar GR1 (9251M)	Mayhem Paintball, Abridge, Essex	
XX745	SEPECAT Jaguar GR1A <ff>	No 1350 Sqn ATC, Fareham	
XX746	SEPECAT Jaguar GR1 (8895M) [S]	DSAE, No 1 SoTT, Cosford	
XX747	SEPECAT Jaguar GR1 (8903M)	Privately owned,	
XX748	SEPECAT Jaguar GR3A [EG]	DSAE, No 1 SoTT, Cosford	
XX751	SEPECAT Jaguar GR1 (8937M) [10]	Privately owned, Bentwaters	
XX752	SEPECAT Jaguar GR3A [EK]	DSAE, No 1 SoTT, Cosford	
XX753	SEPECAT Jaguar GR1 (9087M) <ff>	Newark Air Museum, Winthorpe	
XX756	SEPECAT Jaguar GR1 (8899M) [W]	DSAE, No 1 SoTT, Cosford	
XX761	SEPECAT Jaguar GR1 (8600M) <ff>	Boscombe Down Aviation Collection, Old Sarum	
XX763	SEPECAT Jaguar GR1 (9009M)	Bournemouth Aviation Museum	
XX764	SEPECAT Jaguar GR1 (9010M)	Privately owned, Woodmancote, W Sussex	
XX765	SEPECAT Jaguar ACT	RAF Museum, Cosford	
XX766	SEPECAT Jaguar GR3A [EF]	DSAE, No 1 SoTT, Cosford	
XX767	SEPECAT Jaguar GR3A [FK]	DSAE, No 1 SoTT, Cosford	
XX818	SEPECAT Jaguar GR1 (8945M) [DE]	DSAE, No 1 SoTT, Cosford	

Notes	Serial	Type (code/other identity)	Owner/operator, location or fate
	XX819	SEPECAT Jaguar GR1 (8923M) [CE]	DSAE, No 1 SoTT, Cosford
	XX821	SEPECAT Jaguar GR1 (8896M) [P]	DSAE, No 1 SoTT, Cosford
	XX824	SEPECAT Jaguar GR1 (9019M) [AD]	DSAE, No 1 SoTT, Cosford
	XX825	SEPECAT Jaguar GR1 (9020M) [BN]	DSAE, No 1 SoTT, Cosford
	XX829	SEPECAT Jaguar T2A [GZ]	Newark Air Museum, Winthorpe
	XX830	SEPECAT Jaguar T2 <ff>	City of Norwich Aviation Museum
	XX833	SEPECAT Jaguar T2B	DSAE, No 1 SoTT, Cosford
	XX835	SEPECAT Jaguar T4 [EX]	DSAE, No 1 SoTT, Cosford
	XX836	SEPECAT Jaguar T2A [X]	Privately owned, Sproughton
	XX837	SEPECAT Jaguar T2 (8978M) [Z]	DSAE, No 1 SoTT, Cosford
	XX838	SEPECAT Jaguar T4 [FZ]	Privately owned, Bentwaters
	XX840	SEPECAT Jaguar T4 [EY]	DSAE, No 1 SoTT, Cosford
	XX841	SEPECAT Jaguar T4	Privately owned, Tunbridge Wells
	XX842	SEPECAT Jaguar T2A [FX]	Privately owned, Bentwaters
	XX845	SEPECAT Jaguar T4 [EV]	RN, Predannack Fire School
	XX847	SEPECAT Jaguar T4 [EZ]	DSAE, No 1 SoTT, Cosford
	XX885	HS Buccaneer S2B (9225M/G-HHAA)	Hawker Hunter Aviation, Scampton
	XX888	HS Buccaneer S2B <ff>	Privately owned, Barnstaple
	XX889	HS Buccaneer S2B [T]	Privately owned, Bruntingthorpe
	XX892	HS Buccaneer S2B <ff>	Blue Sky Experiences, Methven, Perth & Kinross
	XX894	HS Buccaneer S2B [020/R]	The Buccaneer Aviation Group, Bruntingthorpe
	XX899	HS Buccaneer S2B <ff>	Midland Air Museum, Coventry
	XX900	HS Buccaneer S2B [900]	Cold War Jets Collection, Bruntingthorpe
	XX901	HS Buccaneer S2B	Yorkshire Air Museum, Elvington
	XX910	WS Lynx HAS2	The Helicopter Museum, Weston-super-Mare
	XX914	BAC VC10/1103 (8777M) <rf>	RAF Defence Movements School, Brize Norton
	XX919	BAC 1-11/402AP (PI-C1121) <ff>	Boscombe Down Aviation Collection, Old Sarum
	XX946	Panavia Tornado (P02) (8883M)	Michael Beetham Conservation Centre, RAFM Cosford
	XX947	Panavia Tornado (P03) (8797M)	Privately owned, Bentwaters
	XX958	SEPECAT Jaguar GR1 (9022M) [BK]	DSAE, No 1 SoTT, Cosford
	XX959	SEPECAT Jaguar GR1 (8953M) [CJ]	DSAE, No 1 SoTT, Cosford
	XX965	SEPECAT Jaguar GR1A (9254M) [C]	DSAE, No 1 SoTT, Cosford
	XX967	SEPECAT Jaguar GR1 (9006M) [AC]	DSAE, No 1 SoTT, Cosford
	XX968	SEPECAT Jaguar GR1 (9007M) [AJ]	DSAE, No 1 SoTT, Cosford
	XX969	SEPECAT Jaguar GR1 (8897M) [01]	DSAE, No 1 SoTT, Cosford
	XX970	SEPECAT Jaguar GR3A [EH]	DSAE, No 1 SoTT, Cosford
	XX975	SEPECAT Jaguar GR1 (8905M) [07]	DSAE, No 1 SoTT, Cosford
	XX976	SEPECAT Jaguar GR1 (8906M) [BD]	DSAE, No 1 SoTT, Cosford
	XX977	SEPECAT Jaguar GR1 (9132M) [DL,05] <rf>	Privately owned, Sproughton
	XX979	SEPECAT Jaguar GR1A (9306M) <ff>	Air Defence Radar Museum, Neatishead
	XZ103	SEPECAT Jaguar GR3A [EF]	DSAE, No 1 SoTT, Cosford
	XZ104	SEPECAT Jaguar GR3A [FM]	DSAE, No 1 SoTT, Cosford
	XZ106	SEPECAT Jaguar GR3A [FR]	RAF Manston History Museum
	XZ107	SEPECAT Jaguar GR3A [FH]	Privately owned, Bentwaters
	XZ109	SEPECAT Jaguar GR3A [EN]	DSAE, No 1 SoTT, Cosford
	XZ112	SEPECAT Jaguar GR3A [GW]	DSAE, No 1 SoTT, Cosford
	XZ113	SEPECAT Jaguar GR3 [FD]	Privately owned, Bentwaters
	XZ114	SEPECAT Jaguar GR3 [EO]	DSAE, No 1 SoTT, Cosford
	XZ115	SEPECAT Jaguar GR3 [ER]	DSAE, No 1 SoTT, Cosford
	XZ117	SEPECAT Jaguar GR3 [ES]	DSAE, No 1 SoTT, Cosford
	XZ119	SEPECAT Jaguar GR1A (9266M) [FG]	Royal Scottish Mus'm of Flight, E Fortune
	XZ130	HS Harrier GR3 (9079M) [A]	Privately owned, Thorpe Wood, Yorkshire
	XZ131	HS Harrier GR3 (9174M) <ff>	Privately owned, Spark Bridge, Cumbria
	XZ132	HS Harrier GR3 (9168M) [C]	Privately owned, Thorpe Wood, N Yorks
	XZ133	HS Harrier GR3 [10]	Imperial War Museum, Duxford
	XZ138	HS Harrier GR3 (9040M) <ff>	RAFC Cranwell, Trenchard Hall
	XZ146	HS Harrier T4 (9281M) [S]	Harrier Heritage Centre, RAF Wittering
	XZ166	WS Lynx HAS2 <ff>	Farnborough Air Sciences Trust, Farnborough
	XZ170	WS Lynx AH9	DSEME SEAE, Arborfield
	XZ171	WS Lynx AH7	Army, Salisbury Plain
	XZ172	WS Lynx AH7	DSEME SEAE, Arborfield

Serial	Type (code/other identity)	Owner/operator, location or fate	Notes
XZ173	WS Lynx AH7 <ff>	Privately owned, Weeton, Lancs	
XZ174	WS Lynx AH7 <ff>	Defence Coll of Policing, Gosport, instructional use	
XZ175	WS Lynx AH7	Privately owned, Fairoaks	
XZ176	WS Lynx AH7	*Sold to the Netherlands, February 2015*	
XZ177	WS Lynx AH7	Privately owned, Sproughton	
XZ178	WS Lynx AH7 <ff>	Privately owned, Hixon, Staffs	
XZ179	WS Lynx AH7 <ff>	Privately owned, Sproughton	
XZ180	WS Lynx AH7 [C]	AAC No 671 Sqn/7 Regt, Middle Wallop	
XZ181	WS Lynx AH1	*Currently not known*	
XZ182	WS Lynx AH7 [Z]	AAC MPSU, Middle Wallop	
XZ183	WS Lynx AH7 <ff>	MoD St Athan, instructional use	
XZ184	WS Lynx AH7 [B]	AAC No 671 Sqn/7 Regt, Middle Wallop	
XZ185	WS Lynx AH7	*Sold to the USA, 2014*	
XZ187	WS Lynx AH7	DSEME SEAE, Arborfield	
XZ188	WS Lynx AH7	DSEME SEAE, Arborfield	
XZ190	WS Lynx AH7 <ff>	Army, Longmoor Camp, Hants, GI use	
XZ191	WS Lynx AH7 [R]	MoD/Vector Aerospace, Fleetlands, RTP	
XZ192	WS Lynx AH7 [H]	AAC No 671 Sqn/7 Regt, Middle Wallop	
XZ193	WS Lynx AH7 <ff>	Privately owned, Weeton, Lancs	
XZ194	WS Lynx AH7 [V]	Imperial War Museum, Duxford	
XZ195	WS Lynx AH7 <ff>	Privately owned, Hixon, Staffs	
XZ196	WS Lynx AH7 [T]	Privately owned, Sproughton	
XZ198	WS Lynx AH7 <ff>	*Currently not known*	
XZ203	WS Lynx AH7 [F]	Army, Salisbury Plain	
XZ206	WS Lynx AH7 <ff>	Privately owned, Hixon, Staffs	
XZ207	WS Lynx AH7 <rf>	DSEME SEAE, Arborfield	
XZ208	WS Lynx AH7	Privately owned, Sproughton	
XZ209	WS Lynx AH7 <ff>	MoD JARTS, Boscombe Down	
XZ211	WS Lynx AH7	AAC MPSU, Middle Wallop	
XZ212	WS Lynx AH7 [X]	AAC Middle Wallop, GI use	
XZ213	WS Lynx AH1 (TAD 213)	Vector Aerospace Fleetlands Apprentice School	
XZ214	WS Lynx AH7	AAC No 657 Sqn, Odiham	
XZ215	WS Lynx AH7	MoD DFTDC, Manston	
XZ216	WS Lynx AH7	DSEME SEAE, Arborfield	
XZ217	WS Lynx AH7	Privately owned, Hixon, Staffs	
XZ218	WS Lynx AH7	Warfighters R6 Centre, Barby, Northants	
XZ219	WS Lynx AH7 <ff>	Army, Bramley, Hants	
XZ220	WS Lynx AH7	Privately owned, Weeton, Lancs	
XZ221	WS Lynx AH7 [Z]	*Sold to Germany, July 2014*	
XZ222	WS Lynx AH7	AAC No 657 Sqn, Odiham	
XZ228	WS Lynx HAS3GMS [313]	RN Historic Flight, stored Culdrose	
XZ229	WS Lynx HAS3GMS [360/MC]	Privately owned, Hixon, Staffs	
XZ230	WS Lynx HAS3GMS	RN Yeovilton, GI use (Wildcat ground trainer)	
XZ232	WS Lynx HAS3GMS [634]	Privately owned, Bentwaters	
XZ233	WS Lynx HAS3S [635] $	RN, stored Culdrose	
XZ234	WS Lynx HAS3S [630]	RN MPSU, Middle Wallop	
XZ235	WS Lynx HAS3S(ICE) [630]	Privately owned, Hixon, Staffs	
XZ236	WS Lynx HMA8 [LST-1]	RN ETS, Yeovilton, GI use	
XZ237	WS Lynx HAS3S [631]	Privately owned, Hixon, Staffs	
XZ238	WS Lynx HAS3S(ICE) [633]	Trinity Marine, Exeter	
XZ239	WS Lynx HAS3GMS [633]	Privately owned, Hixon, Staffs	
XZ245	WS Lynx HAS3GMS	Privately owned, Sproughton	
XZ246	WS Lynx HAS3S(ICE) [434/EE]	Privately owned, Sproughton	
XZ248	WS Lynx HAS3S [666]	SFDO, RNAS Culdrose	
XZ250	WS Lynx HAS3S [426/PO] $	RN, Yeovilton (for display at Portland)	
XZ252	WS Lynx HAS3S	Privately owned, Hixon, Staffs	
XZ254	WS Lynx HAS3S	Privately owned, Hixon, Staffs	
XZ255	WS Lynx HMA8SRU [451/DA]	RN No 815 NAS, *Daring* Flt, Yeovilton	
XZ257	WS Lynx HAS3S	RN Yeovilton, GI use (Wildcat ground trainer)	
XZ287	BAe Nimrod AEW3 (9140M) (fuselage)	RAF TSW, Stafford	
XZ290	WS Gazelle AH1	AAC No 665 Sqn/5 Regt, Aldergrove	
XZ291	WS Gazelle AH1	Privately owned, Stapleford Tawney	

Notes	Serial	Type (code/other identity)	Owner/operator, location or fate
	XZ292	WS Gazelle AH1	Privately owned, Stapleford Tawney
	XZ294	WS Gazelle AH1 [X]	AAC AM&SU, stored Shawbury
	XZ295	WS Gazelle AH1	MoD/ETPS, Boscombe Down
	XZ296	WS Gazelle AH1 [V]	Privately owned, Stapleford Tawney
	XZ298	WS Gazelle AH1 <ff>	AAC Middle Wallop, GI use
	XZ300	WS Gazelle AH1	Army, Bramley, Hants
	XZ303	WS Gazelle AH1	AAC MPSU, Middle Wallop
	XZ304	WS Gazelle AH1	Privately owned, Stapleford Tawney
	XZ305	WS Gazelle AH1 (TAD 020)	DSMarE AESS HMS Sultan, Gosport
	XZ307	WS Gazelle AH1 (G-CHBN)	Sold to Russia, September 2014
	XZ308	WS Gazelle AH1	QinetiQ, Boscombe Down (spares use)
	XZ311	WS Gazelle AH1 [U]	AAC MPSU, Middle Wallop
	XZ312	WS Gazelle AH1	RAF Henlow, instructional use
	XZ313	WS Gazelle AH1 <ff>	Privately owned, Shotton, Durham
	XZ314	WS Gazelle AH1 [A]	Privately owned, Stapleford Tawney
	XZ315	WS Gazelle AH1 <ff>	Privately owned, Babcary, Somerset
	XZ316	WS Gazelle AH1 [B]	DSEME SEAE, Arborfield
	XZ318	WS Gazelle AH1 (fuselage)	Tong Paintball Park, Shropshire
	XZ320	WS Gazelle AH1	AAC No 665 Sqn/5 Regt, Aldergrove
	XZ321	WS Gazelle AH1 (G-CDNS) [D]	Privately owned, Hurstbourne Tarrant, Hants
	XZ322	WS Gazelle AH1 (9283M) [N]	DSAE, No 1 SoTT, Cosford
	XZ323	WS Gazelle AH1	Currently not known
	XZ324	WS Gazelle AH1	Privately owned, Stapleford Tawney
	XZ325	WS Gazelle AH1 [T]	DSEME SEAE, Arborfield
	XZ326	WS Gazelle AH1	No 665 Sqn/5 Regt, Aldergrove
	XZ327	WS Gazelle AH1	AAC Middle Wallop (recruiting aid)
	XZ328	WS Gazelle AH1 [C]	AAC AM&SU, stored Shawbury
	XZ329	WS Gazelle AH1 (G-BZYD) [J]	Privately owned, East Garston, Bucks
	XZ330	WS Gazelle AH1 [Y]	AAC Wattisham, instructional use
	XZ331	WS Gazelle AH1 [D]	Currently not known
	XZ332	WS Gazelle AH1 [O]	DSEME SEAE, Arborfield
	XZ333	WS Gazelle AH1 [A]	DSEME SEAE, Arborfield
	XZ334	WS Gazelle AH1	AAC No 665 Sqn/5 Regt, Aldergrove
	XZ335	WS Gazelle AH1	North-East Aircraft Museum, stored Usworth
	XZ337	WS Gazelle AH1 [Z]	MoD Abbey Wood, Bristol, on display
	XZ338	WS Gazelle AH1 (G-CHZF) [Y]	Privately owned, Stapleford Tawney
	XZ340	WS Gazelle AH1	AAC No 29 Flt, BATUS, Suffield, Canada
	XZ341	WS Gazelle AH1	AAC AM&SU, stored Shawbury
	XZ342	WS Gazelle AH1	AAC No 658 Sqn, Credenhill
	XZ343	WS Gazelle AH1	AAC AM&SU, stored Shawbury
	XZ344	WS Gazelle AH1 [Y]	Privately owned, Stapleford Tawney
	XZ345	WS Gazelle AH1 [M]	AAC No 671 Sqn/7 Regt, Middle Wallop
	XZ345	Aérospatiale SA341G Gazelle (G-SFTA) [T]	North-East Aircraft Museum, Usworth
	XZ346	WS Gazelle AH1	AAC Middle Wallop, at main gate
	XZ347	WS Gazelle AH1	Privately owned, Hurstbourne Tarrant, Hants
	XZ349	WS Gazelle AH1 [G]	QinetiQ, Boscombe Down, spares use
	XZ356	SEPECAT Jaguar GR3A [FU]	Privately owned, Welshpool
	XZ358	SEPECAT Jaguar GR1A (9262M) [L]	DSAE, No 1 SoTT, Cosford
	XZ360	SEPECAT Jaguar GR3 [FN]	Privately owned, Bentwaters
	XZ363	SEPECAT Jaguar GR1A <R> (XX824/BAPC 151) [A]	RAF M&RU, Bottesford
	XZ364	SEPECAT Jaguar GR3A <ff>	Privately owned, Tunbridge Wells
	XZ366	SEPECAT Jaguar GR3A [FC]	Privately owned, Bentwaters
	XZ367	SEPECAT Jaguar GR3 [GP]	DSAE, No 1 SoTT, Cosford
	XZ368	SEPECAT Jaguar GR1 [8900M] [E]	DSAE, No 1 SoTT, Cosford
	XZ369	SEPECAT Jaguar GR3A [EU]	Privately owned, Bentwaters
	XZ370	SEPECAT Jaguar GR1 (9004M) [JB]	DSAE, No 1 SoTT, Cosford
	XZ371	SEPECAT Jaguar GR1 (8907M) [AP]	DSAE, No 1 SoTT, Cosford
	XZ372	SEPECAT Jaguar GR3 [FV]	Aberdeen Airport, on display
	XZ374	SEPECAT Jaguar GR1 (9005M) [JC]	DSAE, stored Cosford
	XZ375	SEPECAT Jaguar GR1A (9255M) <ff>	City of Norwich Aviation Museum
	XZ377	SEPECAT Jaguar GR3A [EP]	DSAE, No 1 SoTT, Cosford
	XZ378	SEPECAT Jaguar GR1A [EP]	Privately owned, Topsham, Devon

Serial	Type (code/other identity)	Owner/operator, location or fate	Notes
XZ382	SEPECAT Jaguar GR1 (8908M)	Cold War Jets Collection, Bruntingthorpe	
XZ383	SEPECAT Jaguar GR1 (8901M) [AF]	DSAE, No 1 SoTT, Cosford	
XZ384	SEPECAT Jaguar GR1 (8954M) [BC]	DSAE, No 1 SoTT, Cosford	
XZ385	SEPECAT Jaguar GR3A [FT]	Privately owned, Bentwaters	
XZ389	SEPECAT Jaguar GR1 (8946M) [BL]	DSAE, No 1 SoTT, Cosford	
XZ390	SEPECAT Jaguar GR1 (9003M) [DM]	DSAE, No 1 SoTT, Cosford	
XZ391	SEPECAT Jaguar GR3A [ET]	DSAE, No 1 SoTT, Cosford	
XZ392	SEPECAT Jaguar GR3A [EM]	DSAE, No 1 SoTT, Cosford	
XZ394	SEPECAT Jaguar GR3 [FG]	Shoreham Airport, on display	
XZ396	SEPECAT Jaguar GR3A [EQ]	Sold to the USA, 2014	
XZ398	SEPECAT Jaguar GR3A [EQ]	DSAE, No 1 SoTT, Cosford	
XZ399	SEPECAT Jaguar GR3A [EJ]	DSAE, No 1 SoTT, Cosford	
XZ400	SEPECAT Jaguar GR3A [FQ]	Privately owned, Bentwaters	
XZ431	HS Buccaneer S2B (9233M) <ff>	Privately owned, Nottingham	
XZ440	BAe Sea Harrier FA2 [40/DD]	SFDO, RNAS Culdrose	
XZ455	BAe Sea Harrier FA2 [001] (wreck)	Privately owned, Thorpe Wood, N Yorks	
XZ457	BAe Sea Harrier FA2 [104/VL]	Boscombe Down Aviation Collection, Old Sarum	
XZ459	BAe Sea Harrier FA2 [126]	Privately owned, Sussex	
XZ493	BAe Sea Harrier FRS1 (comp XV760) [001/N]	FAA Museum, RNAS Yeovilton	
XZ493	BAe Sea Harrier FRS1 <ff>	RN Yeovilton, Fire Section	
XZ494	BAe Sea Harrier FA2 [128]	Privately owned, Wedmore, Somerset	
XZ497	BAe Sea Harrier FA2 [126]	Privately owned, Charlwood	
XZ499	BAe Sea Harrier FA2 [003]	FAA Museum, stored Cobham Hall, RNAS Yeovilton	
XZ559	Slingsby T61F Venture T2 (G-BUEK)	Privately owned, Tibenham	
XZ570	WS61 Sea King HAS5(mod)	RN, Predannack Fire School	
XZ574	WS61 Sea King HAS6	FAA Museum, RNAS Yeovilton	
XZ575	WS61 Sea King HU5	MoD, Boscombe Down (wfu)	
XZ576	WS61 Sea King HAS6	DSMarE AESS, HMS Sultan, Gosport	
XZ578	WS61 Sea King HU5 [30]	RN No 771 NAS, Prestwick	
XZ579	WS61 Sea King HAS6 [707/PW]	DSMarE AESS, HMS Sultan, Gosport	
XZ580	WS61 Sea King HC6 [ZB]	DSMarE AESS, HMS Sultan, Gosport	
XZ581	WS61 Sea King HAS6 [69/CU]	DSMarE AESS, HMS Sultan, Gosport	
XZ585	WS61 Sea King HAR3 [A]	RAF No 202 Sqn, D Flt, Lossiemouth	
XZ586	WS61 Sea King HAR3 [B]	RAF SKAMG, RNAS Yeovilton	
XZ587	WS61 Sea King HAR3 [C]	RAF No 202 Sqn, E Flt, Leconfield	
XZ588	WS61 Sea King HAR3 [D]	RAF No 202 Sqn, D Flt, Lossiemouth	
XZ589	WS61 Sea King HAR3 [E] $	DSMarE AESS, HMS Sultan, Gosport	
XZ590	WS61 Sea King HAR3 [F]	RAF No 202 Sqn, A Flt, Boulmer	
XZ591	WS61 Sea King HAR3 [G]	DSMarE, stored HMS Sultan, Gosport	
XZ592	WS61 Sea King HAR3 [H]	RAF No 22 Sqn, C Flt, Valley	
XZ593	WS61 Sea King HAR3 [I]	RAF No 1564 Flt, Mount Pleasant, FI	
XZ594	WS61 Sea King HAR3	RAF SKAMG, RNAS Yeovilton	
XZ595	WS61 Sea King HAR3 [K]	RAF No 22 Sqn, C Flt, Valley	
XZ596	WS61 Sea King HAR3 [L]	RAF No 202 Sqn, E Flt, Leconfield	
XZ597	WS61 Sea King HAR3 [M]	RAF No 22 Sqn, C Flt, Valley	
XZ598	WS61 Sea King HAR3 [N]	RAF SKAMG, RNAS, Yeovilton	
XZ599	WS61 Sea King HAR3 [P]	DSMarE, stored HMS Sultan, Gosport	
XZ605	WS Lynx AH7	AAC Wattisham, preserved	
XZ606	WS Lynx AH7 [O]	Privately owned, Sproughton	
XZ607	WS Lynx AH7	DSEME SEAE, Arborfield	
XZ608	WS Lynx AH7	AAC No 667 Sqn/7 Regt, Middle Wallop	
XZ608	WS Lynx AH7 (ZD272)[W]	Privately owned, Weeton, Lancs	
XZ609	WS Lynx AH7	AAC MPSU, Middle Wallop	
XZ611	WS Lynx AH7 <ff>	Privately owned, Hixon, Staffs	
XZ612	WS Lynx AH7	Privately owned, Bentwaters	
XZ613	WS Lynx AH7 [F]	AAC Stockwell Hall, Middle Wallop	
XZ615	WS Lynx AH7 <ff>	Privately owned, Hixon, Staffs	
XZ616	WS Lynx AH7	AAC No 657 Sqn, Odiham	
XZ617	WS Lynx AH7	AAC No 9 Regt, Dishforth	
XZ630	Panavia Tornado GR1 (8976M)	RAF Halton, on display	
XZ631	Panavia Tornado GR1	Yorkshire Air Museum, Elvington	
XZ641	WS Lynx AH7 [A]	Privately owned, Bentwaters	

Notes	Serial	Type (code/other identity)	Owner/operator, location or fate
	XZ642	WS Lynx AH7	AAC No 667 Sqn/7 Regt, Middle Wallop
	XZ643	WS Lynx AH7 [C]	Ramco, Croft, Lincs
	XZ645	WS Lynx AH7	Privately owned, Sproughton
	XZ646	WS Lynx AH7 <ff>	MoD JARTS, Boscombe Down
	XZ646	WS Lynx AH7 (really XZ649)	Bristol University, instructional use
	XZ647	WS Lynx AH7 <ff>	Falck Nutec UK, Aberdeen, GI use
	XZ648	WS Lynx AH7 <ff>	Privately owned, Weeton, Lancs
	XZ651	WS Lynx AH7 [O]	AAC No 671 Sqn/7 Regt, Middle Wallop
	XZ652	WS Lynx AH7	IFTC, Durham/Tees Valley
	XZ653	WS Lynx AH7	AAC No 9 Regt, Dishforth
	XZ654	WS Lynx AH7	AAC MPSU, Middle Wallop
	XZ655	WS Lynx AH7 <ff>	Privately owned, Weeton, Lancs
	XZ661	WS Lynx AH7 [V]	Army, Bramley, Hants
	XZ663	WS Lynx AH7 <ff>	Privately owned, Hixon, Staffs
	XZ664	WS Lynx AH7	Warfighters R6 Centre, Barby, Northants
	XZ665	WS Lynx AH7	Privately owned, Barby, Northants
	XZ666	WS Lynx AH7	DSEME SEAE, Arborfield
	XZ669	WS Lynx AH7 [G]	AAC MPSU, Middle Wallop
	XZ670	WS Lynx AH7 [A]	AAC No 671 Sqn/7 Regt, Middle Wallop
	XZ671	WS Lynx AH7 <ff>	AgustaWestland, Yeovil, instructional use
	XZ672	WS Lynx AH7 <ff>	AAC Middle Wallop Fire Section
	XZ673	WS Lynx AH7	Privately owned, Weeton, Lancs
	XZ674	WS Lynx AH7 [T]	AAC MPSU, Middle Wallop
	XZ675	WS Lynx AH7 [H]	Museum of Army Flying, Middle Wallop
	XZ676	WS Lynx AH7 [N]	Privately owned, Weeton, Lancs
	XZ677	WS Lynx AH7 <ff>	Army, Longmoor Camp, Hants, GI use
	XZ678	WS Lynx AH7	AAC MPSU, Middle Wallop
	XZ679	WS Lynx AH7 [W]	AAC No 9 Regt, Dishforth
	XZ680	WS Lynx AH7 <ff>	Privately owned, Weeton, Lancs
	XZ689	WS Lynx HMA8SRU [372/NL]	MoD/Vector Aerospace, Fleetlands
	XZ690	WS Lynx HMA8SRU [640]	RN No 815 NAS, HQ Flt, Yeovilton
	XZ691	WS Lynx HMA8SRU [454]	MoD/Vector Aerospace, Fleetlands
	XZ692	WS Lynx HMA8SRU [641]$	AAC MPSU, Middle Wallop
	XZ693	WS Lynx HAS3S [311]	Privately owned, Hixon, Staffs
	XZ696	WS Lynx HAS3GMS [633]	Privately owned, Hixon, Staffs
	XZ697	WS Lynx HMA8SRU [313]	RN No 815 NAS, HQ Flt, Yeovilton
	XZ698	WS Lynx HMA8SRU [316]	AAC MPSU, Middle Wallop, RTP
	XZ699	WS Lynx HAS2	FAA Museum, stored Cobham Hall, RNAS Yeovilton
	XZ719	WS Lynx HMA8SRU [645]	AAC MPSU, Middle Wallop, RTP
	XZ720	WS Lynx HAS3GMS [410/GC]	FAA Museum, stored Yeovilton
	XZ721	WS Lynx HAS3GMS [322]	Privately owned, Hixon, Staffs
	XZ722	WS Lynx HMA8SRU [671]	RN No 815 NAS, HQ Flt, Yeovilton
	XZ723	WS Lynx HMA8SRU [672/RM]	AAC MPSU, Middle Wallop, RTP
	XZ725	WS Lynx HMA8SRU [337]	RN No 815 NAS, *Cardigan Bay* Flt, Yeovilton
	XZ726	WS Lynx HMA8SRU [316]	RN No 815 NAS, MI Flt, Yeovilton
	XZ727	WS Lynx HAS3S [634]	Privately owned, Hixon, Staffs
	XZ728	WS Lynx HMA8 [326/AW]	RNAS Yeovilton, on display
	XZ729	WS Lynx HMA8SRU [425/DT]	RN No 815 NAS, *Dauntless* Flt, Yeovilton
	XZ730	WS Lynx HAS3S	Privately owned, Hixon, Staffs
	XZ731	WS Lynx HMA8SRU [306]	RN No 815 NAS, HQ Flt, Yeovilton
	XZ732	WS Lynx HMA8SRU [673]	MoD/Vector Aerospace, Fleetlands
	XZ733	WS Lynx HAS3GMS [305]	DSMarE AESS, *HMS Sultan*, Gosport
	XZ735	WS Lynx HAS3GMS <ff>	Privately owned, Sproughton
	XZ736	WS Lynx HMA8SRU [643]	RN No 815 NAS, MI Flt, Yeovilton
	XZ791	Northrop Shelduck D1	Davidstow Airfield & Cornwall at War Museum
	XZ795	Northrop Shelduck D1	Museum of Army Flying, Middle Wallop
	XZ920	WS61 Sea King HU5 [24]	RN No 771 NAS, Culdrose
	XZ921	WS61 Sea King HAS6 [269/N]	Privately owned, Eastleigh, Hants (for scrapping)
	XZ922	WS61 Sea King HC6 [ZA]	DSMarE AESS, *HMS Sultan*, Gosport
	XZ930	WS Gazelle HT3 (A2713) [Q]	DSMarE AESS, *HMS Sultan*, Gosport
	XZ934	WS Gazelle HT3 (G-CBSI) [U]	Privately owned, Babcary, Somerset
	XZ935	WS Gazelle HCC4 (9332M)	DSEME SEAE, Arborfield

Serial	Type (code/other identity)	Owner/operator, location or fate	Notes
XZ936	WS Gazelle HT3 [6]	MoD, stored Boscombe Down (damaged)	
XZ936	WS Gazelle HT3 (XZ933/G-CGJZ)	Privately owned, Hurstbourne Tarrant, Hants	
XZ939	WS Gazelle HT3 [9]	MoD/ETPS, Boscombe Down	
XZ941	WS Gazelle HT3 (9301M) [B]	DSAE, No 1 SoTT, Cosford	
XZ964	BAe Harrier GR3 [D]	Royal Engineers Museum, Chatham	
XZ966	BAe Harrier GR3 (9221M) [G]	MoD DFTDC, Manston	
XZ968	BAe Harrier GR3 (9222M) [3G]	Muckleborough Collection, Weybourne	
XZ969	BAe Harrier GR3 [69/DD]	RN, Predannack Fire School	
XZ971	BAe Harrier GR3 (9219M)	HQ DSDA, Donnington, Shropshire, on display	
XZ987	BAe Harrier GR3 (9185M) [C]	RAF Stafford, at main gate	
XZ990	BAe Harrier GR3 <ff>	*Currently not known*	
XZ990	BAe Harrier GR3 <rf>	RAF Wittering, derelict	
XZ991	BAe Harrier GR3 (9162M) [3A]	DSAE Cosford, on display	
XZ993	BAe Harrier GR3 (9240M) <ff>	Privately owned, Upwood, Cambs	
XZ994	BAe Harrier GR3 (9170M) [U]	RAF Defence Movements School, Brize Norton	
XZ995	BAe Harrier GR3 (9220M/G-CBGK) [3G]	Privately owned, Dunboyne, Eire	
XZ996	BAe Harrier GR3 [96/DD]	Privately owned, Sproughton	
XZ997	BAe Harrier GR3 (9122M) [V]	RAF Museum, HendonPrivately owned,	
ZA101	BAe Hawk 100 (G-HAWK)	BAE Systems Warton, Overseas Customer Training Centre	
ZA105	WS61 Sea King HAR3 [Q]	RAF No 1564 Flt, Mount Pleasant, FI	
ZA110	BAe Jetstream T2 (F-BTMI) [563/CU]	Aberdeen Airport	
ZA111	BAe Jetstream T2 (9Q-CTC) [565/CU]	SFDO, RNAS Culdrose	
ZA126	WS61 Sea King ASaC7 [191]	RN No 849 NAS, Culdrose	
ZA127	WS61 Sea King HAS6 [509/CU]	DSMarE, stored *HMS Sultan*, Gosport	
ZA128	WS61 Sea King HAS6 [010]	DSAE, No 1 SoTT, Cosford	
ZA130	WS61 Sea King HU5 [19]	RN No 771 NAS, Prestwick	
ZA131	WS61 Sea King HAS6 [271/N]	DSAE, No 1 SoTT, Cosford	
ZA133	WS61 Sea King HAS6 [831/CU]	DSMarE, AESS, *HMS Sultan*, Gosport	
ZA134	WS61 Sea King HU5 [25]	RN No 771 NAS, Prestwick	
ZA135	WS61 Sea King HAS6	Privately owned, Woodperry, Oxon	
ZA136	WS61 Sea King HAS6 [018]	Privately owned, Hixon, Staffs	
ZA137	WS61 Sea King HU5 [20]	MoD/AFD/QinetiQ, Boscombe Down	
ZA144	BAe VC10 K2 (G-ARVC) <ff>	MoD JARTS, Boscombe Down	
ZA147	BAe VC10 K3 (5H-MMT) [F]	Privately owned, Bruntingthorpe	
ZA148	BAe VC10 K3 (5Y-ADA) [G]	Classic Air Force, Newquay	
ZA149	BAe VC10 K3 (5X-UVJ) [H]	Privately owned, Bruntingthorpe	
ZA150	BAe VC10 K3 (5H-MOG) [J]	Brooklands Museum, Dunsfold	
ZA166	WS61 Sea King HU5 [16/CU]	RN No 771 NAS, Prestwick	
ZA167	WS61 Sea King HU5 [22/CU]	DSMarE AESS, *HMS Sultan*, stored Gosport	
ZA168	WS61 Sea King HAS6 [830/CU]	DSMarE, AESS, *HMS Sultan*, Gosport	
ZA169	WS61 Sea King HAS6 [515/CW]	DSAE, No 1 SoTT, Cosford	
ZA170	WS61 Sea King HAS5	DSMarE, AESS, *HMS Sultan*, Gosport	
ZA175	BAe Sea Harrier FA2	Norfolk & Suffolk Avn Museum, Flixton	
ZA176	BAe Sea Harrier FA2 [126/R]	Newark Air Museum, Winthorpe	
ZA195	BAe Sea Harrier FA2	Tangmere Military Aviation Museum	
ZA209	Short MATS-B	Museum of Army Flying, Middle Wallop	
ZA220	Short MATS-B	Privately owned, Awbridge, Hants	
ZA250	BAe Harrier T52 (G-VTOL)	Brooklands Museum, Weybridge	
ZA267	Panavia Tornado F2 (9284M)	RAF Marham, instructional use	
ZA291	WS61 Sea King HC4 [N]	Privately owned, Eastleigh, Hants (for scrapping)	
ZA292	WS61 Sea King HC4 [WU]	DSMarE, stored *HMS Sultan*, Gosport	
ZA293	WS61 Sea King HC4 [A]	DSMarE, stored *HMS Sultan*, Gosport	
ZA295	WS61 Sea King HC4 [U]	RN No 845 NAS, Yeovilton	
ZA296	WS61 Sea King HC4 [Q]	RN No 845 NAS, Yeovilton	
ZA297	WS61 Sea King HC4 [W]	DSMarE, stored *HMS Sultan*, Gosport	
ZA298	WS61 Sea King HC4 (G-BJNM) [Y]	RN No 845 NAS, Yeovilton	
ZA299	WS61 Sea King HC4 [D]	RN No 845 NAS, Yeovilton	
ZA310	WS61 Sea King HC4 [B]	Privately owned, Hitchin, Herts (for scrapping)	
ZA312	WS61 Sea King HC4 [E]	RN CHFMU, Yeovilton	
ZA313	WS61 Sea King HC4 [M]	Privately owned, Eastleigh, Hants (for scrapping)	

Notes	Serial	Type (code/other identity)	Owner/operator, location or fate
	ZA314	WS61 Sea King HC4 [WT]	MoD/AFD/QinetiQ, Boscombe Down
	ZA319	Panavia Tornado GR1 (9315M)	DSDA, Arncott, Oxon, on display
	ZA320	Panavia Tornado GR1 (9314M) [TAW]	Privately owned, Thorpe Wood, N Yorks
	ZA323	Panavia Tornado GR1 [TAZ]	DSAE, No 1 SoTT, Cosford
	ZA325	Panavia Tornado GR1 <ff>	RAF Manston History Museum
	ZA325	Panavia Tornado GR1 [TAX] <rf>	RAF AM&SU, stored Shawbury
	ZA326	Panavia Tornado GR1P	Privately owned, Bruntingthorpe
	ZA327	Panavia Tornado GR1 <ff>	BAE Systems, Warton
	ZA328	Panavia Tornado GR1	Marsh Lane Technical School, Preston
	ZA353	Panavia Tornado GR1 [B-53]	Privately owned, Thorpe Wood, N Yorks
	ZA354	Panavia Tornado GR1	Yorkshire Air Museum, Elvington
	ZA355	Panavia Tornado GR1 (9310M) [TAA]	Privately owned, Thorpe Wood, N Yorks
	ZA356	Panavia Tornado GR1 <ff>	RAF Tornado Maintenance School, Marham
	ZA357	Panavia Tornado GR1 [TTV]	DSAE, No 1 SoTT, Cosford
	ZA359	Panavia Tornado GR1	Privately owned, Thorpe Wood, N Yorks
	ZA360	Panavia Tornado GR1 (9318M) <ff>	RAF Marham, instructional use
	ZA361	Panavia Tornado GR1 [TD]	Privately owned, New York, Lincs
	ZA362	Panavia Tornado GR1 [AJ-F]	Highland Aviation Museum, Inverness
	ZA365	Panavia Tornado GR4 [001]	Scrapped at Leeming
	ZA367	Panavia Tornado GR4 [002,KC-N]	Scrapped at Leeming
	ZA369	Panavia Tornado GR4A [003]	RAF CMU, Marham
	ZA370	Panavia Tornado GR4A [004]	RAF No 31 Sqn, Marham
	ZA371	Panavia Tornado GR4A [005]	Scrapped at Leeming
	ZA372	Panavia Tornado GR4A [006]	RAF No 31 Sqn, Marham
	ZA373	Panavia Tornado GR4A [007]	RAF No 9 Sqn, Marham
	ZA375	Panavia Tornado GR1 (9335M) [AJ-W]	RAF Marham, Fire Section
	ZA393	Panavia Tornado GR4 [008] $	RAF No 31 Sqn, Marham
	ZA395	Panavia Tornado GR4A $	RAF Leeming, RTP
	ZA398	Panavia Tornado GR4A $	RAF DFTDC, Manston
	ZA399	Panavia Tornado GR1 (9316M) [AJ-C]	Privately owned, Knutsford, Cheshire
	ZA400	Panavia Tornado GR4A [011]	RAF Marham Wing
	ZA401	Panavia Tornado GR4A [012] $	RAF Leeming, RTP
	ZA402	Panavia Tornado GR4A	Scrapped at Leeming
	ZA404	Panavia Tornado GR4A [013]	Scrapped at Leeming
	ZA405	Panavia Tornado GR4A [014]	RAF No 31 Sqn, Marham
	ZA406	Panavia Tornado GR4 [015]	RAF Marham Wing
	ZA407	Panavia Tornado GR1 (9336M) [AJ-N]	RAF Marham, on display
	ZA409	Panavia Tornado GR1 [VII]	RAF Lossiemouth (wfu)
	ZA410	Panavia Tornado GR4 [016]	MoD/QinetiQ, Boscombe Down (for scrapping)
	ZA411	Panavia Tornado GR1	BAE Systems stored, Warton
	ZA412	Panavia Tornado GR4 $	RAF No 15(R) Sqn, Lossiemouth
	ZA446	Panavia Tornado GR4 [018]	Scrapped at Leeming
	ZA447	Panavia Tornado GR4 [019]	RAF TEF, Lossiemouth
	ZA449	Panavia Tornado GR4 [020]	RAF TEF, Lossiemouth
	ZA450	Panavia Tornado GR1 (9317M) [TH]	DSAE, No 1 SoTT, Cosford
	ZA452	Panavia Tornado GR4 [021]	Midland Air Museum, Coventry
	ZA453	Panavia Tornado GR4 [022]	RAF CMU, Marham
	ZA456	Panavia Tornado GR4 [023]	RAF No 31 Sqn, Marham
	ZA457	Panavia Tornado GR1 [AJ-J]	RAF Museum, Hendon
	ZA458	Panavia Tornado GR4 [024]	RAF TEF, Lossiemouth
	ZA459	Panavia Tornado GR4 [F] $	RAF CMU, Marham
	ZA461	Panavia Tornado GR4 [026]	RAF No 15(R) Sqn, Lossiemouth
	ZA462	Panavia Tornado GR4 [027]	RAF No 15(R) Sqn, Lossiemouth
	ZA463	Panavia Tornado GR4 [028]	RAF CMU, Marham
	ZA465	Panavia Tornado GR1 [FF]	Imperial War Museum, Duxford
	ZA469	Panavia Tornado GR4 [029] $	RAF CMU, Marham
	ZA470	Panavia Tornado GR4	Scrapped at Leeming January 2013
	ZA472	Panavia Tornado GR4 [031]	RAF CMU, Marham
	ZA473	Panavia Tornado GR4 [032]	RAF No 15(R) Sqn, Lossiemouth
	ZA474	Panavia Tornado GR1 (9312M)	RAF, stored Lossiemouth
	ZA475	Panavia Tornado GR1 (9311M) [AJ-G]	RAF Lossiemouth, on display
	ZA492	Panavia Tornado GR4 $	RAF No 15(R) Sqn, Lossiemouth

Serial	Type (code/other identity)	Owner/operator, location or fate	Notes
ZA541	Panavia Tornado GR4 [034]	RAF No 15(R) Sqn, Lossiemouth	
ZA542	Panavia Tornado GR4 [035]	RAF No 31 Sqn, Marham	
ZA543	Panavia Tornado GR4 [036]	RAF TEF, Lossiemouth	
ZA546	Panavia Tornado GR4 [038]	RAF AWC/FJWOEU/No 41(R) Sqn, Coningsby	
ZA547	Panavia Tornado GR4 [039] $	*Scrapped at Leeming, 2014*	
ZA548	Panavia Tornado GR4 [040]	RAF CMU, stored Marham	
ZA549	Panavia Tornado GR4 [041]	RAF Tornado Maintenance School, Marham	
ZA550	Panavia Tornado GR4 [042]	RAF TEF, Lossiemouth	
ZA551	Panavia Tornado GR4 [043]	RAF No 15(R) Sqn, Lossiemouth	
ZA552	Panavia Tornado GR4 [044]	*Scrapped at Leeming*	
ZA553	Panavia Tornado GR4 [045]	RAF TEF, Lossiemouth	
ZA554	Panavia Tornado GR4 [046]	RAF CMU, stored Marham	
ZA556	Panavia Tornado GR4 [047]	RAF AWC/FJWOEU/No 41(R) Sqn, Coningsby	
ZA556	Panavia Tornado GR1 <R> (ZA368/BAPC 155) [Z]	RAF AWC/FJWOEU/No 41(R) Sqn, Coningsby	
ZA557	Panavia Tornado GR4 [048]	RAF Marham Wing	
ZA559	Panavia Tornado GR4 [049]	RAF Marham Wing	
ZA560	Panavia Tornado GR4 [050]	RAF No 31 Sqn, Marham	
ZA562	Panavia Tornado GR4 [051]	RAF No 15(R) Sqn, Lossiemouth	
ZA563	Panavia Tornado GR4	*Scrapped at Leeming*	
ZA564	Panavia Tornado GR4 $	RAF Leeming, RTP	
ZA585	Panavia Tornado GR4 [054]	RAF CMU, Marham	
ZA587	Panavia Tornado GR4 [055]	RAF CMU, Marham	
ZA588	Panavia Tornado GR4 [056]	RAF Marham Wing	
ZA589	Panavia Tornado GR4 [057]	RAF TEF, Lossiemouth	
ZA591	Panavia Tornado GR4 [058]	RAF CMU, Marham	
ZA592	Panavia Tornado GR4 [059]	RAF CMU, Marham	
ZA594	Panavia Tornado GR4 [060]	RAF CMU, stored Marham	
ZA595	Panavia Tornado GR4 [061]	*Scrapped at Leeming, May 2014*	
ZA597	Panavia Tornado GR4 [063]	RAF CMU, Marham	
ZA598	Panavia Tornado GR4 [064]	RAF No 31 Sqn, Marham	
ZA600	Panavia Tornado GR4 [EB-G] $	RAF AWC/FJWOEU/No 41(R) Sqn, Coningsby	
ZA601	Panavia Tornado GR4 [EB-B]	RAF Leeming, RTP	
ZA602	Panavia Tornado GR4 [F] $	RAF CMU, stored Marham	
ZA604	Panavia Tornado GR4 [068]	RAF Leeming, RTP	
ZA606	Panavia Tornado GR4 [069]	RAF Leeming, RTP	
ZA607	Panavia Tornado GR4 [EB-X]	RAF AWC/FJWOEU/No 41(R) Sqn, Coningsby	
ZA608	Panavia Tornado GR4	*Scrapped at Leeming*	
ZA609	Panavia Tornado GR4 [072]	RAF No 15(R) Sqn, Lossiemouth	
ZA611	Panavia Tornado GR4 [073]	RAF TEF, Lossiemouth	
ZA612	Panavia Tornado GR4 [074]	RAF No 12 Sqn, Marham	
ZA613	Panavia Tornado GR4 [075]	RAF TEF, Lossiemouth	
ZA614	Panavia Tornado GR4 [EB-Z] $	RAF CMU, Marham	
ZA630	Slingsby T61F Venture T2 (G-BUGL)	Privately owned, Tibenham	
ZA634	Slingsby T61F Venture T2 (G-BUHA) [C]	Privately owned, Saltby, Leics	
ZA652	Slingsby T61F Venture T2 (G-BUDC)	Privately owned, Enstone	
ZA670	B-V Chinook HC4 (N37010) [AA]	RAF Odiham Wing	
ZA674	B-V Chinook HC4 (N37019)	RAF Odiham Wing	
ZA675	B-V Chinook HC4 (N37020) [AE]	RAF Odiham Wing	
ZA677	B-V Chinook HC4 (N37022) [AF]	RAF Odiham Wing	
ZA678	B-V Chinook HC1 (N37023/9229M) [EZ] (wreck)	RAF Odiham, BDRT	
ZA679	B-V Chinook HC4 (N37025) [AG]	RAF Odiham Wing	
ZA680	B-V Chinook HC4 (N37026) [AH]	RAF Odiham Wing	
ZA681	B-V Chinook HC4 (N37027) [AI]	RAF Odiham Wing	
ZA682	B-V Chinook HC4 (N37029) [AJ]	RAF Odiham Wing	
ZA683	B-V Chinook HC4 (N37030) [AK]	RAF Odiham Wing	
ZA684	B-V Chinook HC4 (N37031) [AL]	MoD/Vector Aerospace, Fleetlands	
ZA704	B-V Chinook HC4 (N37033)	RAF Odiham Wing	
ZA705	B-V Chinook HC4 (N37035) [AN]	RAF Odiham Wing	
ZA707	B-V Chinook HC4 (N37040) [AO]	RAF Odiham Wing	
ZA708	B-V Chinook HC4 (N37042) [AP]	RAF Odiham Wing	
ZA710	B-V Chinook HC4 (N37044) [AR]	RAF Odiham Wing	
ZA711	B-V Chinook HC4 (N37046) $	RAF Odiham Wing	

Notes	Serial	Type (code/other identity)	Owner/operator, location or fate
	ZA712	B-V Chinook HC4 (N37047) [AT]	RAF Odiham Wing
	ZA713	B-V Chinook HC4 (N37048)	RAF Odiham Wing
	ZA714	B-V Chinook HC4 (N37051) [AV]	MoD/Vector Aerospace, Fleetlands (conversion)
	ZA717	B-V Chinook HC1 (N37056/9238M) (wreck)	RAFC Cranwell, instructional use
	ZA718	B-V Chinook HC4 (N37058) [BN]	RAF Odiham Wing
	ZA720	B-V Chinook HC4 (N37060) [AW]	RAF Odiham Wing
	ZA726	WS Gazelle AH1 [F1]	Privately owned, Stapleford Tawney
	ZA726	WS Gazelle AH1 (XW851/G-CIEY)	Privately owned, Breighton
	ZA728	WS Gazelle AH1 [E]	Privately owned, Stapleford Tawney
	ZA729	WS Gazelle AH1	AAC Wattisham, BDRT
	ZA730	WS Gazelle AH1 (G-FUKM)	Privately owned, Hurstbourne Tarrant, Hants
	ZA731	WS Gazelle AH1	AAC No 29 Flt, BATUS, Suffield, Canada
	ZA735	WS Gazelle AH1	DSEME SEAE, Arborfield
	ZA736	WS Gazelle AH1	AAC No 29 Flt, BATUS, Suffield, Canada
	ZA737	WS Gazelle AH1	Museum of Army Flying, Middle Wallop
	ZA766	WS Gazelle AH1	AAC No 665 Sqn/5 Regt, Aldergrove
	ZA768	WS Gazelle AH1 [F] (wreck)	MoD/Vector Aerospace, stored Fleetlands
	ZA769	WS Gazelle AH1 [K]	DSEME SEAE, Arborfield
	ZA771	WS Gazelle AH1	DSAE, No 1 SoTT, Cosford
	ZA772	WS Gazelle AH1	AAC MPSU, Middle Wallop
	ZA773	WS Gazelle AH1 [F]	AAC AM&SU, stored Shawbury
	ZA774	WS Gazelle AH1	Privately owned, Babcary, Somerset
	ZA775	WS Gazelle AH1	AAC AM&SU, stored Shawbury
	ZA776	WS Gazelle AH1 [F]	Privately owned, Stapleford Tawney
	ZA804	WS Gazelle HT3	Privately owned, Solstice Park, Amesbury, Wilts
	ZA935	WS Puma HC2	RAF No 33 Sqn/No 230 Sqn, Benson
	ZA936	WS Puma HC2	RAF No 33 Sqn/No 230 Sqn, Benson
	ZA937	WS Puma HC1	RAF Benson (wfu)
	ZA939	WS Puma HC2	RAF No 33 Sqn/No 230 Sqn, Benson
	ZA940	WS Puma HC2 (F-ZWBZ)	MoD/AFD/QinetiQ, Boscombe Down
	ZA947	Douglas Dakota C3	RAF BBMF, Coningsby
	ZB500	WS Lynx 800 (G-LYNX/ZA500)	The Helicopter Museum, Weston-super-Mare
	ZB506	WS61 Sea King Mk 4X	MoD, Boscombe Down (wfu)
	ZB507	WS61 Sea King HC4 [F]	Privately owned, Eastleigh, Hants (for scrapping)
	ZB601	BAe Harrier T4 (fuselage)	RNAS Yeovilton, Fire Section
	ZB603	BAe Harrier T8 [T03/DD]	SFDO, RNAS Culdrose
	ZB604	BAe Harrier T8 [722]	FAA Museum, stored RNAS Yeovilton
	ZB615	SEPECAT Jaguar T2A	DSAE, No 1 SoTT, Cosford
	ZB625	WS Gazelle HT3 [N]	MoD/AFD/QinetiQ, Boscombe Down
	ZB627	WS Gazelle HT3 (G-CBSK) [A]	Privately owned, Hurstbourne Tarrant, Hants
	ZB646	WS Gazelle HT2 (G-CBGZ) [59/CU]	Privately owned, Knebworth
	ZB647	WS Gazelle HT2 (G-CBSF) [40]	Privately owned, Hurstbourne Tarrant, Hants
	ZB665	WS Gazelle AH1	AAC No 667 Sqn/7 Regt, Middle Wallop
	ZB667	WS Gazelle AH1	AAC AM&SU, stored Shawbury
	ZB668	WS Gazelle AH1 (TAD 015)	DSEME SEAE, Arborfield
	ZB669	WS Gazelle AH1	AAC No 665 Sqn/5 Regt, Aldergrove
	ZB670	WS Gazelle AH1	AAC Dishforth, on display
	ZB671	WS Gazelle AH1	AAC No 29 Flt, BATUS, Suffield, Canada
	ZB672	WS Gazelle AH1	Army Training Regiment, Winchester
	ZB673	WS Gazelle AH1 [P]	Privately owned, Stapleford Tawney
	ZB674	WS Gazelle AH1	AAC No 665 Sqn/5 Regt, Aldergrove
	ZB677	WS Gazelle AH1 [5B]	AAC MPSU, Middle Wallop
	ZB678	WS Gazelle AH1	AAC MPSU, Middle Wallop
	ZB679	WS Gazelle AH1	AAC No 665 Sqn/5 Regt, Aldergrove
	ZB682	WS Gazelle AH1 (G-CIEX)	Privately owned, Deighton, N Yorks
	ZB683	WS Gazelle AH1	AAC No 665 Sqn/5 Regt, Aldergrove
	ZB684	WS Gazelle AH1	RAF Defence Movements School, Brize Norton
	ZB686	WS Gazelle AH1 <ff>	The Helicopter Museum, Weston-super-Mare
	ZB688	WS Gazelle AH1 (G-CHMF)	*To South Africa, 2014*
	ZB689	WS Gazelle AH1	AAC No 665 Sqn/5 Regt, Aldergrove
	ZB690	WS Gazelle AH1	MoD AM&SU, stored Shawbury

Serial	Type (code/other identity)	Owner/operator, location or fate	Notes
ZB691	WS Gazelle AH1 [S]	AAC No 671 Sqn/7 Regt, Middle Wallop	
ZB692	WS Gazelle AH1 [Y]	AAC No 671 Sqn/7 Regt, Middle Wallop	
ZB693	WS Gazelle AH1 [U]	AAC No 671 Sqn/7 Regt, Middle Wallop	
ZD230	BAC Super VC10 K4 (G-ASGA) <ff>	Privately owned, Crondall, Hants	
ZD240	BAC Super VC10 K4 (G-ASGL) <ff>	Privately owned, Crondall, Hants	
ZD241	BAC Super VC10 K4 (G-ASGM) [N]	Privately owned, Bruntingthorpe	
ZD242	BAC Super VC10 K4 (G-ASGP) <ff>	JARTS, Boscombe Down	
ZD249	WS Lynx HAS3S [631]	Privately owned, Sproughton	
ZD250	WS Lynx HAS3S [636]	RAF Henlow, instructional use	
ZD251	WS Lynx HAS3S	Privately owned, Hixon, Staffs	
ZD252	WS Lynx HMA8SRU [455/DN]	RN No 815 NAS, *Dragon* Flt, Yeovilton	
ZD254	WS Lynx HAS3S [305]	DSMarE, AESS, *HMS Sultan*, Gosport	
ZD255	WS Lynx HAS3GMS [635]	Ramco, Croft, Lincs	
ZD257	WS Lynx HMA8SRU [644]	RN No 815 NAS, HQ Flt, Yeovilton	
ZD258	WS Lynx HMA8SRU [365]	Privately owned, Bentwaters	
ZD259	WS Lynx HMA8SRU [474/RM]	RN No 815 NAS, *Richmond* Flt, Yeovilton	
ZD260	WS Lynx HMA8SRU [301]	RN No 815 NAS, HQ Flt, Yeovilton	
ZD261	WS Lynx HMA8SRU [314]	RN No 815 NAS, MI Flt, Yeovilton	
ZD262	WS Lynx HMA8SRU [315]	RN No 815 NAS, MI Flt, Yeovilton	
ZD263	WS Lynx HAS3S [632]	Bawtry Paintball Park, S Yorks	
ZD264	WS Lynx HAS3GMS [407]	Privately owned, Hixon, Staffs	
ZD265	WS Lynx HMA8SRU [302]	RN No 815 NAS, HQ Flt, Yeovilton	
ZD266	WS Lynx HMA8SRU [673]	RN No 815 NAS, HQ Flt, Yeovilton	
ZD267	WS Lynx HMA8 (comp XZ672) [LST-2]	RN ETS, Yeovilton, GI use	
ZD268	WS Lynx HMA8SRU [365]	RN No 815 NAS, *Argyll* Flt, Yeovilton	
ZD272	WS Lynx AH7 [W]	*Painted as XZ608*	
ZD273	WS Lynx AH7 [E]	Territorial Army, Bury St Edmunds, Suffolk	
ZD274	WS Lynx AH7 [M]	AAC No 671 Sqn/7 Regt, Middle Wallop	
ZD276	WS Lynx AH7	Mayhem Paintball, Abridge, Essex	
ZD277	WS Lynx AH7 <ff>	Privately owned, Sproughton	
ZD278	WS Lynx AH7 [F]	Privately owned, Sproughton	
ZD279	WS Lynx AH7 <ff>	Privately owned, Hixon, Staffs	
ZD280	WS Lynx AH7	AAC No 9 Regt, Dishforth	
ZD281	WS Lynx AH7 [K]	MoD/Vector Aerospace, Fleetlands, GI use	
ZD282	WS Lynx AH7	Privately owned, Sproughton	
ZD283	WS Lynx AH7 <ff>	Privately owned,	
ZD284	WS Lynx AH7 [K]	Privately owned, Corley Moor, Warks	
ZD285	WS Lynx AH7	MoD, Boscombe Down (wfu)	
ZD318	BAe Harrier GR7A	Harrier Heritage Centre, RAF Wittering	
ZD353	BAe Harrier GR5 (fuselage)	Privately owned, Sproughton	
ZD412	BAe Harrier GR5 (fuselage)	Privately owned, Sproughton	
ZD433	BAe Harrier GR9A [45A]	FAA Museum, stored Cobham Hall, RNAS Yeovilton	
ZD461	BAe Harrier GR9A [51A]	Imperial War Museum, Lambeth	
ZD462	BAe Harrier GR7 (9302M) [52]	Privately owned, stored Cotswold Airport	
ZD465	BAe Harrier GR9 [55]	DSMarE AESS, *HMS Sultan*, stored Gosport	
ZD469	BAe Harrier GR7A	RAF Wittering, on display	
ZD476	WS61 Sea King HC4 [WZ]	DSMarE AESS, *HMS Sultan*, stored Gosport	
ZD477	WS61 Sea King HC4 [E]	Privately owned, Bruntingthorpe	
ZD478	WS61 Sea King HC4 [J]	Privately owned, Hixon, Staffs	
ZD479	WS61 Sea King HC4 [WQ]	DSMarE AESS, *HMS Sultan*, stored Gosport	
ZD480	WS61 Sea King HC4 [J]	DSMarE AESS, *HMS Sultan*, stored Gosport	
ZD559	WS Lynx AH5X	MoD, Boscombe Down Apprentice School	
ZD560	WS Lynx AH7	MoD, Boscombe Down (wfu)	
ZD565	WS Lynx HMA8SRU [404/IR]	RN No 815 NAS, *Iron Duke* Flt, Yeovilton	
ZD566	WS Lynx HMA8SRU [645]	RN No 815 NAS, HQ Flt, Yeovilton	
ZD574	B-V Chinook HC4 (N37077) [DB]	RAF Odiham Wing	
ZD575	B-V Chinook HC4 (N37078) [DC]	RAF Odiham Wing	
ZD578	BAe Sea Harrier FA2 [000,122]	RNAS Yeovilton, at main gate	
ZD579	BAe Sea Harrier FA2 [79/DD]	SFDO, RNAS Culdrose	
ZD580	BAe Sea Harrier FA2 [710]	Privately owned, Cheshire	
ZD581	BAe Sea Harrier FA2 [124]	RN, Predannack Fire School	

Notes	Serial	Type (code/other identity)	Owner/operator, location or fate
	ZD582	BAe Sea Harrier FA2 [002/N]	Privately owned, Banbury, Oxon
	ZD607	BAe Sea Harrier FA2	DSDA, Arncott, Oxon, preserved
	ZD610	BAe Sea Harrier FA2	Privately owned, Bruntingthorpe
	ZD611	BAe Sea Harrier FA2 [714,002/N]	SFDO, RNAS Culdrose, preserved
	ZD612	BAe Sea Harrier FA2	Privately owned, Topsham, Devon
	ZD613	BAe Sea Harrier FA2 [127/R]	Privately owned, Cross Green, Leeds
	ZD614	BAe Sea Harrier FA2	Privately owned, Lymington, Hants
	ZD620	BAe 125 CC3	RAF No 32(The Royal) Sqn, Northolt
	ZD621	BAe 125 CC3	RAF No 32(The Royal) Sqn, Northolt
	ZD625	WS61 Sea King HC4 [P]	Privately owned, Eastleigh, Hants (for scrapping)
	ZD626	WS61 Sea King HC4 [S]	DSMarE AESS, *HMS Sultan*, stored Gosport
	ZD627	WS61 Sea King HC4 [WO]	Privately owned, Bruntingthorpe
	ZD630	WS61 Sea King HAS6 [012/L]	DSMarE, AESS, *HMS Sultan*, Gosport
	ZD631	WS61 Sea King HAS6 [66] (fuselage)	Privately owned, St Agnes, Cornwall
	ZD633	WS61 Sea King HAS6 [014/L]	Privately owned, Eastleigh, Hants (for scrapping)
	ZD634	WS61 Sea King HAS6 [503]	DSMarE, stored *HMS Sultan*, Gosport
	ZD636	WS61 Sea King ASaC7 [182/CU]	DSMarE AESS, *HMS Sultan*, stored Gosport
	ZD637	WS61 Sea King HAS6 [700/PW]	DSMarE, AESS, *HMS Sultan*, Gosport
	ZD667	BAe Harrier GR3 (9201M) [U]	Bentwaters Cold War Museum
	ZD703	BAe 125 CC3	RAF No 32(The Royal) Sqn, Northolt
	ZD704	BAe 125 CC3	RAF stored, Hawarden (damaged)
	ZD707	Panavia Tornado GR4 [077]	RAF No 15(R) Sqn, Lossiemouth
	ZD708	Panavia Tornado GR4	*Scrapped at Leeming, 2013*
	ZD709	Panavia Tornado GR4 [078]	RAF Marham Wing
	ZD710	Panavia Tornado GR1 <ff>	Privately owned, Ruthin, Denbighshire
	ZD711	Panavia Tornado GR4 [079]	RAF CMU, stored Marham
	ZD712	Panavia Tornado GR4 [080] $	RAF Leeming, RTP
	ZD713	Panavia Tornado GR4 [081]	RAF No 15(R) Sqn, Lossiemouth
	ZD714	Panavia Tornado GR4 [082]	*Scrapped at Leeming, 2014*
	ZD715	Panavia Tornado GR4 [083]	RAF Marham, WLT
	ZD716	Panavia Tornado GR4 [084]	RAF CMU, Marham
	ZD719	Panavia Tornado GR4 [085]	*Scrapped at Leeming, September 2014*
	ZD720	Panavia Tornado GR4 [086]	RAF Leeming, RTP
	ZD739	Panavia Tornado GR4 [087]	RAF TEF, Lossiemouth
	ZD740	Panavia Tornado GR4 [088]	RAF No 15(R) Sqn, Lossiemouth
	ZD741	Panavia Tornado GR4 [089]	RAF CMU, Marham
	ZD742	Panavia Tornado GR4 [090]	MoD/BAE Systems, Warton
	ZD744	Panavia Tornado GR4 [092]	RAF Marham Wing
	ZD745	Panavia Tornado GR4 [093]	RAF Marham Wing
	ZD746	Panavia Tornado GR4 [094]	*Scrapped at Leeming*
	ZD747	Panavia Tornado GR4 [095]	RAF CMU, stored Marham
	ZD748	Panavia Tornado GR4 [096] $	RAF No 31 Sqn, Marham
	ZD749	Panavia Tornado GR4 [097]	RAF No 9 Sqn, Marham
	ZD788	Panavia Tornado GR4 $	RAF No 12 Sqn, Marham
	ZD790	Panavia Tornado GR4 [099]	RAF No 15(R) Sqn, Lossiemouth
	ZD792	Panavia Tornado GR4 [100]	RAF CMU, stored Marham
	ZD793	Panavia Tornado GR4 [101]	RAF Lossiemouth, WLT
	ZD810	Panavia Tornado GR4 [102]	RAF Leeming, RTP
	ZD842	Panavia Tornado GR4 [105]	RAF CMU, Marham
	ZD843	Panavia Tornado GR4 [106]	RAF TEF, Lossiemouth
	ZD844	Panavia Tornado GR4 [107]	RAF Marham Wing
	ZD847	Panavia Tornado GR4 [108]	RAF Leeming, RTP
	ZD848	Panavia Tornado GR4 [109]	RAF CMU, Marham
	ZD849	Panavia Tornado GR4 [110]	RAF CMU, Marham
	ZD850	Panavia Tornado GR4 [111]	*Scrapped at Leeming*
	ZD851	Panavia Tornado GR4 [112]	RAF No 15(R) Sqn, Lossiemouth
	ZD890	Panavia Tornado GR4 [113]	RAF Marham Wing
	ZD892	Panavia Tornado GR4 [TG]	*Scrapped at Leeming*
	ZD895	Panavia Tornado GR4 [115]	RAF Leeming, RTP
	ZD899	Panavia Tornado F2	MoD, Boscombe Down Apprentice School
	ZD902	Panavia Tornado F2A(TIARA)	MoD/QinetiQ, Boscombe Down (wfu)
	ZD906	Panavia Tornado F2 (comp ZE294) <ff>	RAF Leuchars, BDRT

Serial	Type (code/other identity)	Owner/operator, location or fate	Notes
ZD932	Panavia Tornado F2 (comp ZE255) (9308M) (fuselage)	Privately owned, Thorpe Wood, N Yorks	
ZD936	Panavia Tornado F2 (comp ZE251) <ff>	Boscombe Down Aviation Collection, Old Sarum	
ZD938	Panavia Tornado F2 (comp ZE295) <ff>	South Yorkshire Aircraft Museum, Doncaster	
ZD939	Panavia Tornado F2 (comp ZE292) <ff>	DSAE Cosford, instructional use	
ZD948	Lockheed TriStar KC1 (G-BFCA)	*Sold as N304CS, May 2014*	
ZD949	Lockheed TriStar K1 (G-BFCB)	*Scrapped at Cambridge, 27 May 2014*	
ZD950	Lockheed TriStar KC1 (G-BFCC)	*Sold as N405CS, May 2014*	
ZD951	Lockheed TriStar KC1 (G-BFCD) $	*Sold as N309CS, May 2014*	
ZD952	Lockheed TriStar KC1 (G-BFCE)	*Scrapped at Cotswold Airport, 10 September 2014*	
ZD953	Lockheed TriStar KC1 (G-BFCF)	*Sold as N705CS, May 2014*	
ZD980	B-V Chinook HC4 (N37082) [DD]	RAF Odiham Wing	
ZD981	B-V Chinook HC4 (N37083)	RAF Odiham Wing	
ZD982	B-V Chinook HC4 (N37085) [DF]	RAF Odiham Wing	
ZD983	B-V Chinook HC4 (N37086) [DG]	MoD/Vector Aerospace, Fleetlands	
ZD984	B-V Chinook HC4 (N37088) [DH]	RAF Odiham Wing	
ZD990	BAe Harrier T8 [T90/DD]	SFDO, RNAS Culdrose	
ZD992	BAe Harrier T8 [724] (fuselage)	Privately owned, Gr Manchester	
ZD993	BAe Harrier T8 [723/VL]	*Sold to the USA, 2014*	
ZD993	HS Harrier T4 (XZ145)	*Sold to the USA, August 2014*	
ZE114	Panavia Tornado IDS (RSAF 703)	BAE Systems, Warton	
ZE116	Panavia Tornado GR4A [116]	RAF Marham Wing	
ZE119	Panavia Tornado IDS (RSAF 760)	BAE Systems, Warton	
ZE165	Panavia Tornado F3 [GE]	MoD DFTDC, Manston	
ZE168	Panavia Tornado F3 <ff>	Privately owned, Cowes, IoW	
ZE204	Panavia Tornado F3 [FC]	MoD DFTDC, Manston	
ZE256	Panavia Tornado F3 [DZ] (wears ZE343 on port side)	Privately owned, Thorpe Wood, N Yorks	
ZE340	Panavia Tornado F3 (ZE758/9298M) [GO]	DSAE, No 1 SoTT, Cosford	
ZE342	Panavia Tornado F3 [HP]	Rolls Royce, Leeming	
ZE350	McD F-4J(UK) Phantom (9080M) <ff>	Privately owned, Shrewsbury	
ZE352	McD F-4J(UK) Phantom (9086M) <ff>	Privately owned, Hooton Park	
ZE360	McD F-4J(UK) Phantom (9059M) [O]	MoD DFTDC, Manston	
ZE368	WS61 Sea King HAR3 [R]	MoD/Vector Aerospace, Fleetlands	
ZE369	WS61 Sea King HAR3 [S]	RAF No 22 Sqn, C Flt, Valley	
ZE370	WS61 Sea King HAR3 [T]	RAF SKAMG, RNAS Yeovilton	
ZE375	WS Lynx AH9A	MoD/Vector Aerospace stored, Fleetlands	
ZE376	WS Lynx AH9A	MoD/Vector Aerospace, Fleetlands	
ZE378	WS Lynx AH7	AAC No 657 Sqn, Odiham	
ZE379	WS Lynx AH7	AAC Dishforth, BDRT	
ZE380	WS Lynx AH9A	MoD/AgustaWestland, Yeovil	
ZE381	WS Lynx AH7 [X]	DSEME SEAE, Arborfield	
ZE395	BAe 125 CC3	RAF No 32(The Royal) Sqn, Northolt	
ZE396	BAe 125 CC3	RAF No 32(The Royal) Sqn, Northolt	
ZE410	Agusta A109A (AE-334)	Museum of Army Flying, stored Middle Wallop	
ZE412	Agusta A109A	Army, Middle Wallop (for display at Credenhill)	
ZE413	Agusta A109A	Army Whittington Barracks, Lichfield, Gl use	
ZE416	Agusta A109E Power Elite (G-ESLH)	MoD/ETPS, Boscombe Down	
ZE418	WS61 Sea King ASaC7 [186]	MoD/Vector Aerospace, stored Fleetlands	
ZE420	WS61 Sea King ASaC7 [189]	RN No 849 NAS, Culdrose	
ZE422	WS61 Sea King ASaC7 [192]	RN No 849 NAS, Culdrose	
ZE425	WS61 Sea King HC4 [WR]	Privately owned, Eastleigh, Hants (for scrapping)	
ZE426	WS61 Sea King HC4 [WX]	DSMarE, stored *HMS Sultan*, Gosport	
ZE427	WS61 Sea King HC4 [K]	RN No 845 NAS, Yeovilton	
ZE428	WS61 Sea King HC4 [H]	Privately owned, Eastleigh, Hants (for scrapping)	
ZE432	BAC 1-11/479FU (DQ-FBV) <ff>	Bournemouth Aviation Museum	
ZE449	SA330L Puma HC1 (9017M/PA-12)	Army Whittington Barracks, Lichfield, on display	
ZE477	WS Lynx 3	The Helicopter Museum, Weston-super-Mare	
ZE495	Grob G103 Viking T1 (BGA3000) [VA]	RAF No 621 VGS, Hullavington	
ZE496	Grob G103 Viking T1 (BGA3001) [VB]	RAF CGMF, stored Syerston	
ZE498	Grob G103 Viking T1 (BGA3003) [VC]	RAF CGMF, stored Syerston	
ZE499	Grob G103 Viking T1 (BGA3004) [VD]	RAF ACCGS/No 643 VGS, Syerston	
ZE502	Grob G103 Viking T1 (BGA3007) [VF]	RAF No 614 VGS, Wethersfield	

Notes	Serial	Type (code/other identity)	Owner/operator, location or fate
	ZE503	Grob G103 Viking T1 (BGA3008) [VG]	RAF No 615 VGS, Kenley
	ZE504	Grob G103 Viking T1 (BGA3009) [VH]	RAF ACCGS/No 643 VGS, Syerston
	ZE520	Grob G103 Viking T1 (BGA3010) [VJ]	RAF No 661 VGS, Kirknewton
	ZE521	Grob G103 Viking T1 (BGA3011) [VK]	RAF CGMF, stored Syerston
	ZE522	Grob G103 Viking T1 (BGA3012) [VL]	RAF No 621 VGS, Hullavington
	ZE524	Grob G103 Viking T1 (BGA3014) [VM]	RAF CGMF, Syerston
	ZE526	Grob G103 Viking T1 (BGA3016) [VN]	RAF No 626 VGS, Predannack
	ZE527	Grob G103 Viking T1 (BGA3017) [VP]	RAF No 626 VGS, Predannack
	ZE528	Grob G103 Viking T1 (BGA3018) [VQ]	RAF CGMF, stored Syerston
	ZE529	Grob G103 Viking T1 (BGA3019) (comp ZE655) [VR]	RAF ACCGS/No 643 VGS, Syerston
	ZE530	Grob G103 Viking T1 (BGA3020) [VS]	RAF No 621 VGS, Hullavington
	ZE531	Grob G103 Viking T1 (BGA3021) [VT]	RAF CGMF, stored Syerston
	ZE532	Grob G103 Viking T1 (BGA3022) [VU]	RAF CGMF, stored Syerston
	ZE533	Grob G103 Viking T1 (BGA3023) [VV]	RAF No 661 VGS, Kirknewton
	ZE550	Grob G103 Viking T1 (BGA3025) [VX]	RAF CGMF, Syerston (damaged)
	ZE551	Grob G103 Viking T1 (BGA3026) [VY]	RAF No 614 VGS, Wethersfield
	ZE552	Grob G103 Viking T1 (BGA3027) [VZ]	RAF CGMF, stored Syerston
	ZE553	Grob G103 Viking T1 (BGA3028) [WA]	RAF CGMF, stored Syerston
	ZE554	Grob G103 Viking T1 (BGA3029) [WB]	RAF No 615 VGS, Kenley
	ZE555	Grob G103 Viking T1 (BGA3030) [WC]	RAF CGMF, stored Syerston
	ZE556	Grob G103 Viking T1 (BGA3031) <ff>	RAF CGMF, stored Syerston
	ZE557	Grob G103 Viking T1 (BGA3032) [WE]	RAF CGMF, stored Syerston
	ZE558	Grob G103 Viking T1 (BGA3033) [WF]	RAF CGMF, stored Syerston
	ZE559	Grob G103 Viking T1 (BGA3034) [WG]	RAF ACCGS/No 643 VGS, Syerston
	ZE560	Grob G103 Viking T1 (BGA3035) [WH]	RAF ACCGS/No 643 VGS, Syerston
	ZE561	Grob G103 Viking T1 (BGA3036) [WJ]	RAF HQ AC, stored Cranwell
	ZE562	Grob G103 Viking T1 (BGA3037) [WK]	RAF ACCGS/No 643 VGS, Syerston
	ZE563	Grob G103 Viking T1 (BGA3038) [WL]	RAF CGMF, stored Syerston
	ZE564	Grob G103 Viking T1 (BGA3039) [WN]	RAF No 622 VGS, Upavon
	ZE584	Grob G103 Viking T1 (BGA3040) [WP]	RAF CGMF, stored Syerston
	ZE585	Grob G103 Viking T1 (BGA3041) [WQ]	RAF No 614 VGS, Wethersfield
	ZE586	Grob G103 Viking T1 (BGA3042) [WR]	RAF No 661 VGS, Kirknewton
	ZE587	Grob G103 Viking T1 (BGA3043) [WS]	RAF No 622 VGS, Upavon
	ZE590	Grob G103 Viking T1 (BGA3046) [WT]	RAF CGMF, stored Syerston
	ZE591	Grob G103 Viking T1 (BGA3047) [WU]	RAF CGMF, stored Syerston
	ZE592	Grob G103 Viking T1 (BGA3048) <ff>	RAFGSA, stored Halton
	ZE593	Grob G103 Viking T1 (BGA3049) [WW]	RAF No 615 VGS, Kenley
	ZE594	Grob G103 Viking T1 (BGA3050) [WX]	RAF No 662 VGS, Arbroath
	ZE595	Grob G103 Viking T1 (BGA3051) [WY]	RAF No 661 VGS, Kirknewton
	ZE600	Grob G103 Viking T1 (BGA3052) [WZ]	RAF No 615 VGS, Kenley
	ZE601	Grob G103 Viking T1 (BGA3053) [XA]	RAF CGMF, stored Syerston
	ZE602	Grob G103 Viking T1 (BGA3054) [XB]	RAF CGMF, stored Syerston
	ZE603	Grob G103 Viking T1 (BGA3055) [XC]	RAF No 614 VGS, Wethersfield
	ZE604	Grob G103 Viking T1 (BGA3056) [XD]	RAF No 621 VGS, Hullavington
	ZE605	Grob G103 Viking T1 (BGA3057) [XE]	RAF ACCGS/No 643 VGS, Syerston
	ZE606	Grob G103 Viking T1 (BGA3058) [XF]	RAF No 614 VGS, Wethersfield
	ZE607	Grob G103 Viking T1 (BGA3059) [XG]	RAF No 621 VGS, Hullavington
	ZE608	Grob G103 Viking T1 (BGA3060) [XH]	RAF CGMF, Syerston
	ZE609	Grob G103 Viking T1 (BGA3061) [XJ]	RAF ACCGS/No 643 VGS, Syerston
	ZE610	Grob G103 Viking T1 (BGA3062) [XK]	RAF No 614 VGS, Wethersfield
	ZE611	Grob G103 Viking T1 (BGA3063) [XL]	RAF CGMF, Syerston
	ZE613	Grob G103 Viking T1 (BGA3065) [XM]	RAF No 621 VGS, Hullavington
	ZE614	Grob G103 Viking T1 (BGA3066) [XN]	RAF CGMF, Syerston
	ZE625	Grob G103 Viking T1 (BGA3067) [XP]	RAF CGMF, stored Syerston
	ZE626	Grob G103 Viking T1 (BGA3068) [XQ]	RAF CGMF, stored Syerston
	ZE627	Grob G103 Viking T1 (BGA3069) [XR]	RAF ACCGS/No 643 VGS, Syerston
	ZE628	Grob G103 Viking T1 (BGA3070) [XS]	RAF CGMF, Syerston
	ZE629	Grob G103 Viking T1 (BGA3071) [XT]	RAF No 662 VGS, Arbroath
	ZE630	Grob G103 Viking T1 (BGA3072) [XU]	RAF No 662 VGS, Arbroath
	ZE631	Grob G103 Viking T1 (BGA3073) [XV]	RAF No 626 VGS, Predannack
	ZE632	Grob G103 Viking T1 (BGA3074) [XW]	RAF No 662 VGS, Arbroath
	ZE633	Grob G103 Viking T1 (BGA3075) [XX]	RAF CGMF, stored Syerston

Serial	Type (code/other identity)	Owner/operator, location or fate	Notes
ZE636	Grob G103 Viking T1 (BGA3078) [XZ]	RAF No 621 VGS, Hullavington	
ZE637	Grob G103 Viking T1 (BGA3079) [YA]	RAF CGMF, stored Syerston	
ZE650	Grob G103 Viking T1 (BGA3080) [YB]	RAF ACCGS/No 643 VGS, Syerston	
ZE651	Grob G103 Viking T1 (BGA3081) [YC]	RAF No 621 VGS, Hullavington	
ZE652	Grob G103 Viking T1 (BGA3082) [YD]	RAF No 621 VGS, Hullavington	
ZE653	Grob G103 Viking T1 (BGA3083) [YE]	RAF No 621 VGS, Hullavington	
ZE656	Grob G103 Viking T1 (BGA3086) [YH]	RAF No 615 VGS, Kenley	
ZE657	Grob G103 Viking T1 (BGA3087) [YJ]	RAF CGMF, stored Syerston	
ZE658	Grob G103 Viking T1 (BGA3088) [YK]	RAF CGMF, stored Syerston	
ZE677	Grob G103 Viking T1 (BGA3090) [YM]	RAF No 615 VGS, Kenley	
ZE678	Grob G103 Viking T1 (BGA3091) [YN]	RAF ACCGS/No 643 VGS, Syerston	
ZE679	Grob G103 Viking T1 (BGA3092) [YP]	RAF CGMF, stored Syerston	
ZE680	Grob G103 Viking T1 (BGA3093) [YQ]	RAF ACCGS/No 643 VGS, Syerston	
ZE681	Grob G103 Viking T1 (BGA3094) <ff>	RAF Hullavington	
ZE682	Grob G103 Viking T1 (BGA3095) [YS]	RAF CGMF, stored Syerston	
ZE683	Grob G103 Viking T1 (BGA3096) [YT]	RAF ACCGS/No 643 VGS, Syerston	
ZE684	Grob G103 Viking T1 (BGA3097) [YU]	RAF ACCGS/No 643 VGS, Syerston	
ZE685	Grob G103 Viking T1 (BGA3098) [YV]	RAF No 614 VGS, Wethersfield	
ZE686	Grob G103 Viking T1 (BGA3099) <ff>	RAF Museum, Hendon	
ZE690	BAe Sea Harrier FA2 [90/DD]	SFDO, RNAS Culdrose	
ZE691	BAe Sea Harrier FA2 [710]	Classic Autos, Winsford, Cheshire	
ZE692	BAe Sea Harrier FA2 [92/DD]	SFDO, RNAS Culdrose	
ZE694	BAe Sea Harrier FA2 [004]	Midland Air Museum, Coventry	
ZE697	BAe Sea Harrier FA2 [006]	Privately owned, Binbrook	
ZE698	BAe Sea Harrier FA2 [001]	Privately owned, Charlwood	
ZE700	BAe 146 CC2 (G-6-021)	RAF No 32(The Royal) Sqn, Northolt	
ZE701	BAe 146 CC2 (G-6-029)	RAF No 32(The Royal) Sqn, Northolt	
ZE704	Lockheed TriStar C2 (N508PA)	*Sold as N507CS, May 2014*	
ZE705	Lockheed TriStar C2 (N509PA)	*Sold as N703CS, May 2014*	
ZE706	Lockheed TriStar C2A (N503PA)	*Scrapped at Cambridge, 9 June 2014*	
ZE707	BAe 146 C3 (OO-TAZ)	RAF No 32(The Royal) Sqn, Northolt	
ZE708	BAe 146 C3 (OO-TAY)	RAF No 32(The Royal) Sqn, Northolt	
ZE760	Panavia Tornado F3 (MM7206) [AP]	RAF Coningsby, on display	
ZE791	Panavia Tornado F3 [JU-L] $	RAF Leeming, RTP	
ZE887	Panavia Tornado F3 [GF] $	RAF Museum, Hendon	
ZE934	Panavia Tornado F3 [TA]	Royal Scottish Mus'm of Flight, E Fortune	
ZE966	Panavia Tornado F3 [VT]	Science Museum, stored Wroughton	
ZE967	Panavia Tornado F3 [UT]	RAF Leuchars, at main gate	
ZF115	WS61 Sea King HC4 [R,WV]	RN ETS, Yeovilton	
ZF116	WS61 Sea King HC4 [WP]	RN No 845 NAS, Yeovilton	
ZF117	WS61 Sea King HC4 [X]	RN No 845 NAS, Yeovilton	
ZF118	WS61 Sea King HC4 [O]	RN No 845 NAS, Yeovilton	
ZF119	WS61 Sea King HC4 [WY]	DSMarE, stored *HMS Sultan*, Gosport	
ZF120	WS61 Sea King HC4 [Z]	DSMarE, stored *HMS Sultan*, Gosport	
ZF121	WS61 Sea King HC4 [T]	Privately owned, Eastleigh, Hants (for scrapping)	
ZF122	WS61 Sea King HC4 [V]	RN No 845 NAS, Yeovilton	
ZF123	WS61 Sea King HC4 [WW]	Privately owned, Hitchin, Herts (for scrapping)	
ZF124	WS61 Sea King HC4 [L]	DSMarE, stored *HMS Sultan*, Gosport	
ZF135	Shorts Tucano T1 [135]	RAF Linton-on-Ouse, GI use	
ZF137	Shorts Tucano T1 [137]	RAF, stored Linton-on-Ouse	
ZF139	Shorts Tucano T1 [139]	RAF No 1 FTS, Linton-on-Ouse	
ZF140	Shorts Tucano T1 [140] $	RAF No 1 FTS/*72(R) Sqn*, Linton-on-Ouse	
ZF142	Shorts Tucano T1 [142]	RAF No 1 FTS, Linton-on-Ouse	
ZF143	Shorts Tucano T1 [143]	RAF No 1 FTS, Linton-on-Ouse	
ZF144	Shorts Tucano T1 [144]	RAF, stored Linton-on-Ouse	
ZF145	Shorts Tucano T1 [145]	RAF No 1 FTS, Linton-on-Ouse	
ZF160	Shorts Tucano T1 [160]	RAF AM&SU, stored Shawbury	
ZF161	Shorts Tucano T1 [161]	RAF AM&SU, stored Shawbury	
ZF163	Shorts Tucano T1 [163]	RAF AM&SU, stored Shawbury	
ZF166	Shorts Tucano T1 [166]	RAF AM&SU, stored Shawbury	
ZF167	Shorts Tucano T1 (fuselage)	Ulster Aviation Society, Long Kesh	

Notes	Serial	Type (code/other identity)	Owner/operator, location or fate
	ZF169	Shorts Tucano T1 [169]	RAF, stored Linton-on-Ouse
	ZF170	Shorts Tucano T1 [MP-A]	RAF, stored Linton-on-Ouse
	ZF171	Shorts Tucano T1 [171]	RAF No 1 FTS, Linton-on-Ouse
	ZF172	Shorts Tucano T1 [172]	RAF No 1 FTS, Linton-on-Ouse
	ZF202	Shorts Tucano T1 [202]	RAF Linton-on-Ouse, on display
	ZF203	Shorts Tucano T1 [203]	RAF AM&SU, stored Shawbury
	ZF204	Shorts Tucano T1 [204]	RAF No 1 FTS, Linton-on-Ouse
	ZF205	Shorts Tucano T1 [205]	RAF No 1 FTS/72(R) Sqn, Linton-on-Ouse
	ZF210	Shorts Tucano T1 [210]	RAF No 1 FTS, Linton-on-Ouse
	ZF211	Shorts Tucano T1 [211]	RAF AM&SU, stored Shawbury
	ZF212	Shorts Tucano T1 [212]	RAF AM&SU, stored Shawbury
	ZF239	Shorts Tucano T1 [RA-F] $	RAF No 1 FTS, Linton-on-Ouse
	ZF240	Shorts Tucano T1 [240]	RAF No 1 FTS, Linton-on-Ouse
	ZF242	Shorts Tucano T1 [242]	RAF AM&SU, stored Shawbury
	ZF243	Shorts Tucano T1	RAF No 1 FTS, Linton-on-Ouse
	ZF244	Shorts Tucano T1 [244] $	RAF No 1 FTS/72(R) Sqn, Linton-on-Ouse
	ZF263	Shorts Tucano T1 [263]	RAF AM&SU, stored Shawbury
	ZF264	Shorts Tucano T1 [264]	RAF No 1 FTS, Linton-on-Ouse
	ZF268	Shorts Tucano T1 [268]	RAF AM&SU, stored Shawbury
	ZF269	Shorts Tucano T1 $	RAF No 1 FTS, Linton-on-Ouse
	ZF286	Shorts Tucano T1 [286]	RAF AM&SU, stored Shawbury
	ZF287	Shorts Tucano T1 [287]	RAF, stored Linton-on-Ouse
	ZF288	Shorts Tucano T1 [288]	RAF AM&SU, stored Shawbury
	ZF289	Shorts Tucano T1 [289]	RAF No 1 FTS, Linton-on-Ouse
	ZF290	Shorts Tucano T1 [290]	RAF No 1 FTS, Linton-on-Ouse
	ZF291	Shorts Tucano T1 [291]	RAF No 1 FTS, Linton-on-Ouse
	ZF292	Shorts Tucano T1 [292]	RAF, stored Linton-on-Ouse
	ZF293	Shorts Tucano T1 [293]	RAF No 1 FTS/72(R) Sqn, Linton-on-Ouse
	ZF294	Shorts Tucano T1 [294]	RAF, stored Linton-on-Ouse
	ZF295	Shorts Tucano T1 [295] $	RAF No 1 FTS/72(R) Sqn, Linton-on-Ouse
	ZF315	Shorts Tucano T1 [315]	RAF AM&SU, stored Shawbury
	ZF317	Shorts Tucano T1 [317]	RAF No 1 FTS, Linton-on-Ouse
	ZF318	Shorts Tucano T1 [318] $	RAF AM&SU, stored Shawbury
	ZF319	Shorts Tucano T1 [319]	RAF, stored Linton-on-Ouse
	ZF338	Shorts Tucano T1 [338,MP-W]	RAF No 1 FTS, Linton-on-Ouse
	ZF339	Shorts Tucano T1 [339]	RAF No 1 FTS/72(R) Sqn, Linton-on-Ouse
	ZF341	Shorts Tucano T1 [341]	RAF, stored Linton-on-Ouse
	ZF342	Shorts Tucano T1 [342]	RAF, stored Linton-on-Ouse
	ZF343	Shorts Tucano T1 [343]	RAF No 1 FTS/72(R) Sqn, Linton-on-Ouse
	ZF345	Shorts Tucano T1 [345]	RAF AM&SU, stored Shawbury
	ZF347	Shorts Tucano T1 [347]	RAF No 1 FTS, Linton-on-Ouse
	ZF348	Shorts Tucano T1 [348]	RAF, stored Linton-on-Ouse
	ZF349	Shorts Tucano T1 [349]	RAF Linton-on-Ouse (wreck)
	ZF350	Shorts Tucano T1 [350]	RAF AM&SU, stored Shawbury
	ZF372	Shorts Tucano T1 [372]	RAF AM&SU, stored Shawbury
	ZF374	Shorts Tucano T1 $	RAF No 1 FTS, Linton-on-Ouse
	ZF376	Shorts Tucano T1 [376]	RAF AM&SU, stored Shawbury
	ZF377	Shorts Tucano T1 [377]	RAF No 1 FTS, Linton-on-Ouse
	ZF378	Shorts Tucano T1 [378] $	RAF No 1 FTS, Linton-on-Ouse
	ZF379	Shorts Tucano T1 [379]	RAF, stored Linton-on-Ouse
	ZF380	Shorts Tucano T1 [380]	RAF AM&SU, stored Shawbury
	ZF405	Shorts Tucano T1 [405]	RAF AM&SU, stored Shawbury
	ZF406	Shorts Tucano T1 [406]	RAF, stored Linton-on-Ouse
	ZF407	Shorts Tucano T1 [407]	RAF No 1 FTS, Linton-on-Ouse
	ZF408	Shorts Tucano T1 [408]	RAF AM&SU, stored Shawbury
	ZF410	Shorts Tucano T1 [410]	RAF AM&SU, stored Shawbury
	ZF412	Shorts Tucano T1 [412]	RAF AM&SU, stored Shawbury
	ZF414	Shorts Tucano T1 [414]	RAF AM&SU, stored Shawbury
	ZF416	Shorts Tucano T1 [416]	RAF AM&SU, stored Shawbury
	ZF417	Shorts Tucano T1 [417]	RAF No 1 FTS, Linton-on-Ouse
	ZF418	Shorts Tucano T1 [418]	RAF AM&SU, stored Shawbury
	ZF446	Shorts Tucano T1 [446]	RAF AM&SU, stored Shawbury

Serial	Type (code/other identity)	Owner/operator, location or fate	Notes
ZF447	Shorts Tucano T1 [447]	RAF AM&SU, stored Shawbury	
ZF448	Shorts Tucano T1 [448]	RAF, stored Linton-on-Ouse	
ZF449	Shorts Tucano T1 [449]	RAF AM&SU, stored Shawbury	
ZF483	Shorts Tucano T1 [483]	RAF AM&SU, stored Shawbury	
ZF484	Shorts Tucano T1 [484]	RAF AM&SU, stored Shawbury	
ZF485	Shorts Tucano T1 (G-BULU) [485]	RAF No 1 FTS, Linton-on-Ouse	
ZF486	Shorts Tucano T1 [486]	RAF AM&SU, stored Shawbury	
ZF487	Shorts Tucano T1 [487]	RAF AM&SU, stored Shawbury	
ZF488	Shorts Tucano T1 [488]	RAF AM&SU, stored Shawbury	
ZF489	Shorts Tucano T1 [489]	RAF No 1 FTS, Linton-on-Ouse	
ZF490	Shorts Tucano T1 [490]	RAF AM&SU, stored Shawbury	
ZF491	Shorts Tucano T1 [491]	RAF No 1 FTS, Linton-on-Ouse	
ZF492	Shorts Tucano T1 [492]	RAF AM&SU, stored Shawbury	
ZF510	Shorts Tucano T1 [510]	MoD/AFD/QinetiQ, Boscombe Down	
ZF511	Shorts Tucano T1 [511]	MoD/AFD/QinetiQ, Boscombe Down	
ZF512	Shorts Tucano T1 [512]	RAF No 1 FTS/72(R) Sqn, Linton-on-Ouse	
ZF513	Shorts Tucano T1 [513]	RAF AM&SU, stored Shawbury	
ZF514	Shorts Tucano T1 [514]	RAF AM&SU, stored Shawbury	
ZF515	Shorts Tucano T1 [515]	RAF, stored Linton-on-Ouse	
ZF516	Shorts Tucano T1 [516]	RAF AM&SU, stored Shawbury	
ZF534	BAe EAP	RAF Museum, Cosford	
ZF537	WS Lynx AH9A	MoD/Vector Aerospace, Fleetlands	
ZF538	WS Lynx AH9A	MoD/Vector Aerospace, Fleetlands	
ZF539	WS Lynx AH9A	MoD/Vector Aerospace, stored Fleetlands	
ZF540	WS Lynx AH9A	Crashed 26 April 2014, Southern Afghanistan	
ZF557	WS Lynx HMA8SRU [425/KT]	RN No 815 NAS, Westminster Flt, Yeovilton	
ZF558	WS Lynx HMA8SRU [642/PD]	RN No 815 NAS, HQ Flt, Yeovilton	
ZF560	WS Lynx HMA8SRU [456]	RN No 815 NAS, Duncan Flt, Yeovilton	
ZF562	WS Lynx HMA8SRU [452/DT]	RN No 815 NAS, Dauntless Flt, Yeovilton	
ZF563	WS Lynx HMA8SRU [453/DM]	MoD/Vector Aerospace, stored Fleetlands	
ZF573	PBN 2T Islander CC2 (G-SRAY)	RAF Northolt Station Flight	
ZF579	BAC Lightning F53	Gatwick Aviation Museum, Charlwood	
ZF580	BAC Lightning F53	Classic Air Force, Newquay	
ZF581	BAC Lightning F53	Bentwaters Cold War Museum	
ZF582	BAC Lightning F53 <ff>	Bournemouth Aviation Museum	
ZF583	BAC Lightning F53	Solway Aviation Society, Carlisle	
ZF584	BAC Lightning F53	Dumfries & Galloway Avn Mus, Dumfries	
ZF587	BAC Lightning F53 <ff>	Lashenden Air Warfare Museum, Headcorn	
ZF588	BAC Lightning F53 [L]	East Midlands Airport Aeropark	
ZF590	BAC Lightning F53 <ff>	Privately owned, stored Bruntingthorpe	
ZF595	BAC Lightning T55 (fuselage)	Privately owned, Binbrook	
ZF596	BAC Lightning T55 <ff>	Lakes Lightnings, Spark Bridge, Cumbria	
ZF622	Piper PA-31 Navajo Chieftain 350 (N3548Y)	MoD/AFD/QinetiQ, Boscombe Down	
ZF641	EHI-101 [PP1]	SFDO, RNAS Culdrose	
ZF649	EHI-101 Merlin (A2714) [PP5]	DSMarE AESS, HMS Sultan, Gosport	
ZG101	EHI-101 (mock-up) [GB]	AgustaWestland, Yeovil	
ZG347	Northrop Chukar D2	Davidstow Airfield & Cornwall At War Museum	
ZG477	BAe Harrier GR9 $	RAF Museum, Cosford	
ZG478	BAe Harrier GR9 (fuselage)	Privately owned, Sproughton	
ZG509	BAe Harrier GR7 [80]	Privately owned, Sproughton	
ZG631	Northrop Chukar D2	Farnborough Air Sciences Trust, Farnborough	
ZG705	Panavia Tornado GR4A [118]	RAF No 31 Sqn, Marham	
ZG707	Panavia Tornado GR4A [EB-Z]	RAF TEF, Lossiemouth	
ZG709	Panavia Tornado GR4A [120]	RAF Leeming, RTP	
ZG712	Panavia Tornado GR4A [122]	Scrapped at Leeming, January 2014	
ZG713	Panavia Tornado GR4A [123]	Scrapped at Leeming	
ZG714	Panavia Tornado GR4A [124]	RAF Marham Wing	
ZG726	Panavia Tornado GR4A [125]	Scrapped at Leeming	
ZG727	Panavia Tornado GR4A [126]	RAF Leeming, RTP	
ZG729	Panavia Tornado GR4A [127]	RAF TEF, Lossiemouth	
ZG750	Panavia Tornado GR4 [128]	RAF No 31 Sqn, Marham	

Notes	Serial	Type (code/other identity)	Owner/operator, location or fate
	ZG752	Panavia Tornado GR4 [129]	RAF No 12 Sqn, Marham
	ZG754	Panavia Tornado GR4 [130]	*Scrapped at Leeming*
	ZG756	Panavia Tornado GR4 [131]	RAF Leeming, RTP
	ZG771	Panavia Tornado GR4 [133]	RAF No 15(R) Sqn, Lossiemouth
	ZG773	Panavia Tornado GR4	MoD/BAE Systems, Warton
	ZG775	Panavia Tornado GR4 [EB-Z]	RAF AWC/FJWOEU/No 41(R) Sqn, Coningsby
	ZG777	Panavia Tornado GR4 [EB-Q]	RAF CMU, Marham
	ZG779	Panavia Tornado GR4 [136]	RAF CMU stored, Marham
	ZG791	Panavia Tornado GR4 [137]	RAF CMU stored, Marham
	ZG794	Panavia Tornado GR4 [TN]	*Scrapped at Leeming, 2013*
	ZG816	WS61 Sea King HAS6 [014/L]	DSMarE, stored *HMS Sultan*, Gosport
	ZG817	WS61 Sea King HAS6 [702/PW]	DSMarE AESS, *HMS Sultan*, Gosport
	ZG818	WS61 Sea King HAS6 [707/PW]	DSMarE AESS, *HMS Sultan*, Gosport
	ZG819	WS61 Sea King HAS6 [265/N]	DSMarE AESS, *HMS Sultan*, Gosport
	ZG820	WS61 Sea King HC4 [I]	Privately owned, Hitchin, Herts (for scrapping)
	ZG821	WS61 Sea King HC4 [G]	RN No 845 NAS, Yeovilton
	ZG822	WS61 Sea King HC4 [WS]	Privately owned, Bruntingthorpe
	ZG844	PBN 2T Islander AL1 (G-BLNE)	AAC AM&SU, stored Shawbury
	ZG845	PBN 2T Islander AL1 (G-BLNT)	AAC No 651 Sqn/5 Regt, Aldergrove
	ZG846	PBN 2T Islander AL1 (G-BLNU)	AAC No 651 Sqn/5 Regt, Aldergrove
	ZG847	PBN 2T Islander AL1 (G-BLNV)	AAC AM&SU, stored Shawbury
	ZG848	PBN 2T Islander AL1 (G-BLNY)	AAC No 651 Sqn/5 Regt, Aldergrove
	ZG875	WS61 Sea King HAS6 [013] <ff>	Privately owned, Market Drayton, Shrops
	ZG875	WS61 Sea King HAS6 <rf>	Mayhem Paintball, Abridge, Essex
	ZG884	WS Lynx AH9A	MoD/AgustaWestland, Yeovil
	ZG885	WS Lynx AH9A	AAC No 657 Sqn, Odiham
	ZG886	WS Lynx AH9A	AAC No 9 Regt, Dishforth
	ZG887	WS Lynx AH9A	AAC No 9 Regt, Dishforth
	ZG888	WS Lynx AH9A	MoD/AgustaWestland, Yeovil
	ZG889	WS Lynx AH9A	AAC No 667 Sqn/7 Regt, Middle Wallop
	ZG914	WS Lynx AH9A	MoD/AgustaWestland, Yeovil
	ZG915	WS Lynx AH9A	AAC No 9 Regt, Dishforth
	ZG916	WS Lynx AH9A	MoD/AgustaWestland, Yeovil
	ZG917	WS Lynx AH9A	AAC No 9 Regt, Dishforth
	ZG918	WS Lynx AH9A	AAC No 657 Sqn, Odiham
	ZG919	WS Lynx AH9A	AAC No 9 Regt, Dishforth
	ZG920	WS Lynx AH9A	MoD/Vector Aerospace, Fleetlands
	ZG921	WS Lynx AH9A	MoD/Vector Aerospace, Fleetlands
	ZG922	WS Lynx AH9	Privately owned, Staverton, instructional use
	ZG923	WS Lynx AH9A	AAC No 657 Sqn, Odiham
	ZG969	Pilatus PC-9 (HB-HQE)	*Currently not known*
	ZG989	PBN 2T Islander ASTOR (G-DLRA)	AAC AM&SU, stored Shawbury
	ZG993	PBN 2T Islander AL1 (G-BOMD)	AAC AM&SU, stored Shawbury
	ZG994	PBN 2T Islander AL1 (G-BPLN) (fuselage)	Britten-Norman, stored Bembridge
	ZG995	PBN 2T Defender AL1 (G-SURV)	AAC No 651 Sqn/5 Regt, Aldergrove
	ZG996	PBN 2T Defender AL2 (G-BWPR)	AAC No 651 Sqn/5 Regt, Aldergrove
	ZG997	PBN 2T Defender AL2 (G-BWPV)	AAC No 651 Sqn/5 Regt, Aldergrove
	ZG998	PBN 2T Defender AL1 (G-BWPX)	AAC No 651 Sqn/5 Regt, Aldergrove
	ZH001	PBN 2T Defender AL2 (G-CEIO)	AAC No 651 Sqn/5 Regt, Aldergrove
	ZH002	PBN 2T Defender AL2 (G-CEIP)	AAC No 651 Sqn/5 Regt, Aldergrove
	ZH003	PBN 2T Defender AL2 (G-CEIR)	AAC No 651 Sqn/5 Regt, Aldergrove
	ZH004	PBN 2T Defender T3 (G-BWPO)	AAC No 651 Sqn/5 Regt, Aldergrove
	ZH005	PBN 2T Defender AL2 (G-CGVB)	AAC No 651 Sqn/5 Regt, Aldergrove
	ZH101	Boeing E-3D Sentry AEW1 [01]	RAF No 8 Sqn, Waddington
	ZH102	Boeing E-3D Sentry AEW1 [02]	RAF No 8 Sqn, Waddington
	ZH103	Boeing E-3D Sentry AEW1 [03]	RAF No 8 Sqn, Waddington
	ZH104	Boeing E-3D Sentry AEW1 [04]	RAF No 8 Sqn, Waddington
	ZH105	Boeing E-3D Sentry AEW1 [05]	RAF, stored Waddington
	ZH106	Boeing E-3D Sentry AEW1 [06]	RAF No 8 Sqn, Waddington
	ZH107	Boeing E-3D Sentry AEW1 [07]	RAF No 8 Sqn, Waddington
	ZH115	Grob G109B Vigilant T1 [TA]	RAF ACCGS/No 644 VGS, Syerston

Serial	Type (code/other identity)	Owner/operator, location or fate	Notes
ZH116	Grob G109B Vigilant T1 [TB]	RAF No 618 VGS, Odiham	
ZH117	Grob G109B Vigilant T1 [TC]	RAF CGMF, Syerston	
ZH118	Grob G109B Vigilant T1 [TD]	RAF No 664 VGS, Newtownards	
ZH119	Grob G109B Vigilant T1 [TE]	RAF No 631 VGS, Woodvale	
ZH120	Grob G109B Vigilant T1 [TF]	RAF No 635 VGS, Topcliffe	
ZH121	Grob G109B Vigilant T1 [TG]	RAF No 642 VGS, Linton-on-Ouse	
ZH122	Grob G109B Vigilant T1 [TH]	RAF CGMF, Syerston	
ZH123	Grob G109B Vigilant T1 [TJ]	RAF No 637 VGS, Little Rissington	
ZH124	Grob G109B Vigilant T1 [TK]	RAF No 631 VGS, Woodvale	
ZH125	Grob G109B Vigilant T1 [TL]	RAF ACCGS/No 644 VGS, Syerston	
ZH126	Grob G109B Vigilant T1 (D-KGRA) [TM]	RAF No 613 VGS, Halton	
ZH127	Grob G109B Vigilant T1 (D-KEEC) [TN]	RAF CGMF, Syerston	
ZH128	Grob G109B Vigilant T1 [TP]	RAF No 632 VGS, Ternhill	
ZH129	Grob G109B Vigilant T1 [TQ]	RAF CGMF, Syerston	
ZH139	BAe Harrier GR7 <R> (BAPC 191/*ZD472*)	RAF M&RU, Bottesford (wfu)	
ZH144	Grob G109B Vigilant T1 [TR]	RAF No 635 VGS, Topcliffe	
ZH145	Grob G109B Vigilant T1 [TS]	RAF No 624 VGS, Chivenor RMB	
ZH146	Grob G109B Vigilant T1 [TT]	RAF No 642 VGS, Linton-on-Ouse	
ZH147	Grob G109B Vigilant T1 [TU]	RAF No 613 VGS, Halton	
ZH148	Grob G109B Vigilant T1 [TV]	RAF No 645 VGS, Topcliffe	
ZH184	Grob G109B Vigilant T1 [TW]	RAF No 631 VGS, Woodvale	
ZH185	Grob G109B Vigilant T1 [TX]	RAF No 632 VGS, Ternhill	
ZH186	Grob G109B Vigilant T1 [TY]	RAF ACCGS/No 644 VGS, Syerston	
ZH187	Grob G109B Vigilant T1 [TZ]	RAF No 663 VGS, Kinloss	
ZH188	Grob G109B Vigilant T1 [UA]	RAF ACCGS/No 644 VGS, Syerston	
ZH189	Grob G109B Vigilant T1 [UB]	RAF No 636 VGS, Swansea	
ZH190	Grob G109B Vigilant T1 [UC]	RAF No 645 VGS, Topcliffe	
ZH191	Grob G109B Vigilant T1 [UD]	RAF No 612 VGS, Abingdon	
ZH192	Grob G109B Vigilant T1 [UE]	RAF No 633 VGS, Cosford	
ZH193	Grob G109B Vigilant T1 [UF]	RAF No 631 VGS, Woodvale	
ZH194	Grob G109B Vigilant T1 [UG]	RAF No 637 VGS, Little Rissington	
ZH195	Grob G109B Vigilant T1 [UH]	RAF No 624 VGS, Chivenor RMB	
ZH196	Grob G109B Vigilant T1 [UJJ]	RAF ACCGS/No 644 VGS, Syerston	
ZH197	Grob G109B Vigilant T1 [UK]	RAF No 618 VGS, Odiham	
ZH200	BAe Hawk 200	Loughborough University	
ZH205	Grob G109B Vigilant T1 [UL]	RAF No 642 VGS, Linton-on-Ouse	
ZH206	Grob G109B Vigilant T1 [UM]	RAF CGMF, Syerston	
ZH207	Grob G109B Vigilant T1 [UN]	RAF No 612 VGS, Abingdon	
ZH208	Grob G109B Vigilant T1 [UP]	RAF No 645 VGS, Topcliffe	
ZH209	Grob G109B Vigilant T1 [UQ]	RAF No 634 VGS, St Athan	
ZH211	Grob G109B Vigilant T1 [UR]	RAF No 636 VGS, Swansea	
ZH247	Grob G109B Vigilant T1 [US]	RAF No 616 VGS, Henlow	
ZH248	Grob G109B Vigilant T1 [UT]	RAF No 618 VGS, Odiham	
ZH249	Grob G109B Vigilant T1 [UU]	RAF CGMF, Syerston	
ZH257	B-V CH-47C Chinook (9217M) (fuselage)	RAF Odiham, BDRT	
ZH263	Grob G109B Vigilant T1 [UV]	RAF ACCGS/No 644 VGS, Syerston	
ZH264	Grob G109B Vigilant T1 [UW]	RAF CGMF, Syerston	
ZH265	Grob G109B Vigilant T1 [UX]	RAF No 618 VGS, Odiham	
ZH266	Grob G109B Vigilant T1 [UY]	RAF No 612 VGS, Abingdon	
ZH267	Grob G109B Vigilant T1 [UZ]	RAF No 635 VGS, Topcliffe	
ZH268	Grob G109B Vigilant T1 [SA]	RAF No 612 VGS, Abingdon	
ZH269	Grob G109B Vigilant T1 [SB]	RAF No 663 VGS, Kinloss	
ZH270	Grob G109B Vigilant T1 [SC]	RAF No 616 VGS, Henlow	
ZH271	Grob G109B Vigilant T1 [SD]	RAF No 637 VGS, Little Rissington	
ZH278	Grob G109B Vigilant T1 (D-KAIS) [SF]	RAF CGMF, Syerston	
ZH279	Grob G109B Vigilant T1 (D-KNPS) [SG]	RAF CGMF, Syerston	
ZH536	PBN 2T Islander CC2 (G-BSAH)	RAF Northolt Station Flight	
ZH537	PBN 2T Islander CC2 (G-SELX)	RAF Northolt Station Flight	
ZH540	WS61 Sea King HAR3A	RAF No 22 Sqn, A Flt, Chivenor RMB	
ZH541	WS61 Sea King HAR3A [V]	RAF No 22 Sqn, B Flt, Wattisham	
ZH542	WS61 Sea King HAR3A	RAF No 22 Sqn, B Flt, Wattisham	
ZH543	WS61 Sea King HAR3A [X]	RAF No 22 Sqn, B Flt, Wattisham	

Notes	Serial	Type (code/other identity)	Owner/operator, location or fate
	ZH544	WS61 Sea King HAR3A	RAF No 22 Sqn, A Flt, Chivenor RMB
	ZH545	WS61 Sea King HAR3A [Z]	RAF Valley (on repair)
	ZH552	Panavia Tornado F3 [HW]	*Scrapped at Leeming*
	ZH553	Panavia Tornado F3 [RT]	MoD DFTDC, Manston
	ZH554	Panavia Tornado F3 [HX,JU-C]	*Scrapped at Leeming*
	ZH588	Eurofighter Typhoon (DA2)	RAF Museum, Hendon
	ZH590	Eurofighter Typhoon (DA4)	Imperial War Museum, Duxford
	ZH654	BAe Harrier T10 <ff>	*Currently not known*
	ZH655	BAe Harrier T10 <ff>	Privately owned, Bentwaters
	ZH658	BAe Harrier T10 (fuselage)	Privately owned, Sproughton
	ZH763	BAC 1-11/539GL (G-BGKE)	Classic Air Force, Newquay
	ZH775	B-V Chinook HC4 (N7424J) [HB]	RAF Odiham Wing
	ZH776	B-V Chinook HC4 (N7424L) [HC]	RAF Odiham Wing
	ZH777	B-V Chinook HC4 (N7424M) [HE]	RAF Odiham Wing
	ZH796	BAe Sea Harrier FA2 [001/L]	DSAE No 1 SoTT, Cosford
	ZH797	BAe Sea Harrier FA2 [97/DD]	SFDO, RNAS Culdrose
	ZH798	BAe Sea Harrier FA2 [98/DD]	SFDO, RNAS Culdrose
	ZH799	BAe Sea Harrier FA2 [730]	Privately owned, Tunbridge Wells
	ZH800	BAe Sea Harrier FA2 (ZH801) [123]	RNAS Yeovilton, stored
	ZH801	BAe Sea Harrier FA2 (ZH800) [001]	RNAS Yeovilton, stored
	ZH802	BAe Sea Harrier FA2 [02/DD]	SFDO, RNAS Culdrose
	ZH803	BAe Sea Harrier FA2 [03/DD]	SFDO, RNAS Culdrose
	ZH804	BAe Sea Harrier FA2 [003/L]	RN, stored Culdrose
	ZH806	BAe Sea Harrier FA2 [007]	Privately owned, Bentwaters
	ZH807	BAe Sea Harrier FA2 <ff>	Privately owned, Thorpe Wood, N Yorks
	ZH810	BAe Sea Harrier FA2 [125]	Privately owned, Sproughton
	ZH811	BAe Sea Harrier FA2 [002/L]	RN, stored Culdrose
	ZH812	BAe Sea Harrier FA2 [005/L]	Privately owned, Bentwaters
	ZH813	BAe Sea Harrier FA2 [13/DD]	SFDO, RNAS Culdrose
	ZH814	Bell 212HP AH1 (G-BGMH)	AAC No 7 Flt, Brunei
	ZH815	Bell 212HP AH1 (G-BGCZ)	AAC No 7 Flt, Brunei
	ZH816	Bell 212HP AH1 (G-BGMG)	AAC No 7 Flt, Brunei
	ZH821	EHI-101 Merlin HM1	RN AM&SU, stored Shawbury
	ZH822	EHI-101 Merlin HM1	RN AM&SU, stored Shawbury
	ZH823	EHI-101 Merlin HM1	RN AM&SU, stored Shawbury
	ZH824	EHI-101 Merlin HM2 [14}	MoD/AgustaWestland, Yeovil (conversion)
	ZH825	EHI-101 Merlin HM1 [583]	RN AM&SU, stored Shawbury
	ZH826	EHI-101 Merlin HM2	MoD/AFD/QinetiQ, Boscombe Down
	ZH827	EHI-101 Merlin HM2 [15]	RN No 820 NAS, Culdrose
	ZH828	EHI-101 Merlin HM2	RN No 829 NAS, Culdrose
	ZH829	EHI-101 Merlin HM2 [CU]	MoD/AgustaWestland, Yeovil
	ZH830	EHI-101 Merlin HM1	RN No 829 NAS, Culdrose
	ZH831	EHI-101 Merlin HM2	MoD/AFD/QinetiQ, Boscombe Down
	ZH832	EHI-101 Merlin HM2 [85]	RN No 824 NAS, Culdrose
	ZH833	EHI-101 Merlin HM2	RN No 829 NAS, Culdrose
	ZH834	EHI-101 Merlin HM2 [84]	RN No 824 NAS, Culdrose
	ZH835	EHI-101 Merlin HM2	RN No 814 NAS, Culdrose
	ZH836	EHI-101 Merlin HM2	RN No 829 NAS, Culdrose
	ZH837	EHI-101 Merlin HM2 [503]	MoD/AFD/QinetiQ, Boscombe Down
	ZH838	EHI-101 Merlin HM1 [70]	RN No 814 NAS, Culdrose
	ZH839	EHI-101 Merlin HM2 [88]	MoD/AgustaWestland, Yeovil (conversion)
	ZH840	EHI-101 Merlin HM2 [99]	RN No 820 NAS, Culdrose
	ZH841	EHI-101 Merlin HM2	RN No 814 NAS, Culdrose
	ZH842	EHI-101 Merlin HM2 [86]	RN No 824 NAS, Culdrose
	ZH843	EHI-101 Merlin HM2	MoD/AFD/QineriQ, Boscombe Down
	ZH845	EHI-101 Merlin HM2 [81]	RN No 824 NAS, Culdrose
	ZH846	EHI-101 Merlin HM2 [13/CU]	MoD/AgustaWestland, Yeovil (conversion)
	ZH847	EHI-101 Merlin HM2 [12]	MoD/AgustaWestland, Yeovil (conversion)
	ZH848	EHI-101 Merlin HM1	RN No 829 NAS, Culdrose
	ZH849	EHI-101 Merlin HM1 [67]	RN No 814 NAS, Culdrose
	ZH850	EHI-101 Merlin HM2 [80]	RN No 820 NAS, Culdrose
	ZH851	EHI-101 Merlin HM2 [82]	RN No 824 NAS, Culdrose

Serial	Type (code/other identity)	Owner/operator, location or fate	Notes
ZH852	EHI-101 Merlin HM1(mod)	MoD/AgustaWestland, Yeovil	
ZH853	EHI-101 Merlin HM2 [83]	RN No 824 NAS, Culdrose	
ZH854	EHI-101 Merlin HM2 [84]	MoD/AgustaWestland, Yeovil (conversion)	
ZH855	EHI-101 Merlin HM1 [68]	RN No 814 NAS, Culdrose	
ZH856	EHI-101 Merlin HM2 [10]	RN No 820 NAS, Culdrose	
ZH857	EHI-101 Merlin HM2 [14]	RN No 829 NAS, Culdrose	
ZH858	EHI-101 Merlin HM1	RN No 829 NAS, Culdrose	
ZH860	EHI-101 Merlin HM2	RN No 820 NAS, Culdrose	
ZH861	EHI-101 Merlin HM2 [NL]	MoD/AgustaWestland, Yeovil (conversion)	
ZH862	EHI-101 Merlin HM2 [82]	RN No 824 NAS, Culdrose	
ZH863	EHI-101 Merlin HM1	RN stored, QinetiQ Boscombe Down	
ZH864	EHI-101 Merlin HM2 [12]	RN No 820 NAS, Culdrose	
ZH865	Lockheed C-130J-30 Hercules C4 (N130JA) [865]	RAF No 24 Sqn/No 30 Sqn/No 47 Sqn, Brize Norton	
ZH866	Lockheed C-130J-30 Hercules C6 (N130JE) [866]	MoD/AFD/QineriQ, Boscombe Down	
ZH867	Lockheed C-130J-30 Hercules C4 (N130JJ) [867]	RAF No 24 Sqn/No 30 Sqn/No 47 Sqn, Brize Norton	
ZH868	Lockheed C-130J-30 Hercules C4 (N130JN) [868]	RAF No 24 Sqn/No 30 Sqn/No 47 Sqn, Brize Norton	
ZH869	Lockheed C-130J-30 Hercules C4 (N130JV) [869]	RAF No 24 Sqn/No 30 Sqn/No 47 Sqn, Brize Norton	
ZH870	Lockheed C-130J-30 Hercules C4 (N78235) [870]	RAF No 24 Sqn/No 30 Sqn/No 47 Sqn, Brize Norton	
ZH871	Lockheed C-130J-30 Hercules C4 (N73238) [871]	RAF No 24 Sqn/No 30 Sqn/No 47 Sqn, Brize Norton	
ZH872	Lockheed C-130J-30 Hercules C4 (N4249Y) [872]	RAF No 24 Sqn/No 30 Sqn/No 47 Sqn, Brize Norton	
ZH873	Lockheed C-130J-30 Hercules C4 (N4242N) [873]	RAF No 24 Sqn/No 30 Sqn/No 47 Sqn, Brize Norton	
ZH874	Lockheed C-130J-30 Hercules C4 (N41030) [874]	RAF No 24 Sqn/No 30 Sqn/No 47 Sqn, Brize Norton	
ZH875	Lockheed C-130J-30 Hercules C4 (N4099R) [875]	RAF No 24 Sqn/No 30 Sqn/No 47 Sqn, Brize Norton	
ZH877	Lockheed C-130J-30 Hercules C4 (N4081M) [877]	MoD/AFD/QinetiQ, Boscombe Down	
ZH878	Lockheed C-130J-30 Hercules C4 (N73232) [878]	RAF No 24 Sqn/No 30 Sqn/No 47 Sqn, Brize Norton	
ZH879	Lockheed C-130J-30 Hercules C4 (N4080M) [879]	RAF No 24 Sqn/No 30 Sqn/No 47 Sqn, Brize Norton	
ZH880	Lockheed C-130J Hercules C5 (N73238) [880]	RAF No 24 Sqn/No 30 Sqn/No 47 Sqn, Brize Norton	
ZH881	Lockheed C-130J Hercules C5 (N4081M) [881]	RAF No 24 Sqn/No 30 Sqn/No 47 Sqn, Brize Norton	
ZH882	Lockheed C-130J Hercules C5 (N4099R) [882]	RAF No 24 Sqn/No 30 Sqn/No 47 Sqn, Brize Norton	
ZH883	Lockheed C-130J Hercules C5 (N4242N) [883]	RAF No 24 Sqn/No 30 Sqn/No 47 Sqn, Brize Norton	
ZH884	Lockheed C-130J Hercules C5 (N4249Y) [884]	RAF No 24 Sqn/No 30 Sqn/No 47 Sqn, Brize Norton	
ZH885	Lockheed C-130J Hercules C5 (N41030) [885]	RAF No 24 Sqn/No 30 Sqn/No 47 Sqn, Brize Norton	
ZH886	Lockheed C-130J Hercules C5 (N73235) [886]	RAF No 24 Sqn/No 30 Sqn/No 47 Sqn, Brize Norton	
ZH887	Lockheed C-130J Hercules C5 (N4187W) [887]	RAF No 1312 Flt, Mount Pleasant, FI	
ZH888	Lockheed C-130J Hercules C5 (N4187) [888]	RAF No 24 Sqn/No 30 Sqn/No 47 Sqn, Brize Norton	
ZH889	Lockheed C-130J Hercules C5 (N4099R) [889]	RAF No 24 Sqn/No 30 Sqn/No 47 Sqn, Brize Norton	
ZH890	Grob G109B Vigilant T1 [SE]	RAF No 613 VGS, Halton	
ZH891	B-V Chinook HC2A (N20075) [HF]	RAF Odiham Wing	
ZH892	B-V Chinook HC4 (N2019V) [HG]	MoD/Vector Aerospace, Fleetlands (conversion)	
ZH893	B-V Chinook HC2A (N2025L) [HH]	RAF Odiham Wing	
ZH894	B-V Chinook HC4 (N2026E) [HI]	RAF Odiham Wing	
ZH895	B-V Chinook HC2A (N2034K) [HJ] $	RAF Odiham Wing	
ZH896	B-V Chinook HC2A (N2038G) [HK]	RAF Odiham Wing	
ZH897	B-V Chinook HC3R (N2045G)	RAF Odiham Wing	
ZH898	B-V Chinook HC5 (N2057Q)	RAF Odiham Wing	
ZH899	B-V Chinook HC5 (N2057R)	MoD/Vector Aerospace, Fleetlands (conversion)	
ZH900	B-V Chinook HC5 (N2060H)	MoD/Vector Aerospace, Fleetlands (conversion)	
ZH901	B-V Chinook HC3R (N2060M)	RAF Odiham Wing	
ZH902	B-V Chinook HC3R (N2064W)	RAF Odiham Wing	
ZH903	B-V Chinook HC3R (N20671) [HR]	RAF Odiham Wing	
ZH904	B-V Chinook HC3R (N2083K)	RAF Odiham Wing	
ZJ100	BAe Hawk 102D	BAe Systems National Training Academy, Humberside	
ZJ117	EHI-101 Merlin HC3	MoD/AFD/QinetiQ, Boscombe Down	
ZJ118	EHI-101 Merlin HC3 [B]	RN MDMF, Culdrose	
ZJ119	EHI-101 Merlin HC3 [C]	RN MDMF, Culdrose	
ZJ120	EHI-101 Merlin HC3 [D]	RAF No 28 Sqn, Benson	
ZJ121	EHI-101 Merlin HC3 [E]	RAF No 28 Sqn/RN No 846 NAS, Benson	
ZJ122	EHI-101 Merlin HC3 [F]	RAF No 28 Sqn/RN No 846 NAS, Benson	
ZJ123	EHI-101 Merlin HC3 [G]	RAF No 28 Sqn/RN No 846 NAS, Benson	
ZJ124	EHI-101 Merlin HC3 [H]	RAF No 28 Sqn/RN No 846 NAS, Benson	
ZJ125	EHI-101 Merlin HC3 [J]	RAF No 28 Sqn/RN No 846 NAS, Benson	

Notes	Serial	Type (code/other identity)	Owner/operator, location or fate
	ZJ126	EHI-101 Merlin HC3 [K]	MoD/AgustaWestland, Yeovil
	ZJ127	EHI-101 Merlin HC3 [L]	RAF No 28 Sqn/RN No 846 NAS, Benson
	ZJ128	EHI-101 Merlin HC3 [M]	RAF No 28 Sqn/RN No 846 NAS, Benson
	ZJ129	EHI-101 Merlin HC3 [N]	RAF No 28 Sqn/RN No 846 NAS, Benson
	ZJ130	EHI-101 Merlin HC3 [O]	RAF No 28 Sqn/RN No 846 NAS, Benson
	ZJ131	EHI-101 Merlin HC3 [P]	RAF No 28 Sqn/RN No 846 NAS, Benson
	ZJ132	EHI-101 Merlin HC3 [Q]	RN MDMF, Culdrose
	ZJ133	EHI-101 Merlin HC3 [R]	RN MDMF, Culdrose
	ZJ134	EHI-101 Merlin HC3 [S]	RAF No 28 Sqn/RN No 846 NAS, Benson
	ZJ135	EHI-101 Merlin HC3 [T]	RAF No 28 Sqn/RN No 846 NAS, Benson
	ZJ136	EHI-101 Merlin HC3 [U]	RAF No 28 Sqn/RN No 846 NAS, Benson
	ZJ137	EHI-101 Merlin HC3 [W]	RN MDMF, Culdrose
	ZJ138	EHI-101 Merlin HC3 [X]	MoD, stored Boscombe Down
	ZJ164	AS365N-2 Dauphin 2 (G-BTLC)	RN/Bond Helicopters, Newquay
	ZJ165	AS365N-2 Dauphin 2 (G-NTOO)	RN/Bond Helicopters, Newquay
	ZJ166	WAH-64 Apache AH1 (N9219G)	AAC No 673 Sqn/7 Regt, Middle Wallop
	ZJ167	WAH-64 Apache AH1 (N3266B)	AAC No 3 Regt, Wattisham
	ZJ168	WAH-64 Apache AH1 (N3123T)	AAC No 4 Regt, Wattisham
	ZJ169	WAH-64 Apache AH1 (N3114H)	AAC No 673 Sqn/7 Regt, Middle Wallop
	ZJ170	WAH-64 Apache AH1 (N3065U)	AAC No 673 Sqn/7 Regt, Middle Wallop
	ZJ171	WAH-64 Apache AH1 (N3266T)	AAC No 3 Regt, Wattisham
	ZJ172	WAH-64 Apache AH1	AAC No 3 Regt, Wattisham
	ZJ173	WAH-64 Apache AH1 (N3266W)	AAC No 4 Regt, Wattisham
	ZJ174	WAH-64 Apache AH1	AAC No 3 Regt, Wattisham
	ZJ175	WAH-64 Apache AH1 (N3218V)	AAC No 673 Sqn/7 Regt, Middle Wallop
	ZJ176	WAH-64 Apache AH1	AAC No 3 Regt, Wattisham
	ZJ177	WAH-64 Apache AH1	AAC, stored Wattisham (damaged)
	ZJ178	WAH-64 Apache AH1	AAC No 4 Regt, Wattisham
	ZJ179	WAH-64 Apache AH1	AAC No 4 Regt, Wattisham
	ZJ180	WAH-64 Apache AH1	AAC No 673 Sqn/7 Regt, Middle Wallop
	ZJ181	WAH-64 Apache AH1	AAC No 4 Regt, Wattisham
	ZJ182	WAH-64 Apache AH1	AAC No 4 Regt, Wattisham
	ZJ183	WAH-64 Apache AH1	AAC No 4 Regt, Wattisham
	ZJ184	WAH-64 Apache AH1	AAC No 3 Regt, Wattisham
	ZJ185	WAH-64 Apache AH1	AAC No 673 Sqn/7 Regt, Middle Wallop
	ZJ186	WAH-64 Apache AH1	AAC No 4 Regt, Wattisham
	ZJ187	WAH-64 Apache AH1	AAC No 4 Regt, Wattisham
	ZJ188	WAH-64 Apache AH1	AAC No 4 Regt, Wattisham
	ZJ189	WAH-64 Apache AH1	AAC No 3 Regt, Wattisham
	ZJ190	WAH-64 Apache AH1	AAC No 4 Regt, Wattisham
	ZJ191	WAH-64 Apache AH1	AAC No 4 Regt, Wattisham
	ZJ192	WAH-64 Apache AH1	AAC No 3 Regt, Wattisham
	ZJ193	WAH-64 Apache AH1	AAC No 673 Sqn/7 Regt, Middle Wallop
	ZJ194	WAH-64 Apache AH1	AAC No 3 Regt, Wattisham
	ZJ195	WAH-64 Apache AH1	AAC No 3 Regt, Wattisham
	ZJ196	WAH-64 Apache AH1	AAC No 3 Regt, Wattisham
	ZJ197	WAH-64 Apache AH1	AAC No 4 Regt, Wattisham
	ZJ198	WAH-64 Apache AH1	AAC No 673 Sqn/7 Regt, Middle Wallop
	ZJ199	WAH-64 Apache AH1	AAC No 3 Regt, Wattisham
	ZJ200	WAH-64 Apache AH1	AAC No 3 Regt, Wattisham
	ZJ202	WAH-64 Apache AH1	AAC No 3 Regt, Wattisham
	ZJ203	WAH-64 Apache AH1	AAC No 4 Regt, Wattisham
	ZJ204	WAH-64 Apache AH1	AAC No 4 Regt, Wattisham
	ZJ205	WAH-64 Apache AH1	AAC No 4 Regt, Wattisham
	ZJ206	WAH-64 Apache AH1	AAC No 673 Sqn/7 Regt, Middle Wallop
	ZJ207	WAH-64 Apache AH1	AAC No 673 Sqn/7 Regt, Middle Wallop
	ZJ208	WAH-64 Apache AH1	AAC No 4 Regt, Wattisham
	ZJ209	WAH-64 Apache AH1	AAC No 3 Regt, Wattisham
	ZJ210	WAH-64 Apache AH1	AAC No 673 Sqn/7 Regt, Middle Wallop
	ZJ211	WAH-64 Apache AH1	AAC No 3 Regt, Wattisham
	ZJ212	WAH-64 Apache AH1	AAC No 673 Sqn/7 Regt, Middle Wallop
	ZJ213	WAH-64 Apache AH1	AAC No 673 Sqn/7 Regt, Middle Wallop

Serial	Type (code/other identity)	Owner/operator, location or fate	Notes
ZJ214	WAH-64 Apache AH1	AAC No 673 Sqn/7 Regt, Middle Wallop	
ZJ215	WAH-64 Apache AH1	AAC No 3 Regt, Wattisham	
ZJ216	WAH-64 Apache AH1	AAC No 4 Regt, Wattisham	
ZJ217	WAH-64 Apache AH1	AAC No 4 Regt, Wattisham	
ZJ218	WAH-64 Apache AH1	AAC No 4 Regt, Wattisham	
ZJ219	WAH-64 Apache AH1	AAC No 673 Sqn/7 Regt, Middle Wallop	
ZJ220	WAH-64 Apache AH1	AAC No 673 Sqn/7 Regt, Middle Wallop	
ZJ221	WAH-64 Apache AH1	AAC No 3 Regt, Wattisham	
ZJ222	WAH-64 Apache AH1	AAC No 3 Regt, Wattisham	
ZJ223	WAH-64 Apache AH1	AAC No 4 Regt, Wattisham	
ZJ224	WAH-64 Apache AH1	AAC No 4 Regt, Wattisham	
ZJ225	WAH-64 Apache AH1	AAC No 3 Regt, Wattisham	
ZJ226	WAH-64 Apache AH1	AAC No 4 Regt, Wattisham	
ZJ227	WAH-64 Apache AH1	AAC No 3 Regt, Wattisham	
ZJ228	WAH-64 Apache AH1	AAC No 3 Regt, Wattisham	
ZJ229	WAH-64 Apache AH1	AAC No 4 Regt, Wattisham	
ZJ230	WAH-64 Apache AH1	AAC No 3 Regt, Wattisham	
ZJ231	WAH-64 Apache AH1	AAC No 4 Regt, Wattisham	
ZJ232	WAH-64 Apache AH1	AAC No 3 Regt, Wattisham	
ZJ233	WAH-64 Apache AH1	AAC No 673 Sqn/7 Regt, Middle Wallop	
ZJ234	Bell 412EP Griffin HT1 (G-BWZR) [S]	DHFS No 60(R) Sqn, RAF Shawbury	
ZJ235	Bell 412EP Griffin HT1 (G-BXBF) [I]	DHFS No 60(R) Sqn/SARTU, RAF Valley	
ZJ236	Bell 412EP Griffin HT1 (G-BXBE) [X]	DHFS No 60(R) Sqn, RAF Shawbury	
ZJ237	Bell 412EP Griffin HT1 (G-BXFF) [T]	DHFS No 60(R) Sqn, RAF Shawbury	
ZJ238	Bell 412EP Griffin HT1 (G-BXHC) [Y]	DHFS No 60(R) Sqn, RAF Shawbury	
ZJ239	Bell 412EP Griffin HT1 (G-BXFH) [R]	DHFS No 60(R) Sqn/SARTU, RAF Valley	
ZJ240	Bell 412EP Griffin HT1 (G-BXIR) [U] $	DHFS No 60(R) Sqn, RAF Shawbury	
ZJ241	Bell 412EP Griffin HT1 (G-BXIS) [L]	DHFS No 60(R) Sqn/SARTU, RAF Valley	
ZJ242	Bell 412EP Griffin HT1 (G-BXDK) [E]	DHFS No 60(R) Sqn/SARTU, RAF Valley	
ZJ243	AS350BA Squirrel HT2 (G-BWZS) [43]	DHFS, RAF Shawbury	
ZJ244	AS350BA Squirrel HT2 (G-BXMD) [44]	School of Army Aviation/No 670 Sqn, Middle Wallop	
ZJ245	AS350BA Squirrel HT2 (G-BXME) [45]	School of Army Aviation/No 670 Sqn, Middle Wallop	
ZJ246	AS350BA Squirrel HT2 (G-BXMJ) [46]	School of Army Aviation/No 670 Sqn, Middle Wallop	
ZJ248	AS350BA Squirrel HT2 (G-BXNE) [48]	School of Army Aviation/No 670 Sqn, Middle Wallop	
ZJ249	AS350BA Squirrel HT2 (G-BXNJ) [49]	School of Army Aviation/No 670 Sqn, Middle Wallop	
ZJ250	AS350BA Squirrel HT2 (G-BXNY) [50]	School of Army Aviation/No 670 Sqn, Middle Wallop	
ZJ251	AS350BA Squirrel HT2 (G-BXOG) [51]	DHFS, RAF Shawbury	
ZJ252	AS350BA Squirrel HT2 (G-BXOK) [52]	School of Army Aviation/No 670 Sqn, Middle Wallop	
ZJ253	AS350BA Squirrel HT2 (G-BXPG) [53]	School of Army Aviation/No 670 Sqn, Middle Wallop	
ZJ254	AS350BA Squirrel HT2 (G-BXPJ) [54]	School of Army Aviation/No 670 Sqn, Middle Wallop	
ZJ255	AS350BB Squirrel HT1 (G-BXAG) [55]	DHFS, RAF Shawbury	
ZJ256	AS350BB Squirrel HT1 (G-BXCE) [56]	DHFS, RAF Shawbury	
ZJ257	AS350BB Squirrel HT1 (G-BXDJ) [57]	DHFS, RAF Shawbury	
ZJ260	AS350BB Squirrel HT1 (G-BXGB) [60]	DHFS, RAF Shawbury	
ZJ261	AS350BB Squirrel HT1 (G-BXGJ) [61]	DHFS, RAF Shawbury	
ZJ262	AS350BB Squirrel HT1 (G-BXHB) [62]	DHFS, RAF Shawbury	
ZJ264	AS350BB Squirrel HT1 (G-BXHW) [64]	DHFS, RAF Shawbury	
ZJ265	AS350BB Squirrel HT1 (G-BXHX) [65]	DHFS, RAF Shawbury	
ZJ266	AS350BB Squirrel HT1 (G-BXIL) [66]	DHFS, RAF Shawbury	
ZJ267	AS350BB Squirrel HT1 (G-BXIP) [67]	DHFS, RAF Shawbury	
ZJ268	AS350BB Squirrel HT1 (G-BXJE) [68]	DHFS, RAF Shawbury	
ZJ269	AS350BB Squirrel HT1 (G-BXJN) [69]	DHFS, RAF Shawbury	
ZJ270	AS350BB Squirrel HT1 (G-BXJR) [70]	DHFS, RAF Shawbury	
ZJ271	AS350BB Squirrel HT1 (G-BXKE) [71]	DHFS, RAF Shawbury	
ZJ272	AS350BB Squirrel HT1 (G-BXKN) [72]	DHFS, RAF Shawbury	
ZJ273	AS350BB Squirrel HT1 (G-BXKP) [73]	DHFS, RAF Shawbury	
ZJ274	AS350BB Squirrel HT1 (G-BXKR) [74]	DHFS, RAF Shawbury	
ZJ275	AS350BB Squirrel HT1 (G-BXLB) [75]	DHFS, RAF Shawbury	
ZJ276	AS350BB Squirrel HT1 (G-BXLE) [76]	DHFS, RAF Shawbury	
ZJ277	AS350BB Squirrel HT1 (G-BXLH) [77]	DHFS, RAF Shawbury	
ZJ278	AS350BB Squirrel HT1 (G-BXMB) [78]	DHFS, RAF Shawbury	
ZJ279	AS350BB Squirrel HT1 (G-BXMC) [79]	DHFS, RAF Shawbury	

Notes	Serial	Type (code/other identity)	Owner/operator, location or fate
	ZJ280	AS350BB Squirrel HT1 (G-BXMI) [80]	DHFS, RAF Shawbury
	ZJ369	GEC Phoenix RPAS	Defence Academy of the UK, Shrivenham
	ZJ385	GEC Phoenix RPAS	Muckleburgh Collection, Weybourne, Norfolk
	ZJ449	GEC Phoenix RPAS	REME Museum, Arborfield
	ZJ452	GEC Phoenix RPAS	Science Museum, Wroughton
	ZJ469	GEC Phoenix RPAS	Army, Larkhill, on display
	ZJ489	GAF Jindivik 104AL (A92-808)	Caernarfon Airworld
	ZJ493	GAF Jindivik 104AL (A92-814)	RAF Museum Reserve Collection, Stafford
	ZJ496	GAF Jindivik 104AL (A92-901)	Farnborough Air Sciences Trust, Farnborough
	ZJ515	BAE Systems Nimrod MRA4 (XV258) <ff>	Cranfield University, instructional use
	ZJ645	D-BD Alpha Jet (98+62) [45]	MoD/AFD/QinetiQ, Boscombe Down
	ZJ646	D-BD Alpha Jet (98+55) [46]	MoD/AFD/QinetiQ, Boscombe Down
	ZJ647	D-BD Alpha Jet (98+71) [47]	MoD/AFD/QinetiQ, Boscombe Down
	ZJ648	D-BD Alpha Jet (98+09) [48]	MoD/QinetiQ Boscombe Down, spares use
	ZJ649	D-BD Alpha Jet (98+73) [49]	MoD/ETPS, Boscombe Down
	ZJ650	D-BD Alpha Jet (98+35)	MoD/QinetiQ, stored Boscombe Down
	ZJ651	D-BD Alpha Jet (41+42) [51]	MoD/ETPS, Boscombe Down
	ZJ652	D-BD Alpha Jet (41+09)	MoD/QinetiQ Boscombe Down, spares use
	ZJ653	D-BD Alpha Jet (40+22)	MoD/QinetiQ Boscombe Down, spares use
	ZJ654	D-BD Alpha Jet (41+02)	MoD/QinetiQ Boscombe Down, spares use
	ZJ655	D-BD Alpha Jet (41+19)	MoD/QinetiQ Boscombe Down, spares use
	ZJ656	D-BD Alpha Jet (41+40)	MoD/QinetiQ Boscombe Down, spares use
	ZJ690	Bombardier Sentinel R1 (C-GJRG)	RAF No 5 Sqn, Waddington
	ZJ691	Bombardier Sentinel R1 (C-FZVM)	RAF No 5 Sqn, Waddington
	ZJ692	Bombardier Sentinel R1 (C-FZWW)	RAF No 5 Sqn, Waddington
	ZJ693	Bombardier Sentinel R1 (C-FZXC)	RAF No 5 Sqn, Waddington
	ZJ694	Bombardier Sentinel R1 (C-FZYL)	RAF No 5 Sqn, Waddington
	ZJ699	Eurofighter Typhoon (PT001)	MoD/BAE Systems, Warton
	ZJ700	Eurofighter Typhoon (PS002)	MoD/BAE Systems, Warton
	ZJ703	Bell 412EP Griffin HAR2 (G-CBST) [Spades,3]	RAF No 84 Sqn, Akrotiri
	ZJ704	Bell 412EP Griffin HAR2 (G-CBWT) [Clubs,4]	RAF/FBS, Shawbury
	ZJ705	Bell 412EP Griffin HAR2 (G-CBXL) [Hearts,5]	DHFS No 60(R) Sqn, RAF Shawbury
	ZJ706	Bell 412EP Griffin HAR2 (G-CBYR) [Diamonds,6]	RAF No 84 Sqn, Akrotiri
	ZJ707	Bell 412EP Griffin HT1 (G-CBUB) [O]	DHFS No 60(R) Sqn, RAF Shawbury
	ZJ708	Bell 412EP Griffin HT1 (G-CBVP) [K]	DHFS No 60(R) Sqn, RAF Shawbury
	ZJ780	AS365N-3 Dauphin AH1 (G-CEXT)	AAC No 658 Sqn, Credenhill
	ZJ781	AS365N-3 Dauphin AH1 (G-CEXU)	AAC No 658 Sqn, Credenhill
	ZJ782	AS365N-3 Dauphin AH1 (G-CEXV)	AAC No 658 Sqn, Credenhill
	ZJ783	AS365N-3 Dauphin AH1 (G-CEXW)	AAC No 671 Sqn/7 Regt, Middle Wallop
	ZJ785	AS365N-3 Dauphin AH1 (G-CFFW)	AAC No 658 Sqn, Credenhill
	ZJ787	AS365N-3 Dauphin AH1 (G-CHNJ)	AAC No 658 Sqn, Credenhill
	ZJ800	Eurofighter Typhoon T3 [BC]	RAF No 29(R) Sqn, Coningsby
	ZJ801	Eurofighter Typhoon T3 [BJ]	RAF No 29(R) Sqn, Coningsby
	ZJ802	Eurofighter Typhoon T3 [BP]	RAF No 29(R) Sqn, Coningsby
	ZJ803	Eurofighter Typhoon T3 [BA]	RAF No 29(R) Sqn, Coningsby
	ZJ804	Eurofighter Typhoon T3 [BM]	RAF No 29(R) Sqn, Coningsby
	ZJ805	Eurofighter Typhoon T3 [BD]	RAF No 29(R) Sqn, Coningsby
	ZJ806	Eurofighter Typhoon T3 [BE]	RAF No 29(R) Sqn, Coningsby
	ZJ807	Eurofighter Typhoon T3 [BF]	RAF No 29(R) Sqn, Coningsby
	ZJ808	Eurofighter Typhoon T3 [DW]	RAF No 11 Sqn, Coningsby
	ZJ809	Eurofighter Typhoon T3 [EY]	RAF No 6 Sqn, Lossiemouth
	ZJ810	Eurofighter Typhoon T3 [BI]	RAF No 29(R) Sqn, Coningsby
	ZJ811	Eurofighter Typhoon T3 [QO-B]	RAF No 3 Sqn, Coningsby
	ZJ812	Eurofighter Typhoon T3 [BK]	RAF No 29(R) Sqn, Coningsby
	ZJ813	Eurofighter Typhoon T3 [BL]	RAF No 29(R) Sqn, Coningsby
	ZJ814	Eurofighter Typhoon T3 [BH]	RAF No 29(R) Sqn, Coningsby
	ZJ815	Eurofighter Typhoon T3 [EB-H]	RAF AWC/FJWOEU/No 41(R) Sqn, Coningsby
	ZJ910	Eurofighter Typhoon FGR4 [BV]	RAF Coningsby, WLT
	ZJ911	Eurofighter Typhoon FGR4 [QO-Z]	RAF No 3 Sqn, Coningsby
	ZJ912	Eurofighter Typhoon FGR4 [DR]	RAF No 2 Sqn, Lossiemouth
	ZJ913	Eurofighter Typhoon FGR4 [QO-M]	RAF No 3 Sqn, Coningsby
	ZJ914	Eurofighter Typhoon FGR4 [QO-X]	RAF No 3 Sqn, Coningsby

Serial	Type (code/other identity)	Owner/operator, location or fate	Notes
ZJ915	Eurofighter Typhoon FGR4 [DP]	RAF No 11 Sqn, Coningsby	
ZJ916	Eurofighter Typhoon FGR4 [QO-S]	RAF No 3 Sqn, Coningsby	
ZJ917	Eurofighter Typhoon FGR4 [QO-G]	RAF No 3 Sqn, Coningsby	
ZJ918	Eurofighter Typhoon FGR4 [QO-L]	RAF No 3 Sqn, Coningsby	
ZJ919	Eurofighter Typhoon FGR4 [DC]	RAF No 11 Sqn, Coningsby	
ZJ920	Eurofighter Typhoon FGR4 [QO-A]	RAF No 3 Sqn, Coningsby	
ZJ921	Eurofighter Typhoon FGR4 [BY]	RAF No 29(R) Sqn, Coningsby	
ZJ922	Eurofighter Typhoon FGR4 [QO-C]	RAF, stored Coningsby	
ZJ923	Eurofighter Typhoon FGR4 [DM]	RAF No 11 Sqn, Coningsby	
ZJ924	Eurofighter Typhoon FGR4 [DD]	RAF No 11 Sqn, Coningsby	
ZJ925	Eurofighter Typhoon FGR4 [QO-R]	RAF No 3 Sqn, Coningsby	
ZJ926	Eurofighter Typhoon FGR4 [QO-Y]	RAF No 3 Sqn, Coningsby	
ZJ927	Eurofighter Typhoon FGR4 [BO]	RAF No 29(R) Sqn, Coningsby	
ZJ928	Eurofighter Typhoon FGR4 [FQ]	RAF No 1 Sqn, Lossiemouth	
ZJ929	Eurofighter Typhoon FGR4 [DL]	RAF No 11 Sqn, Coningsby	
ZJ930	Eurofighter Typhoon FGR4 [EB-R]	RAF AWC/FJWOEU/No 41(R) Sqn, Coningsby	
ZJ931	Eurofighter Typhoon FGR4 [DA]	RAF No 11 Sqn, Coningsby	
ZJ932	Eurofighter Typhoon FGR4 [DB]	RAF No 11 Sqn, Coningsby	
ZJ933	Eurofighter Typhoon FGR4 [DF]	RAF No 11 Sqn, Coningsby	
ZJ934	Eurofighter Typhoon FGR4 [QO-T]	RAF No 3 Sqn, Coningsby	
ZJ935	Eurofighter Typhoon FGR4 [DJ]	RAF No 11 Sqn, Coningsby	
ZJ936	Eurofighter Typhoon FGR4 [QO-C]	RAF No 6 Sqn, Lossiemouth	
ZJ937	Eurofighter Typhoon FGR4 [EG]	RAF No 6 Sqn, Lossiemouth	
ZJ938	Eurofighter Typhoon FGR4	MoD/BAE Systems, Warton	
ZJ939	Eurofighter Typhoon FGR4 [DXI]	RAF No 11 Sqn, Coningsby	
ZJ940	Eurofighter Typhoon FGR4	RAF, stored Coningsby	
ZJ941	Eurofighter Typhoon FGR4 [QO-J]	RAF No 3 Sqn, Coningsby	
ZJ942	Eurofighter Typhoon FGR4 [DH]	RAF No 11 Sqn, Coningsby	
ZJ943	Eurofighter Typhoon FGR4 [DK]	RAF, stored Coningsby (wreck)	
ZJ944	Eurofighter Typhoon FGR4 [F]	RAF No 1435 Flt, Mount Pleasant, FI	
ZJ945	Eurofighter Typhoon FGR4	RAF, stored Coningsby	
ZJ946	Eurofighter Typhoon FGR4 [EB-A]	RAF AWC/FJWOEU/No 41(R) Sqn, Coningsby	
ZJ947	Eurofighter Typhoon FGR4 [EB-L] $	RAF AWC/FJWOEU/No 41(R) Sqn, Coningsby	
ZJ948	Eurofighter Typhoon FGR4	RAF, stored Coningsby	
ZJ949	Eurofighter Typhoon FGR4 [H]	RAF No 1435 Flt, Mount Pleasant, FI	
ZJ950	Eurofighter Typhoon FGR4 [C]	RAF No 1435 Flt, Mount Pleasant, FI	
ZJ951	BAE Systems Hawk 120D	MoD/BAE Systems, Warton (for Loughborough University)	
ZJ954	SA330H Puma HC2 (SAAF 144)	RAF No 33 Sqn/No 230 Sqn, Benson	
ZJ955	SA330H Puma HC2 (SAAF 148) [P] $	RAF No 33 Sqn/No 230 Sqn, Benson	
ZJ956	SA330H Puma HC2 (SAAF 172/F-ZWCC)	RAF No 33 Sqn/No 230 Sqn, Benson	
ZJ957	SA330H Puma HC2 (SAAF 169)	RAF No 33 Sqn/No 230 Sqn, Benson	
ZJ958	SA330H Puma (SAAF 173)	DE&S, stored Bicester	
ZJ959	SA330H Puma (SAAF 184)	DE&S, stored Bicester	
ZJ960	Grob G109B Vigilant T1 (D-KSMU) [SH]	RAF ACCGS/No 644 VGS, Syerston	
ZJ961	Grob G109B Vigilant T1 (D-KLCW) [SJ]	RAF No 637 VGS, Little Rissington	
ZJ962	Grob G109B Vigilant T1 (D-KBEU) [SK]	RAF No 634 VGS, St Athan	
ZJ963	Grob G109B Vigilant T1 (D-KMSN) [SL]	RAF No 611 VGS, Honington	
ZJ964	Bell 212HP AH2 (G-BJGV) [D]	AAC No 25 Flt, Kenya	
ZJ966	Bell 212HP AH2 (G-BJJO) [C]	AAC No 25 Flt, Kenya	
ZJ967	Grob G109B Vigilant T1 (G-DEWS) [SM]	RAF CGMF, Syerston	
ZJ968	Grob G109B Vigilant T1 (N109BT) [SN]	RAF No 631 VGS, Woodvale	
ZJ969	Bell 212HP AH1 (G-BGLJ) [K]	AAC JHC/No 7 Regt, Middle Wallop	
ZJ990	EHI-101 Merlin HC3A (M-501) [AA]	RAF No 28 Sqn/RN No 846 NAS, Benson	
ZJ992	EHI-101 Merlin HC3A (M-503) [AB]	RAF MDMF, RNAS Culdrose	
ZJ994	EHI-101 Merlin HC3A (M-505) [AC]	RAF MDMF, RNAS Culdrose	
ZJ995	EHI-101 Merlin HC3A (M-506) [AD]	RAF MDMF, RNAS Culdrose	
ZJ998	EHI-101 Merlin HC3A (M-509) [AE]	RAF No 28 Sqn/RN No 846 NAS, Benson	
ZK001	EHI-101 Merlin HC3A (M-511) [AF]	RAF No 28 Sqn/RN No 846 NAS, Benson	
ZK005	Grob G109B Vigilant T1 (OH-797) [SP]	RAF No 645 VGS, Topcliffe	
ZK010	BAE Systems Hawk T2 [A]	RAF No 4 FTS/4(R) Sqn, Valley	
ZK011	BAE Systems Hawk T2 [B]	RAF No 4 FTS/4(R) Sqn, Valley	

Notes	Serial	Type (code/other identity)	Owner/operator, location or fate
	ZK012	BAE Systems Hawk T2 [C]	RAF No 4 FTS/4(R) Sqn, Valley
	ZK013	BAE Systems Hawk T2 [D]	RAF No 4 FTS/4(R) Sqn, Valley
	ZK014	BAE Systems Hawk T2 [E]	RAF No 4 FTS/4(R) Sqn, Valley
	ZK015	BAE Systems Hawk T2 [F]	RAF No 4 FTS/4(R) Sqn, Valley
	ZK016	BAE Systems Hawk T2 [G]	RAF No 4 FTS/4(R) Sqn, Valley
	ZK017	BAE Systems Hawk T2 [H]	RAF No 4 FTS/4(R) Sqn, Valley
	ZK018	BAE Systems Hawk T2 [I] $	RAF No 4 FTS/4(R) Sqn, Valley
	ZK019	BAE Systems Hawk T2 [J]	RAF No 4 FTS/4(R) Sqn, Valley
	ZK020	BAE Systems Hawk T2 [K] $	RAF No 4 FTS/4(R) Sqn, Valley
	ZK021	BAE Systems Hawk T2 [L]	RAF No 4 FTS/4(R) Sqn, Valley
	ZK022	BAE Systems Hawk T2 [M]	RAF No 4 FTS/4(R) Sqn, Valley
	ZK023	BAE Systems Hawk T2 [N]	RAF No 4 FTS/4(R) Sqn, Valley
	ZK024	BAE Systems Hawk T2 [O]	RAF No 4 FTS/4(R) Sqn, Valley
	ZK025	BAE Systems Hawk T2 [P]	RAF No 4 FTS/4(R) Sqn, Valley
	ZK026	BAE Systems Hawk T2 [Q]	RAF No 4 FTS/4(R) Sqn, Valley
	ZK027	BAE Systems Hawk T2 [R]	RAF No 4 FTS/4(R) Sqn, Valley
	ZK028	BAE Systems Hawk T2 [S]	RAF No 4 FTS/4(R) Sqn, Valley
	ZK029	BAE Systems Hawk T2 [T]	RAF No 4 FTS/4(R) Sqn, Valley
	ZK030	BAE Systems Hawk T2 [U]	RAF No 4 FTS/4(R) Sqn, Valley
	ZK031	BAE Systems Hawk T2 [V]	RAF No 4 FTS/4(R) Sqn, Valley
	ZK032	BAE Systems Hawk T2 [W]	RAF No 4 FTS/4(R) Sqn, Valley
	ZK033	BAE Systems Hawk T2 [X]	RAF No 4 FTS/4(R) Sqn, Valley
	ZK034	BAE Systems Hawk T2 [Y]	RAF No 4 FTS/4(R) Sqn, Valley
	ZK035	BAE Systems Hawk T2 [Z]	RAF No 4 FTS/4(R) Sqn, Valley
	ZK036	BAE Systems Hawk T2 [AA]	RAF No 4 FTS/4(R) Sqn, Valley
	ZK037	BAE Systems Hawk T2 [AB]	RAF No 4 FTS/4(R) Sqn, Valley
	ZK067	Bell 212HP AH3 (G-BFER) [B]	AAC JHC/No 7 Regt, Middle Wallop
	ZK113	Panavia Tornado IDS (RSAF 6606)	*To R Saudi AF as 6606, 15 December 2014*
	ZK114	M2370 RPAS	QinetiQ
	ZK150*	Lockheed Martin Desert Hawk 3 RPAS	Army 47 Regt Royal Artillery, Thorney Island
	ZK150	Lockheed Martin Desert Hawk 3 RPAS (ZK150/617)	Imperial War Museum, Lambeth
	ZK155*	Honeywell T-Hawk RPAS	Army 32 Regt Royal Artillery, Larkhill
	ZK191	AgustaWestland Super Lynx Mk.140	AgustaWestland, for Algeria
	ZK192	AgustaWestland Super Lynx Mk.140	AgustaWestland, for Algeria
	ZK193	AgustaWestland Super Lynx Mk.140	AgustaWestland, for Algeria
	ZK194	AgustaWestland Super Lynx Mk.140	AgustaWestland, for Algeria
	ZK195	AgustaWestland Super Lynx Mk.140	AgustaWestland, for Algeria
	ZK196	AgustaWestland Super Lynx Mk.140	AgustaWestland, for Algeria
	ZK205	Grob G109B Vigilant T1 (D-KBRU) [SS]	RAF No 624 VGS, Chivenor RMB
	ZK206	Bell 212EP AH2 (G-CFXE) [A]	AAC No 25 Flt, Kenya
	ZK210	BAE Systems Mantis RPAS	MoD/BAE Systems, Warton
	ZK300	Eurofighter Typhoon FGR4 [DG]	RAF No 11 Sqn, Coningsby
	ZK301	Eurofighter Typhoon FGR4 [D]	RAF No 1435 Flt, Mount Pleasant, FI
	ZK302	Eurofighter Typhoon FGR4 [EC]	RAF No 6 Sqn, Lossiemouth
	ZK303	Eurofighter Typhoon T3 [AX]	MoD/BAE Systems, Warton
	ZK304	Eurofighter Typhoon FGR4 [FM]	RAF No 1 Sqn, Lossiemouth
	ZK305	Eurofighter Typhoon FGR4 [DE]	RAF No 1 Sqn, Lossiemouth
	ZK306	Eurofighter Typhoon FGR4 [BT]	RAF No 29(R) Sqn, Coningsby
	ZK307	Eurofighter Typhoon FGR4 [O]	RAF No 2 Sqn, Lossiemouth
	ZK308	Eurofighter Typhoon FGR4 [TP-V] $	RAF No 6 Sqn, Lossiemouth
	ZK309	Eurofighter Typhoon FGR4 [QO-P]	RAF No 3 Sqn, Coningsby
	ZK310	Eurofighter Typhoon FGR4 [FL]	RAF No 1 Sqn, Lossiemouth
	ZK311	Eurofighter Typhoon FGR4 [EK]	RAF No 2 Sqn, Lossiemouth
	ZK312	Eurofighter Typhoon FGR4 [EM]	RAF No 6 Sqn, Lossiemouth
	ZK313	Eurofighter Typhoon FGR4 [W]	RAF No 2 Sqn, Lossiemouth
	ZK314	Eurofighter Typhoon FGR4 [EO]	RAF No 6 Sqn, Lossiemouth
	ZK315	Eurofighter Typhoon FGR4	RAF TMU, Coningsby
	ZK316	Eurofighter Typhoon FGR4 [FA]	RAF No 1 Sqn, Lossiemouth
	ZK317	Eurofighter Typhoon FGR4 [ES]	RAF No 6 Sqn, Lossiemouth
	ZK318	Eurofighter Typhoon FGR4 [ET]	RAF No 6 Sqn, Lossiemouth
	ZK319	Eurofighter Typhoon FGR4 [QO-D]	RAF No 3 Sqn, Coningsby
	ZK320	Eurofighter Typhoon FGR4 [EV]	RAF No 29(R) Sqn, Coningsby

Serial	Type (code/other identity)	Owner/operator, location or fate	Notes
ZK321	Eurofighter Typhoon FGR4 [EU]	RAF No 6 Sqn, Lossiemouth	
ZK322	Eurofighter Typhoon FGR4 [BR]	RAF No 6 Sqn, Lossiemouth	
ZK323	Eurofighter Typhoon FGR4 [DN]	RAF No 6 Sqn, Lossiemouth	
ZK324	Eurofighter Typhoon FGR4 [EI]	RAF No 6 Sqn, Lossiemouth	
ZK325	Eurofighter Typhoon FGR4 [FK]	RAF No 1 Sqn, Lossiemouth	
ZK326	Eurofighter Typhoon FGR4 [FB]	RAF, stored Coningsby	
ZK327	Eurofighter Typhoon FGR4 [FR]	RAF No 1 Sqn, Lossiemouth	
ZK328	Eurofighter Typhoon FGR4 [BS]	RAF No 6 Sqn, Lossiemouth	
ZK329	Eurofighter Typhoon FGR4 [FH]	RAF No 1 Sqn, Lossiemouth	
ZK330	Eurofighter Typhoon FGR4 [FT]	RAF No 1 Sqn, Lossiemouth	
ZK331	Eurofighter Typhoon FGR4 [FE]	RAF No 1 Sqn, Lossiemouth	
ZK332	Eurofighter Typhoon FGR4 [EB-J]	RAF AWC/FJWOEU/No 41(R) Sqn, Coningsby	
ZK333	Eurofighter Typhoon FGR4 [FS]	RAF No 1 Sqn, Lossiemouth	
ZK334	Eurofighter Typhoon FGR4 [FB]	RAF No 1 Sqn, Lossiemouth	
ZK335	Eurofighter Typhoon FGR4 [FC]	RAF No 1 Sqn, Lossiemouth	
ZK336	Eurofighter Typhoon FGR4 [FD]	RAF No 1 Sqn, Lossiemouth	
ZK337	Eurofighter Typhoon FGR4 [FP]	RAF No 1 Sqn, Lossiemouth	
ZK338	Eurofighter Typhoon FGR4 [FF]	RAF No 1 Sqn, Lossiemouth	
ZK339	Eurofighter Typhoon FGR4 [EB-B]	RAF AWC/FJWOEU/No 41(R) Sqn, Coningsby	
ZK340	Eurofighter Typhoon FGR4 [FI]	RAF No 1 Sqn, Lossiemouth	
ZK341	Eurofighter Typhoon FGR4 [FJ]	RAF No 1 Sqn, Lossiemouth	
ZK342	Eurofighter Typhoon FGR4 [ED] $	RAF No 6 Sqn, Lossiemouth	
ZK343	Eurofighter Typhoon FGR4 [BX] $	RAF No 29(R) Sqn, Coningsby	
ZK344	Eurofighter Typhoon FGR4 [II]	RAF No 2 Sqn, Lossiemouth	
ZK345	Eurofighter Typhoon FGR4 [EP]	RAF No 6 Sqn, Lossiemouth	
ZK346	Eurofighter Typhoon FGR4 [ER]	RAF No 6 Sqn, Lossiemouth	
ZK347	Eurofighter Typhoon FGR4 [EF]	RAF No 6 Sqn, Lossiemouth	
ZK348	Eurofighter Typhoon FGR4 [FN]	RAF No 1 Sqn, Lossiemouth	
ZK349	Eurofighter Typhoon FGR4 [BZ]	RAF No 29(R) Sqn, Coningsby	
ZK350	Eurofighter Typhoon FGR4	MoD/BAE Systems, Warton	
ZK351	Eurofighter Typhoon FGR4	RAF TMU, Coningsby	
ZK352	Eurofighter Typhoon FGR4	RAF TMU, Coningsby	
ZK353	Eurofighter Typhoon FGR4 [BQ]	RAF No 29(R) Sqn, Coningsby	
ZK354	Eurofighter Typhoon FGR4	RAF TMU, Coningsby	
ZK355	Eurofighter Typhoon FGR4	MoD/BAE Systems, Warton	
ZK356	Eurofighter Typhoon FGR4	MoD/BAE Systems, Warton	
ZK357	Eurofighter Typhoon FGR4	MoD/BAE Systems, Warton	
ZK358	Eurofighter Typhoon FGR4	MoD/BAE Systems, Warton	
ZK359	Eurofighter Typhoon FGR4	MoD/BAE Systems, Warton	
ZK360	Eurofighter Typhoon FGR4	MoD/BAE Systems, Warton	
ZK361	Eurofighter Typhoon FGR4	MoD/BAE Systems, Warton	
ZK362	Eurofighter Typhoon FGR4	MoD/BAE Systems, Warton	
ZK363	Eurofighter Typhoon FGR4	MoD/BAE Systems, Warton	
ZK364	Eurofighter Typhoon FGR4	MoD/BAE Systems, Warton	
ZK365	Eurofighter Typhoon FGR4	DE&S/BAE Systems, Warton, for RAF	
ZK366	Eurofighter Typhoon FGR4	MoD/BAE Systems, Warton	
ZK367	Eurofighter Typhoon FGR4	MoD/BAE Systems, Warton	
ZK368	Eurofighter Typhoon FGR4	MoD/BAE Systems, Warton	
ZK369	Eurofighter Typhoon FGR4	DE&S/BAE Systems, Warton, for RAF	
ZK370	Eurofighter Typhoon FGR4	DE&S/BAE Systems, Warton, for RAF	
ZK371	Eurofighter Typhoon FGR4	DE&S/BAE Systems, Warton, for RAF	
ZK372	Eurofighter Typhoon FGR4	DE&S/BAE Systems, Warton, for RAF	
ZK373	Eurofighter Typhoon FGR4	DE&S/BAE Systems, Warton, for RAF	
ZK374	Eurofighter Typhoon FGR4	DE&S/BAE Systems, Warton, for RAF	
ZK375	Eurofighter Typhoon FGR4	DE&S/BAE Systems, Warton, for RAF	
ZK376	Eurofighter Typhoon FGR4	DE&S/BAE Systems, Warton, for RAF	
ZK377	Eurofighter Typhoon FGR4	DE&S/BAE Systems, Warton, for RAF	
ZK378	Eurofighter Typhoon FGR4	DE&S/BAE Systems, Warton, for RAF	
ZK379	Eurofighter Typhoon T3 [BB]	RAF No 29(R) Sqn, Coningsby	
ZK380	Eurofighter Typhoon T3 [BG]	RAF No 2 Sqn, Lossiemouth	
ZK381	Eurofighter Typhoon T3 [EX]	RAF No 6 Sqn, Lossiemouth	
ZK382	Eurofighter Typhoon T3 [FX]	RAF No 1 Sqn, Lossiemouth	

Notes	Serial	Type (code/other identity)	Owner/operator, location or fate
	ZK383	Eurofighter Typhoon T3 [BN]	RAF No 29(R) Sqn, Coningsby
	ZK385	Eurofighter Typhoon	*To R Saudi AF as 1012, 27 February 2014*
	ZK386	Eurofighter Typhoon	BAe Systems, for R Saudi AF as 1013
	ZK387	Eurofighter Typhoon T	*To R Saudi AF as 1010, 27 February 2014*
	ZK389	Eurofighter Typhoon	*To R Saudi AF as 1014, 16 May 2014*
	ZK390	Eurofighter Typhoon	*To R Saudi AF as 1015, 16 May 2014*
	ZK391	Eurofighter Typhoon	*To R Saudi AF as 1018, 7 August 2014*
	ZK392	Eurofighter Typhoon	*To R Saudi AF as 1019, 5 September 2014*
	ZK393	Eurofighter Typhoon	BAE Systems, for R Saudi AF as 1022
	ZK394	Eurofighter Typhoon	*To R Saudi AF as 1023, 8 December 2014*
	ZK395	Eurofighter Typhoon	BAE Systems, for R Saudi AF as 1024
	ZK396	Eurofighter Typhoon T	*To R Saudi AF as 1016, 7 August 2014*
	ZK397	Eurofighter Typhoon T	*To R Saudi AF as 1017, 5 September 2014*
	ZK398	Eurofighter Typhoon T	*To R Saudi AF as 1020, 21 December 2014*
	ZK399	Eurofighter Typhoon T	*To R Saudi AF as 1021, 8 December 2014*
	ZK450	Beech King Air B200 (G-RAFJ) [J]	*To G-RAFJ, 26 March 2014*
	ZK451	Beech King Air B200 (G-RAFK) [K]	SERCO/RAF No 3 FTS/45(R) Sqn, Cranwell
	ZK452	Beech King Air B200 (G-RAFL) [L]	SERCO/RAF No 3 FTS/45(R) Sqn, Cranwell
	ZK453	Beech King Air B200 (G-RAFM) [M]	*To G-RAFM, 26 March 2014*
	ZK454	Beech King Air B200 (G-RAFN) [N]	*To G-RAFN, 13 June 2014*
	ZK455	Beech King Air B200 (G-RAFO) [O]	SERCO/RAF No 3 FTS/45(R) Sqn, Cranwell
	ZK456	Beech King Air B200 (G-RAFP) [P]	SERCO/RAF No 3 FTS/45(R) Sqn, Cranwell
	ZK457	Beech King Air B200 (G-ROWN)	*To G-ROWN, 7 March 2014*
	ZK458	Hawker Beechcraft King Air B200GT (G-RAFD) [D]	SERCO/RAF No 3 FTS/45(R) Sqn, Cranwell
	ZK459	Hawker Beechcraft King Air B200GT (G-RAFX) [X]	SERCO/RAF No 3 FTS/45(R) Sqn, Cranwell
	ZK460	Hawker Beechcraft King Air B200GT (G-RAFU) [U]	SERCO/RAF No 3 FTS/45(R) Sqn, Cranwell
	ZK501	Elbit Hermes 450 RPAS	Army 32 Regt Royal Artillery, Larkhill
	ZK502	Elbit Hermes 450 RPAS	Army 32 Regt Royal Artillery, Larkhill
	ZK503	Elbit Hermes 450 RPAS	Army 32 Regt Royal Artillery, Larkhill
	ZK504	Elbit Hermes 450 RPAS	Army 32 Regt Royal Artillery, Larkhill
	ZK505	Elbit Hermes 450 RPAS	Army 32 Regt Royal Artillery, Larkhill
	ZK506	Elbit Hermes 450 RPAS	Army 43 Battery Royal Artillery, Boscombe Down
	ZK507	Elbit Hermes 450 RPAS	Army 32 Regt Royal Artillery, Larkhill
	ZK508	Elbit Hermes 450 RPAS	Army 32 Regt Royal Artillery, Larkhill
	ZK509	Elbit Hermes 450 RPAS	Army 32 Regt Royal Artillery, Larkhill
	ZK510	Elbit Hermes 450 RPAS	Army 32 Regt Royal Artillery, Larkhill
	ZK511	Elbit Hermes 450 RPAS	Army 32 Regt Royal Artillery, Larkhill
	ZK512	Elbit Hermes 450 RPAS	Army 32 Regt Royal Artillery, Larkhill
	ZK513	Elbit Hermes 450 RPAS	Army 32 Regt Royal Artillery, Larkhill
	ZK514	Elbit Hermes 450 RPAS	Army 32 Regt Royal Artillery, Larkhill
	ZK515	Elbit Hermes 450 RPAS	Army 32 Regt Royal Artillery, Larkhill
	ZK516	Elbit Hermes 450 RPAS	Army 32 Regt Royal Artillery, Larkhill
	ZK517	Elbit Hermes 450 RPAS	Army 32 Regt Royal Artillery, Larkhill
	ZK518	Elbit Hermes 450 RPAS	Army 32 Regt Royal Artillery, Larkhill
	ZK519	Elbit Hermes 450 RPAS	Army 32 Regt Royal Artillery, Larkhill
	ZK520	Elbit Hermes 450 RPAS	Army 32 Regt Royal Artillery, Larkhill
	ZK521	Elbit Hermes 450 RPAS	Army 32 Regt Royal Artillery, Larkhill
	ZK522	Elbit Hermes 450 RPAS	Army 32 Regt Royal Artillery, Larkhill
	ZK523	Elbit Hermes 450 RPAS	Army 32 Regt Royal Artillery, Larkhill
	ZK524	Elbit Hermes 450 RPAS	Army 32 Regt Royal Artillery, Larkhill
	ZK525	Elbit Hermes 450 RPAS	Army 32 Regt Royal Artillery, Larkhill
	ZK531	BAe Hawk T53 (LL-5306)	MoD/BAE Systems, Warton
	ZK532	BAe Hawk T53 (LL-5315)	MoD/BAE Systems, Warton
	ZK533	BAe Hawk T53 (LL-5317)	BAE Systems National Training Academy, Humberside
	ZK534	BAe Hawk T53 (LL-5319)	Privately owned, Bentwaters
	ZK535	BAe Hawk T53 (LL-5320)	BAE Systems National Training Academy, Humberside
	ZK550	Boeing Chinook HC6 (N701UK)	RAF Odiham Wing
	ZK551	Boeing Chinook HC6 (N702UK)	RAF Odiham Wing
	ZK552	Boeing Chinook HC6 (N703UK)	RAF Odiham Wing
	ZK553	Boeing Chinook HC6 (N700UK)	RAF Odiham Wing
	ZK554	Boeing Chinook HC6 (N705UK)	RAF Odiham Wing
	ZK555	Boeing Chinook HC6 (N706UK)	RAF Odiham Wing

Serial	Type (code/other identity)	Owner/operator, location or fate	Notes
ZK556	Boeing Chinook HC6 (N707UK)	Boeing, for RAF	
ZK557	Boeing Chinook HC6 (N708UK)	Boeing, for RAF	
ZK558	Boeing Chinook HC6 (N709UK)	Boeing, for RAF	
ZK559	Boeing Chinook HC6 (N710UK)	Boeing, for RAF	
ZK560	Boeing Chinook HC6 (N711UK)	Boeing, for RAF	
ZK561	Boeing Chinook HC6 (N712UK)	Boeing, for RAF	
ZK562	Boeing Chinook HC6 (N713UK)	Boeing, for RAF	
ZK563	Boeing Chinook HC6 (N714UK)	Boeing, for RAF	
ZK600	Eurofighter Typhoon	BAE Systems, for R Saudi AF	
ZM135	Lockheed Martin F-35B Lightning II (BK-1)	RAF No 17(R) Sqn, Edwards AFB, California	
ZM136	Lockheed Martin F-35B Lightning II (BK-2)	RAF No 17(R) Sqn, Edwards AFB, California	
ZM137	Lockheed Martin F-35B Lightning II (BK-3)	RAF/VMFAT-501, MCAS Beaufort, S Carolina	
ZM138	Lockheed Martin F-35B Lightning II	Reservation for RAF/RN	
ZM139	Lockheed Martin F-35B Lightning II	Reservation for RAF/RN	
ZM140	Lockheed Martin F-35B Lightning II	Reservation for RAF/RN	
ZM141	Lockheed Martin F-35B Lightning II	Reservation for RAF/RN	
ZM142	Lockheed Martin F-35B Lightning II	Reservation for RAF/RN	
ZM143	Lockheed Martin F-35B Lightning II	Reservation for RAF/RN	
ZM144	Lockheed Martin F-35B Lightning II	Reservation for RAF/RN	
ZM145	Lockheed Martin F-35B Lightning II	Reservation for RAF/RN	
ZM146	Lockheed Martin F-35B Lightning II	Reservation for RAF/RN	
ZM147	Lockheed Martin F-35B Lightning II	Reservation for RAF/RN	
ZM148	Lockheed Martin F-35B Lightning II	Reservation for RAF/RN	
ZM149	Lockheed Martin F-35B Lightning II	Reservation for RAF/RN	
ZM150	Lockheed Martin F-35B Lightning II	Reservation for RAF/RN	
ZM151	Lockheed Martin F-35B Lightning II	Reservation for RAF/RN	
ZM152	Lockheed Martin F-35B Lightning II	Reservation for RAF/RN	
ZM153	Lockheed Martin F-35B Lightning II	Reservation for RAF/RN	
ZM154	Lockheed Martin F-35B Lightning II	Reservation for RAF/RN	
ZM155	Lockheed Martin F-35B Lightning II	Reservation for RAF/RN	
ZM156	Lockheed Martin F-35B Lightning II	Reservation for RAF/RN	
ZM157	Lockheed Martin F-35B Lightning II	Reservation for RAF/RN	
ZM158	Lockheed Martin F-35B Lightning II	Reservation for RAF/RN	
ZM159	Lockheed Martin F-35B Lightning II	Reservation for RAF/RN	
ZM160	Lockheed Martin F-35B Lightning II	Reservation for RAF/RN	
ZM161	Lockheed Martin F-35B Lightning II	Reservation for RAF/RN	
ZM162	Lockheed Martin F-35B Lightning II	Reservation for RAF/RN	
ZM163	Lockheed Martin F-35B Lightning II	Reservation for RAF/RN	
ZM164	Lockheed Martin F-35B Lightning II	Reservation for RAF/RN	
ZM165	Lockheed Martin F-35B Lightning II	Reservation for RAF/RN	
ZM166	Lockheed Martin F-35B Lightning II	Reservation for RAF/RN	
ZM167	Lockheed Martin F-35B Lightning II	Reservation for RAF/RN	
ZM168	Lockheed Martin F-35B Lightning II	Reservation for RAF/RN	
ZM169	Lockheed Martin F-35B Lightning II	Reservation for RAF/RN	
ZM170	Lockheed Martin F-35B Lightning II	Reservation for RAF/RN	
ZM171	Lockheed Martin F-35B Lightning II	Reservation for RAF/RN	
ZM172	Lockheed Martin F-35B Lightning II	Reservation for RAF/RN	
ZM173	Lockheed Martin F-35B Lightning II	Reservation for RAF/RN	
ZM174	Lockheed Martin F-35B Lightning II	Reservation for RAF/RN	
ZM175	Lockheed Martin F-35B Lightning II	Reservation for RAF/RN	
ZM176	Lockheed Martin F-35B Lightning II	Reservation for RAF/RN	
ZM177	Lockheed Martin F-35B Lightning II	Reservation for RAF/RN	
ZM178	Lockheed Martin F-35B Lightning II	Reservation for RAF/RN	
ZM179	Lockheed Martin F-35B Lightning II	Reservation for RAF/RN	
ZM180	Lockheed Martin F-35B Lightning II	Reservation for RAF/RN	
ZM181	Lockheed Martin F-35B Lightning II	Reservation for RAF/RN	
ZM182	Lockheed Martin F-35B Lightning II	Reservation for RAF/RN	
ZM183	Lockheed Martin F-35B Lightning II	Reservation for RAF/RN	
ZM184	Lockheed Martin F-35B Lightning II	Reservation for RAF/RN	
ZM185	Lockheed Martin F-35B Lightning II	Reservation for RAF/RN	
ZM186	Lockheed Martin F-35B Lightning II	Reservation for RAF/RN	

Notes	Serial	Type (code/other identity)	Owner/operator, location or fate
	ZM187	Lockheed Martin F-35B Lightning II	Reservation for RAF/RN
	ZM188	Lockheed Martin F-35B Lightning II	Reservation for RAF/RN
	ZM189	Lockheed Martin F-35B Lightning II	Reservation for RAF/RN
	ZM190	Lockheed Martin F-35B Lightning II	Reservation for RAF/RN
	ZM191	Lockheed Martin F-35B Lightning II	Reservation for RAF/RN
	ZM192	Lockheed Martin F-35B Lightning II	Reservation for RAF/RN
	ZM193	Lockheed Martin F-35B Lightning II	Reservation for RAF/RN
	ZM194	Lockheed Martin F-35B Lightning II	Reservation for RAF/RN
	ZM195	Lockheed Martin F-35B Lightning II	Reservation for RAF/RN
	ZM196	Lockheed Martin F-35B Lightning II	Reservation for RAF/RN
	ZM197	Lockheed Martin F-35B Lightning II	Reservation for RAF/RN
	ZM198	Lockheed Martin F-35B Lightning II	Reservation for RAF/RN
	ZM199	Lockheed Martin F-35B Lightning II	Reservation for RAF/RN
	ZM200	Lockheed Martin F-35B Lightning II	Reservation for RAF/RN
	ZM400	Airbus A400M Atlas C1 (EC-405)	RAF No 24 Sqn, Brize Norton
	ZM401	Airbus A400M Atlas C1 (EC-406)	Airbus Defence & Space, Seville, for RAF
	ZM402	Airbus A400M Atlas C1	Airbus Defence & Space, Seville, for RAF
	ZM403	Airbus A400M Atlas C1	Airbus Defence & Space, Seville, for RAF
	ZM404	Airbus A400M Atlas C1	Airbus Defence & Space, Seville, for RAF
	ZM405	Airbus A400M Atlas C1	Reservation for RAF
	ZM406	Airbus A400M Atlas C1	Reservation for RAF
	ZM407	Airbus A400M Atlas C1	Reservation for RAF
	ZM408	Airbus A400M Atlas C1	Reservation for RAF
	ZM409	Airbus A400M Atlas C1	Reservation for RAF
	ZM410	Airbus A400M Atlas C1	Reservation for RAF
	ZM411	Airbus A400M Atlas C1	Reservation for RAF
	ZM412	Airbus A400M Atlas C1	Reservation for RAF
	ZM413	Airbus A400M Atlas C1	Reservation for RAF
	ZM414	Airbus A400M Atlas C1	Reservation for RAF
	ZM415	Airbus A400M Atlas C1	Reservation for RAF
	ZM416	Airbus A400M Atlas C1	Reservation for RAF
	ZM417	Airbus A400M Atlas C1	Reservation for RAF
	ZM418	Airbus A400M Atlas C1	Reservation for RAF
	ZM419	Airbus A400M Atlas C1	Reservation for RAF
	ZM420	Airbus A400M Atlas C1	Reservation for RAF
	ZM421	Airbus A400M Atlas C1	Reservation for RAF
	ZR283	AgustaWestland AW139 (G-FBHA)	FBS Helicopters/RAF DHFS, Valley
	ZR322	Agusta A109E Power Elite (G-CDVC)	RAF No 32(The Royal) Sqn, Northolt
	ZR324	Agusta A109E Power (G-EMHB)	DHFS, RAF Shawbury
	ZR325	Agusta A109E Power (G-BZEI)	DHFS, RAF Shawbury
	ZR335	AgustaWestland AW101 Mk.640	*To Saudi Arabia as HMH-2, 3 November 20014*
	ZR339	AgustaWestland AW101 Mk.641 (ZW-4302)	AgustaWestland, Yeovil
	ZR342	AgustaWestland AW101 Mk.641 (ZW-4305)	AgustaWestland, Yeovil
	ZR343	AgustaWestland AW101 Mk.641 (ZW-4306)	AgustaWestland, Yeovil
	ZR344	AgustaWestland AW101 Mk.641 (ZW-4307)	*To Nigeria as NAF 280, 30 September 2014*
	ZR345	AgustaWestland AW101 Mk.641 (ZW-4308)	*To Nigeria as NAF 281, 30 October 2014*
	ZR346	AgustaWestland AW101 Mk.641 (ZW-4309)	AgustaWestland, Yeovil
	ZR347	AgustaWestland AW101 Mk.641 (ZW-4310)	AgustaWestland, for Azerbaijan as 4K-AI010
	ZR348	AgustaWestland AW101 Mk.641 (ZW-4311)	AgustaWestland, for Azerbaijan
	ZR349	AgustaWestland AW101 Mk.641 (ZW-4312)	AgustaWestland, for Azerbaijan
	ZR352	AgustaWestland AW101 Mk.611 [15-01]	AgustaWestland, Yeovil, for Italy
	ZR353	AgustaWestland AW101 Mk.611 [15-03]	AgustaWestland, Yeovil, for Italy
	ZR354	AgustaWestland AW101 Mk.611 [15-03]	AgustaWestland, Yeovil, for Italy
	ZS782	WS WG25 Sharpeye	The Helicopter Museum, Weston-super-Mare
	ZT800	WS Super Lynx Mk 300	MoD/AgustaWestland, Yeovil
	ZZ171	Boeing C-17A Globemaster III (00-201/N171UK)	RAF No 99 Sqn, Brize Norton
	ZZ172	Boeing C-17A Globemaster III (00-202/N172UK)	RAF No 99 Sqn, Brize Norton
	ZZ173	Boeing C-17A Globemaster III (00-203/N173UK)	RAF No 99 Sqn, Brize Norton

Serial	Type (code/other identity)	Owner/operator, location or fate	Notes
ZZ174	Boeing C-17A Globemaster III (00-204/N174UK)	RAF No 99 Sqn, Brize Norton	
ZZ175	Boeing C-17A Globemaster III (06-0205/N9500Z)	RAF No 99 Sqn, Brize Norton	
ZZ176	Boeing C-17A Globemaster III (08-0206/N9500B)	RAF No 99 Sqn, Brize Norton	
ZZ177	Boeing C-17A Globemaster III (09-8207/N9500B)	RAF No 99 Sqn, Brize Norton	
ZZ178	Boeing C-17A Globemaster III (12-0208/N9500N)	RAF No 99 Sqn, Brize Norton	
ZZ190	Hawker Hunter F58 (J-4066/G-HHAE)	Hawker Hunter Aviation, Yeovilton	
ZZ191	Hawker Hunter F58 (J-4058/G-HHAD)	Hawker Hunter Aviation, Yeovilton	
ZZ192	Grob G109B Vigilant T1 (D-KLVI) [SQ]	RAF CGMF, Syerston	
ZZ193	Grob G109B Vigilant T1 (D-KBLO) [SR]	RAF CGMF, Syerston	
ZZ194	Hawker Hunter F58 (J-4021/G-HHAC)	Hawker Hunter Aviation, Scampton	
ZZ201	General Atomics Reaper RPAS (07-111)	RAF No 13 Sqn/No 39 Sqn, Creech AFB, Nevada, USA	
ZZ202	General Atomics Reaper RPAS (07-117)	RAF No 13 Sqn/No 39 Sqn, Creech AFB, Nevada, USA	
ZZ203	General Atomics Reaper RPAS (08-133)	RAF No 13 Sqn/No 39 Sqn, Creech AFB, Nevada, USA	
ZZ204	General Atomics Reaper RPAS (10-0157)	RAF No 13 Sqn/No 39 Sqn, Creech AFB, Nevada, USA	
ZZ205	General Atomics Reaper RPAS (10-0162)	RAF No 13 Sqn/No 39 Sqn, Creech AFB, Nevada, USA	
ZZ206	General Atomics Reaper RPAS (12-0707)	RAF No 13 Sqn/No 39 Sqn, Creech AFB, Nevada, USA	
ZZ207	General Atomics Reaper RPAS (12-0708)	RAF No 13 Sqn/No 39 Sqn, Creech AFB, Nevada, USA	
ZZ208	General Atomics Reaper RPAS (12-0709)	RAF No 13 Sqn/No 39 Sqn, Creech AFB, Nevada, USA	
ZZ209	General Atomics Reaper RPAS (12-0710)	RAF No 13 Sqn/No 39 Sqn, Creech AFB, Nevada, USA	
ZZ210	General Atomics Reaper RPAS (12-0711)	RAF No 13 Sqn/No 39 Sqn, Creech AFB, Nevada, USA	
ZZ211	General Atomics Reaper RPAS	General Atomics, for RAF	
ZZ212	General Atomics Reaper RPAS	General Atomics, for RAF	
ZZ213	General Atomics Reaper RPAS	General Atomics, for RAF	
ZZ250	BAE Systems Taranis RPAS	BAE Systems, Warton	
ZZ251	BAE Systems HERTI RPAS	BAE Systems, Warton	
ZZ252	BAE Systems HERTI RPAS	BAE Systems, Warton	
ZZ253	BAE Systems HERTI RPAS	BAE Systems, Warton	
ZZ254	BAE Systems HERTI RPAS	BAE Systems, Warton	
ZZ330	Airbus A330 Voyager KC2 (MRTT017/EC-337/G-VYGA)	RAF No 10 Sqn/No 101 Sqn, Brize Norton	
ZZ331	Airbus A330 Voyager KC2 (MRTT018/EC-331/G-VYGB)	RAF No 10 Sqn/No 101 Sqn, Brize Norton	
ZZ332	Airbus A330 Voyager KC3 (MRTT019/EC-330/G-VYGC)	RAF No 1312 Flt, Mount Pleasant, FI	
ZZ333	Airbus A330 Voyager KC3 (MRTT020/G-VYGD)	RAF No 10 Sqn/No 101 Sqn, Brize Norton	
ZZ334	Airbus A330 Voyager KC3 (MRTT016/EC-335/G-VYGE)	RAF No 10 Sqn/No 101 Sqn, Brize Norton	
ZZ335	Airbus A330 Voyager KC3 (MRTT021/EC-338/G-VYGF)	RAF No 10 Sqn/No 101 Sqn, Brize Norton	
ZZ336	Airbus A330 Voyager KC3 (EC-333/G-VYGG)	MoD/Airbus, Getafe (conversion)	
ZZ337	Airbus A330 Voyager KC3 (MRTT023/EC-336/G-VYGH)	RAF No 10 Sqn/No 101 Sqn, Brize Norton	
ZZ338	Airbus A330 Voyager KC3 (MRTT024/EC-331/G-VYGI)	RAF No 10 Sqn/No 101 Sqn, Brize Norton	
ZZ339	Airbus A330-243 (EC-333/G-VYGJ)	Airtanker Ltd, Brize Norton [flies as G-VYGJ]	
ZZ340	Airbus A330 Voyager KC2 (MRTT026/EC-330/G-VYGK)	DE&S/Airbus, for Airtanker Ltd	
ZZ341	Airbus A330 Voyager KC2	Reservation for Airtanker Ltd	
ZZ342	Airbus A330 Voyager KC2/KC3	Reservation for Airtanker Ltd	
ZZ343	Airbus A330 Voyager KC2/KC3	Reservation for Airtanker Ltd	
ZZ349	AgustaWestland AW159 Wildcat	DE&S/AgustaWestland, Yeovil, for RN/AAC	
ZZ350	AgustaWestland AW159 Wildcat	DE&S/AgustaWestland, Yeovil, for RN/AAC	
ZZ351	AgustaWestland AW159 Wildcat	DE&S/AgustaWestland, Yeovil, for RN/AAC	
ZZ352	AgustaWestland AW159 Wildcat	DE&S/AgustaWestland, Yeovil, for RN/AAC	
ZZ353	AgustaWestland AW159 Wildcat	DE&S/AgustaWestland, Yeovil, for RN/AAC	
ZZ354	AgustaWestland AW159 Wildcat	DE&S/AgustaWestland, Yeovil, for RN/AAC	
ZZ355	AgustaWestland AW159 Wildcat	DE&S/AgustaWestland, Yeovil, for RN/AAC	
ZZ356	AgustaWestland AW159 Wildcat	DE&S/AgustaWestland, Yeovil, for RN/AAC	
ZZ357	AgustaWestland AW159 Wildcat	DE&S/AgustaWestland, Yeovil, for RN/AAC	
ZZ358	AgustaWestland AW159 Wildcat	DE&S/AgustaWestland, Yeovil, for RN/AAC	
ZZ359	AgustaWestland AW159 Wildcat	DE&S/AgustaWestland, Yeovil, for RN/AAC	

Notes	Serial	Type (code/other identity)	Owner/operator, location or fate
	ZZ360	AgustaWestland AW159 Wildcat	DE&S/AgustaWestland, Yeovil, for RN/AAC
	ZZ361	AgustaWestland AW159 Wildcat	DE&S/AgustaWestland, Yeovil, for RN/AAC
	ZZ362	AgustaWestland AW159 Wildcat	DE&S/AgustaWestland, Yeovil, for RN/AAC
	ZZ363	AgustaWestland AW159 Wildcat	DE&S/AgustaWestland, Yeovil, for RN/AAC
	ZZ364	AgustaWestland AW159 Wildcat	DE&S/AgustaWestland, Yeovil, for RN/AAC
	ZZ365	AgustaWestland AW159 Wildcat	DE&S/AgustaWestland, Yeovil, for RN/AAC
	ZZ366	AgustaWestland AW159 Wildcat	DE&S/AgustaWestland, Yeovil, for RN/AAC
	ZZ367	AgustaWestland AW159 Wildcat	DE&S/AgustaWestland, Yeovil, for RN/AAC
	ZZ368	AgustaWestland AW159 Wildcat	DE&S/AgustaWestland, Yeovil, for RN/AAC
	ZZ369	AgustaWestland AW159 Wildcat	DE&S/AgustaWestland, Yeovil, for RN/AAC
	ZZ370	AgustaWestland AW159 Wildcat	DE&S/AgustaWestland, Yeovil, for RN/AAC
	ZZ371	AgustaWestland AW159 Wildcat	DE&S/AgustaWestland, Yeovil, for RN/AAC
	ZZ372	AgustaWestland AW159 Wildcat	DE&S/AgustaWestland, Yeovil, for RN/AAC
	ZZ373	AgustaWestland AW159 Wildcat	DE&S/AgustaWestland, Yeovil, for RN/AAC
	ZZ374	AgustaWestland AW159 Wildcat	DE&S/AgustaWestland, Yeovil, for RN/AAC
	ZZ375	AgustaWestland AW159 Wildcat HMA2	RN No 825 NAS, Yeovilton
	ZZ376	AgustaWestland AW159 Wildcat HMA2	RN No 825 NAS, Yeovilton
	ZZ377	AgustaWestland AW159 Wildcat HMA2	RN No 825 NAS, Yeovilton
	ZZ378	AgustaWestland AW159 Wildcat HMA2	RN No 825 NAS, Yeovilton
	ZZ379	AgustaWestland AW159 Wildcat HMA2	RN No 825 NAS, Yeovilton
	ZZ380	AgustaWestland AW159 Wildcat HMA2	RN No 825 NAS, Yeovilton
	ZZ381	AgustaWestland AW159 Wildcat HMA2	RN No 825 NAS, Yeovilton
	ZZ382	AgustaWestland AW159 Wildcat AH1	AAC No 652 Sqn, Yeovilton
	ZZ383	AgustaWestland AW159 Wildcat AH1	AAC No 652 Sqn, Yeovilton
	ZZ384	AgustaWestland AW159 Wildcat AH1	AAC No 652 Sqn, Yeovilton
	ZZ385	AgustaWestland AW159 Wildcat AH1	AAC No 652 Sqn, Yeovilton
	ZZ386	AgustaWestland AW159 Wildcat AH1	AAC, stored, Yeovilton
	ZZ387	AgustaWestland AW159 Wildcat AH1	AAC, stored Yeovilton
	ZZ388	AgustaWestland AW159 Wildcat AH1	AAC No 652 Sqn, Yeovilton
	ZZ389	AgustaWestland AW159 Wildcat AH1	AAC No 652 Sqn, Yeovilton
	ZZ390	AgustaWestland AW159 Wildcat AH1	AAC No 652 Sqn, Yeovilton
	ZZ391	AgustaWestland AW159 Wildcat AH1	AAC, stored, Yeovilton
	ZZ392	AgustaWestland AW159 Wildcat AH1	AAC No 652 Sqn, Yeovilton
	ZZ393	AgustaWestland AW159 Wildcat AH1	AAC No 652 Sqn, Yeovilton
	ZZ394	AgustaWestland AW159 Wildcat AH1	AAC No 652 Sqn, Yeovilton
	ZZ395	AgustaWestland AW159 Wildcat AH1	AAC No 652 Sqn, Yeovilton
	ZZ396	AgustaWestland AW159 Wildcat HMA2	MoD/AgustaWestland, Yeovil
	ZZ397	AgustaWestland AW159 Wildcat HMA2	RN No 825 NAS, Yeovilton
	ZZ398	AgustaWestland AW159 Wildcat AH1	AAC No 652 Sqn, Yeovilton
	ZZ399	AgustaWestland AW159 Wildcat AH1	AAC, stored Yeovilton
	ZZ400	AgustaWestland AW159 Wildcat (TI01)	MoD/AgustaWestland, Yeovil
	ZZ401	AgustaWestland AW159 Wildcat (TI02)	MoD/AgustaWestland, Yeovil
	ZZ402	AgustaWestland AW159 Wildcat (TI03)	MoD/AgustaWestland, Yeovil
	ZZ403	AgustaWestland AW159 Wildcat AH1	MoD/AgustaWestland, Yeovil
	ZZ404	AgustaWestland AW159 Wildcat AH1	AAC, stored, Yeovilton
	ZZ405	AgustaWestland AW159 Wildcat AH1	AAC, stored, Yeovilton
	ZZ406	AgustaWestland AW159 Wildcat AH1	AAC, stored, Yeovilton
	ZZ407	AgustaWestland AW159 Wildcat AH1	MoD/AgustaWestland, Yeovil
	ZZ408	AgustaWestland AW159 Wildcat AH1	AAC, stored, Yeovilton
	ZZ409	AgustaWestland AW159 Wildcat AH1	AAC No 652 Sqn, Yeovilton
	ZZ410	AgustaWestland AW159 Wildcat AH1	AAC No 652 Sqn, Yeovilton
	ZZ411	AgustaWestland AW159 Wildcat	DE&S/AgustaWestland, Yeovil, for RN/AAC
	ZZ412	AgustaWestland AW159 Wildcat	DE&S/AgustaWestland, Yeovil, for RN/AAC
	ZZ413	AgustaWestland AW159 Wildcat HMA2	RN No 825 NAS, Yeovilton
	ZZ414	AgustaWestland AW159 Wildcat HMA2	MoD/AgustaWestland, Yeovil
	ZZ415	AgustaWestland AW159 Wildcat HMA2	MoD/AgustaWestland, Yeovil
	ZZ416	Hawker Beechcraft Shadow R1 (G-JENC)	RAF No 14 Sqn, Waddington
	ZZ417	Hawker Beechcraft Shadow R1 (G-NICY)	RAF No 14 Sqn, Waddington
	ZZ418	Hawker Beechcraft Shadow R1 (G-JIMG)	RAF No 14 Sqn, Waddington
	ZZ419	Hawker Beechcraft Shadow R1 (G-OTCS)	RAF No 14 Sqn, Waddington
	ZZ500	Hawker Beechcraft Avenger T1 (G-MFTA)	RN No 750 NAS, Culdrose
	ZZ501	Hawker Beechcraft Avenger T1 (G-MFTB)	RN No 750 NAS, Culdrose

Serial	Type (code/other identity)	Owner/operator, location or fate	Notes
ZZ502	Hawker Beechcraft Avenger T1 (G-MFTC)	RN No 750 NAS, Culdrose	
ZZ503	Hawker Beechcraft Avenger T1 (G-MFTD)	RN No 750 NAS, Culdrose	
ZZ504	Hawker Beechcraft Shadow R1 (G-CGUM)	RAF No 14 Sqn, Waddington	
ZZ510	AgustaWestland AW159 Wildcat AH1	AAC, stored Yeovilton	
ZZ511	AgustaWestland AW159 Wildcat AH1	AAC No 652 Sqn, Yeovilton	
ZZ512	AgustaWestland AW159 Wildcat AH1	MoD/AgustaWestland, Yeovil	
ZZ513	AgustaWestland AW159 Wildcat HMA2	MoD/AgustaWestland, Yeovil	
ZZ514	AgustaWestland AW159 Wildcat HMA2	DE&S/AgustaWestland, Yeovil, for RN	
ZZ515	AgustaWestland AW159 Wildcat HMA2	DE&S/AgustaWestland, Yeovil, for RN	
ZZ516	AgustaWestland AW159 Wildcat HMA2	DE&S/AgustaWestland, Yeovil, for RN	
ZZ517	AgustaWestland AW159 Wildcat HMA2	DE&S/AgustaWestland, Yeovil, for RN	
ZZ518	AgustaWestland AW159 Wildcat HMA2	DE&S/AgustaWestland, Yeovil, for RN	
ZZ519	AgustaWestland AW159 Wildcat HMA2	DE&S/AgustaWestland, Yeovil, for RN	
ZZ520	AgustaWestland AW159 Wildcat AH1	DE&S/AgustaWestland, Yeovil, for AAC	
ZZ521	AgustaWestland AW159 Wildcat AH1	DE&S/AgustaWestland, Yeovil, for AAC	
ZZ522	AgustaWestland AW159 Wildcat HMA2	DE&S/AgustaWestland, Yeovil, for RN	
ZZ523	AgustaWestland AW159 Wildcat AH1	DE&S/AgustaWestland, Yeovil, for AAC	
ZZ524	AgustaWestland AW159 Wildcat AH1	DE&S/AgustaWestland, Yeovil, for AAC	
ZZ525	AgustaWestland AW159 Wildcat AH1	DE&S/AgustaWestland, Yeovil, for AAC	
ZZ526	AgustaWestland AW159 Wildcat AH1	DE&S/AgustaWestland, Yeovil, for AAC	
ZZ527	AgustaWestland AW159 Wildcat AH1	DE&S/AgustaWestland, Yeovil, for AAC	
ZZ528	AgustaWestland AW159 Wildcat HMA2	DE&S/AgustaWestland, Yeovil, for RN	
ZZ529	AgustaWestland AW159 Wildcat HMA2	DE&S/AgustaWestland, Yeovil, for RN	
ZZ530	AgustaWestland AW159 Wildcat HMA2	DE&S/AgustaWestland, Yeovil, for RN	
ZZ531	AgustaWestland AW159 Wildcat HMA2	DE&S/AgustaWestland, Yeovil, for RN	
ZZ532	AgustaWestland AW159 Wildcat HMA2	DE&S/AgustaWestland, Yeovil, for RN	
ZZ533	AgustaWestland AW159 Wildcat HMA2	DE&S/AgustaWestland, Yeovil, for RN	
ZZ534	AgustaWestland AW159 Wildcat HMA2	DE&S/AgustaWestland, Yeovil, for RN	
ZZ535	AgustaWestland AW159 Wildcat HMA2	DE&S/AgustaWestland, Yeovil, for RN	
ZZ541	AgustaWestland AW159 Mk.210	AgustaWestland, Yeovil for Rep of Korea Navy as 15-0601	
ZZ542	AgustaWestland AW159 Mk.210	AgustaWestland, Yeovil for Rep of Korea Navy as 15-0602	
ZZ543	AgustaWestland AW159 Mk.210	AgustaWestland, Yeovil for Rep of Korea Navy as 15-0603	
ZZ544	AgustaWestland AW159 Mk.210	AgustaWestland, Yeovil for Rep of Korea Navy as 15-0604	
ZZ545	AgustaWestland AW159 Mk.210	AgustaWestland, Yeovil for Rep of Korea Navy as 15-0605	
ZZ546	AgustaWestland AW159 Mk.210	AgustaWestland, Yeovil for Rep of Korea Navy as 15-0606	
ZZ547	AgustaWestland AW159 Mk.210	AgustaWestland, Yeovil for Rep of Korea Navy as 15-0607	
ZZ548	AgustaWestland AW159 Mk.210	AgustaWestland, Yeovil for Rep of Korea Navy as 15-0608	
ZZ664	Boeing RC-135W Airseeker R1 (64-14833)	RAF No 51 Sqn, Waddington/Mildenhall	
ZZ665	Boeing RC-135W Airseeker R1 (64-14838)	Boeing, for RAF	
ZZ666	Boeing RC-135W Airseeker R1 (64-14830)	Boeing, for RAF	

Notes	Serial	Type (code/other identity)	Owner/operator, location or fate
	G-BYUA	Grob G.115E Tutor T1	VT Aerospace/No 1 EFTS, Barkston Heath
	G-BYUB	Grob G.115E Tutor T1	VT Aerospace/East Midlands Universities AS/
			No 16(R) Sqn/No 57(R) Sqn, No 115(R) Sqn, Cranwell
	G-BYUC	Grob G.115E Tutor T1	VT Aerospace/East Midlands Universities AS/
			No 16(R) Sqn/No 57(R) Sqn, No 115(R) Sqn, Cranwell
	G-BYUD	Grob G.115E Tutor T1	VT Aerospace/No 1 EFTS, Barkston Heath
	G-BYUE	Grob G.115E Tutor T1	VT Aerospace/East Midlands Universities AS/
			No 16(R) Sqn/No 57(R) Sqn, No 115(R) Sqn, Cranwell
	G-BYUF	Grob G.115E Tutor T1	VT Aerospace/Northumbrian Universities AS, Leeming
	G-BYUG	Grob G.115E Tutor T1	VT Aerospace/Cambridge UAS/University of London
			AS, Cranwell
	G-BYUH	Grob G.115E Tutor T1	VT Aerospace/Southampton UAS, Boscombe Down
	G-BYUI	Grob G.115E Tutor T1	VT Aerospace/East Midlands Universities AS/
			No 16(R) Sqn/No 57(R) Sqn, No 115(R) Sqn, Cranwell
	G-BYUJ	Grob G.115E Tutor T1	VT Aerospace/Oxford UAS, Benson
	G-BYUK	Grob G.115E Tutor T1	VT Aerospace/No 1 EFTS, Barkston Heath
	G-BYUL	Grob G.115E Tutor T1	VT Aerospace/No 1 EFTS/676 Sqn, Middle Wallop
	G-BYUM	Grob G.115E Tutor T1	VT Aerospace/No 1 EFTS, Barkston Heath
	G-BYUN	Grob G.115E Tutor T1	VT Aerospace/No 1 EFTS, Barkston Heath
	G-BYUO	Grob G.115E Tutor T1	VT Aerospace/Cambridge UAS/University of London
			AS, Wyton
	G-BYUP	Grob G.115E Tutor T1	VT Aerospace/East Midlands Universities AS/
			No 16(R) Sqn/No 57(R) Sqn, No 115(R) Sqn, Cranwell
	G-BYUR	Grob G.115E Tutor T1	VT Aerospace/No 1 EFTS, Barkston Heath
	G-BYUS	Grob G.115E Tutor T1	VT Aerospace/Cambridge UAS/University of London
			AS, Cranwell
	G-BYUU	Grob G.115E Tutor T1	VT Aerospace/Cambridge UAS/University of London
			AS, Wyton
	G-BYUV	Grob G.115E Tutor T1	VT Aerospace/Bristol UAS, Colerne
	G-BYUW	Grob G.115E Tutor T1	VT Aerospace/Cambridge UAS/University of London
			AS, Cranwell
	G-BYUX	Grob G.115E Tutor T1	VT Aerospace/No 1 EFTS, Barkston Heath
	G-BYUY	Grob G.115E Tutor T1	VT Aerospace/East Midlands Universities AS/
			No 16(R) Sqn/No 57(R) Sqn, No 115(R) Sqn, Cranwell
	G-BYUZ	Grob G.115E Tutor T1	VT Aerospace/No 1 EFTS, Barkston Heath
	G-BYVA	Grob G.115E Tutor T1	VT Aerospace/No 1 EFTS/676 Sqn, Middle Wallop
	G-BYVB	Grob G.115E Tutor T1	VT Aerospace/Oxford UAS, Benson
	G-BYVC	Grob G.115E Tutor T1	VT Aerospace/East Midlands Universities AS/
			No 16(R) Sqn/No 57(R) Sqn, No 115(R) Sqn, Cranwell
	G-BYVD	Grob G.115E Tutor T1	VT Aerospace/No 1 EFTS, Barkston Heath
	G-BYVE	Grob G.115E Tutor T1	VT Aerospace/Cambridge UAS/University of London
			AS, Wittering
	G-BYVF	Grob G.115E Tutor T1	VT Aerospace/RN No 727 NAS, Yeovilton
	G-BYVG	Grob G.115E Tutor T1	VT Aerospace/East Midlands Universities AS/
			No 16(R) Sqn/No 57(R) Sqn, No 115(R) Sqn, Cranwell
	G-BYVH	Grob G.115E Tutor T1	VT Aerospace/No 1 EFTS, Barkston Heath
	G-BYVI	Grob G.115E Tutor T1	VT Aerospace/Cambridge UAS/University of London
			AS, Wyton
	G-BYVJ	Grob G.115E Tutor T1	VT Aerospace/Yorkshire Universities AS,
			Linton-on-Ouse
	G-BYVK	Grob G.115E Tutor T1	VT Aerospace/RN No 727 NAS, Yeovilton
	G-BYVL	Grob G.115E Tutor T1	VT Aerospace/Oxford UAS, Benson
	G-BYVM	Grob G.115E Tutor T1	VT Aerospace/No 1 EFTS, Barkston Heath
	G-BYVO	Grob G.115E Tutor T1	VT Aerospace/No 1 EFTS, Barkston Heath
	G-BYVP	Grob G.115E Tutor T1	VT Aerospace/Cambridge UAS/University of London
			AS, Wyton
	G-BYVR	Grob G.115E Tutor T1	VT Aerospace/East Midlands Universities AS/
			No 16(R) Sqn/No 57(R) Sqn, No 115(R) Sqn, Cranwell

Serial	Type (code/other identity)	Owner/operator, location or fate	Notes
G-BYVS	Grob G.115E Tutor T1	VT Aerospace/Bristol UAS, Colerne	
G-BYVT	Grob G.115E Tutor T1	VT Aerospace/Cambridge UAS/University of London AS, Wyton	
G-BYVU	Grob G.115E Tutor T1	VT Aerospace/No 1 EFTS/676 Sqn, Middle Wallop	
G-BYVV	Grob G.115E Tutor T1	VT Aerospace/Northumbrian Universities AS, Leeming	
G-BYVW	Grob G.115E Tutor T1	VT Aerospace/University of Wales AS, St Athan	
G-BYVX	Grob G.115E Tutor T1	VT Aerospace/Yorkshire Universities AS, Linton-on-Ouse	
G-BYVY	Grob G.115E Tutor T1	VT Aerospace/No 1 EFTS/676 Sqn, Middle Wallop	
G-BYVZ	Grob G.115E Tutor T1	VT Aerospace/No 1 EFTS, Barkston Heath	
G-BYWA	Grob G.115E Tutor T1	VT Aerospace/University of Wales AS, St Athan	
G-BYWB	Grob G.115E Tutor T1	VT Aerospace/East Midlands Universities AS/ No 16(R) Sqn/No 57(R) Sqn, No 115(R) Sqn, Cranwell	
G-BYWC	Grob G.115E Tutor T1	VT Aerospace/Bristol UAS, Colerne	
G-BYWD	Grob G.115E Tutor T1	VT Aerospace/University of Wales AS, St Athan	
G-BYWE	Grob G.115E Tutor T1	VT Aerospace/Bristol UAS, Colerne	
G-BYWF	Grob G.115E Tutor T1	VT Aerospace/Cambridge UAS/University of London AS, Cranwell	
G-BYWG	Grob G.115E Tutor T1	VT Aerospace/East Midlands Universities AS/ No 16(R) Sqn/No 57(R) Sqn, No 115(R) Sqn, Cranwell	
G-BYWH	Grob G.115E Tutor T1	VT Aerospace/Cambridge UAS/University of London AS, Wyton	
G-BYWI	Grob G.115E Tutor T1	VT Aerospace/No 1 EFTS, Barkston Heath	
G-BYWJ	Grob G.115E Tutor T1	VT Aerospace/No 1 EFTS, Barkston Heath	
G-BYWK	Grob G.115E Tutor T1	VT Aerospace/Bristol UAS, Colerne	
G-BYWL	Grob G.115E Tutor T1	VT Aerospace/East Midlands Universities AS/ No 16(R) Sqn/No 57(R) Sqn, No 115(R) Sqn, Cranwell	
G-BYWM	Grob G.115E Tutor T1	VT Aerospace/RN No 727 NAS, Yeovilton	
G-BYWN	Grob G.115E Tutor T1	VT Aerospace/No 1 EFTS, Barkston Heath	
G-BYWO	Grob G.115E Tutor T1	VT Aerospace/Cambridge UAS/University of London AS, Wyton	
G-BYWP	Grob G.115E Tutor T1	VT Aerospace/Yorkshire Universities AS, Linton-on-Ouse	
G-BYWR	Grob G.115E Tutor T1	VT Aerospace/East Midlands Universities AS/ No 16(R) Sqn/No 57(R) Sqn, No 115(R) Sqn, Cranwell	
G-BYWS	Grob G.115E Tutor T1	VT Aerospace/No 1 EFTS, Barkston Heath	
G-BYWT	Grob G.115E Tutor T1	VT Aerospace/Northumbrian Universities AS, Leeming	
G-BYWU	Grob G.115E Tutor T1	VT Aerospace/Oxford UAS, Benson	
G-BYWV	Grob G.115E Tutor T1	VT Aerospace/Yorkshire Universities AS, Linton-on-Ouse	
G-BYWW	Grob G.115E Tutor T1	VT Aerospace/Southampton UAS, Boscombe Down	
G-BYWX	Grob G.115E Tutor T1	VT Aerospace/Cambridge UAS/University of London AS, Wittering	
G-BYWY	Grob G.115E Tutor T1	VT Aerospace/Oxford UAS, Benson	
G-BYWZ	Grob G.115E Tutor T1	VT Aerospace/East Midlands Universities AS/ No 16(R) Sqn/No 57(R) Sqn, No 115(R) Sqn, Cranwell	
G-BYXA	Grob G.115E Tutor T1	VT Aerospace/Oxford UAS, Benson	
G-BYXB	Grob G.115E Tutor T1	VT Aerospace/Bristol UAS, Colerne	
G-BYXC	Grob G.115E Tutor T1	VT Aerospace/East Midlands Universities AS/ No 16(R) Sqn/No 57(R) Sqn, No 115(R) Sqn, Cranwell	
G-BYXD	Grob G.115E Tutor T1	VT Aerospace/Southampton UAS, Boscombe Down	
G-BYXE	Grob G.115E Tutor T1	VT Aerospace/Southampton UAS, Boscombe Down	
G-BYXF	Grob G.115E Tutor T1	VT Aerospace/No 1 EFTS/676 Sqn, Middle Wallop	
G-BYXG	Grob G.115E Tutor T1	VT Aerospace/Yorkshire Universities AS, Linton-on-Ouse	
G-BYXH	Grob G.115E Tutor T1	VT Aerospace/East Midlands Universities AS/ No 16(R) Sqn/No 57(R) Sqn, No 115(R) Sqn, Cranwell	
G-BYXI	Grob G.115E Tutor T1	VT Aerospace/Southampton UAS, Boscombe Down	

Notes	Serial	Type (code/other identity)	Owner/operator, location or fate
	G-BYXJ	Grob G.115E Tutor T1	VT Aerospace/Southampton UAS, Boscombe Down
	G-BYXK	Grob G.115E Tutor T1	VT Aerospace/RN No 727 NAS, Yeovilton
	G-BYXL	Grob G.115E Tutor T1	VT Aerospace/Oxford UAS, Benson
	G-BYXM	Grob G.115E Tutor T1 $	VT Aerospace/East Midlands Universities AS/ No 16(R) Sqn/No 57(R) Sqn, No 115(R) Sqn, Cranwell
	G-BYXN	Grob G.115E Tutor T1	VT Aerospace/East Midlands Universities AS/ No 16(R) Sqn/No 57(R) Sqn, No 115(R) Sqn, Cranwell
	G-BYXO	Grob G.115E Tutor T1	VT Aerospace/East Midlands Universities AS/ No 16(R) Sqn/No 57(R) Sqn, No 115(R) Sqn, Cranwell
	G-BYXP	Grob G.115E Tutor T1	VT Aerospace/Cambridge UAS/University of London AS, Cranwell
	G-BYXS	Grob G.115E Tutor T1	VT Aerospace/RN No 727 NAS, Yeovilton
	G-BYXT	Grob G.115E Tutor T1	VT Aerospace/East Midlands Universities AS/ No 16(R) Sqn/No 57(R) Sqn, No 115(R) Sqn, Cranwell
	G-BYXX	Grob G.115E Tutor T1	VT Aerospace/No 1 EFTS, Barkston Heath
	G-BYXY	Grob G.115E Tutor T1	VT Aerospace/Cambridge UAS/University of London AS, Wyton
	G-BYXZ	Grob G.115E Tutor T1	VT Aerospace/East Midlands Universities AS/ No 16(R) Sqn/No 57(R) Sqn, No 115(R) Sqn, Cranwell
	G-BYYA	Grob G.115E Tutor T1	VT Aerospace/Northumbrian Universities AS, Leeming
	G-BYYB	Grob G.115E Tutor T1	VT Aerospace/No 1 EFTS, Barkston Heath
	G-CEYO	AS350BB Squirrel HT3 [00]	DHFS stored, RAF Shawbury
	G-CGKA	Grob G.115E Tutor T1EA	VT Aerospace/No 3 FTS/45(R) Sqn, D Flt, Cranwell
	G-CGKB	Grob G.115E Tutor T1EA	VT Aerospace/No 3 FTS/45(R) Sqn, D Flt, Cranwell
	G-CGKC	Grob G.115E Tutor T1EA	VT Aerospace/No 3 FTS/45(R) Sqn, D Flt, Cranwell
	G-CGKD	Grob G.115E Tutor T1EA	VT Aerospace/University of Birmingham AS, Cosford
	G-CGKE	Grob G.115E Tutor T1EA	VT Aerospace/University of Birmingham AS, Cosford
	G-CGKF	Grob G.115E Tutor T1EA	VT Aerospace/Liverpool UAS/Manchester and Salford Universities AS, Woodvale
	G-CGKG	Grob G.115E Tutor T1EA	VT Aerospace/University of Birmingham AS, Cosford
	G-CGKH	Grob G.115E Tutor T1EA	VT Aerospace/University of Birmingham AS, Cosford
	G-CGKI	Grob G.115E Tutor T1EA	VT Aerospace/East of Scotland UAS, Leuchars
	G-CGKJ	Grob G.115E Tutor T1EA	VT Aerospace/East of Scotland UAS, Leuchars
	G-CGKK	Grob G.115E Tutor T1EA	VT Aerospace/East of Scotland UAS, Leuchars
	G-CGKL	Grob G.115E Tutor T1EA	VT Aerospace/East of Scotland UAS, Leuchars
	G-CGKM	Grob G.115E Tutor T1EA	VT Aerospace/East of Scotland UAS, Leuchars
	G-CGKN	Grob G.115E Tutor T1EA	VT Aerospace/Universities of Glasgow & Strathclyde AS, Glasgow
	G-CGKO	Grob G.115E Tutor T1EA	VT Aerospace/East of Scotland UAS, Leuchars
	G-CGKP	Grob G.115E Tutor T1EA	VT Aerospace/East of Scotland UAS, Leuchars
	G-CGKR	Grob G.115E Tutor T1EA	VT Aerospace/University of Birmingham AS, Cosford
	G-CGKS	Grob G.115E Tutor T1EA	VT Aerospace/Liverpool UAS/Manchester and Salford Universities AS, Woodvale
	G-CGKT	Grob G.115E Tutor T1EA	VT Aerospace/Liverpool UAS/Manchester and Salford Universities AS, Woodvale
	G-CGKU	Grob G.115E Tutor T1EA	VT Aerospace/Liverpool UAS/Manchester and Salford Universities AS, Woodvale
	G-CGKV	Grob G.115E Tutor T1EA	VT Aerospace/Liverpool UAS/Manchester and Salford Universities AS, Woodvale
	G-CGKW	Grob G.115E Tutor T1EA	VT Aerospace/No 3 FTS/45(R) Sqn, D Flt, Cranwell
	G-CGKX	Grob G.115E Tutor T1EA	VT Aerospace/Liverpool UAS/Manchester and Salford Universities AS, Woodvale
	G-DOIT	AS350BB Squirrel HT3 [99]	DHFS stored, RAF Shawbury
	G-FFRA	Dassault Falcon 20DC (N902FR)	Cobham Leasing Ltd, Durham/Tees Valley
	G-FRAD	Dassault Falcon 20E	Cobham Leasing Ltd, Bournemouth
	G-FRAF	Dassault Falcon 20E (N911FR)	Cobham Leasing Ltd, Bournemouth
	G-FRAH	Dassault Falcon 20DC (N900FR)	Cobham Leasing Ltd, Durham/Tees Valley
	G-FRAI	Dassault Falcon 20E (N901FR)	Cobham Leasing Ltd, Bournemouth

Serial	Type (code/other identity)	Owner/operator, location or fate	Notes
G-FRAJ	Dassault Falcon 20E (N903FR)	Cobham Leasing Ltd, Bournemouth	
G-FRAK	Dassault Falcon 20DC (N905FR)	Cobham Leasing Ltd, Durham/Tees Valley	
G-FRAL	Dassault Falcon 20DC (N904FR)	Cobham Leasing Ltd, Bournemouth	
G-FRAO	Dassault Falcon 20DC (N906FR)	Cobham Leasing Ltd, Bournemouth	
G-FRAP	Dassault Falcon 20DC (N908FR)	Cobham Leasing Ltd, Durham/Tees Valley	
G-FRAR	Dassault Falcon 20DC (N909FR)	Cobham Leasing Ltd, Durham/Tees Valley	
G-FRAS	Dassault Falcon 20C (117501)	Cobham Leasing Ltd, Durham/Tees Valley	
G-FRAT	Dassault Falcon 20C (117502)	Cobham Leasing Ltd, Durham/Tees Valley	
G-FRAU	Dassault Falcon 20C (117504)	Cobham Leasing Ltd, Durham/Tees Valley	
G-FRAW	Dassault Falcon 20ECM (117507)	Cobham Leasing Ltd, Durham/Tees Valley	

G-BYVY is a Tutor T1 operated by 676 Squadron Army Air Corps as part of No 1 EFTS and seen here at its base at Middle Wallop. In late 2014 it was announced that these machines would eventually be replaced by the Grob G120TP.

Tutor T1 G-BYXE is operated by Southampton University Air Squadron and No 2 Air Experience Flight from the QinetiQ base at Boscombe Down.

1764M/K4972	7491M/WT569	7829M/XH992	8019M/WZ869
2015M/K5600	7496M/WT612	7839M/WV781	8021M/XL824
2292M/K8203	7499M/WT555	7841M/WV783	8022M/XN341
2361M/K6035	7510M/WT694	7851M/WZ706	8027M/XM555
3118M/H5199/(BK892)	7525M/WT619	7854M/XM191	8032M/XH837
3858M/X7688	7530M/WT648	7855M/XK416	8034M/*XL554*/(XL703)
4354M/BL614	7532M/WT651	7859M/XP283	8041M/XF690
4552M/T5298	7533M/WT680	7861M/*XL738*/(XM565)	8043M/XF836
4887M/JN768	7544M/WN904	7862M/XR246	8046M/XL770
5377M/EP120	7548M/PS915	7863M/*XP248*	8049M/WE168
5405M/LF738	7556M/WK584	7864M/XP244	8050M/XG329
5466M/*BN230*/(LF751)	7564M/XE982	7865M/TX226	8052M/WH166
5690M/MK356	7570M/XD674	7866M/XH278	8054AM/XM410
5718M/BM597	7582M/WP190	7868M/WZ736	8054BM/XM417
5758M/DG202	7583M/WP185	7869M/WK935	8055AM/XM402
6457M/ML427	7602M/WE600	7872M/*WZ826*/(XD826)	8055BM/XM404
6490M/LA255	7605M/WS692	7881M/WD413	8056M/XG337
6640M/RM694	7606M/*XF688*/(WV562)	7882M/XD525	8057M/XR243
6850M/TE184	7607M/TJ138	7883M/XT150	8063M/WT536
6946M/RW388	7615M/WV679	7887M/XD375	8070M/EP120
6948M/DE673	7616M/WW388	7891M/XM693	8072M/PK624
6960M/MT847	7618M/WW442	7894M/XD818	8073M/TB252
7008M/EE549	7622M/WV606	7895M/WF784	8075M/RW382
7014M/N6720	7631M/VX185	7898M/XP854	8078M/XM351
7015M/NL985	7641M/XA634	7900M/WA576	8080M/XM480
7035M/*K2567*/(DE306)	7645M/WD293	7906M/WH132	8081M/XM468
7060M/VF301	7648M/XF785	7917M/WA591	8082M/XM409
7090M/EE531	7673M/WV332	7930M/WH301	8086M/TB752
7118M/LA198	7689M/*WW421*/	7931M/RD253	8092M/WK654
7119M/LA226	(WW450)	7932M/WZ744	8094M/WT520
7150M/PK683	7696M/WV493	7933M/XR220	8097M/XN492
7154M/WB188	7698M/WV499	7937M/WS843	8101M/WH984
7174M/VX272	7704M/TW536	7938M/XH903	8102M/WT486
7175M/VV106	7705M/WL505	7939M/XD596	8103M/WR985
7200M/VT812	7706M/WB584	7940M/XL764	8106M/WR982
7241M/TE311/*(MK178)*	7709M/WT933	7955M/XH767	8114M/WL798
7243M/TE462	7711M/PS915	7957M/XF545	8117M/WR974
7245M/RW382	7712M/WK281	7960M/WS726	8118M/WZ549
7246M/TD248	7715M/XK724	7961M/WS739	8119M/WR971
7256M/TB752	7716M/WS776	7964M/WS760	8121M/XM474
7257M/TB252	7718M/WA577	7965M/WS792	8124M/WZ572
7279M/TB752	7719M/WK277	7967M/WS788	8128M/WH775
7281M/TB252	7726M/XM373	7969M/WS840	8130M/WH798
7288M/PK724	7741M/VZ477	7971M/XK699	8131M/WT507
7293M/*TB675*/(RW393)	7750M/*WK864*/(WL168)	7973M/WS807	8140M/XJ571
7323M/VV217	7751M/WL131	7979M/XM529	8142M/XJ560
7325M/R5868	7755M/WG760	7980M/XM561	8147M/XR526
7326M/VN485	7759M/PK664	7982M/XH892	8151M/WV795
7362M/475081/(VP546)	7761M/XH318	7983M/XD506	8153M/WV903
7416M/WN907	7762M/XE670	7984M/XN597	8154M/WV908
7421M/WT660	7764M/XH318	7986M/WG777	8155M/WV797
7422M/WT684	7770M/*XF506*/(WT746)	7988M/XL149	8156M/XE339
7428M/WK198	7793M/XG523	7990M/XD452	8158M/XE369
7432M/WZ724	7798M/XH783	7997M/XG452	8160M/XD622
7438M/*18671*/(WP905)	7806M/TA639	7998M/*XM515*/(XD515)	8162M/WM913
7443M/WX853	7809M/XA699	8005M/WG768	8164M/*WN105*/(WF299)
7458M/WX905	7816M/WG763	8009M/XG518	8165M/WH791
7464M/XA564	7817M/TX214	8010M/XG547	8169M/WH644
7470M/XA553	7825M/WK991	8017M/XL762	8173M/XN685
7473M/XE946	7827M/XA917	8018M/XN344	8176M/WH791

8177M/WM224	8394M/WG422	8545M/XN726	8703M/VW453
8179M/XN928	8395M/WF408	8546M/XN728	8706M/XF383
8183M/*XN972*/(XN962)	8396M/XK740	8548M/WT507	8708M/XF509
8184M/WT520	8399M/WR539	8549M/WT534	8709M/XG209
8186M/WR977	8401M/XP686	8554M/TG511	8710M/XG274
8187M/WH791	8406M/XP831	8561M/XS100	8711M/XG290
8189M/*WD615*/(WD646)	8407M/XP585	8563M/*XX822*/(XW563)	8713M/XG225
8190M/XJ918	8408M/XS186	8565M/*WT720*/(E-408)	8718M/XX396
8192M/XR658	8409M/XS209	8566M/XV279	8719M/XT257
8198M/WT339	8410M/XR662	8573M/XM708	8721M/XP354
8203M/XD377	8413M/XM192	8575M/XP542	8724M/XW923
8205M/XN819	8414M/XM173	8576M/XP502	8726M/XP299
8206M/WG419	8422M/XM169	8578M/XR534	8727M/XR486
8208M/WG303	8427M/XM172	8581M/WJ775	8728M/WT532
8209M/WG418	8429M/XH592	8582M/XE874	8729M/WJ815
8210M/WG471	8434M/XM411	8583M/BAPC 94	8732M/XJ729
8211M/WK570	8436M/XN554	8585M/XE670	8733M/XL318
8213M/WK626	8437M/WG362	8586M/XE643	8736M/XF375
8215M/WP869	8439M/WZ846	8588M/XR681	8739M/XH170
8216M/WP927	8440M/WD935	8589M/XR700	8740M/WE173
8229M/XM355	8442M/XP411	8590M/XM191	8741M/XW329
8230M/XM362	8452M/XK885	8591M/XA813	8743M/WD790
8234M/XN458	8453M/XP745	8595M/XH278	8746M/XH171
8235M/XN549	8458M/XP672	8598M/WP270	8749M/XH537
8236M/XP573	8459M/XR650	8600M/XX761	8751M/XT255
8237M/XS179	8460M/XP680	8602M/*PF179*/(XR541)	8753M/WL795
8238M/XS180	8462M/XX477	8604M/XS104	8762M/WH740
8338M/XS180	8463M/XP355	8606M/XP530	8764M/XP344
8342M/WP848	8464M/XJ758	8608M/XP540	8768M/A-522
8344M/WH960	8465M/W1048	8610M/XL502	8769M/A-528
8350M/WH840	8466M/L-866	8611M/WF128	8770M/XL623
8352M/XN632	8467M/WP912	8618M/*XS111*/(XP504)	8771M/XM602
8355M/KN645	8468M/MM5701/(BT474)	8620M/XP534	8772M/WR960
8357M/WK576	8469M/100503	8621M/XR538	8777M/XX914
8359M/WF825	8470M/584219	8624M/*XR991*/(XS102)	8778M/XM598
8361M/WB670	8471M/701152	8627M/XP558	8779M/XM607
8362M/WG477	8472M/120227/(VN679)	8628M/XJ380	8780M/WK102
8364M/WG464	8473M/WP190	8630M/WG362	8781M/WE982
8365M/XK421	8474M/494083	8631M/XR574	8782M/XH136
8366M/XG454	8475M/360043/(PJ876)	8633M/3W-17/MK732	8783M/XW272
8367M/XG474	8476M/24	8634M/WP314	8785M/XS642
8368M/XF926	8477M/4101/(DG200)	8640M/XR977	8789M/XK970
8369M/WE139	8478M/10639	8642M/XR537	8791M/XP329
8370M/N1671	8479M/730301	8645M/XD163	8792M/XP345
8371M/XA847	8481M/191614	8648M/XK526	8793M/XP346
8372M/K8042	8482M/112372/(VK893)	8653M/XS120	8794M/XP398
8373M/P2617	8483M/420430	8655M/XN126	8796M/XK943
8375M/NX611	8484M/5439	8656M/XP405	8797M/XX947
8376M/RF398	8485M/997	8657M/VZ634	8799M/WV787
8377M/R9125	8486M/BAPC 99	8661M/XJ727	8800M/XG226
8378M/*T9707*	8487M/J-1172	8666M/XE793	8805M/XV722
8379M/DG590	8488M/WL627	8671M/XJ435	8807M/XL587
8380M/Z7197	8491M/WJ880	8672M/XP351	8810M/XJ825
8382M/VR930	8493M/XR571	8673M/XD165	8816M/XX734
8383M/K9942	8494M/XP557	8676M/XL577	8818M/XK527
8384M/X4590	8501M/XP640	8679M/XF526	8820M/WP952
8385M/N5912	8502M/XP686	8680M/XF527	8821M/XX115
8386M/NV778	8508M/XS218	8681M/XG164	8822M/VP957
8387M/T6296	8509M/XT141	8682M/XP404	8828M/XS587
8388M/XL993	8514M/XS176	8693M/WH863	8830M/*N-294*/(XF515)
8389M/VX573	8535M/XN776	8696M/WH773	8831M/XG160
8392M/SL674	8538M/XN781	8702M/XG196	8832M/*XG168*/(XG172)

8833M/XL569	8943M/XE799	9056M/XS488	9173M/XW418
8834M/*XL571*	8944M/WZ791	9059M/ZE360	9174M/XZ131
8836M/XL592	8945M/XX818	9066M/XV582	9175M/P1344
8838M/*34037*/(429356)	8946M/XZ389	9067M/XV586	9176M/XW430
8839M/XG194	8947M/XX726	9070M/XV581	9179M/XW309
8841M/XE606	8949M/XX743	9072M/XW768	9180M/XW311
8853M/XT277	8951M/XX727	9073M/XW924	9181M/XW358
8855M/XT284	8953M/XX959	9075M/XV752	9185M/XZ987
8857M/XW544	8954M/XZ384	9076M/XV808	9187M/XW405
8858M/XW541	8955M/XX110	9078M/XV753	9188M/XW364
8863M/XG154	8957M/XN582	9079M/XZ130	9191M/XW416
8867M/XK532	8961M/XS925	9080M/ZE350	9193M/XW367
8868M/WH775	8967M/XV263	9086M/ZE352	9194M/XW420
8869M/WH957	8969M/XR753	9087M/XV753	9195M/XW330
8870M/WH964	8972M/XR754	9090M/XW353	9196M/XW370
8871M/WJ565	8973M/XS922	9091M/XW434	9197M/*XX530*/(XX637)
8873M/XR453	8974M/XM473	9092M/XH669	9198M/XS641
8874M/XE597	8975M/XW917	9093M/WK124	9199M/XW290
8875M/XE624	8976M/XZ630	9095M/XW547	9201M/ZD667
8876M/*VM791*/(XA312)	8978M/XX837	9096M/WV322	9203M/*3066*
8880M/XF435	8984M/XN551	9098M/XV406	9205M/*E449*
8881M/XG254	8985M/WK127	9101M/WL756	9206M/F6314
8883M/XX946	8986M/XV261	9103M/XV411	9207M/8417/18
8884M/VX275	8987M/XM358	9110M/XX736	9208M/F938
8885M/XW922	8990M/XM419	9111M/XW421	9210M/MF628
8886M/XA243	8995M/XM425	9115M/XV863	9211M/733682
8888M/XA231	8996M/XM414	9117M/XV161	9212M/*KL216*/(45-49295)
8889M/XN239	8997M/XX669	9119M/XW303	9213M/N5182
8890M/WT532	8998M/XT864	9120M/XW419	9215M/XL164
8892M/XL618	9002M/XW763	9122M/XZ997	9216M/XL190
8895M/XX746	9003M/XZ390	9123M/XT773	9217M/ZH257
8896M/XX821	9004M/XZ370	9125M/XW410	9218M/XL563
8897M/XX969	9005M/XZ374	9127M/XW432	9219M/XZ971
8898M/XX119	9006M/XX967	9130M/XW327	9221M/XZ966
8899M/XX756	9007M/XX968	9131M/*DD931*	9222M/XZ968
8900M/XZ368	9008M/XX140	9132M/XX977	9224M/XL568
8901M/XZ383	9009M/XX763	9133M/*413573*	9225M/XX885
8902M/XX739	9010M/XX764	9134M/XT288	9226M/XV865
8903M/XX747	9011M/XM412	9136M/XT891	9227M/XB812
8905M/XX975	9012M/XN494	9137M/XN579	9229M/ZA678
8906M/XX976	9014M/XN584	9139M/XV863	9233M/XZ431
8907M/XZ371	9015M/XW320	9140M/XZ287	9234M/XV864
8908M/XZ382	9017M/ZE449	9141M/XV118	9236M/WV318
8909M/XV784	9019M/XX824	9143M/XN589	9237M/XF445
8910M/XL160	9020M/XX825	9145M/XV863	9238M/ZA717
8911M/XH673	9022M/XX958	9146M/XW299	9239M/7198/18
8918M/XX109	9026M/XP629	9148M/XW436	9241M/XS639
8919M/XT486	9027M/XP556	9149M/XW375	9242M/XH672
8920M/XT469	9028M/XP563	9150M/*FX760*	9246M/XS714
8921M/XT466	9032M/XR673	9151M/XT907	9248M/WB627
8922M/XT467	9033M/XS181	9152M/XV424	9249M/WV396
8923M/XX819	9036M/XM350	9153M/XW360	9251M/XX744
8924M/XP701	9038M/XV810	9154M/XW321	9252M/XX722
8925M/XP706	9039M/XN586	9155M/WL679	9254M/XX965
8931M/XV779	9040M/XZ138	9162M/XZ991	9255M/XZ375
8932M/XR718	9041M/XW763	9163M/XV415	9257M/XX962
8934M/XR749	9042M/XL954	9166M/XW323	9258M/XW265
8935M/XR713	9044M/XS177	9167M/XV744	9259M/XS710
8937M/XX751	9047M/XW409	9168M/XZ132	9260M/XS734
8938M/WV746	9048M/XM403	9169M/XW547	9261M/*W2068*
8941M/XT456	9049M/XW404	9170M/XZ994	9262M/XZ358
8942M/XN185	9052M/WJ717	9172M/XW304	9263M/XW267

9264M/XS735	9284M/ZA267	9306M/XX979	9328M/ZD607
9265M/WK585	9285M/XR806	9308M/ZD932	9329M/ZD578
9266M/XZ119	9286M/XT905	9310M/ZA355	9330M/ZB684
9267M/XW269	9287M/WP962	9311M/ZA475	9331M/XW852
9268M/XR529	9288M/XX520	9312M/ZA474	9332M/XZ935
9269M/XT914	9289M/XX665	9314M/ZA320	9335M/ZA375
9270M/XZ145	9290M/XX626	9315M/ZA319	9336M/ZA407
9272M/XS486	9292M/XW892	9316M/ZA399	9337M/ZA774
9273M/XS726	9293M/XX830	9317M/ZA450	9338M/ZA325
9274M/XS738	9294M/XX655	9318M/ZA360	9339M/ZA323
9275M/XS729	9295M/XV497	9319M/XR516	9340M/XX745
9276M/XS733	9298M/ZE340	9320M/XX153	9341M/ZA357
9277M/XT601	9299M/XW870	9321M/XZ367	9342M/XR498
9278M/XS643	9300M/XX431	9322M/ZB686	9343M/XR506
9279M/XT681	9301M/XZ941	9323M/XV643	9344M/XV706
9281M/XZ146	9302M/ZD462	9324M/XV659	
9283M/XZ322	9303M/XV709	9326M/XV653	

	0	1	2	3	4	5	6	7	8	9
30	LA									
33								CB		
36					AY					
37			NL			SM				
40					IR					
41						MM				
42			SU			KT	PD	SB		
44					MR					
45		DA	DT	DM	DF	DN	DU			
46			WM							
47					RM					

Deck/Base Code Numbers	Letters	Unit	Location	Aircraft Type(s)
010 — 015	CU	820 NAS	Culdrose	Merlin HM2
180 — 190	CU	849 NAS	Culdrose	Sea King ASaC7
264 — 274	CU	814 NAS	Culdrose	Merlin HM2
300 — 308	VL	815 NAS	Yeovilton	Lynx HMA8
311 — 316	VL	815 NAS, MI Flight	Yeovilton	Lynx HMA8
321 — 474	*	815 NAS	Yeovilton	Lynx HMA8
500 — 515	CU	829 NAS	Culdrose	Merlin HM1
580 — 588	CU	824 NAS	Culdrose	Merlin HM2
630 — 674	VL	815 NAS	Yeovilton	Lynx HMA8
817 — 831	CU	771 NAS	Culdrose	Sea King HU5/HAS6

*See foregoing separate ships' Deck Letters Analysis
Note that only the 'last two' digits of the Code are worn by some aircraft types, especially helicopters.

RN LANDING PLATFORM AND SHORE STATION CODE-LETTERS

Code	Deck Letters	Vessel Name and Pennant No.	Vessel Type and Unit
—	AB	HMS *Albion* (L14)	Assault
—	AS	RFA *Argus* (A135)	Aviation Training ship
365	AY	HMS *Argyll* (F231)	Type 23 (815 NAS/211 Flt)
—	BK	HMS *Bulwark* (L15)	Assault
—	BV	RFA *Black Rover* (A273)	Fleet tanker
337	CB	RFA *Cardigan Bay* (L3009)	Landing ship(815 NAS/229 Flt)
—	CU	RNAS Culdrose (HMS *Seahawk*)	
451	DA	HMS *Daring* (D32)	Type 45 (815 NAS/200 Flt)
—	DF	HMS *Defender* (D36)	Type 45 (815 NAS)
—	DG	RFA *Diligence* (A132)	Maintenance
453	DM	HMS *Diamond* (D34)	Type 45 (815 NAS/208 Flt)
455	DN	HMS *Dragon* (D35)	Type 45 (815 NAS/219 Flt/226/Flt)
452	DT	HMS *Dauntless* (D33)	Type 45 (815 NAS)
456	DU	HMS *Duncan* (D37)	Type 45 (815 NAS)
—	FA	RFA *Fort Austin* (A386)	Support ship
—	FE	RFA *Fort Rosalie* (A385)	Support ship
—	GV	RFA *Gold Rover* (A271)	Fleet tanker
404	IR	HMS *Iron Duke* (F234)	Type 23 (815 NAS/207 Flt)
425	KT	HMS *Kent* (F78)	Type 23 (815 NAS)
300	LA	HMS *Lancaster* (F229)	Type 23 (829 NAS/202 Flt)
—	MB	RFA *Mounts Bay* (L3008)	Landing ship
415	MM	HMS *Monmouth* (F235)	Type 23 (829 NAS/215 Flt)
444	MR	HMS *Montrose* (F236)	Type 23 (815 NAS/214 Flt)
372	NL	HMS *Northumberland* (F238)	Type 23 (829 NAS)
—	O	HMS *Ocean* (L12)	Helicopter carrier
426	PD	HMS *Portland* (F79)	Type 23 (815 NAS/210 Flt)
474	RM	HMS *Richmond* (F239)	Type 23 (815 NAS/206 Flt)
427	SB	HMS *St Albans* (F83)	Type 23 (815 NAS)
375	SM	HMS *Somerset* (F82)	Type 23 (815 NAS/203 Flt)
422	SU	HMS *Sutherland* (F81)	Type 23 (815 NAS)
—	VL	RNAS Yeovilton (HMS *Heron*)	
462	WM	HMS *Westminster* (F237)	Type 23 (829 NAS/234 Flt)
—	—	HMS *Protector* (A173)	Ice patrol
—	—	RFA *Fort Victoria* (A387)	Auxiliary Oiler
—	—	RFA *Lyme Bay* (L3007)	Landing ship
—	—	RFA *Wave Knight* (A389)	Fleet tanker
—	—	RFA *Wave Ruler* (A390)	Fleet tanker

This table gives brief details of the markings worn by aircraft of RAF squadrons. While this may help to identify the operator of a particular machine, it may not always give the true picture. For example, from time to time aircraft are loaned to other units while others wear squadron marks but are actually operated on a pool basis. Squadron badges are usually located on the front fuselage.

Squadron	Type(s) operated	Base(s)	Distinguishing marks & other comments
No 1 Sqn	Typhoon T3/FGR4	RAF Lossiemouth	Badge (on tail): A red 1 with yellow wings on a white background, flanked in red. Roundel is flanked by two white chevrons, edged in red. Aircraft are coded F*.
No 2 Sqn	Typhoon T3/FGR4	RAF Lossiemouth	Badge: A wake knot on a white circular background flanked on either side by black and white triangles. Tail fin has a black stripe with white triangles and the badge repeated on it.
No 3 Sqn	Typhoon T3/FGR4	RAF Coningsby	Badge: A blue cockatrice on a white circular background flanked by two green bars edged with yellow. Tail fin as a green stripe edged with yellow. Aircraft are coded QO-*
No 4(R) Sqn	Hawk T2	RAF Valley	Badge (on nose): A yellow lightning flash on a red background with IV superimposed. Aircraft carry a yellow lightning flash on a red and black background on the tail and this is repeated in bars either side of the roundel on the fuselage. Part of No 4 FTS.
No 5 Sqn	Sentinel R1	RAF Waddington	Badge (on tail): A green maple leaf on a white circle over a red horizontal band.
No 6 Sqn	Typhoon T3/FGR4	RAF Lossiemouth	Badge (on tail): A red, winged can opener on a blue shield, edged in red. The roundel is flanked by a red zigzag on a blue background. Aircraft are coded E*.
No 7 Sqn	Chinook HC6	RAF Odiham	Badge (on tail): A blue badge containing the seven stars of Ursa Major ('The Plough') in yellow.
No 8 Sqn	Sentry AEW1	RAF Waddington	Badge (on tail): A grey, sheathed, Arabian dagger. Aircraft pooled with No 54(R) Sqn.
No 9 Sqn	Tornado GR4/GR4A	RAF Marham	Badge: A green bat on a black circular background, flanked by yellow and green horizontal stripes. The green bat also appears on the tail, edged in yellow.
No 10 Sqn	Voyager KC2/KC3	RAF Brize Norton	No markings worn.
No 11 Sqn	Typhoon T3/FGR4	RAF Coningsby	Badge (on tail): Two eagles in flight on a white shield. The roundel is flanked by yellow and black triangles. Aircraft are coded D*
No 12 Sqn	Tornado GR4	RAF Marham	
No 13 Sqn	Reaper	RAF Waddington	No markings worn.
No 14 Sqn	Shadow R1	RAF Waddington	No markings worn.
No 15(R) Sqn	Tornado GR4/GR4A	RAF Lossiemouth	Roman numerals XV appear in white on the tail.
No 16(R) Sqn	Tutor T1	RAF Cranwell	No markings carried. Aircraft pooled with No 57(R) Sqn. Part of No 1 EFTS.
No 18 Sqn	Chinook HC2A/HC3R/HC4	RAF Odiham	Badge (on tail): A red winged horse on a black circle. Aircraft pooled with No 27 Sqn.

Squadron	Type(s) operated	Base(s)	Distinguishing marks & other comments
No 22 Sqn	Sea King HAR3/HAR3A	A Flt: RMB Chivenor B Flt: Wattisham C Flt: RAF Valley	Badge: A black pi symbol in front of a white Maltese cross on a red circle.
No 24 Sqn	Hercules C4/C5/Atlas C1	RAF Brize Norton	No squadron markings carried. Aircraft pooled with No 30 Sqn and No 47 Sqn.
No 27 Sqn	Chinook HC2A/HC3R/HC4	RAF Odiham	Badge (on tail): A dark green elephant on a green circle, flanked by green and dark green stripes. Aircraft pooled with No 18 Sqn.
No 28 Sqn	Merlin HC3/HC3A	RAF Benson	Badge: A winged horse above two white crosses on a red shield. Aircraft currently pooled with 846 NAS.
No 29(R) Sqn	Typhoon T3/FGR4	RAF Coningsby	Badge (on tail): An eagle in flight, preying on a buzzard, with three red Xs across the top. The roundel is flanked by two white bars outlined by a red line, each containing three red Xs. Aircraft are coded B*.
No 30 Sqn	Hercules C4/C5	RAF Brize Norton	No squadron markings carried. Aircraft pooled with No 24 Sqn and No 47 Sqn.
No 31 Sqn	Tornado GR4/GR4A	RAF Marham	Badge: A gold, five-pointed star on a yellow circle flanked by yellow and green checks. The star is repeated on the tail.
No 32(The Royal) Sqn	BAe 125 CC3/146 CC2/ 146 C3/Agusta 109	RAF Northolt	No squadron markings carried but aircraft carry a distinctive livery with a red stripe, edged in blue along the middle of the fuselage and a red tail.
No 33 Sqn	Puma HC2	RAF Benson	Badge: A stag's head.
No 39 Sqn	Predator/ Reaper	Nellis AFB Creech AFB	No markings worn
No 41(R) Test & Evaluation Sqn [FJWOEU]	Tornado GR4/ Typhoon T3/FGR4	RAF Coningsby	Badge: A red, double armed cross on the tail with a gold crown above. White and red horizontal bars flanking the roundel on the fuselage. Aircraft are coded EB-*.
No 45(R) Sqn	Super King Air 200/200GT/Tutor T1EA	RAF Cranwell	The King Airs carry a dark blue stripe on the tail superimposed with red diamonds. Part of No 3 FTS.
No 47 Sqn	Hercules C4/C5	RAF Brize Norton	No squadron markings usually carried. Aircraft pooled with No 24 Sqn and No 30 Sqn.
No 51 Sqn	Airseeker R1	RAF Waddington	Badge (on tail): A red goose in flight.
No 54(R) Sqn [ISTAR OCU]	Sentry AEW1	RAF Waddington	Based aircraft as required.
No 56(R) Sqn [ISTAR Test & Evaluation Sqn]	Shadow R1/ Sentry AEW1/ Sentinel R1	RAF Waddington	Based aircraft as required.
No 57(R) Sqn	Tutor T1	RAF Cranwell	No markings carried. Aircraft pooled with No 16(R) Sqn. Part of No 1 EFTS.

Squadron	Type(s) operated	Base(s)	Distinguishing marks & other comments
No 60(R) Sqn	Griffin HT1	RAF Shawbury [DHFS] & RAF Valley [SARTU]	No squadron markings usually carried.
No 70 Sqn	Atlas C1	RAF Brize Norton	No squadron markings usually carried.
No 72(R) Sqn	Tucano T1	RAF Linton-on-Ouse	Badge: A black swift in flight on a red disk, flanked by blue bars edged with red. The blue bars edged with red also flank the roundel on the fuselage; part of No 1 FTS
No 84 Sqn	Griffin HAR2	RAF Akrotiri	Badge (on tail): A scorpion on a playing card symbol (diamonds, clubs etc.). Aircraft carry a vertical light blue stripe through the roundel on the fuselage.
No 99 Sqn	Globemaster III	RAF Brize Norton	Badge (on tail): A black puma leaping.
No 100 Sqn	Hawk T1A	RAF Leeming	Badge (on tail): A skull in front of two bones crossed. Aircraft are usually coded C*. Incorporates the Joint Forward Air Control Training and Standards Unit (JFACTSU)
No 101 Sqn	Voyager KC2/KC3	RAF Brize Norton	No markings worn.
No 115(R) Sqn	Tutor T1	Wittering	No markings carried. Part of the CFS.
No 202 Sqn	Sea King HAR3	A Flt: RAF Boulmer D Flt: RAF Lossiemouth E Flt: RAF Leconfield	Badge: A mallard alighting on a white circle.
No 206(R) Sqn [HAT&ES]	Hercules C4/C5	RAF Brize Norton/ Boscombe Down	No squadron markings usually carried.
No 208(R) Sqn	Hawk T1/T1A/T1W	RAF Valley	Badge (on tail): A Sphinx inside a white circle, flanked by flashes of yellow. Aircraft also carry blue and yellow bars either side of the roundel on the fuselage and a blue and yellow chevron on the nose. Part of No 4 FTS.
No 230 Sqn	Puma HC2	RAF Benson	Badge: A tiger in front of a palm tree on a black pentagon.
No 1312 Flt	Hercules C5	Mount Pleasant, FI	Badge (on tail): A red Maltese cross on a white circle, flanked by red and white horizontal bars.
No 1435 Flt	Typhoon FGR4	Mount Pleasant, FI	Badge (on tail): A red Maltese cross on a white circle, flanked by red and white horizontal bars.
No 1564 Flt	Sea King HAR3	Mount Pleasant, FI	No markings known.

Some UAS aircraft carry squadron badges and markings, usually on the tail. Squadron crests all consist of a white circle surrounded by a blue circle, topped with a red crown and having a yellow scroll beneath. Each differs by the motto on the scroll, the UAS name running around the blue circle & by the contents at the centre and it is the latter which are described below.

* All AEFs come under the administration of local UASs and these are listed here.

UAS	Base	Marks
Bristol UAS/ No 3 AEF	Colerne	A sailing ship on water.
Cambridge UAS/ No 5 AEF	RAF Wittering	A heraldic lion in front of a red badge. Aircraft pooled with University of London AS.
East Midlands Universities AS/ No 7 AEF	Cranwell	A yellow quiver, full of arrows. [To move to Wittering in 2015.]
East of Scotland UAS/ No 12 AEF	RAF Leuchars	An open book in front of a white diagonal cross edged in blue.
Liverpool UAS	RAF Woodvale	A bird atop an open book, holding a branch in its beak. Aircraft pooled with Manchester and Salford Universities AS
Manchester and Salford Universities AS/ No 10 AEF	RAF Woodvale	A bird of prey with a green snake in its beak. Aircraft pooled with Liverpool UAS
Northumbrian Universities AS/ No 11 AEF	RAF Leeming	A white cross on a blue background.
Oxford UAS/ No 6 AEF	RAF Benson	An open book in front of crossed swords.
Southampton UAS/ No 2 AEF	Boscombe Down	A red stag in front of a stone pillar.
Universities of Glasgow and Strathclyde AS/ No 4 AEF	Glasgow	A bird of prey in flight, holding a branch in its beak, in front of an upright sword.
University of Birmingham AS/ No 8 AEF	DCAE Cosford	A blue griffon with two heads.
University of London AS	RAF Wittering	A globe superimposed over an open book. Aircraft pooled with Cambridge UAS.
University of Wales AS/ No 1 AEF	MoD St Athan	A red Welsh dragon in front of an open book, clasping a sword. Some aircraft have the dragon in front of white and green squares.
Yorkshire Universities AS/ No 9 AEF	RAF Linton-on-Ouse	An open book in front of a Yorkshire rose with leaves.

This table gives brief details of the markings worn by aircraft of FAA squadrons. Squadron badges, when worn, are usually located on the front fuselage. All FAA squadron badges comprise a crown atop a circle edged in gold braid and so the badge details below list only what appears in the circular part.

Squadron	Type(s) operated	Base(s)	Distinguishing marks & other comments
No 727 NAS	Tutor T1	RNAS Yeovilton	Badge: The head of Britannia wearing a gold helmet on a background of blue and white waves.
No 736 NAS	Hawk T1A	RNAS Culdrose & RNAS Yeovilton	A white lightning bolt on the tail.
No 750 NAS	Avenger T1	RNAS Culdrose	Badge: A Greek runner bearing a torch & sword on a background of blue and white waves.
No 771 NAS	Sea King HU5/HAS6	RNAS Culdrose & Prestwick	Badge: Three bees on a background of blue and white waves.
No 814 NAS	Merlin HM2	RNAS Culdrose	Badge: A winged tiger mask on a background of dark blue and white waves.
No 815 NAS	Lynx HMA8	RNAS Yeovilton	Badge: A winged, gold harpoon on a background of blue and white waves.
No 820 NAS	Merlin HM2	RNAS Culdrose	Badge: A flying fish on a background of blue and white waves.
No 824 NAS	Merlin HM2	RNAS Culdrose	Badge: A heron on a background of blue and white waves.
No 825 NAS	Wildcat HMA2	RNAS Yeovilton	Badge: An eagle over a Maltese cross.
No 829 NAS	Merlin HM1	RNAS Culdrose	Badge: A kingfisher hovering on a background of blue and white waves.
No 845 NAS	Sea King HC4	RNAS Yeovilton	Badge: A dragonfly on a background of blue and white waves.
No 846 NAS	Merlin HC3	RAF Benson	Badge: A swordsman riding a winged horse whilst attacking a serpent on a background of blue and white waves. Aircraft currently pooled with 28 Sqn. [To move to RNAS Yeovilton in Easter 2015.]
No 847 NAS	Wildcat AH1	RNAS Yeovilton	Badge: A gold sea lion on a blue background.
No 849 NAS	Sea King ASaC7	RNAS Culdrose	Badge: A winged streak of lightning with an eye in front on a background of blue and white waves.

This section lists the codes worn by some UK military aircraft and, alongside, the Registration of the aircraft currently wearing this code. It should be pointed out that in some cases more than one aircraft wears the same code but the aircraft listed is the one believed to be in service with the unit concerned at the time of going to press. This list will be updated regularly and those with Internet access can download the latest version via the 'Military Aircraft Markings' Web Site, www.militaryaircraftmarkings.co.uk.

Code	Serial	Code	Serial	Code	Serial	Code	Serial
ROYAL AIR FORCE				D	ZJ120	BV	ZJ910
BAE Hawk T1/T2		**Bell 412EP Griffin HT1**		E	ZJ121	BX	ZK343
95-Y	XX246 & XX318	E	ZJ242	F	ZJ122	BY	ZJ921
A	ZK010	I	ZJ235	G	ZJ123	BZ	ZK349
B	ZK011	K	ZJ708	H	ZJ124	DA	ZJ931
C	ZK012	L	ZJ241	J	ZJ125	DB	ZJ932
D	ZK013	O	ZJ707	K	ZJ126	DC	ZJ919
E	ZK014	R	ZJ239	L	ZJ127	DD	ZJ924
F	ZK015	S	ZJ234	M	ZJ128	DE	ZK305
G	ZK016	T	ZJ237	N	ZJ129	DF	ZJ933
H	ZK017	U	ZJ240	O	ZJ130	DG	ZK300
I	ZK018	X	ZJ236	P	ZJ131	DH	ZJ942
J	ZK019	Y	ZJ238	Q	ZJ132	DJ	ZJ935
K	ZK020			R	ZJ133	DL	ZJ929
L	ZK021	**B-V Chinook**		S	ZJ134	DM	ZJ923
M	ZK022	AA	ZA670	T	ZJ135	DN	ZK323
N	ZK023	AB	ZA671	U	ZJ136	DP	ZJ915
O	ZK024	AD	ZA674	W	ZJ137	DR	ZJ912
P	ZK025	AE	ZA675	X	ZJ138	DW	ZJ808
Q	ZK026	AF	ZA677	AA	ZJ990	DXI	ZJ939
R	ZK027	AG	ZA679	AB	ZJ992	EC	ZK302
S	ZK028	AH	ZA680	AC	ZJ994	ED	ZK342
T	ZK029	AI	ZA681	AD	ZJ995	EF	ZK347
U	ZK030	AJ	ZA682	AE	ZJ998	EG	ZJ937
V	ZK031	AK	ZA683	AF	ZK001	EI	ZK324
W	ZK032	AL	ZA684			EK	ZK311
X	ZK033	AN	ZA705	**Eurofighter Typhoon**		EM	ZK312
Y	ZK034	AO	ZA707	C	ZJ950	EO	ZK314
Z	ZK035	AP	ZA708	D	ZK301	EP	ZK345
AA	ZK036	AR	ZA710	F	ZJ944	ER	ZK346
AB	ZK037	AT	ZA712	H	ZJ949	ES	ZK317
CC	XX203	AV	ZA714	O	ZK307	ET	ZK318
CD	XX332	AW	ZA720	W	ZK313	EU	ZK321
CE	XX258	BN	ZA718	AX	ZK303	EV	ZK320
CF	XX202	DB	ZD574	BA	ZJ803	EX	ZK381
CG	XX198	DC	ZD575	BB	ZK379	EY	ZJ809
CH	XX346	DD	ZD980	BC	ZJ800	FA	ZK316
CI	XX321	DF	ZD982	BD	ZJ805	FB	ZK334
CJ	XX329	DG	ZD983	BE	ZJ806	FC	ZK335
CK	XX339	DH	ZD984	BF	ZJ807	FD	ZK336
CL	XX255	HB	ZH775	BG	ZK380	FE	ZK331
CM	XX280	HC	ZH776	BH	ZJ814	FF	ZK338
CO	XX200	HE	ZH777	BI	ZJ810	FH	ZK329
CQ	XX184	HF	ZH891	BJ	ZJ801	FI	ZK340
CR	XX189	HG	ZH892	BK	ZJ812	FJ	ZK341
		HH	ZH893	BL	ZJ813	FK	ZK325
Beech King Air 200		HI	ZH894	BM	ZJ804	FL	ZK310
D	ZK458	HJ	ZH895	BN	ZK383	FM	ZK304
K	ZK451	HK	ZH896	BO	ZJ927	FN	ZK348
L	ZK452	HR	ZH903	BP	ZJ802	FP	ZK337
O	ZK455			BQ	ZK353	FQ	ZJ928
P	ZK456	**EHI-101 Merlin**		BR	ZK322	FR	ZK327
U	ZK460	B	ZJ118	BS	ZK328	FS	ZK333
X	ZK459	C	ZJ119	BT	ZK306	FT	ZK330

Code	Serial	Code	Serial	Code	Serial	Code	Serial
...Eurofighter Typhoon		055	ZA587	D	XZ588	642	ZF558
FX	ZK382	056	ZA588	F	XZ590	643	XZ736
EB-A	ZJ946	057	ZA589	H	XZ592	644	ZD257
EB-B	ZK339	058	ZA591	I	XZ593	645	ZD566
EB-H	ZJ815	059	ZA592	K	XZ595	671	XZ722
EB-J	ZK332	060	ZA594	L	XZ596	673	ZD266
EB-L	ZJ947	062	ZA596	M	XZ597		
EB-R	ZJ930	063	ZA597	N	XZ598	**WS61 Sea King**	
IT	ZK344	064	ZA598	P	XZ599	16	ZA166
QO-A	ZJ920	072	ZA609	Q	ZA105	17	XV670
QO-B	ZJ811	073	ZA611	S	ZE369	18	XV648
QO-C	ZJ936	074	ZA612	T	ZE370	19	ZA130
QO-D	ZK319	075	ZA613	V	ZH541	20	ZA137
QO-G	ZJ917	077	ZD707	X	ZH543	21	XV666
QO-J	ZJ941	078	ZD709	Z	ZH545	22	ZA167
QO-L	ZJ918	079	ZD711			24	XZ920
QO-M	ZJ913	081	ZD713	**ROYAL NAVY**		25	ZA134
QO-P	ZK309	083	ZD715	**BAE Hawk T1**		26	XV661
QO-R	ZJ925	084	ZD716	840	XX240	27	XV673
QO-S	ZJ916	087	ZD739	846	XX205	28	XV647
QO-T	ZJ934	088	ZD740	849	XX316	29	XV705
QO-X	ZJ914	089	ZD741			30	XZ578
QO-Y	ZJ926	090	ZD742	**EHI-101 Merlin**		180	XV649
QO-Z	ZJ911	092	ZD744	10	ZH856	181	XV697
TP-V	ZK308	093	ZD745	12	ZH864	183	XV671
		095	ZD747	14	ZH857	184	XV707
Panavia Tornado GR4		096	ZD748	15	ZH827	185	XV656
003	ZA369	097	ZD749	67	ZH849	186	ZE418
004	ZA370	099	ZD790	68	ZH855	187	XV672
006	ZA372	100	ZD792	70	ZH838	188	XV714
007	ZA373	101	ZD793	80	ZH850	189	ZE420
008	ZA393	105	ZD842	81	ZH845	190	XV664
011	ZA400	106	ZD843	82	ZH851	191	ZA126
014	ZA405	107	ZD844	83	ZH853	192	ZE422
015	ZA406	109	ZD848	84	ZH834	D	ZA299
019	ZA447	110	ZD849	85	ZH832	G	ZG821
020	ZA449	112	ZD851	86	ZH842	K	ZE427
022	ZA453	113	ZD890	503	ZH837	O	ZF118
023	ZA456	116	ZE116			Q	ZA296
024	ZA458	118	ZG705	**WS Lynx**		U	ZA295
026	ZA461	127	ZG719	301	ZD260	V	ZF122
027	ZA462	128	ZG750	302	ZD265	X	ZF117
028	ZA463	129	ZG752	306	XZ731	Y	ZA298
029	ZA469	133	ZG771	313	XZ697	WP	ZF116
031	ZA472	136	ZG779	314	ZD261	WT	ZA314
032	ZA473	137	ZG791	315	ZD262		
034	ZA541	F	ZA602	316	XZ726	**ARMY AIR CORPS**	
035	ZA542	EB-B	ZA601	337	XZ725	**WS Gazelle AH1**	
036	ZA543	EB-G	ZA600	365	ZD268	S	ZB691
038	ZA546	EB-Q	ZG777	372	XZ689	U	ZB693
040	ZA548	EB-X	ZA607	375	XZ255	Y	ZB692
041	ZA549	EB-Z	ZG775	404	ZD565		
042	ZA550			417	XZ690	**WS Lynx AH7**	
043	ZA551	**Shorts Tucano T1**		425	ZF557	A	XZ670
045	ZA553	MP-A	ZF170	444	XZ725	B	XZ184
046	ZA554	MP-W	ZF338	451	XZ255	C	XZ180
047	ZA556	RA-F	ZF239	452	XZ729	H	XZ192
048	ZA557			454	XZ691	M	ZD274
049	ZA559	**WS61 Sea King**		455	ZD252	O	XZ651
050	ZA560	A	XZ585	456	ZF560	S	ZB691
051	ZA562	B	XZ586	474	ZD259	T	XZ674
054	ZA585	C	XZ587	640	XZ690		

Some *historic, classic and warbird* aircraft carry the markings of overseas air arms and can be seen in the UK, mainly preserved in museums and collections or taking part in air shows.

Serial	Type (code/other identity)	Owner/operator, location or fate	Notes
AFGHANISTAN			
–	Hawker Afghan Hind (BAPC 82)	RAF Museum, Cosford	
ARGENTINA			
–	Bell UH-1H Iroquois (AE-406/*998-8888*) [Z]	RAF Valley, instructional use	
0729	Beech T-34C Turbo Mentor	FAA Museum, stored Cobham Hall, RNAS Yeovilton	
0767	Aermacchi MB339AA	South Yorkshire Aircraft Museum, Doncaster	
A-515	FMA IA58 Pucara (ZD485)	RAF Museum, Cosford	
A-517	FMA IA58 Pucara (G-BLRP)	Privately owned, Channel Islands	
A-522	FMA IA58 Pucara (8768M)	FAA Museum, at NE Aircraft Museum, Usworth	
A-528	FMA IA58 Pucara (8769M)	Norfolk & Suffolk Avn Museum, Flixton	
A-533	FMA IA58 Pucara (ZD486) <ff>	Privately owned, Cheltenham	
A-549	FMA IA58 Pucara (ZD487)	Imperial War Museum, Duxford	
AE-331	Agusta A109A (ZE411)	FAA Museum, stored Cobham Hall, RNAS Yeovilton	
AE-409	Bell UH-1H Iroquois [656]	Museum of Army Flying, Middle Wallop	
AE-422	Bell UH-1H Iroquois	FAA Museum, stored Cobham Hall, RNAS Yeovilton	
AUSTRALIA			
369	Hawker Fury ISS (F-AZXL) [D]	Privately owned, Avignon, France	
A2-4	Supermarine Seagull V (VH-ALB)	RAF Museum, Hendon	
A11-301	Auster J/5G (G-ARKG) [931-NW]	Privately owned, Spanhoe	
A16-199	Lockheed Hudson IIIA (G-BEOX) [SF-R]	RAF Museum, Hendon	
A17-48	DH82A Tiger Moth (G-BPHR)	Privately owned, Wanborough, Wilts	
A17-376	DH82A Tiger Moth (T6830/G-ANJI) [376]	Privately owned, Gloucester	
A19-144	Bristol 156 Beaufighter XIc (JM135/A8-324)	The Fighter Collection, Duxford	
A92-255	GAF Jindivik 102	DPA/QinetiQ, Boscombe Down, apprentice use	
A92-664	GAF Jindivik 103A	Boscombe Down Aviation Collection, Old Sarum	
A92-708	GAF Jindivik 103BL	Bristol Aero Collection, stored Filton	
A92-740	GAF Jindivik 103BL	Caernarfon Airworld	
A92-908	GAF Jindivik 104AL (ZJ503)	No 2445 Sqn ATC, Llanbedr	
N16-114	WS61 Sea King Mk.50A	Privately owned, Horsham	
N16-125	WS61 Sea King Mk.50A	Privately owned, Horsham	
N16-238	WS61 Sea King Mk.50A	Privately owned, Horsham	
N16-239	WS61 Sea King Mk.50A	Privately owned, Horsham	
N16-918	WS61 Sea King Mk.50B	Privately owned, Horsham	
WH589	Hawker Fury ISS (F-AZXJ) [115-NW]	Privately owned, Dijon, France	
BELGIUM			
A-41	SA318C Alouette II	The Helicopter Museum, Weston-super-Mare	
FT-36	Lockheed T-33A Shooting Star	Dumfries & Galloway Avn Mus, Dumfries	
H-01	Agusta A109HO	Privately owned, Woodmancote, W Sussex	
H-02	Agusta A109HO	Privately owned, Cotswold Airport	
H-05	Agusta A109HO	Privately owned, Cotswold Airport	
H-50	Noorduyn AT-16 Harvard IIB (OO-DAF)	Privately owned, Brasschaat, Belgium	
IF-68	Hawker Hunter F6 <ff>	Privately owned, Kings Lynn	
L-44	Piper L-18C Super Cub (OO-SPQ)	Royal Aéro Para Club de Spa, Belgium	
L-47	Piper L-18C Super Cub (OO-SPG)	Aeroclub Brasschaat VZW, Brasschaat, Belgium	
OL-L49	Piper L-18C Super Cub (L-156/OO-LGB)	Aeroclub Brasschaat VZW, Brasschaat, Belgium	
V-4	SNCAN Stampe SV-4B (OO-EIR)	Antwerp Stampe Centre, Antwerp-Deurne, Belgium	
V-18	SNCAN Stampe SV-4B (OO-GWD)	Antwerp Stampe Centre, Antwerp-Deurne, Belgium	
V-29	SNCAN Stampe SV-4B (OO-GWB)	Antwerp Stampe Centre, Antwerp-Deurne, Belgium	
V-66	SNCAN Stampe SV-4C (OO-GWA)	Antwerp Stampe Centre, Antwerp-Deurne, Belgium	
BOLIVIA			
FAB184	SIAI-Marchetti SF.260W (G-SIAI)	Privately owned, Booker	

Notes	Serial	Type (code/other identity)	Owner/operator, location or fate
	BRAZIL		
	1317	Embraer T-27 Tucano	Shorts, Belfast (engine test bed)
	BURKINA FASO		
	BF8431	SIAI-Marchetti SF.260 (G-NRRA) [31]	Privately owned, Oaksey Park
	CANADA		
	–	Lockheed T-33A Shooting Star (17473)	Midland Air Museum, Coventry
	622	Piasecki HUP-3 Retriever (51-16622/N6699D)	The Helicopter Museum, Weston-super-Mare
	920	VS Stranraer (CF-BXO) [Q-N]	RAF Museum, Hendon
	3091	NA81 Harvard II (3019/G-CPPM)	Beech Restorations, Bruntingthorpe
	3349	NA64 Yale (G-BYNF)	Privately owned, Duxford
	5487	Hawker Hurricane XII (G-CBOE)	*Repainted in Finnish marks as HC-465, July 2014*
	9041	Bristol 149 Bolingbroke IV <ff>	Manx Aviation Museum, Ronaldsway
	9048	Bristol 149 Bolingbroke IV	Bristol Aero Collection, Filton
	9754	Consolidated PBY-5A Catalina (VP-BPS) [P]	Privately owned, Lee-on-Solent
	9893	Bristol 149 Bolingbroke IVT	Imperial War Museum store, Duxford
	9940	Bristol 149 Bolingbroke IVT	Royal Scottish Mus'm of Flight, E Fortune
	15252	Fairchild PT-19A Cornell (comp 15195)	RAF Museum Reserve Collection, Stafford
	16693	Auster J/1N Alpha (G-BLPG) [693]	Privately owned, Clacton
	18393	Avro Canada CF-100 Canuck 4B (G-BCYK)	Imperial War Museum, Duxford
	18671	DHC1 Chipmunk 22 (WP905/7438M/G-BNZC) [671]	The Shuttleworth Collection, Old Warden
	20249	Noorduyn AT-16 Harvard IIB (PH-KLU) [XS-249]	Privately owned, Texel, The Netherlands
	20310	CCF T-6J Texan (G-BSBG) [310]	Privately owned, Tatenhill
	21417	Canadair CT-133 Silver Star	Yorkshire Air Museum, Elvington
	23140	Canadair CL-13 Sabre [AX] <ff>	Midland Air Museum, Coventry
	23380	Canadair CL-13 Sabre <rf>	Privately owned, Haverigg
	FJ777	Boeing-Stearman PT-17D Kaydet(41-8689/G-BIXN)	Privately owned, Rendcomb
	KN448	Douglas Dakota IV <ff>	Science Museum, South Kensington
	CHILE		
	H225	Aérospatiale SA 330H Puma	Holmbush Paintball, Faygate
	CHINA		
	68 r	Nanchang CJ-6A Chujiao (2751219/G-BVVG)	Privately owned, White Waltham
	61367	Nanchang CJ-6A Chujiao (4532009/G-CGHB) [37]	Privately owned, Headcorn
	61762	Nanchang CJ-6A Chujiao (4532008/G-CGFS) [72]	Privately owned, Seething
	CZECH REPUBLIC		
	3677	Letov S-103 (MiG-15bisSB) (613677)	Royal Scottish Mus'm of Flight, E Fortune
	3794	Letov S-102 (MiG-15) (623794)	Norfolk & Suffolk Avn Museum, Flixton
		(starboard side only, painted in Polish marks as 1972 on port side)	
	9147	Mil Mi-4	The Helicopter Museum, Weston-super-Mare
	DENMARK		
	A-011	SAAB A-35XD Draken	Privately owned, Westhoughton, Lancs
	AR-107	SAAB S-35XD Draken	Newark Air Museum, Winthorpe
	E-419	Hawker Hunter F51 (G-9-441)	North-East Aircraft Museum, Usworth
	E-420	Hawker Hunter F51 (G-9-442) <ff>	Privately owned, Walton-on-Thames
	E-421	Hawker Hunter F51 (G-9-443)	Brooklands Museum, Weybridge
	E-424	Hawker Hunter F51 (G-9-445)	South Yorkshire Aircraft Museum, Doncaster
	ET-272	Hawker Hunter T7 <ff>	Norfolk & Suffolk Avn Museum, Flixton
	K-682	Douglas C-47A Skytrain (OY-BPB)	Foreningen For Flyvende Mus, Vaerløse, Denmark
	L-866	Consolidated PBY-6A Catalina (8466M)	RAF Museum, Cosford
	P-129	DHC-1 Chipmunk 22 (OY-ATO)	Privately owned, Roskilde, Denmark
	R-756	Lockheed F-104G Starfighter	Midland Air Museum, Coventry
	S-881	Sikorsky S-55C	The Helicopter Museum, Weston-super-Mare
	S-882	Sikorsky S-55C	Skirmish Paintball, Portishead
	S-886	Sikorsky S-55C	Hamburger Hill Paintball, Marksbury, Somerset
	S-887	Sikorsky S-55C	The Helicopter Museum, Weston-super-Mare

Serial	Type (code/other identity)	Owner/operator, location or fate	Notes
ECUADOR			
FAE 259	BAC Strikemaster Mk.80A (G-UPPI) [T59]	Privately owned, St Athan	
EGYPT			
356	Heliopolis Gomhouria Mk 6	Privately owned, Breighton	
764	Mikoyan MiG-21SPS <ff>	Privately owned, Northampton	
773	WS61 Sea King Mk.47 (WA.823)	RNAS Yeovilton Fire Section	
774	WS61 Sea King Mk.47 (WA.822)	Privately owned,	
775	WS61 Sea King Mk.47 (WA.824)	Privately owned,	
776	WS61 Sea King Mk.47 (WA.825) <ff>	Mayhem Paintball, Aybridge, Essex	
0446	Mikoyan MiG-21UM <ff>	Thameside Aviation Museum, Tilbury	
7907	Sukhoi Su-7 <ff>	Robertsbridge Aviation Society, Mayfield	
FINLAND			
GA-43	Gloster Gamecock II (G-CGYF)	Privately owned, Dursley, Glos	
GN-101	Folland Gnat F1 (XK741)	Midland Air Museum, Coventry	
HC-465	Hawker Hurricane XII (RCAF 5487/G-CBOE)	Privately owned, Thruxton	
SZ-12	Focke-Wulf Fw44J Stieglitz (D-EXWO)	Privately owned, Bienenfarm, Germany	
VI-3	Valtion Viima 2 (OO-EBL)	Privately owned, Brasschaat, Belgium	
FRANCE			
1/4513	Spad XIII <R> (G-BFYO/*S3398*)	American Air Museum, Duxford	
37	Nord 3400 (G-ZARA) [MAB]	Privately owned, Swanton Morley	
54	SNCAN NC856A Norvigie (G-CGWR) [AOM]	Privately owned, Spanhoe	
67	SNCAN 1101 Noralpha (F-GMCY) [CY]	Privately owned, la Ferté-Alais, France	
70	Dassault Mystère IVA	Midland Air Museum, Coventry	
78	Nord 3202B-1 (G-BIZK)	Privately owned, Norfolk	
79	Dassault Mystère IVA [2-EG]	Norfolk & Suffolk Avn Museum, Flixton	
82	Curtiss H75-C1 Hawk (G-CCVH) [X-881]	The Fighter Collection, Duxford	
82	NA T-28D Fennec (F-AZKG)	Privately owned, Strasbourg, France	
83	Dassault Mystère IVA [8-MS]	Newark Air Museum, Winthorpe	
83	Morane-Saulnier MS733 Alcyon (F-AZKS)	Privately owned, Montlucon, France	
84	Dassault Mystère IVA [8-NF]	Lashenden Air Warfare Museum, Headcorn	
85	Dassault Mystère IVA [8-MV]	Cold War Jets Collection, Bruntingthorpe	
104	MH1521M Broussard (F-GHFG) [307-FG]	Privately owned, Montceau-les-Mines, France	
105	Nord N2501F Noratlas (F-AZVM) [62-SI]	Le Noratlas de Provence, Marseilles, France	
106	MH1521M Broussard (F-GKJT) [33-JT]	Privately owned, Montceau-les-Mines, France	
108	MH1521M Broussard (F-BNEX) [50S9]	Privately owned, Lelystad, The Netherlands	
108	SO1221 Djinn (FR108) [CDL]	The Helicopter Museum, Weston-super-Mare	
121	Dassault Mystère IVA [8-MY]	City of Norwich Aviation Museum	
128	Morane-Saulnier MS733 Alcyon (F-BMMY)	Privately owned, St Cyr, France	
143	Morane-Saulnier MS733 Alcyon (G-MSAL)	Privately owned, Spanhoe	
146	Dassault Mystère IVA [8-MC]	North-East Aircraft Museum, Usworth	
156	SNCAN Stampe SV-4B (G-NIFE)	Privately owned, Gloucester	
158	Dassault MD312 Flamant (F-AZGE) [12-XA]	Privately owned, Albert, France	
208	MH1521C1 Broussard (G-YYYY) [IR]	Privately owned, Eggesford	
255	MH1521M Broussard (G-CIGH) [5-ML]	Privately owned, Breighton	
260	Dassault MD311 Flamant (F-AZKT) [316-KT]	Privately owned, Albert, France	
261	MH1521M Broussard (F-GIBN) [30-QA]	Privately owned, Rotterdam, The Netherlands	
282	Dassault MD311 Flamant (F-AZFX) [316-KY]	Memorial Flt Association, la Ferté-Alais, France	
290	Dewoitine D27 (F-AZJD)	Les Casques de Cuir, la Ferté-Alais, France	
316	MH1521M Broussard (F-GGKR) [315-SN]	Privately owned, Lognes, France	
318	Dassault Mystère IVA [8-NY]	Dumfries & Galloway Avn Mus, Dumfries	
319	Dassault Mystère IVA [8-ND]	Rebel Air Museum, Andrewsfield	
319	Grumman TBM-3E Avenger (HB-RDG) [4F.6]	Privately owned, Lausanne, Switzerland	
351	Morane-Saulnier MS317 (G-MOSA) [HY22]	Privately owned, Barton	
354	Morane-Saulnier MS315E-D2 (G-BZNK)	Privately owned, Wickenby	
394	SNCAN Stampe SV-4C (G-BIMO)	*Crashed 10 July 2010, Oxon*	
538	Dassault Mirage IIIE [3-QH]	Yorkshire Air Museum, Elvington	
569	Fouga CM170R Magister [F-AZZP]	Privately owned, Le Havre, France	
42157	NA F-100D Super Sabre [11-ML]	North-East Aircraft Museum, Usworth	
54439	Lockheed T-33A Shooting Star (55-4439) [WI]	North-East Aircraft Museum, Usworth	

Notes	Serial	Type (code/other identity)	Owner/operator, location or fate
	63938	NA F-100F Super Sabre [11-EZ]	Lashenden Air Warfare Museum, Headcorn
	125716	Douglas AD-4N Skyraider (F-AZFN) [22-DG]	Privately owned, Mélun, France
	127002	Douglas AD-4NA Skyraider (F-AZHK) [20-LN]	Privately owned, Avignon, France
	517692	NA T-28S Fennec (G-TROY) [142]	Privately owned, Duxford
18-5395	Piper L-18C Super Cub (52-2436/G-CUBJ) [CDG]	Privately owned, Old Warden
	51-7545	NA T-28S Fennec (N14113)	Privately owned, Duxford
C850	Salmson 2A2 <R>	Barton Aviation Heritage Society, Barton
MS824	Morane-Saulnier Type N <R> (G-AWBU)	Privately owned, Booker
N856	SNCAN NC856 (G-CDWE)	Privately owned, Wickenby
N1977	Nieuport Scout 17/23 <R> (N1723/G-BWMJ) [8]	Privately owned, Duxford
	GERMANY		
	–	Fieseler Fi103R-IV (V-1) (BAPC 91)	Lashenden Air Warfare Museum, Headcorn
	–	Fokker Dr1 Dreidekker <R> (BAPC 88)	FAA Museum, stored Cobham Hall, RNAS Yeovilton
	–	Messerschmitt Bf109 <R> (6357/BAPC 74) [6]	Kent Battle of Britain Museum, Hawkinge
	–	Messerschmitt Bf109 <R> [<-]	Battle of Britain Experience, Canterbury
1	Hispano HA 1.112M1L Buchón (C.4K-31/G-AWHE)	Spitfire Ltd, Humberside
1	Messerschmitt Bf109G <R> (BAPC 240)	Yorkshire Air Museum, Elvington
3	Messerschmitt Bf109G-10 (D-FDME)	Messerschmitt Stiftung, Manching, Germany
3	SNCAN 1101 Noralpha (G-BAYV)	Barton Aviation Heritage Society, Barton
6	Messerschmitt Bf109G-2/Trop (10639/8478M/G-USTV)	RAF Museum, Hendon
7	Messerschmitt Bf109G-4 (D-FWME)	Messerschmitt Stiftung, Manching, Germany
8	Focke-Wulf Fw190 <R> (G-WULF)	Privately owned, Halfpenny Green
9	Focke-Wulf Fw190 <R> (G-CCFW)	Privately owned, Gloucester
10	Hispano HA 1.112M1L Buchón (C4K-102/G-BWUE)	Historic Flying Ltd, Duxford
14	Messerschmitt Bf109 <R> (BAPC 67)	Kent Battle of Britain Museum, Hawkinge
14	Nord 1002 (G-ETME)	Privately owned, White Waltham
33/15	Fokker EIII <R> (G-CHAW)	Privately owned, Membury
87	Heinkel He111 <R> <ff>	Privately owned, East Kirkby
105/15	Fokker EIII <R> (G-UDET)	Privately owned, Horsham
139	SNCAN 1101 Noralpha (G-BSMD)	Privately owned, Perth
152/17	Fokker Dr1 Dreidekker <R> (F-AZPQ)	Les Casques de Cuir, la Ferté-Alais, France
152/17	Fokker Dr1 Dreidekker <R> (G-BVGZ)	Privately owned, Breighton
157/18	Fokker D.VIII <R> (BAPC 239)	Norfolk & Suffolk Air Museum, Flixton
210/16	Fokker EIII (BAPC 56)	Science Museum, South Kensington
403/17	Fokker Dr1 Dreidekker <R> (G-CDXR)	Privately owned, Popham
416/15	Fokker EIII <R> (G-GSAL)	Privately owned, Aston Down
422/15	Fokker EIII <R> (G-AVJO)	Privately owned, Booker
422/15	Fokker EIII <R> (G-FOKR)	Privately owned, Eshott
425/17	Fokker Dr1 Dreidekker <R> (BAPC 133)	Kent Battle of Britain Museum, Hawkinge
477/17	Fokker Dr1 Dreidekker <R> (G-FOKK)	Privately owned, Sywell
556/17	Fokker Dr1 Dreidekker <R> (G-CFHY)	Privately owned, Tibenham
626/8	Fokker DVII <R> (N6268)	Privately owned, Booker
764	Mikoyan MiG-21SPS <ff>	Privately owned, Norfolk
959	Mikoyan MiG-21SPS	Midland Air Museum, Coventry
1160	Dornier Do17Z-2	Michael Beetham Conservation Centre, Cosford
1190	Messerschmitt Bf109E-3 [4]	Imperial War Museum, Duxford
1480	Messerschmitt Bf109 <R> (BAPC 66) [6]	Kent Battle of Britain Museum, Hawkinge
1801/18	Bowers Fly Baby 1A (G-BNPV)	Privately owned, Chessington
1803/18	Bowers Fly Baby 1A (G-BUYU)	Privately owned, Chessington
1983	Messerschmitt Bf109E-3 (G-EMIL)	Privately owned, Colchester
2100	Focke-Wulf Fw189A-1 (G-BZKY) [V7+1H]	Privately owned, Sandown
3579	Messerschmitt Bf109E-7 (CF-EML) [14]	Privately owned
4101	Messerschmitt Bf109E-3 (DG200/8477M) [12]	RAF Museum, Hendon
4477	CASA 1.131E Jungmann (G-RETA) [GD+EG]	The Shuttleworth Collection, Old Warden
7198/18	LVG CVI (G-AANJ/9239M)	Michael Beetham Conservation Centre, Cosford
8417/18	Fokker DVII (9207M)	RAF Museum, Hendon
12802	Antonov An-2T (D-FOFM)	Historische Flugzeuge, Grossenhain, Germany
100143	Focke-Achgelis Fa330A-1 Bachstelze	Imperial War Museum, Duxford
100502	Focke-Achgelis Fa330A-1 Bachstelze	Privately owned, Millom
100503	Focke-Achgelis Fa330A-1 Bachstelze (8469M)	RAF Museum, Cosford
100509	Focke-Achgelis Fa330A-1 Bachstelze	Science Museum, stored Wroughton

Serial	Type (code/other identity)	Owner/operator, location or fate	Notes
100545	Focke-Achgelis Fa330A-1 Bachstelze	FAA Museum, RNAS Yeovilton	
100549	Focke-Achgelis Fa330A-1 Bachstelze	Lashenden Air Warfare Museum, Headcorn	
112372	Messerschmitt Me262A-2a (AM.51/VK893/8482M) [4]	RAF Museum, Hendon	
120227	Heinkel He162A-2 Salamander (VN679/AM.65/8472M) [2]	RAF Museum, Hendon	
120235	Heinkel He162A-1 Salamander (AM.68)	Imperial War Museum, Duxford	
191316	Messerschmitt Me163B Komet	Science Museum, South Kensington	
191454	Messerschmitt Me163B Komet <R> (BAPC 271)	The Shuttleworth Collection, Old Warden	
191461	Messerschmitt Me163B Komet (191614/8481M) [14]	RAF Museum, Cosford	
191659	Messerschmitt Me163B Komet (8480M) [15]	Royal Scottish Mus'm of Flight, E Fortune	
280020	Flettner Fl282/B-V20 Kolibri (frame only)	Midland Air Museum, Coventry	
360043	Junkers Ju88R-1 (PJ876/8475M) [D5+EV]	RAF Museum, Hendon	
420430	Messerschmitt Me410A-1/U2 (AM.72/8483M) [3U+CC]	RAF Museum, Cosford	
475081	Fieseler Fi156C-7 Storch (VP546/AM.101/7362M)[GM+AK]	RAF Museum, Cosford	
494083	Junkers Ju87D-3 (8474M) [RI+JK]	RAF Museum, Hendon	
502074	Heliopolis Gomhouria Mk 6 (158/G-CGEV) [CG+EV]	Privately owned, Breighton	
584219	Focke-Wulf Fw190F-8/U1 (AM.29/8470M) [38]	RAF Museum, Hendon	
701152	Heinkel He111H-23 (8471M) [NT+SL]	RAF Museum, Hendon	
730301	Messerschmitt Bf110G-4 (AM.34/8479M) [D5+RL]	RAF Museum, Hendon	
733682	Focke-Wulf Fw190A-8/R7 (AM.75/9211M)	RAF Museum, Cosford	
980554	Flug Werk FW190A-8/N (G-FWAB)	Meier Motors, Bremgarten, Germany	
2+1	Focke-Wulf Fw190 <R> (G-SYFW) [7334]	Privately owned, Empingham, Rutland	
17+TF	CASA 1.133C Jungmeister (G-BZTJ)	Privately owned, Turweston	
22+35	Lockheed F-104G Starfighter	Privately owned, Bruntingthorpe	
23.02	Albatros B.II <R> (D-EKGH)	Historischer Flugzeugbau, Fürstenwalde, Germany	
28+08	Aero L-39ZO Albatros (142/28+04)	Pinewood Studios, Bucks	
2E+RA	Fieseler Fi156C-3 Storch (F-AZRA)	Amicale J-B Salis, la Ferté-Alais, France	
4+1	Focke-Wulf Fw190 <R> (G-BSLX)	Privately owned, Norwich	
6G+ED	Slepcev Storch (G-BZOB) [5447]	Privately owned, Breighton	
37+86	McD F-4F Phantom II <ff>	Privately owned, Bruntingthorpe	
58+89	Dornier Do28D-2 Skyservant (D-ICDY)	Privately owned, Uetersen, Germany	
80+39	MBB Bo.105M	Privately owned, Coney Park, Leeds	
80+40	MBB Bo.105M	Privately owned, Coney Park, Leeds	
80+55	MBB Bo.105M	Lufthansa Resource Technical Training, Cotswold Airport	
80+77	MBB Bo.105M	Lufthansa Resource Technical Training, Cotswold Airport	
81+00	MBB Bo.105M (D-HZYR)	The Helicopter Museum, Weston-super-Mare	
96+26	Mil Mi-24D (421)	The Helicopter Museum, Weston-super-Mare	
98+14	Sukhoi Su-22M-4	Hawker Hunter Aviation Ltd, stored Scampton	
99+18	NA OV-10B Bronco (G-ONAA)	Bronco Demo Team, Wevelgem, Belgium	
AZ+JU	CASA 3.52L (F-AZJU)	Amicale J-B Salis, la Ferté-Alais, France	
BG+KM	Nord 1002 Pingouin (G-ASTG)	Privately owned, Duxford	
BU+CC	CASA 1.131E Jungmann (G-BUCC)	Privately owned, Sandown	
D5397/17	Albatros DVA <R> (G-BFXL)	FAA Museum, stored Cobham Hall, RNAS Yeovilton	
D7343/17	Albatros DVA <R> (ZK-TVD)	RAF Museum, Hendon	
DM+BK	Morane-Saulnier MS505 (G-BPHZ)	Aero Vintage, Westfield, Sussex	
ES+BH	Messerschmitt Bf108B-2 (D-ESBH)	Messerschmitt Stiftung, Manching, Germany	
FI+S	Morane-Saulnier MS505 (G-BIRW)	Royal Scottish Mus'm of Flight, E Fortune	
FM+BB	Messerschmitt Bf109G-6 (D-FMBB)	Messerschmitt Stiftung, Manching, Germany	
GM+AI	Fieseler Fi156A Storch (2088/G-STCH)	Privately owned, Old Warden	
KB+GB	CASA 1.131E Jungmann (G-BHSL)	Privately owned, Oaksey Park	
LG+03	Bücker Bü133C Jungmeister (G-AEZX)	Privately owned, Milden	
NJ+C11	Nord 1002 (G-ATBG)	Privately owned, Duxford	
NQ+NR	Klemm Kl35D (D-EQXD)	Quax Flieger, Hamm, Germany	
NV+KG	Focke-Wulf Fw44J Stieglitz (D-ENAY)	Quax Flieger, Hamm, Germany	
S4-A07	CASA 1.131E Jungmann (G-BWHP)	Privately owned, Yarcombe, Devon	
S5+B06	CASA 1.131E Jungmann 2000 (G-BSFB)	Privately owned, Old Buckenham	
TP+WX	Heliopolis Gomhouria Mk 6 (G-TPWX)	Privately owned, Swanborough	

GHANA

| G360 | PBN 2T Islander (G-BRSR) | Privately owned, Biggin Hill | |

Notes	Serial	Type (code/other identity)	Owner/operator, location or fate
	G361	PBN 2T Islander (G-BRPB)	Privately owned, Biggin Hill
	G362	PBN 2T Islander (G-BRPC)	Privately owned, Biggin Hill
	G363	PBN 2T Islander (G-BRSV)	Privately owned, Biggin Hill
	GREECE		
	52-6541	Republic F-84F Thunderflash [541]	North-East Aircraft Museum, Usworth
	63-8418	Northrop F-5A <ff>	Martin-Baker Ltd, Chalgrove, Fire Section
	HONG KONG		
	HKG-5	SA128 Bulldog (G-BULL)	Privately owned, Cotswold Airport
	HKG-6	SA128 Bulldog (G-BPCL)	Privately owned, North Weald
	HKG-11	Slingsby T.67M Firefly 200 (G-BYRY)	Privately owned, Antwerp, Belgium
	HKG-13	Slingsby T.67M Firefly 200 (G-BXKW)	Privately owned, St Ghislain, Belgium
	HUNGARY		
	335	Mil Mi-24D (3532461715415)	Privately owned, Harlow
	501	Mikoyan MiG-21PF	Imperial War Museum, Duxford
	503	Mikoyan MiG-21SMT (G-BRAM)	RAF Museum, Cosford
	INDIA		
	E296	Hindustan Gnat F1 (G-SLYR)	Privately owned, North Weald
	Q497	EE Canberra T4 (WE191) <ff>	Privately owned, Stoneykirk, D&G
	HA561	Hawker Tempest II (MW743)	Privately owned, stored Wickenby
	HT291	Noorduyn AT-16 Harvard IIB (G-CGYM)	*To India, July 2014*
	INDONESIA		
	LL-5313	BAe Hawk T53	BAE Systems, Brough, on display
	IRAQ		
	333	DH115 Vampire T55 <ff>	South Yorkshire Aircraft Mus'm, stored Doncaster
	ITALY		
	MM5701	Fiat CR42 (BT474/8468M) [13-95]	RAF Museum, Hendon
	MM53211	Fiat G46-1B (MM52799) [ZI-4]	Privately owned, Shipdham
	MM53692	CCF T-6G Texan	RAeS Medway Branch, Rochester
	MM53774	Fiat G59-4B (I-MRSV) [181]	Privately owned, Parma, Italy
	MM54099	NA T-6G Texan (G-BRBC) [RR-56]	Privately owned, Chigwell
	MM54532	SIAI-Marchetti SF.260AM (G-ITAF) [70-42]	Privately owned, Leicester
	MM80270	Agusta-Bell AB204B (MM80279)	London Motor Museum, Hayes
	MM80927	Agusta-Bell AB206A-1 JetRanger [CC-49]	The Helicopter Museum, Weston-super-Mare
	MM81205	Agusta A109A-2 SEM [GF-128]	The Helicopter Museum, Weston-super-Mare
	MM54-2372	Piper L-21B Super Cub	Privately owned, Foxhall Heath, Suffolk
	JAPAN		
	–	Kawasaki Ki100-1B (8476M/BAPC 83)	RAF Museum, Cosford
	–	Yokosuka MXY 7 Ohka II (BAPC 159)	Imperial War Museum, stored Duxford
	997	Yokosuka MXY 7 Ohka II (8485M/BAPC 98)	Museum of Science & Industry, Manchester
	5439	Mitsubishi Ki46-III (8484M/BAPC 84)	RAF Museum, Cosford
	15-1585	Yokosuka MXY 7 Ohka II (BAPC 58)	Science Museum, at FAA Museum, RNAS Yeovilton
	I-13	Yokosuka MXY 7 Ohka II (8486M/BAPC 99)	RAF Museum, Cosford
	Y2-176	Mitsubishi A6M3-2 Zero (3685) [76]	Imperial War Museum, Lambeth
	JORDAN		
	408	SA125 Bulldog (G-BDIN)	South Yorkshire Aircraft Musm, stored Doncaster
	KENYA		
	115	Dornier Do28D-2 Skyservant	Privately owned, stored Hibaldstow
	117	Dornier Do28D-2 Skyservant	Privately owned, stored Hibaldstow
	KUWAIT		
	113	BAC Strikemaster Mk.80A (G-CFBK) [K167-113/D]	Privately owned, Bentwaters

Serial	Type (code/other identity)	Owner/operator, location or fate	Notes
MYANMAR			
UB441	VS361 Spitfire IX (ML119/G-SDNI)	Privately owned, Sandown	
THE NETHERLANDS			
16-218	Consolidated PBY-5A Catalina (2459/PH-PBY)	Neptune Association, Lelystad, The Netherlands	
174	Fokker S-11 Instructor (E-31/G-BEPV) [K]	Privately owned, Spanhoe	
179	Fokker S-11 Instructor (PH-ACG) [K]	Privately owned, Lelystad, The Netherlands	
197	Fokker S-11 Instructor (PH-GRY) [K]	KLu Historic Flt, Gilze-Rijen, The Netherlands	
204	Lockheed SP-2H Neptune [V]	RAF Museum, Cosford	
A-57	DH82A Tiger Moth (PH-TYG)	KLu Historic Flt, Gilze-Rijen, The Netherlands	
B-64	Noorduyn AT-16 Harvard IIB (PH-LSK)	KLu Historic Flt, Gilze-Rijen, The Netherlands	
B-71	Noorduyn AT-16 Harvard IIB (PH-MLM)	KLu Historic Flt, Gilze-Rijen, The Netherlands	
B-118	Noorduyn AT-16 Harvard IIB (PH-IIB)	KLu Historic Flt, Gilze-Rijen, The Netherlands	
B-182	Noorduyn AT-16 Harvard IIB (PH-TBR)	KLu Historic Flt, Gilze-Rijen, The Netherlands	
E-14	Fokker S-11 Instructor (PH-AFS)	Privately owned, Lelystad, The Netherlands	
E-15	Fokker S-11 Instructor (G-BIYU)	Privately owned, RAF Topcliffe	
E-20	Fokker S-11 Instructor (PH-GRB)	Privately owned, Gilze-Rijen, The Netherlands	
E-27	Fokker S-11 Instructor (PH-HOL)	Privately owned, Lelystad, The Netherlands	
E-32	Fokker S-11 Instructor (PH-HOI)	Privately owned, Gilze-Rijen, The Netherlands	
E-39	Fokker S-11 Instructor (PH-HOG)	Privately owned, Lelystad, The Netherlands	
G-29	Beech D18S (PH-KHV)	KLu Historic Flt, Gilze-Rijen, The Netherlands	
MH424	VS361 Spitfire LFIXC (MJ271/H-53)	Privately owned, Duxford	
MK732	VS361 Spitfire LFIXC (8633M/PH-OUQ) [3W-17]	KLu Historic Flt, Gilze-Rijen, The Netherlands	
N-202	Hawker Hunter F6 [10] <ff>	Privately owned, Stockport	
N-250	Hawker Hunter F6 (G-9-185) <ff>	Imperial War Museum, Duxford	
N-268	Hawker Hunter FGA78 (Qatar QA-10)	Yorkshire Air Museum, Elvington	
N-294	Hawker Hunter F6A (XF515/G-KAXF)	Stichting Hawker Hunter Foundation, Leeuwarden, The Netherlands	
N-302	Hawker Hunter T7 (ET-273/G-9-431) <ff>	South Yorkshire Aircraft Museum, Doncaster	
N-315	Hawker Hunter T7 (comp XM121)	Privately owned, Netherley, Aberdeenshire	
N-321	Hawker Hunter T8C (G-BWGL)	Stichting Hawker Hunter Foundation, Leeuwarden, The Netherlands	
N5-149	NA B-25J Mitchell (44-29507/HD346/PH-XXV) [232511]	KLu Historic Flt, Gilze-Rijen, The Netherlands	
R-18	Auster III (PH-NGK)	KLu Historic Flt, Gilze-Rijen, The Netherlands	
R-55	Piper L-18C Super Cub (52-2466/G-BLMI)	Privately owned, Antwerp, Belgium	
R-109	Piper L-21B Super Cub (54-2337/PH-GAZ)	KLu Historic Flt, Gilze-Rijen, The Netherlands	
R-122	Piper L-21B Super Cub (54-2412/PH-PPW)	KLu Historic Flt, Gilze-Rijen, The Netherlands	
R-124	Piper L-21B Super Cub (54-2414/PH-APA)	Privately owned, Eindhoven, The Netherlands	
R-137	Piper L-21B Super Cub (54-2427/PH-PSC)	Privately owned, Gilze-Rijen, The Netherlands	
R-151	Piper L-21B Super Cub (54-2441/G-BIYR)	Privately owned, Yarcombe, Devon	
R-156	Piper L-21B Super Cub (54-2446/G-ROVE)	Privately owned, Headcorn	
R-167	Piper L-21B Super Cub (54-2457/G-LION)	Privately owned, Turweston, Bucks	
R-170	Piper L-21B Super Cub (52-6222/PH-ENJ)	Privately owned, Midden Zealand, The Netherlands	
R-177	Piper L-21B Super Cub (54-2467/PH-KNR)	KLu Historic Flt, Gilze-Rijen, The Netherlands	
R-181	Piper L-21B Super Cub (54-2471/PH-GAU)	Privately owned, Gilze-Rijen, The Netherlands	
R-213	Piper L-21A Super Cub (51-15682/PH-RED)	Privately owned, The Netherlands	
R-345	Piper J-3C Cub (PH-UCS)	Privately owned, Hilversum, The Netherlands	
S-9	DHC2 L-20A Beaver (55-4585/PH-DHC)	KLu Historic Flt, Gilze-Rijen, The Netherlands	
NEW ZEALAND			
NZ3909	WS Wasp HAS1 (XT782/G-KANZ)	Kennet Aviation, stored North Weald	
NORTH KOREA			
–	WSK Lim-2 (MiG-15) (01420/G-BMZF)	FAA Museum, RNAS Yeovilton	
NORTH VIETNAM			
1211	WSK Lim-5 (MiG-17F) (G-MIGG)	Privately owned, North Weald	
NORWAY			
145	DH82A Tiger Moth II (DE248/LN-BDM)	Privately owned, Kjeller, Norway	
163	Fairchild PT-19A Cornell (42-83641/LN-BIF)	Privately owned, Kjeller, Norway	
599	Canadair CT-133AUP Silver Star Mk.3 (133599/NX865SA)	Royal Norwegian Historical Sqn, Ørland, Norway	

Notes	Serial	Type (code/other identity)	Owner/operator, location or fate
	848	Piper L-18C Super Cub (LN-ACL) [FA-N]	Privately owned, Norway
	56321	SAAB S91B Safir (G-BKPY)	Newark Air Museum, Winthorpe
PX-K	DH100 Vampire FB6 (SE-DXS)	Privately owned, Rygge, Norway
PX-M	DH115 Vampire T55 (LN-DHZ)	Privately owned, Rygge, Norway

OMAN

Notes	Serial	Type (code/other identity)	Owner/operator, location or fate
	425	BAC Strikemaster Mk.82A (G-SOAF)	Privately owned, Hawarden
	801	Hawker Hunter T66B <ff>	Privately owned, St Athan
	801	Hawker Hunter T66B <rf>	Privately owned, Hawarden
	853	Hawker Hunter FR10 (XF426)	RAF Museum, Hendon
XF688	Percival P56 Provost T1 (WV562/7606M)	RAF Museum, Hendon
XL554	SAL Pioneer CC1 (XL703/8034M)	RAF Museum, Hendon

POLAND

Notes	Serial	Type (code/other identity)	Owner/operator, location or fate
05	WSK SM-2 (Mi-2) (S2-03006)	The Helicopter Museum, Weston-super-Mare
309	WSK SBLim-2A (MiG-15UTI) <ff>	R Scottish Mus'm of Flight, stored Granton
458	Mikoyan MiG-23ML (04 red/024003607)	Newark Air Museum, Winthorpe
618	Mil Mi-8P (10618)	The Helicopter Museum, Weston-super-Mare
1018	WSK-PZL Mielec TS-11 Iskra (1H-1018/G-ISKA)	Cold War Jets Collection, Bruntingthorpe
1120	WSK Lim-2 (MiG-15bis)	RAF Museum, Cosford
1706	WSK-PZL Mielec TS-11 Iskra (1H-0408)	Midland Air Museum, Coventry
1972	Letov S-102 (MiG-15) (623794)	Norfolk & Suffolk Avn Museum, Flixton
		(port side only, painted in Czech marks as 3794 on starboard side)	

PORTUGAL

Notes	Serial	Type (code/other identity)	Owner/operator, location or fate
85	Isaacs Fury II (G-BTPZ)	Privately owned, Ormskirk
1350	OGMA/DHC1 Chipmunk T20 (G-CGAO)	Privately owned, Spanhoe
1360	OGMA/DHC1 Chipmunk T20 (G-BYYU) (fuselage)	Sold to Poland, 2015
1365	OGMA/DHC1 Chipmunk T20 (G-DHPM)	Privately owned, Sywell
1367	OGMA/DHC1 Chipmunk T20 (G-UANO)	Privately owned, Sherburn-in-Elmet
1372	OGMA/DHC1 Chipmunk T20 (HB-TUM)	Privately owned, Switzerland
1373	OGMA/DHC1 Chipmunk T20 (G-CBJG)	Privately owned, Winwick, Cambs
1375	OGMA/DHC1 Chipmunk T20 (F-AZJV)	Privately owned, Valenciennes, France
1377	DHC1 Chipmunk Mk.22 (G-BARS)	Privately owned, Yeovilton
1741	CCF T-6J Texan (G-HRVD)	Privately owned, Bruntingthorpe
1747	CCF T-6J Texan (20385/G-BGPB)	The Aircraft Restoration Co, Duxford
1765	CCF T-6J Texan (G-CHYN)	Privately owned, Cotswold Airport
3303	MH1521M Broussard (G-CBGL)	Privately owned, Rochester

QATAR

Notes	Serial	Type (code/other identity)	Owner/operator, location or fate
QA12	Hawker Hunter FGA78 <ff>	Privately owned, New Inn, Torfaen
QP30	WS Lynx Mk.28 (G-BFDV/TD 013)	DSEME SEAE, Arborfield
QP31	WS Lynx Mk.28	Vector Aerospace Fleetlands Apprentice School
QP32	WS Lynx Mk.28 (TAD 016)	DSEME SEAE, Arborfield

ROMANIA

Notes	Serial	Type (code/other identity)	Owner/operator, location or fate
29	LET L-29 Delfin	Privately owned, Pocklington
42	LET L-29 Delfin <ff>	Privately owned, Catshill, Worcs
42	LET L-29 Delfin <rf>	Privately owned, Hinstock, Shrops
47	LET L-29 Delfin <ff>	Top Gun Flight Simulator Centre, Stalybridge
53	LET L-29 Delfin (99954)	Privately owned, Bruntingthorpe

RUSSIA (& FORMER SOVIET UNION)

Notes	Serial	Type (code/other identity)	Owner/operator, location or fate
–	Mil Mi-24D (3532464505029)	Midland Air Museum, Coventry
1 w	SPP Yak C-11 (171314/G-BZMY)	Privately owned, Coventry
01 y	Yakovlev Yak-52 (9311709/G-YKSZ)	Privately owned, White Waltham
03 bl	Yakovlev Yak-55M (910103/RA-01274)	Privately owned, Halfpenny Green
03 w	Yakovlev Yak-18A (1160403/G-CEIB)	Privately owned, Breighton
03 w	Yakovlev Yak-52 (899803/G-YAKR)	Crashed 29 March 2014, North Weald
5 w	Yakovlev Yak-3UA (0470204/D-FYGJ)	Privately owned, Bremgarten, Germany
06 y	Yakovlev Yak-9UM (HB-RYA)	Flying Fighter Association, Bex, Switzerland

Serial	Type (code/other identity)	Owner/operator, location or fate	Notes
07 y	WSK SM-1 (Mi-1) (Polish AF 2007)	The Helicopter Museum, Weston-super-Mare	
07 y	Yakovlev Yak-18M (G-BMJY)	Privately owned, East Garston, Bucks	
07 r	Yakovlev Yak-52 (9011107/G-HYAK)	Privately owned, Exeter	
09 y	Yakovlev Yak-52 (9411809/G-BVMU)	Privately owned, Lille, France	
9 w	SPP Yak C-11 (1701139/G-OYAK)	Privately owned, Little Gransden	
10 r	Yakovlev Yak-50 (801810/G-BTZB)	Privately owned, Lee-on-Solent	
10 r	Yakovlev Yak-50 (877610/G-YAKE)	Privately owned, Henstridge	
11 y	SPP Yak C-11 (G-YCII)	Privately owned, Woodchurch, Kent	
12 r	LET L-29 Delfin (194555/ES-YLM/G-DELF)	Privately owned, St Athan	
15 w	SPP Yak C-11 (170103/D-FYAK)	Classic Aviation Company, Hannover, Germany	
20 r	Yakovlev Yak-50 (812003/G-YAAK)	Privately owned, Henstridge	
20 w	Lavochkin La-11	The Fighter Collection, Duxford	
21 w	Yakovlev Yak-3UA (0470203/G-CDBJ)	Privately owned, Headcorn	
21 w	Yakovlev Yak-9UM (0470403/D-FENK)	Privately owned, Magdeburg, Germany	
23 y	Bell P-39Q Airacobra (44-2911)	Privately owned, Sussex	
23 w	Mikoyan MiG-27D (83712515040)	Privately owned, Hawarden	
26 bl	Yakovlev Yak-52 (9111306/G-BVXK)	Privately owned, White Waltham	
27 w	Yakovlev Yak-3UTI-PW (9/04623/F-AZIM)	Privately owned, la Ferté-Alais, France	
27 r	Yakovlev Yak-52 (9111307/G-YAKX)	Privately owned, Popham	
28 w	Polikarpov Po-2 (0094/G-BSSY)	The Shuttleworth Collection, Old Warden	
31 gy	Yakovlev Yak-52 (9111311/G-YAKV)	Privately owned, Rendcomb	
33 r	Yakovlev Yak-50 (853206/G-YAKZ)	Privately owned, Henstridge	
33 w	Yakovlev Yak-52 (899915/G-YAKH)	Privately owned, White Waltham	
35 r	Sukhoi Su-17M-3 (25102)	Privately owned, Hawarden	
36 w	LET/Yak C-11 (171101/G-KYAK)	Privately owned, North Weald	
36 r	Yakovlev Yak-52 (9111604/G-IUII)	Privately owned, North Weald	
43 bl	Yakovlev Yak-52 (877601/G-BWSV)	Privately owned, North Weald	
48 bl	Yakovlev Yak-52 (9111413/G-CBSN)	Privately owned, Manston	
49 r	Yakovlev Yak-50 (822305/G-YAKU)	Privately owned, Henstridge	
50 bk	Yakovlev Yak-50 (812101/G-CBPM)	Repainted as G-CBPM	
50 gy	Yakovlev Yak-52 (9111415/G-CBRW)	Meier Motors, Bremgarten, Germany	
51 r	LET L-29 Delfin (893019/G-BZNT)	Privately owned, St Athan	
51 r	LET L-29S Delfin (491273/YL-PAG)	Privately owned, Breighton	
51 y	Yakovlev Yak-50 (812004/G-BWYK)	Privately owned, West Meon, Hants	
52 w	Yakovlev Yak C-11 (171312/G-BTZE)	Privately owned, Booker	
52 w	Yakovlev Yak-52 (9612001/G-CCJK)	Privately owned, White Waltham	
52 y	Yakovlev Yak-52 (878202/G-BWVR)	Privately owned, Eshott	
54 r	Sukhoi Su-17M (69004)	Privately owned, Hawarden	
55 y	Yakovlev Yak-52 (9111505/G-BVOK)	Privately owned, Shoreham	
61 r	Yakovlev Yak-50 (842710/G-YAKM)	Privately owned, Henstridge	
66 r	Yakovlev Yak-52 (855905/G-YAKN)	Privately owned, Henstridge	
67 r	Yakovlev Yak-52 (822013/G-CBSL)	Privately owned, Church Fenton	
69 bl	Yakovlev Yak-52 (899413/G-XYAK)	Repainted as G-XYAK	
69 y	Yakovlev Yak-52 (888712/G-CCSU)	Privately owned, Germany	
71 r	Mikoyan MiG-27K (61912507006)	Newark Air Museum, Winthorpe	
74 w	Yakovlev Yak-52 (877404/G-LAOK) [JA-74, IV-62]	Privately owned, Tollerton	
93 w	Yakovlev Yak-50 (853001/G-JYAK) [R]	Privately owned, North Weald	
100 bl	Yakovlev Yak-52 (866904/G-YAKI)	Privately owned, Popham	
100 w	Yakovlev Yak-3M (0470107/G-CGXG)	Privately owned, Bentwaters	
139 y	Yakovlev Yak-52 (833810/G-BWOD)	Privately owned, Sywell	
526 bk	Mikoyan MiG-29 (2960725887) <f>	Fenland & West Norfolk Aviation Museum, Wisbech	
1342	Yakovlev Yak-1 (G-BTZD)	Privately owned, Westfield, Sussex	
1870710	Ilyushin Il-2 (G-BZVW)	Privately owned, Wickenby	
1878576	Ilyushin Il-2 (G-BZVX)	Privately owned, Wickenby	
(PT879)	VS361 Spitfire FIX (G-BYDE)	Hangar 11 Collection, North Weald	
(RK858)	VS361 Spitfire LFIX (G-CGJE)	The Fighter Collection, Duxford	
(SM639)	VS361 Spitfire LFIX	Privately owned, Catfield	

SAUDI ARABIA

1104	BAC Strikemaster Mk.80 (G-SMAS)	Privately owned, Hawarden	
1112	BAC Strikemaster Mk.80 (G-FLYY)	Privately owned, Cotswold Airport	
1115	BAC Strikemaster Mk.80A	Global Aviation, Humberside	

Notes	Serial	Type (code/other identity)	Owner/operator, location or fate
	1120	BAC Strikemaster Mk.80A (G-RSAF)	Privately owned, Hawarden
	1129	BAC Strikemaster Mk.80A	Global Aviation, Humberside
	1133	BAC Strikemaster Mk.80A (G-BESY)	Imperial War Museum, Duxford
	53-686	BAC Lightning F53 (G-AWON/ZF592)	City of Norwich Aviation Museum
	55-713	BAC Lightning T55 (ZF598) [C]	Midland Air Museum, Coventry
	SINGAPORE		
	311	BAC Strikemaster Mk.84 (G-MXPH)	Privately owned, North Weald
	323	BAC Strikemaster Mk.81 (N21419)	Privately owned, stored Hawarden
	SOUTH AFRICA		
	91	Westland Wasp HAS1 (pod)	Privately owned, Oaksey Park
	92	Westland Wasp HAS1 (G-BYCX)	Privately owned, Babcary, Somerset
	6130	Lockheed Ventura II (AJ469)	RAF Museum, stored Cosford
	7429	NA AT-6D Harvard III (D-FASS)	Privately owned, Aachen, Germany
	SOUTH VIETNAM		
	24550	Cessna L-19E Bird Dog (G-PDOG) [GP]	Privately owned, Fenland
	SPAIN		
	B.2I-27	CASA 2.111B (He111H-16) (B.2I-103)	Imperial War Museum, stored Duxford
	E.1-9	Bücker Bü133C Jungmeister (G-BVXJ)	Privately owned, Old Warden
	E.3B-143	CASA 1.131E Jungmann (G-JUNG)	Privately owned, White Waltham
	E.3B-153	CASA 1.131E Jungmann (G-BPTS) [781-75]	Privately owned, Egginton
	E.3B-350	CASA 1.131E Jungmann (G-BHPL) [05-97]	Privately owned, Henstridge
	(E.3B-369)	CASA 1.131E Jungmann (G-BPDM) [781-32]	Privately owned, Heighington
	(E.3B-379)	CASA 1.131E Jungmann (G-CDJU) [72-36]	Privately owned, Abbeyshrule, Eire
	E.3B-494	CASA 1.131E Jungmann (G-CDLC) [81-47]	Privately owned, Chiseldon
	E.3B-521	CASA 1.131E Jungmann [781-3]	RAF Museum, Hendon
	E.3B-599	CASA 1.131E Jungmann (G-CGTX) [31]	Privately owned, Archerfield, Lothian
	E.18-2	Piper PA-31P Navajo 425 [42-71]	Privately owned, Stamford
	EM-01	DH60G Moth (G-AAOR)	Privately owned, Rendcomb
	ES.1-4	Bücker Bü133C Jungmeister (G-BUTX)	Privately owned, Breighton
	ES.1-16	CASA 1.133L Jungmeister	Privately owned, Stretton, Cheshire
	SRI LANKA		
	CT180	Nanchang CJ-6A Chujiao (2632016/G-BXZB)	Privately owned, White Waltham
	SWEDEN		
	–	Thulin A/Bleriot XI (SE-XMC)	Privately owned, Loberod, Sweden
	087	CFM 01 Tummelisa <R> (SE-XIL)	Privately owned, Loberod, Sweden
	2542	Fiat CR42 (G-CBLS)	The Fighter Collection, Duxford
	5033	Klemm Kl35D (SE-BPT) [78]	Privately owned, Barkaby, Sweden
	5060	Klemm Kl35D (SE-BPU) [174]	Privately owned, Barkaby, Sweden
	05108	DH60 Moth	Privately owned, Langham
	17239	SAAB B-17A (SE-BYH) [7-J]	Flygvapenmuseum, Linköping, Sweden
	28693	DH100 Vampire FB6 (J-1184/SE-DXY) [9-G]	Scandinavian Historic Flight, Oslo, Norway
	29640	SAAB J-29F [20-08]	Midland Air Museum, Coventry
	29670	SAAB J-29F (SE-DXB) [10-R]	Flygvapenmuseum/F10 Wing, Angelholm, Sweden
	32028	SAAB 32A Lansen (G-BMSG)	Privately owned, Willenhall, Staffs
	35075	SAAB J-35A Draken [40]	Dumfries & Galloway Aviation Museum
	35515	SAAB J-35F Draken [49]	Airborne Systems, Llangeinor
	35810	SAAB Sk-35C Draken (SE-DXP) [810]	Swedish Air Force Historic Flight, Såtenäs, Sweden
	37098	SAAB AJSF-37 Viggen (SE-DXN) [52]	Swedish Air Force Historic Flight, Såtenäs, Sweden
	37918	SAAB AJSH-37 Viggen [57]	Newark Air Museum, Winthorpe
	60140	SAAB 105 (SE-DXG) [140-5]	Swedish Air Force Historic Flight, Såtenäs, Sweden
	91130	SAAB S91A Safir (SE-BNN) [10-30]	Privately owned, Barkaby, Sweden
	SWITZERLAND		
	–	DH112 Venom FB54 (J-1758/N203DM)	Grove Technology Park, Wantage, Oxon
	A-10	CASA 1.131E Jungmann (G-BECW)	Privately owned, Rochester
	A-57	CASA 1.131E Jungmann (G-BECT)	Privately owned, Goodwood

Serial	Type (code/other identity)	Owner/operator, location or fate	Notes
A-125	Pilatus P2-05 (G-BLKZ) (fuselage)	Privately owned, Newquay	
A-701	Junkers Ju52/3m (HB-HOS)	Ju-Air, Dubendorf, Switzerland	
A-702	Junkers Ju52/3m (HB-HOT)	Ju-Air, Dubendorf, Switzerland	
A-703	Junkers Ju52/3m (HB-HOP)	Ju-Air, Dubendorf, Switzerland	
A-806	Pilatus P3-03 (G-BTLL)	Privately owned, Guist, Norfolk	
A-815	Pilatus P3-03 (HB-RCQ)	Privately owned, Locarno, Switzerland	
A-818	Pilatus P3-03 (HB-RCH)	Privately owned, Ambri, Switzerland	
A-829	Pilatus P3-03 (HB-RCJ)	Privately owned, Locarno, Switzerland	
A-873	Pilatus P3-03 (HB-RCL)	Privately owned, Locarno, Switzerland	
C-552	EKW C-3605 (G-DORN)	Currently not known	
C-558	EKW C-3605 (G-CCYZ)	Privately owned, Wickenby	
J-143	Morane-Saulnier MS406 (HB-RCF)	Privately owned, Bex, Switzerland	
J-901	NA P-51D Mustang (44-72773/D-FPSI)	Meier Motors, Bremgarten, Germany	
J-1008	DH100 Vampire FB6	DHAHC, London Colney	
J-1169	DH100 Vampire FB6	Privately owned, Henley-on-Thames	
J-1172	DH100 Vampire FB6 (8487M)	RAF Museum Reserve Collection, Stafford	
J-1605	DH112 Venom FB50 (G-BLID)	Gatwick Aviation Museum, Charlwood, Surrey	
J-1629	DH112 Venom FB50	Privately owned, Shropshire	
J-1632	DH112 Venom FB50 (G-VNOM) <ff>	Privately owned, Cantley, Norfolk	
J-1649	DH112 Venom FB50 <ff>	Classic Air Force, Newquay	
J-1704	DH112 Venom FB54	RAF Museum, Cosford	
J-1712	DH112 Venom FB54 <ff>	Privately owned, Connah's Quay, Flintshire	
J-4015	Hawker Hunter F58 (J-4040/HB-RVS)	Privately owned, St Stephan, Switzerland	
J-4064	Hawker Hunter F58 (HB-RVQ)	Fliegermuseum Altenrhein, Switzerland	
J-4083	Hawker Hunter F58 (G-EGHH)	Privately owned, St Athan	
J-4086	Hawker Hunter F58 (HB-RVU)	Privately owned, Altenrhein, Switzerland	
J-4201	Hawker Hunter T68 (HB-RVR)	Amici dell'Hunter, Sion, Switzerland	
J-4205	Hawker Hunter T68 (HB-RVP)	Fliegermuseum Altenrhein, Switzerland	
J-4206	Hawker Hunter T68 (HB-RVV)	Fliegermuseum Altenrhein, Switzerland	
U-80	Bücker Bü133D Jungmeister (G-BUKK)	Privately owned, Kirdford, W Sussex	
U-95	Bücker Bü133C Jungmeister (G-BVGP)	Privately owned, Booker	
U-99	Bücker Bü133C Jungmeister (G-AXMT)	Privately owned, Italy	
U-1215	DH115 Vampire T11 (XE998)	Solent Sky, Southampton	
U-1229	DH115 Vampire T55 (G-HATD)	Aviation Heritage Ltd, Coventry	
V-54	SE3130 Alouette II (G-BVSD)	Privately owned, Glos	

UNITED ARAB EMIRATES

DU-103	Bell 206B JetRanger II	Privately owned, Sparkford, Somerset	

USA

–	Noorduyn AT-16 Harvard IIB (KLu B-168)	American Air Museum, Duxford	
001	Ryan ST-3KR Recruit (G-BYPY)	Privately owned, Goodwood	
14	Boeing-Stearman A75N-1 Kaydet (G-ISDN)	Privately owned, Oaksey Park	
23	Fairchild PT-23 (N49272)	Privately owned, Sleap	
26	Boeing-Stearman A75N-1 Kaydet (G-BAVO)	Privately owned, Turweston	
27	NA SNJ-7 Texan (90678/G-BRVG)	Privately owned, Rochester	
43	Noorduyn AT-16 Harvard IIB (43-13064/G-AZSC) [SC]	Privately owned, North Weald	
44	Boeing-Stearman D75N-1 Kaydet (42-15852/G-RJAH)	Privately owned, Duxford	
85	WAR P-47 Thunderbolt <R> (G-BTBI)	Privately owned, Perth	
104	Boeing-Stearman PT-13D Kaydet (42-16931/N4712V) [W]	Privately owned, Hardwick, Norfolk	
112	Boeing-Stearman PT-13D Kaydet (42-17397/G-BSWC)	Privately owned, Gloucester	
164	Boeing-Stearman PT-13B Kaydet (N60320)	Privately owned, Fenland	
379	Boeing-Stearman PT-13D Kaydet (42-14865/G-ILLE)	Privately owned, Hohenems, Austria	
399	Boeing-Stearman N2S-5 Kaydet (38495/N67193)	Privately owned, Gelnhausen, Germany	
441	Boeing-Stearman N2S-4 Kaydet (30010/G-BTFG)	Privately owned, Manston	
443	Boeing-Stearman E75 Kaydet (N43YP) [6018]	Privately owned, Bicester	
466	Boeing-Stearman PT-13A Kaydet (37-0089/N731)	Privately owned, Gloucester	
540	Piper L-4H Grasshopper (43-29877/G-BCNX)	Privately owned, Monewden	
560	Bell UH-1H Iroquois (73-22077/G-HUEY)	Privately owned, North Weald	
578	Boeing-Stearman N2S-5 Kaydet (N1364V)	Privately owned, North Weald	
586	Boeing-Stearman N2S-3 Kaydet (07874/N74650)	Privately owned, Seppe, The Netherlands	
628	Beech D17S (44-67761/N18V)	Privately owned, stored East Garston, Bucks	

Notes	Serial	Type (code/other identity)	Owner/operator, location or fate
	669	Boeing-Stearman A75N-1 Kaydet (37869/G-CCXA)	Privately owned, Old Buckenham
	671	Boeing-Stearman PT-13D Kaydet (61181/G-CGPY)	Privately owned, Breighton
	699	Boeing-Stearman N2S-3 Kaydet (38233/G-CCXB)	Privately owned, Goodwood
	716	Boeing-Stearman PT-13D Kaydet (42-17553/N1731B)	Privately owned, Compton Abbas
	718	Boeing-Stearman PT-13D Kaydet (42-17555/N5345N)	Privately owned, Tibenham
	744	Boeing-Stearman A75N-1 Kaydet (42-16532/OO-USN)	Privately owned, Wevelgem, Belgium
	805	Boeing-Stearman PT-17 Kaydet (42-17642/N3922B)	Privately owned, Tibenham
	854	Ryan PT-22 Recruit (41-20854/G-BTBH)	Privately owned, Goodwood
	855	Ryan PT-22 Recruit (41-15510/N56421)	Privately owned, Sleap
	897	Aeronca 11AC Chief (G-BJEV) [E]	Privately owned, English Bicknor, Glos
	985	Boeing-Stearman PT-13D Kaydet (42-16930/OO-OPS)	Privately owned, Antwerp, Belgium
	1102	Boeing-Stearman N2S-5 Kaydet (G-AZLE) [102]	Privately owned, Tongham
	1164	Beech D18S (G-BKGL)	The Aircraft Restoration Co, Duxford
	1180	Boeing-Stearman N2S-3 Kaydet (3403/G-BRSK)	Privately owned, Morley
	3072	NA T-6G Texan (49-3072/G-TEXN) [72]	Privately owned, Shoreham
	3397	Boeing-Stearman N2S-3 Kaydet (G-OBEE) [174]	Privately owned, Old Buckenham
	3403	Boeing-Stearman N2S-3 Kaydet (N75TQ) [180]	Privately owned, Tibenham
	3583	Piper L-4B Grasshopper (45-0583/G-FINT) [44-D]	Privately owned, Eggesford
	3681	Piper L-4J Grasshopper (44-80248/G-AXGP)	Privately owned, Biggin Hill
	4406	Naval Aircraft Factory N3N-3 (G-ONAF) [12]	Privately owned, Sandown
	6136	Boeing-Stearman A75N-1 Kaydet(42-16136/G-BRUJ) [205]	Privately owned, Liverpool
	6771	Republic F-84F Thunderstreak (BAF FU-6)	RAF Museum, stored Cosford
	7797	Aeronca L-16A Grasshopper (47-0797/G-BFAF)	Privately owned, Finmere
	8084	NA AT-6D Texan (42-85068/G-KAMY)	Kennet Aviation, North Weald
	8178	NA F-86A Sabre (48-0178/G-SABR) [FU-178]	*Sold to the USA as N528YH, November 2014*
	8242	NA F-86A Sabre (48-0242) [FU-242]	Midland Air Museum, Coventry
	01532	Northrop F-5E Tiger II <R>	RAF Alconbury on display
	02538	Fairchild PT-19B (N33870)	Privately owned, Mendlesham, Suffolk
	07539	Boeing-Stearman N2S-3 Kaydet (N63590) [143]	Privately owned, Billericay
	14286	Lockheed T-33A Shooting Star (51-4286)	American Air Museum, Duxford
	O-14419	Lockheed T-33A Shooting Star (51-4419)	Midland Air Museum, Coventry
	14863	NA AT-6D Harvard III (41-33908/G-BGOR)	Privately owned, Rednal
	15154	Bell OH-58A Kiowa (70-15154)	RAF Molesworth (for disposal)
	15372	Piper L-18C Super Cub (51-15372/N123SA) [372-A]	Privately owned, Anwick, Lincs
	15979	Hughes OH-6A Cayuse (69-15979)	RAF Mildenhall, instructional use
	15990	Bell AH-1F Hueycobra (70-15990)	Museum of Army Flying, Middle Wallop
	16011	Hughes OH-6A Cayuse (69-16011/G-OHGA)	Privately owned, Wesham, Lancs
	16037	Piper J-3C Cub 65 (G-BSFD)	Privately owned, Sleap
	16171	NA F-86D Sabre (51-6171)	North-East Aircraft Museum, Usworth
	16445	Bell AH-1F Hueycobra (69-16445)	Defence Academy of the UK, Shrivenham
	16506	Hughes OH-6A Cayuse (67-16506)	The Helicopter Museum, Weston-super-Mare
	16544	NA AT-6A Texan (41-16544/N13FY) [FY]	Privately owned, Hilversum, The Netherlands
	16579	Bell UH-1H Iroquois (66-16579)	The Helicopter Museum, Weston-super-Mare
	16718	Lockheed T-33A Shooting Star (51-6718)	City of Norwich Aviation Museum
	17962	Lockheed SR-71A Blackbird (61-7962)	American Air Museum, Duxford
	18263	Boeing-Stearman PT-17 Kaydet (41-8263/N38940) [822]	Privately owned, Tibenham
	19252	Lockheed T-33A Shooting Star (51-9252)	Tangmere Military Aviation Museum
	21509	Bell UH-1H Iroquois (72-21509/G-UHIH)	Privately owned, Wesham, Lancs
	21605	Bell UH-1H Iroquois (72-21605)	American Air Museum, Duxford
	24538	Kaman HH-43F Huskie (62-4535)	Midland Air Museum, Coventry
	24541	Cessna L-19E Bird Dog (N134TT)	Privately owned, Cotswold Airport
	24568	Cessna L-19E Bird Dog (LN-WNO)	Army Aviation Norway, Kjeller, Norway
	24582	Cessna L-19E Bird Dog (G-VDOG)	Privately owned, Chattis Hill, Hants
	28521	CCF T-6J Texan(G-TVIJ) [TA-521]	Privately owned, Woodchurch, Kent
	30274	Piper AE-1 Cub Cruiser (N203SA)	Privately owned, Nangis, France
	30861	NA TB-25J Mitchell (44-30861/N9089Z)	Privately owned, Booker
	31145	Piper L-4B Grasshopper (43-1145/G-BBLH) [26-G]	Privately owned, Biggin Hill
	31171	NA B-25J Mitchell (44-31171/N7614C)	American Air Museum, Duxford
	31430	Piper L-4B Grasshopper (43-1430/G-BHVV)	Privately owned, Perranporth
	31952	Aeronca O-58B Defender (G-BRPR)	Privately owned, Belchamp Water
	34037	NA TB-25N Mitchell (44-29366/N9115Z/8838M)	RAF Museum, Hendon
	37414	McD F-4C Phantom II (63-7414)	Midland Air Museum, stored Coventry

Serial	Type (code/other identity)	Owner/operator, location or fate	Notes
39624	Wag Aero Sport Trainer (G-BVMH) [39-D]	Privately owned, Abbeyshrule, Eire	
40467	Grumman F6F-5K Hellcat (80141/G-BTCC) [19]	Sold to the USA, February 2015	
41386	Thomas-Morse S4 Scout <R> (G-MJTD)	Privately owned, Lutterworth	
42165	NA F-100D Super Sabre (54-2165) [VM]	American Air Museum, Duxford	
42196	NA F-100D Super Sabre (54-2196)	Norfolk & Suffolk Avn Museum, Flixton	
43517	Boeing-Stearman N2S-5 Kaydet (G-NZSS) [227]	Privately owned, Woodchurch, Kent	
46214	Grumman TBM-3E Avenger (69327/CF-KCG) [X-3]	American Air Museum, Duxford	
51970	NA AT-6D Texan (41-33888/G-TXAN) [V-970]	Privately owned, Thruxton	
54433	Lockheed T-33A Shooting Star (55-4433) [TR-433]	Norfolk & Suffolk Avn Museum, Flixton	
54884	Piper L-4J Grasshopper (45-4884/N61787) [57-D]	Privately owned, Sywell	
56498	Douglas C-54Q Skymaster (N44914)	Privately owned, stored North Weald	
60312	McD F-101F Voodoo (56-0312)	Midland Air Museum, Coventry	
60344	Ryan Navion (N4956C)	Privately owned, Earls Colne	
60689	Boeing B-52D Stratofortress (56-0689)	American Air Museum, Duxford	
63000	NA F-100D Super Sabre (54-2212) [FW-000]	USAF Croughton, Oxon, at gate	
63319	NA F-100D Super Sabre (54-2269) [FW-319]	RAF Lakenheath, on display	
66692	Lockheed U-2CT (56-6692)	American Air Museum, Duxford	
70270	McD F-101B Voodoo (57-270) (fuselage)	Midland Air Museum, Coventry	
80105	Replica SE5a <R> (PH-WWI/G-CCBN) [19]	Privately owned, Thruxton	
80995	Cessna 337D Super Skymaster (F-BRPQ)	Privately owned, Strasbourg, France	
82062	DHC U-6A Beaver (58-2062)	Midland Air Museum, Coventry	
85061	NA SNJ-5 Texan (G-CHIA) [61]	Privately owned, Duxford	
86690	Grumman FM-2 Wildcat (G-CHPN) [F-2]	The Shuttleworth Collection, Old Warden	
93542	CCF T-6J Harvard IV (G-BRLV) [LTA-542]	Privately owned, North Weald	
96995	CV F4U-4 Corsair (OE-EAS) [BR-37]	Flying Bulls, Salzburg, Austria	
111836	NA AT-6C Harvard IIA (41-33262/G-TSIX) [JZ-6]	Privately owned, Linton-on-Ouse	
111989	Cessna L-19A Bird Dog (51-11989/N33600)	Museum of Army Flying, Middle Wallop	
114700	NA T-6G Texan (51-14700/G-TOMC)	Privately owned, Netherthorpe	
115042	NA T-6G Texan (51-15042/G-BGHU) [TA-042]	Privately owned, Headcorn	
115227	NA T-6G Texan (51-15227/G-BKRA)	Privately owned, Gloucester	
115302	Piper L-18C Super Cub (51-15302/G-BJTP) [TP]	Privately owned, Defford	
115373	Piper L-18C Super Cub (51-15373/G-AYPM) [A-373]	Privately owned, Leicester	
115684	Piper L-21A Super Cub (51-15684/G-BKVM) [DC]	Privately owned, Strubby	
117415	Canadair CT-133 Silver Star (G-BYOY) [TR-415]	RAF Manston History Museum	
121714	Grumman F8F-2P Bearcat (G-RUMM) [201-B]	The Fighter Collection, Duxford	
124143	Douglas AD-4NA Skyraider (F-AZDP) [205-RM]	Amicale J-B Salis, la Ferté-Alais, France	
124485	Boeing B-17G Flying Fortress (44-85784/G-BEDF)[DF-A]	B-17 Preservation Ltd, Duxford	
124541	CV F4U-5NL Corsair (F-AZYS) [16-WF]	Privately owned, Avignon, France	
124724	CV F4U-5NL Corsair (F-AZEG) [22]	Les Casques de Cuir, la Ferté-Alais, France	
126922	Douglas AD-4NA Skyraider (G-RADR) [503-H]	Kennet Aviation, North Weald	
134076	NA AT-6D Texan (41-34671/F-AZSC) [TA076]	Privately owned, Yvetot, France	
138179	NA T-28A Trojan (OE-ESA) [BA]	The Flying Bulls, Salzburg, Austria	
138266	NA T-28B Trojan (HB-RCT) [266-CT]	Jet Alpine Fighter, Sion, Switzerland	
138343	NA T-28B Trojan (N343NA) [212]	Privately owned, Antwerp, Belgium	
140547	NA T-28C Trojan (F-AZHN) [IF-28]	Privately owned, Toussus le Noble, France	
140566	NA T-28C Trojan (N556EB) [252]	Privately owned, la Ferté-Alais, France	
146289	NA T-28C Trojan (N99153) [2W]	Norfolk & Suffolk Avn Museum, Flixton	
150225	WS58 Wessex 60 (G-AWOX) [123]	Privately owned, Lulsgate	
155454	NA OV-10B Bronco (158300/99+24/F-AZKM) [26]	Privately owned, Montelimar, France	
155529	McD F-4J(UK) Phantom II (ZE359) [AJ-114]	American Air Museum, Duxford	
155848	McD F-4S Phantom II [WT-11]	Royal Scottish Mus'm of Flight, E Fortune	
159233	HS AV-8A Harrier [CG-33]	Imperial War Museum North, Salford Quays	
162068	McD AV-8B Harrier II <ff>	Privately owned, Thorpe Wood, N Yorks	
162071	McD AV-8B Harrier II (fuselage)	Rolls-Royce, Filton	
162074	McD AV-8B Harrier II <ff>	Privately owned, South Molton, Devon	
162730	McD AV-8B Harrier II <ff>	Privately owned, Liverpool	
162737	McD AV-8B Harrier II (fuselage) [38]	MoD, Boscombe Down	
162958	McD AV-8B Harrier II <ff>	QinetiQ, Farnborough	
162964	McD AV-8B Harrier II <ff>	Harrier Heritage Centre, RAF Wittering	
162964	McD AV-8B Harrier II <r>	Privately owned, Charlwood, Surrey	
163205	McD AV-8B Harrier II (fuselage)	Privately owned, Thorpe Wood, N Yorks	
163423	McD AV-8B Harrier II <ff>	QinetiQ, Boscombe Down	

Notes	Serial	Type (code/other identity)	Owner/operator, location or fate
	163423	McD AV-8B Harrier II <rf>	Privately owned, Sproughton
	217786	Boeing-Stearman PT-17 Kaydet (41-8169/CF-EQS) [25]	American Air Museum, Duxford
224319		Douglas C-47B Skytrain (44-77047/G-AMSN) <ff>	Privately owned, Sussex
	226413	Republic P-47D Thunderbolt (45-49192/N47DD) [ZU-N]	American Air Museum, Duxford
	231983	Boeing B-17G Flying Fortress (44-83735/F-BDRS)[IY-G]	American Air Museum, Duxford
	234539	Fairchild PT-19B Cornell (42-34539/N50429) [63]	Privately owned, Dunkeswell
	236657	Piper L-4A Grasshopper (42-36657/G-BGSJ) [72-D]	Privately owned, Langport
	238410	Piper L-4A Grasshopper (42-38410/G-BHPK) [44-A]	Privately owned, Tibenham
	241079	Waco CG-4A Hadrian <R>	RAF Museum, stored Cosford
243809		Waco CG-4A Hadrian (BAPC 185)	Museum of Army Flying, Middle Wallop
	252983	Schweizer TG-3A (42-52983/N66630)	Imperial War Museum, stored Duxford
	298177	Stinson L-5A Sentinel (42-98177/N6438C) [8-R]	Privately owned, Tibenham
313048		NA AT-6D Texan (44-81506/G-TDJN)	Privately owned, Gloucester
	314887	Fairchild Argus III (43-14887/G-AJPI)	Privately owned, Eelde, The Netherlands
	315509	Douglas C-47A Skytrain (43-15509/G-BHUB) [W7-S]	American Air Museum, Duxford
319764		Waco CG-4A Hadrian (237123/BAPC 157) (fuselage)	Yorkshire Air Museum, Elvington
	329282	Piper J-3C Cub 65 (N46779)	Privately owned, Abbots Bromley
	329405	Piper L-4H Grasshopper (43-29405/G-BCOB) [23-A]	Privately owned, Turweston
	329417	Piper L-4A Grasshopper (42-38400/G-BDHK)	Privately owned, English Bicknor, Glos
	329471	Piper L-4H Grasshopper (43-29471/G-BGXA) [44-F]	Privately owned, Martley, Worcs
	329601	Piper L-4H Grasshopper (43-29601/G-AXHR) [44-D]	Privately owned, Nayland
	329707	Piper L-4H Grasshopper (43-29707/G-BFBY) [44-S]	Privately owned, Old Buckenham
	329854	Piper L-4H Grasshopper (43-29854/G-BMKC) [44-R]	Privately owned, Biggin Hill
	329934	Piper L-4H Grasshopper (43-29934/G-BCPH) [72-B]	Privately owned, Garford, Oxon
	330238	Piper L-4H Grasshopper (43-30238/G-LIVH) [24-A]	Privately owned, Yarcombe, Devon
	330244	Piper L-4H Grasshopper (43-30244/G-CGIY) [46-C]	Privately owned, Leeds
	330372	Piper L-4H Grasshopper (43-30372/G-AISX)	Privately owned, Booker
330426		Piper L-4J Grasshopper (45-4884/N61787) [53-K]	Privately owned, Sywell
	330485	Piper L-4H Grasshopper (43-30485/G-AJES) [44-C]	Privately owned, Shifnal
379994		Piper J-3L Cub 65 (G-BPUR) [52-J]	Privately owned, Popham
411631		NA P-51D Mustang (44-73979) [MX-V]	American Air Museum, Duxford
411622		NA P-51D Mustang (44-74427/F-AZSB) [G4-C]	Amicale J-B Salis, la Ferté-Alais, France
413317		NA P-51D Mustang (44-74409/N51RT) [VF-B]	RAF Museum, Hendon
413357		NA P-51D Mustang <R> [B7-R]	Privately owned, Byfleet, Surrey
413521		NA P-51D Mustang (44-13521/G-MRLL) [5Q-B]	Privately owned, Hardwick, Norfolk
413573		NA P-51D Mustang (44-73415/9133M/N6526D) [B6-V]	RAF Museum, Cosford
413578		NA P-51D Mustang (44-74923/PH-JAT) [C5-W]	Privately owned, Lelystad, The Netherlands
413704		NA P-51D Mustang (44-73149/G-BTCD) [B7-H]	The Old Flying Machine Company, Duxford
414419		NA P-51D Mustang (45-15118/G-MSTG) [LH-F]	Privately owned, Hardwick, Norfolk
414237		NA P-51D Mustang (44-73656/F-AZXS) [HO-W]	Privately owned, Avignon, France
414450		NA P-51D Mustang (44-73877/N167F) [B6-S]	Repainted in RAF markings as KH774, July 2014
414673		Bonsall Mustang <R> (G-BDWM) [LH-I]	Privately owned, Gamston
414907		Titan T-51 Mustang (G-DHYS) [CY-S]	Privately owned, Pershore
433915		Consolidated PBV-1A Canso A (RCAF 11005/G-PBYA)	Privately owned, Duxford
434602		Douglas A-26B Invader (44-34602/LN-IVA) [B]	Nordic Warbirds, Västerås, Sweden
436021		Piper J/3C Cub 65 (G-BWEZ)	Privately owned, Strathaven, Strathclyde
442268		Noorduyn AT-16 Harvard IIB (KF568/LN-TEX) [TA-268]	Scandinavian Historic Flight, Oslo, Norway
	454467	Piper L-4J Grasshopper (45-4467/G-BILI) [44-J]	Privately owned, White Waltham
	454537	Piper L-4J Grasshopper (45-4537/G-BFDL) [04-J]	Privately owned, Shempston Farm, Lossiemouth
	461748	Boeing B-29A Superfortress (44-61748/G-BHDK) [Y]	American Air Museum, Duxford
	463209	NA P-51D Mustang <R> (BAPC 255) [WZ-S]	American Air Museum, Duxford
	472035	NA P-51D Mustang (44-72035/G-SIJJ)	Hangar 11 Collection, North Weald
	472216	NA P-51D Mustang (44-72216/G-BIXL) [HO-M]	Privately owned, East Garston, Bucks
	472218	NA P-51D Mustang (44-73979) [WZ-I]	Repainted as 411631, 2014
	472218	Titan T-51 Mustang (G-MUZY) [WZ-I]	Privately owned, Damyns Hall, Essex
472773		NA P-51D Mustang (44-72773/D-FPSI) [QP-M]	Repainted in Swiss marks as J-901, 2014
473871		NA TF-51D Mustang (44-73871/D-FTSI) [TF-871]	Meier Motors, Bremgarten, Germany
474008		Jurca MJ77 Gnatsum (G-PSIR) [VF-R]	Privately owned, Fishburn
474425		NA P-51D Mustang (44-74425/PH-PSI) [OC-G]	Privately owned, Lelystad, The Netherlands
	479712	Piper L-4H Grasshopper (44-79826/G-AHIP) [8-R]	Privately owned, Coleford
	479744	Piper L-4H Grasshopper (44-79744/G-BGPD) [49-M]	Privately owned, Marsh, Bucks
	479766	Piper L-4H Grasshopper (44-79766/G-BKHG) [63-D]	Privately owned, Frogland Cross

Serial	Type (code/other identity)	Owner/operator, location or fate	Notes
479781	Piper L-4H Grasshopper (44-79781/G-AISS)	Privately owned, Insch (under restoration)	
479897	Piper L-4H Grasshopper (44-79897/G-BOXJ) [JD]	Privately owned, Rochester	
480015	Piper L-4H Grasshopper (44-80015/G-AKIB) [44-M]	Privately owned, Perranporth	
480133	Piper L-4J Grasshopper (44-80133/G-BDCD) [44-B]	Privately owned, Slinfold	
480173	Piper L-4J Grasshopper (44-80609/G-RRSR) [57-H]	Privately owned, Wellesbourne Mountford	
480321	Piper L-4J Grasshopper (44-80321/G-FRAN) [44-H]	Privately owned, Rayne, Essex	
480480	Piper L-4J Grasshopper (44-80480/G-BECN) [44-E]	Privately owned, Rayne, Essex	
480551	Piper L-4J Grasshopper (44-80551/LN-KLT) [43-S]	Scandinavian Historic Flight, Oslo, Norway	
480636	Piper L-4J Grasshopper (44-80636/G-AXHP) [58-A]	Privately owned, Spanhoe	
480723	Piper L-4J Grasshopper (44-80723/G-BFZB) [E5-J]	Privately owned, Egginton	
480752	Piper L-4J Grasshopper (44-80752/G-BCXJ) [39-E]	Privately owned, Old Sarum	
483868	Boeing B-17G Flying Fortress (44-83868/N5237V) [A-N]	RAF Museum, Hendon	
493209	NA T-6G Texan (49-3209/G-DDMV/41)	Privately owned, Headcorn	
511701A	Beech C-45H (51-11701/G-BSZC) [AF258]	Privately owned, Bryngwyn Bach	
779465	Hiller UH-12C (N5315V)	Privately owned, Lower Upham, Hants	
2106638	Titan T-51 Mustang (G-CIFD)	Privately owned, Hereford	
2100882	Douglas C-47A Skytrain (42-100882/N473DC) [3X-P]	Privately owned, East Kirkby	
2100884	Douglas C-47A Skytrain (42-100884/N147DC) [L4-D]	Privately owned, Dunsfold	
2104590	Curtiss P-40M Kittyhawk (43-5802/G-KITT) [44]	Hangar 11 Collection, North Weald	
2105915	Curtiss P-40N Kittyhawk (42-105915/F-AZKU) [12]	Privately owned, la Ferté-Alais, France	
03-08003	B-V CH-47F Chinook [DT]	RAF Odiham, at main gate	
3-1923	Aeronca O-58B Defender (43-1923/G-BRHP)	Privately owned, Chiseldon	
18-2001	Piper L-18C Super Cub (52-2401/G-BIZV)	Privately owned, Wicklow, Ireland	
39-139	Beech YC-43 Traveler (N295BS)	Duke of Brabant AF, Eindhoven, The Netherlands	
39-0285	Curtiss P-40B Warhawk	The Fighter Collection, stored Duxford (wreck)	
39-0287	Curtiss P-40B Warhawk	The Fighter Collection, stored Duxford (wreck)	
40-2538	Fairchild PT-19A Cornell (N33870)	Privately owned, Mendlesham	
41-13357	Curtiss P-40C Warhawk (G-CIIO) [160,10AB]	The Fighter Collection, Duxford	
41-19393	Douglas A-20C Havoc (wreck)	WWII Remembrance Museum, Handcross, W Sussex	
41-33275	NA AT-6C Texan (G-BICE) [CE]	Privately owned, Great Oakley, Essex	
42-66841	Lockheed P-38H Lightning [153]	Privately owned, Bentwaters	
42-12417	Noorduyn AT-16 Harvard IIB (KLu. B-163)	Newark Air Museum, Winthorpe	
42-35870	Taylorcraft DCO-65 (G-BWLJ) [129]	Privately owned, Nayland	
42-58678	Taylorcraft DF-65 (G-BRIY) [IY]	Privately owned, Carlisle	
42-78044	Aeronca 11AC Chief (G-BRXL)	Privately owned, Andrewsfield	
42-84555	NA AT-6D Harvard III (FAP.1662/G-ELMH) [EP-H]	Privately owned, Hardwick, Norfolk	
42-93510	Douglas C-47A Skytrain [CM] <ff>	Privately owned, Kew	
43-9628	Douglas A-20G Havoc <ff>	Privately owned, Hinckley, Leics	
43-11137	Bell P-63C Kingcobra (wreck)	WWII Remembrance Museum, Handcross, W Sussex	
43-21664	Douglas A-20G Havoc (wreck)	WWII Remembrance Museum, Handcross, W Sussex	
43-35943	Beech 3N (G-BKRN) [943]	Beech Restorations, Bruntingthorpe	
43-36140	NA B-25J Mitchell <ff>	WWII Remembrance Museum, Handcross, W Sussex	
44-4315	Bell P-63C Kingcobra	WWII Remembrance Museum, Handcross, W Sussex	
44-4368	Bell P-63C Kingcobra	Privately owned, Surrey	
44-13954	NA P-51D Mustang (G-UAKE)	Mustang Restoration Co Ltd, Coventry	
44-14574	NA P-51D Mustang (fuselage)	East Essex Aviation Museum, Clacton	
44-42914	Douglas DC-4 (N31356)	Privately owned, stored North Weald	
44-51228	Consolidated B-24M Liberator [EC-493]	American Air Museum, Duxford	
44-79609	Piper L-4H Grasshopper (G-BHXY) [PR]	Privately owned, Bealbury, Cornwall	
44-80594	Piper L-4J Grasshopper (G-BEDJ)	Privately owned, White Waltham	
44-80647	Piper L-4J Grasshopper (D-EGAF)	The Vintage Aircraft Co, Fürstenwalde, Germany	
44-83184	Fairchild UC-61K Argus III (G-RGUS)	Privately owned, Sibson	
51-9036	Lockheed T-33A Shooting Star	Newark Air Museum, Winthorpe	
51-15319	Piper L-18C Super Cub (G-FUZZ) [A-319]	Privately owned, Elvington	
51-15555	Piper L-18C Super Cub (G-OSPS)	Privately owned, Weston, Eire	
52-8543	CCF T-6J Texan (G-BUKY) [66]	Privately owned, Duxford	
54-005	NA F-100D Super Sabre (54-2163)	Dumfries & Galloway Avn Mus, Dumfries	

Notes	Serial	Type (code/other identity)	Owner/operator, location or fate
	54-174	NA F-100D Super Sabre (54-2174) [SM]	Midland Air Museum, Coventry
	54-2223	NA F-100D Super Sabre	Newark Air Museum, Winthorpe
	54-2445	Piper L-21B Super Cub (G-OTAN) [A-445]	Privately owned, Hawarden
	54-2447	Piper L-21B Super Cub (G-SCUB)	Privately owned, Anwick
	55-138354	NA T-28B Trojan (138354/N1328B) [TL-354]	Privately owned, Antwerp, Belgium
	62-428	Republic F-105G Thunderchief (62-4428) [WW]	USAF Croughton, Oxon, at gate
	63-699	McD F-4C Phantom II (63-7699) [CG]	Midland Air Museum, Coventry
	64-17657	Douglas B-26K Counter Invader (N99218) <ff>	WWII Remembrance Museum, Handcross, W Sussex
	65-777	McD F-4C Phantom II (63-7419) [SA]	RAF Lakenheath, on display
	65-10450	Northrop AT-38B Talon <rf>	Martin-Baker Ltd, Chalgrove, Fire Section
	67-120	GD F-111E Aardvark (67-0120) [UH]	American Air Museum, Duxford
	68-0060	GD F-111E Aardvark <ff>	Dumfries & Galloway Avn Mus, Dumfries
	68-8284	Sikorsky MH-53M Pave Low IV	RAF Museum, Cosford
	70-0389	GD F-111F Aardvark (68-0011) [LN]	RAF Lakenheath, on display
	72-1447	GD F-111F Aardvark <ff>	American Air Museum, Duxford
	74-0177	GD F-111F Aardvark [FN]	RAF Museum, Cosford
	76-020	McD F-15A Eagle (76-0020)	American Air Museum, Duxford
	76-124	McD F-15B Eagle (76-0124) [LN]	RAF Lakenheath, instructional use
	77-259	Fairchild A-10A Thunderbolt (77-0259) [AR]	American Air Museum, Duxford
	80-219	Fairchild GA-10A Thunderbolt (80-0219) [AR]	RAF Alconbury, on display
	82-23762	B-V CH-47D Chinook <ff>	RAF Odiham, instructional use
	83-24104	B-V CH-47D Chinook [BN] <ff>	RAF Museum, Hendon
	86-01677	B-V CH-47D Chinook	RAF Odiham, instructional use
	89-00159	B-V CH-47D Chinook	Army, Bramley, GI use
	92-048	McD F-15A Eagle (74-0131) [LN]	RAF Lakenheath, on display
	108-1601	Stinson 108-1 Voyager (G-CFGE) [H]	Privately owned, Spanhoe
	146-11042	Wolf WII Boredom Fighter (G-BMZX) [7]	Privately owned, Popham
	146-11083	Wolf WII Boredom Fighter (G-BNAI) [5]	Privately owned, Haverfordwest
	G-57	Piper L-4A Grasshopper (42-36375/G-AKAZ)	Privately owned, Duxford
	X-17	Curtiss P-40F Warhawk (41-19841/G-CGZP)	The Fighter Collection, Duxford
	CY-D	NA TF-51D Mustang (44-84847/G-TFSI)	The Fighter Collection, Duxford
	CY-G	Titan T-51 Mustang (G-TSIM)	Privately owned, Hereford
	YEMEN		
	104	BAC Jet Provost T52A (G-PROV)	Privately owned, North Weald
	YUGOSLAVIA		
	30131	Soko P-2 Kraguj [131]	Privately owned, stored Minskip, N Yorks
	30139	Soko P-2 Kraguj [139]	Privately owned, Biggin Hill
	30140	Soko P-2 Kraguj (G-RADA) [140]	Privately owned, Linton-on-Ouse
	30146	Soko P-2 Kraguj (G-BSXD) [146]	Privately owned, Linton-on-Ouse
	30149	Soko P-2 Kraguj (G-SOKO) [149]	Privately owned, Fenland
	30151	Soko P-2 Kraguj [151]	Privately owned, Sopley, Hants
	51109	UTVA-66 (YU-DLG)	Privately owned, Morpeth

Bristol Scout D A1742 is a replica and is also BAPC 38. It is owned by the Shuttleworth Trust at Old Warden.

F8010/Z is a 1978-vintage replica SE5A with civil registration G-BDWJ. Operated from Langport in Somerset as part of the Great War Display Team, it saw considerable action in 2014 in commemoration of the 100th anniversary of the start of World War I.

P6382/C is a Miles M14A Hawk Trainer 3. Allocated the civil registration G-AJRS, this 1939-vintage machine is owned by the Shuttleworth trust at Old Warden, where it flies regularly.

KF388 is a Noorduyn AT-16 Harvard IIB composite, on display at the Bournemouth Aviation Museum, complete with Bournemouth Airport as a backdrop.

2014 saw the commemoration of the 70th Anniversary of D-Day and the Battle of Britain Memorial Flight's Spitfire LF IXC MK356/5J-K wore D-Day markings as a consequence. Seen here at RIAT 2014, it did a pairs display with a suitably marked RAF Typhoon FGR4.

Chipmunk WK517, appropriately registered G-ULAS, wears the University of London Air Squadron markings and is seen here at its home base at Turweston.

Midair Squadron's silver painted Canberra PR9 XH134 (G-OMHD) was one of the stars of the 2014 air show season, often displaying with a pair of similarly painted Hunter trainers.

2014 saw the 40th anniversary of the first flight of the Hawk and T1 XX154 was the machine in question, first taking to the air from Dunsfold on 21 August 1974. Still flown by ETPS at Boscombe Down and painted in a stylish black and gold scheme, it is seen here in the static at RIAT, Fairford.

Bulldog T1 XX614 (G-GGRR) is seen here in the sun at its Turweston base not long after emerging from here following an expensive repaint and rebuild.

Lynx AH7 XZ180/C belongs to 671 Squadron AAC, part of 7 Regiment, based at Middle Wallop. Due to be retired in 2015, the Lynx AH7 still provides a surprisingly agile display item, as indeed it did here, at Farnborough.

Some Sea Kings still in service celebrated 45 years since their first flight in 2014; the RAF's HAR3 ZA105/Q is a mere baby, at just 34 years old. It wears the badge of 203(R) Squadron on the nose, a unit which celebrated its centenary in 2014, although ZA105 has since been dispatched to operate with 1564 Flight in the Falkland Islands.

Coded W, ZA297 is a Sea King HC4 and is seen here at its Yeovilton base in February 2014 shortly before being retired from service. A dwindling number of 'Junglies' remain, all operated by 845 NAS, and all will be retired in the next year, replaced by the Merlin HC3.

Seen here in the static display at RIAT, Viking ZE636/XZ is operated by 621 VGS at RAF Hullavington.

For 2014 Tucano T1 ZF244 of 1 FTS at Church Fenton was painted up in special markings to commemorate the 100th anniversary of the start of World War I, while still carrying 72(R) Squadron markings on the fuselage and tail. By the end of 2014 the fate of these machines was becoming clearer, following the announcement that they will be replaced by the Beechcraft T-6C Texan.

ZF377 is a Shorts Tucano T1 of No 1 FTS at Linton-on-Ouse. Unlike others in the fleet, this one does not carry squadron markings.

ZF560 is a Lynx HMA8SRU and is seen here wearing the code 455/DN and a Dragon on the nose, all of which reveal the operator as HMS Dragon Flight. The Lynx HMA8 fleet was 'condensed' into one squadron, 815 NAS, when 702 NAS decommissioned on 1 August 2014 and ZF560 is now assigned to 815 NAS Duncan Flight.

ZH004 is the sole Defender T3 in the Army Air Corps fleet, being operated by 651 Squadron as part of 5 Regiment at Aldergrove.

ZH883 is a Hercules C5, operated by 24, 30 and 47 Squadrons of the RAF from a pool at RAF Brize Norton. The fleet is due to undergo a mid-life update in the next few years with C4s becoming C6s and C5s becoming C7s. First in line is C6 ZH866.

Seen here in the static at RIAT is Merlin HC3 ZJ122/F. The badge on the nose is of 28 Squadron, the last remaining RAF Merlin squadron, based at RAF Benson. The other unit, 78 Squadron, disbanded in 2014 as the Navy began to operate these machines in the guise of 846 NAS.

It is sometimes said that if you can read the markings on an Army Apache you are probably about to bang your head on it and AH1 ZJ172 here is fairly typical! Seen here at Fairford, it is operated by 662 Squadron, part of 3 Regiment at Wattisham.

Alpha Jet ZJ647 is one of a small fleet operated by QinetiQ from its base at Boscombe Down.

Puma HC2 ZJ955/P is one of several converted from the HC1 variant and now building up numbers with 33 and 230 Squadrons at RAF Benson. 2014 saw the 95th anniversary of 230 Squadron and this one was specially marked in this 'Black Peter' colour scheme, similar to that worn by the unit's Sunderland, JM673, in November 1944.

ZK308/TP-V is a Typhoon FGR4 specially marked with D-Day stripes to commemorate the 70th anniversary in 2014. TP-V was the code worn by Typhoon IB MN526 of 198 Squadron on D-Day and this one performed at many air shows with Spitfire LFIXC MK356.

2014 was an eventful year for 6 Squadron as it celebrated its 100th anniversary and also moved from RAF Leuchars to RAF Lossiemouth during the year. To commemorate its centenary, Typhoon FGR4 ZK342/ED was painted in special markings with a sand camouflage spine, a large, red flying can opener and battle honours on the tail, along with additional markings on the canards depicting initial and current types operated.

29(R) Squadron at RAF Coningsby had the honour of forming the RAF's Typhoon Display Team for 2014 and its Typhoon FGR4, ZK343/BX wore special markings to depict this.

ZZ383 is a Wildcat AH1 of 652 Squadron Army Air Corps, the first AAC unit to operate the type and now doing so in large numbers from RNAS Yeovilton, where the whole Wildcat fleet is currently located.

ZZ501 is one of five Hawker Beechcraft Super Kingair 350s operated by the Royal Navy's 750 NAS at RNAS Culdrose. It is known as the Avenger T1 in RN service.

G-BYWM is one of a small number of Tutor T1s wearing 'Fly Navy' stickers and operated by 727 NAS from RNAS Yeovilton.

Painted in false marks as (18)671, Chipmunk G-BNZC is actually an ex RAF machine. It is based at Old Warden with The Shuttleworth Collection.

18-5395/CDG is a Piper L-18C Super Cub painted in false French Army colours. Registered G-CUBJ, it is based at Old Warden.

99+18 is a North American OV-10B Bronco, registered as G-ONAA and operated by the Bronco Demonstration Team based at Wevelgem in Belgium. This one replaces the team's previous mount which was written off in a crash in 2012.

556/17 is a Fokker Dr.1 Replica, based at Tibenham and registered as G-CFHY. Flown as part of the Great War Display Team, the centenary of the start of World War I meant that this was in great demand at air shows in 2014. The other Fokker Dr.1 Replica used by GWDT, 403/17 (G-CDXR), was flown regularly by Iron Maiden front man Bruce Dickinson.

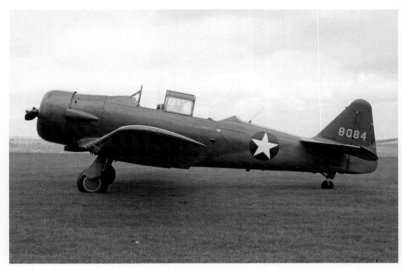

8084 is a North American AT-6D Texan, registered as G-KAMY and operated by Kennet Aviation at North Weald, but frequently flown from Yeovilton, which has become a second base for this machine.

Basking in the sun at Turweston, 329405/23-A, Piper L-4H Grasshopper G-BCOB, wears USAAF markings from World War II.

262 is a Pilatus PC-9M of the Irish Air Corps' Flying Training School at Baldonnel, which operates seven of the type.

Wearing Algerian Air Force markings, C-130H Hercules 7T-WHE also wears its construction number (4935) on the tail. It is operated by 2 Escadre de Transport at Boufarik.

Wearing a red H, SAAB 105ÖE 1128 (construction number 105428) belongs to the Austrian Air Force's Düsentrainerstaffel at Linz.

Transall 51+06 is one of a dwindling fleet of these transports operated by the Luftwaffe, in this case with LTG 63 at Hohn. These should disappear from service completely in the next few years as they are replaced by the Airbus A400M.

Wearing the code 6-22 of 154 Gruppo Italian Air Force based at Ghedi, this Panavia A-200A Tornado, MM7029, was actually being flown rather well by a pilot from the RSV at the time!

C-27J Spartan 08 is one of three operated by the Lithuanian Air Force's Transporto Eskadrile, based at Siauliai-Zokniai.

Look carefully on the tail and you can just about make out the serial 671 of this Norwegian Air Force F-16A, wearing special Tiger markings to reflect 338 Skv's participation at the 2014 Tiger Meet.

506 is a C-130J Hercules II belonging to 16 Squadron of the Royal Air Force of Oman, based at Seeb. It was delivered in 2014.

040 is a PZL-130 TC-II Turbo Orlik and forms part of the Polish Air Force's 'Team Orlik', based at Radom.

For many the highlight of RIAT 2014 was seeing a pair of Polish Air Force Su-22M-4Ks in the flying display. 3612 was one of three brought over and is operated by 40.elt at Swidwin.

Swiss Air Force F/A-18C Hornet J-5009 touches down after another flying display. Wearing no external unit markings, this one is believed to be based at Payerne.

T-314 is an AS.332M-1 Super Puma, operated by Lufttransport Staffel 6 of the Swiss Air Force at Alpnach.

Turkish Air Force F-16C 91-0011 is operated by 141 Filo at Akinci and is one of a pair wearing special 'Solo Turk' markings. Its flying display is as equally stunning as its colour scheme!

The Boeing P-8A Poseidon is slowly entering service with more and more former US Navy P-3 Orion units. This one, 167955, wears the JA tail code of VX-1, based at Patuxent River in Maryland and is seen here with its weapons bay open - not something you'd see on the Boeing 737-800 that these airframes are based upon!

Serial	Type (code/other identity)	Owner/operator, location or fate	Notes
C7	Avro 631 Cadet (EI-AGO)	IAC Museum, Baldonnel	
34	Miles M14A Magister I (N5392)	National Museum of Ireland, Dublin	
141	Avro 652A Anson C19	IAC Museum, Baldonnel	
164	DHC1 Chipmunk T20	IAC Museum, Baldonnel	
168	DHC1 Chipmunk T20	IAC Museum, Baldonnel	
169	DHC1 Chipmunk 22 (G-ARGG)	Privately owned, Ballyboy	
170	DHC1 Chipmunk 22 (G-BDRJ)	Privately owned, Kilkerran	
172	DHC1 Chipmunk T20	IAC, stored Baldonnel	
173	DHC1 Chipmunk T20	South East Aviation Enthusiasts, Dromod	
176	DH104 Dove 4 (VP-YKF)	South East Aviation Enthusiasts, Waterford	
183	Percival P56 Provost T51	IAC Museum, Baldonnel	
184	Percival P56 Provost T51	South East Aviation Enthusiasts, Dromod	
187	DH115 Vampire T55 <ff>	South East Aviation Enthusiasts, Dromod	
191	DH115 Vampire T55	IAC Museum, Baldonnel	
192	DH115 Vampire T55 <ff>	South East Aviation Enthusiasts, Dromod	
195	Sud SA316 Alouette III (F-WJDH)	National Museum of Ireland, Dublin	
198	DH115 Vampire T11 (XE977)	National Museum of Ireland, Dublin	
199	DHC1 Chipmunk T22	IAC, stored Baldonnel	
202	Sud SA316 Alouette III	Ulster Aviation Society, Long Kesh	
203	Reims-Cessna FR172H	IAC No 104 Sqn/1 Operations Wing, Baldonnel	
205	Reims-Cessna FR172H	IAC No 104 Sqn/1 Operations Wing, Baldonnel	
206	Reims-Cessna FR172H	IAC No 104 Sqn/1 Operations Wing, Baldonnel	
207	Reims-Cessna FR172H	IAC, stored Waterford	
208	Reims-Cessna FR172H	IAC No 104 Sqn/1 Operations Wing, Baldonnel	
210	Reims-Cessna FR172H	IAC No 104 Sqn/1 Operations Wing, Baldonnel	
215	Fouga CM170R Super Magister	Dublin Institute of Technology	
216	Fouga CM170R Super Magister	IAC Museum, Baldonnel	
218	Fouga CM170R Super Magister	Shannon Aerospace, Shannon Airport	
219	Fouga CM170R Super Magister	IAC Museum, Baldonnel	
220	Fouga CM170R Super Magister	Cork University, instructional use	
221	Fouga CM170R Super Magister [3-KE]	IAC Museum, stored Baldonnel	
231	SIAI SF-260WE Warrior	IAC Museum, stored Baldonnel	
240	Beech Super King Air 200MR	IAC No 102 Sqn/1 Operations Wing, Baldonnel	
251	Grumman G1159C Gulfstream IV (N17584)	IAC No 102 Sqn/1 Operations Wing, Baldonnel	
252	Airtech CN.235 MPA Persuader	IAC No 101 Sqn/1 Operations Wing, Baldonnel	
253	Airtech CN.235 MPA Persuader	IAC No 101 Sqn/1 Operations Wing, Baldonnel	
254	PBN-2T Defender 4000 (G-BWPN)	IAC No 106 Sqn/1 Operations Wing, Baldonnel	
256	Eurocopter EC135T-1 (G-BZRM)	IAC No 106 Sqn/1 Operations Wing, Baldonnel	
258	Gates Learjet 45 (N5009T)	IAC No 102 Sqn/1 Operations Wing, Baldonnel	
260	Pilatus PC-9M (HB-HQS)	IAC Flying Training School, Baldonnel	
261	Pilatus PC-9M (HB-HQT)	IAC Flying Training School, Baldonnel	
262	Pilatus PC-9M (HB-HQU)	IAC Flying Training School, Baldonnel	
263	Pilatus PC-9M (HB-HQV)	IAC Flying Training School, Baldonnel	
264	Pilatus PC-9M (HB-HQW)	IAC Flying Training School, Baldonnel	
266	Pilatus PC-9M (HB-HQY)	IAC Flying Training School, Baldonnel	
267	Pilatus PC-9M (HB-HQZ)	IAC Flying Training School, Baldonnel	
270	Eurocopter EC135P-2	IAC No 302 Sqn/3 Operations Wing, Baldonnel	
271	Eurocopter EC135P-2	IAC No 302 Sqn/3 Operations Wing, Baldonnel	
272	Eurocopter EC135T-2 (G-CECT)	IAC No 106 Sqn/1 Operations Wing, Baldonnel	
274	AgustaWestland AW139	IAC No 301 Sqn/3 Operations Wing, Baldonnel	
275	AgustaWestland AW139	IAC No 301 Sqn/3 Operations Wing, Baldonnel	
276	AgustaWestland AW139	IAC No 301 Sqn/3 Operations Wing, Baldonnel	
277	AgustaWestland AW139	IAC No 301 Sqn/3 Operations Wing, Baldonnel	
278	AgustaWestland AW139	IAC No 301 Sqn/3 Operations Wing, Baldonnel	
279	AgustaWestland AW139	IAC No 301 Sqn/3 Operations Wing, Baldonnel	

Aircraft included in this section include those likely to be seen visiting UK civil and military airfields on transport flights, exchange visits, exercises and for air shows. It is not a comprehensive list of *all* aircraft operated by the air arms concerned.

ALGERIA
Force Aérienne Algérienne/
Al Quwwat al Jawwiya al Jaza'eriya
Airbus A.340-541
Ministry of Defence, Boufarik
7T-VPP

Ilyushin
Il-76MD/Il-76TD/Il-78
347 Escadron de Transport
  Strategique,
  Boufarik;
357 Escadron de
  Ravitaillement en Vol,
  Boufarik

7T-WIA	Il-76MD	347 Esc
7T-WIB	Il-76MD	347 Esc
7T-WIC	Il-76MD	347 Esc
7T-WID	Il-76TD	347 Esc
7T-WIE	Il-76TD	347 Esc
7T-WIF	Il-78	357 Esc
7T-WIG	Il-76TD	347 Esc
7T-WIL	Il-78	357 Esc
7T-WIM	Il-76TD	347 Esc
7T-WIN	Il-78	357 Esc
7T-WIP	Il-76TD	347 Esc
7T-WIQ	Il-78	357 Esc
7T-WIR	Il-76TD	347 Esc
7T-WIS	Il-78	357 Esc
7T-WIT	Il-76TD	347 Esc
7T-WIU	Il-76TD	347 Esc
7T-WIV	Il-76TD	347 Esc

Lockheed
C-130H/C-130H-30 Hercules
2 Escadre de Transport
  Tactique et Logistique,
  Boufarik

7T-WHB	(5224)	C-130H-30
7T-WHD	(4987)	C-130H-30
7T-WHE	(4935)	C-130H
7T-WHF	(4934)	C-130H
7T-WHI	(4930)	C-130H
7T-WHJ	(4928)	C-130H
7T-WHL	(4989)	C-130H-30
7T-WHN	(4894)	C-130H-30
7T-WHO	(4897)	C-130H-30
7T-WHP	(4921)	C-130H-30
7T-WHQ	(4926)	C-130H
7T-WHR	(4924)	C-130H
7T-WHS	(4912)	C-130H
7T-WHT	(4911)	C-130H
7T-WHY	(4913)	C-130H
7T-WHZ	(4914)	C-130H

Grumman
G.1159C Gulfstream IVSP
Ministry of Defence, Boufarik

7T-VPC	(1418)
7T-VPM	(1421)
7T-VPR	(1288)
7T-VPS	(1291)

Gulfstream Aerospace
Gulfstream V
Ministry of Defence, Boufarik
7T-VPG   (617)

ANGOLA
Angolan Government
Bombardier Global 5000
Angola Government, Luanda
D2-ANG

ARMENIA
Armenian Government
Airbus A.319CJ-132
Armenian Government, Yerevan
EK-RA01

AUSTRALIA
Royal Australian Air Force
Airbus A.330-203 MRTT (KC-30A)
33 Sqn, Amberley
A39-001
A39-002
A39-003
A39-004
A39-005

Boeing
737-7DF/-7DT/-7ES AEW&C (E-7A)
34 Sqn, Canberra

A30-001	737-7ES (E-7A)
A30-002	737-7ES (E-7A)
A30-003	737-7ES (E-7A)
A30-004	737-7ES (E-7A)
A30-005	737-7ES (E-7A)
A30-006	737-7ES (E-7A)
A36-001	737-7DT
A36-002	737-7DF

Boeing
C-17A Globemaster III
36 Sqn, Amberley
A41-206
A41-207
A41-208
A41-209
A41-210
A41-211

Canadair
CL.604 Challenger
34 Sqn, Canberra
A37-001
A37-002
A37-003

Lockheed
C-130J-30 Hercules II
37 Sqn, Richmond, NSW
A97-440
A97-441
A97-442
A97-447
A97-448
A97-449
A97-450
A97-464
A97-465
A97-466
A97-467
A97-468

Lockheed
AP-3C Orion/EAP-3C Orion*
10/11 Sqns, 92 Wing,
  Edinburgh, NSW

A9-656	11 Sqn
A9-657*	11 Sqn
A9-658	11 Sqn
A9-659	11 Sqn
A9-660*	11 Sqn
A9-661	11 Sqn
A9-662	11 Sqn
A9-664	11 Sqn
A9-665	11 Sqn
A9-751	10 Sqn
A9-752	10 Sqn
A9-753	10 Sqn
A9-756	10 Sqn
A9-757	10 Sqn
A9-759	10 Sqn
A9-760	10 Sqn

AUSTRIA
Öesterreichische
Luftstreitkräfte
Agusta-Bell AB.212/
Bell 212*
1. & 2. leichte Transporthubschrauberstaffel,
  Linz
5D-HB
5D-HC
5D-HD
5D-HF
5D-HG
5D-HH
5D-HI
5D-HJ

5D-HK
5D-HL
5D-HN
5D-HO
5D-HP
5D-HQ
5D-HR
5D-HS
5D-HT
5D-HU
5D-HV
5D-HW
5D-HX
5D-HY*
5D-HZ $

Bell
OH-58B Kiowa
Mehrzweckhubschrauberstaffel,
   Tulln
3C-OA
3C-OB
3C-OC
3C-OE
3C-OG
3C-OH
3C-OI
3C-OJ
3C-OK $
3C-OL

Eurofighter
EF.2000
Überwachungsgeschwader:
   1.Staffel & 2.Staffel, Zeltweg
7L-WA
7L-WB
7L-WC $
7L-WD
7L-WE
7L-WF
7L-WG
7L-WH
7L-WI
7L-WJ
7L-WK
7L-WL
7L-WM
7L-WN
7L-WO

Lockheed
C-130K Hercules
Lufttransportstaffel, Linz
8T-CA
8T-CB
8T-CC

Pilatus
PC-6B/B2-H2 Turbo Porter/
PC-6B/B2-H4 Turbo Porter*
leichte Lufttransportstaffel, Tulln
3G-EB
3G-ED
3G-EE
3G-EF
3G-EG
3G-EH
3G-EL
3G-EN*

Pilatus
PC-7 Turbo Trainer
Lehrabteilung Fläche, Zeltweg
3H-FA
3H-FB
3H-FC $
3H-FD
3H-FE
3H-FF
3H-FG $
3H-FH
3H-FJ
3H-FK
3H-FL
3H-FM
3H-FO

SAAB 105ÖE
Überwachungsgeschwader:
Düsentrainerstaffel, Linz
(yellow)		
D	(1104)	
I	(1109)	
J	(1110)	
(green)		
D	(1114)	
GG-17	(1117)	
(red)		
B	(1122)	
C	(1123)	
D	(1124)	
E	(1125)	
RF-26	(1126) $	
G	(1127)	
H	(1128)	
I	(1129)	
J	(1130)	
(blue)		
A	(1131)	
B	(1132)	
C	(1133)	
D	(1134)	
E	(1135)	
F	(1136)	
G	(1137)	
I	(1139)	
J	(1140)	

Sikorsky S-70A
mittlere Transporthubschrauberstaffel, Tulln
6M-BA
6M-BB
6M-BC
6M-BD
6M-BE
6M-BF
6M-BG
6M-BH
6M-BI

AZERBAIJAN
Airbus A.319-115LR
Azerbaijan Govt, Baku
4K-AI02

Boeing 767-32LER
Azerbaijan Govt, Baku
4K-AI01

BAHRAIN
BAE RJ.85/RJ.100*
Bahrain Defence Force
A9C-AWL*
A9C-BDF*
A9C-HWR

Boeing 747SP-Z5
Bahrain Amiri Flt
A9C-HAK

Boeing 747-4P8
Bahrain Amiri Flt
A9C-HMK

Boeing 767-4FSER
Bahrain Amiri Flt
A9C-HMH

Grumman
G.1159 Gulfstream IITT/
G.1159C Gulfstream IV-SP
Govt of Bahrain
A9C-BG   Gulfstream IITT
A9C-BRF Gulfstream IV-SP

Gulfstream Aerospace
G.450
Govt of Bahrain
A9C-BHR

Gulfstream Aerospace
G.550
Govt of Bahrain
A9C-BRN

Gulfstream Aerospace
G.650
Govt of Bahrain
A9C-BAH

BELGIUM

BELGIUM
Composante Aérienne Belge/
Belgische Luchtcomponent
D-BD Alpha Jet E
11 Smaldeel (1 Wg),
    Cazaux, France (ET 02.008)

AT-01
AT-02
AT-03
AT-05
AT-06
AT-08
AT-10
AT-11
AT-12
AT-13
AT-14
AT-15
AT-17
AT-18
AT-19
AT-20
AT-21
AT-22
AT-23
AT-24
AT-25
AT-26
AT-27
AT-28
AT-29 $
AT-30
AT-31
AT-32 $
AT-33

Airbus A.321-231
21 Smaldeel (15 Wg), Melsbroek
CS-TRJ

Dassault
Falcon 900B
21 Smaldeel (15 Wg), Melsbroek
CD-01

Embraer
ERJ.135LR/ERJ.145LR*
21 Smaldeel (15 Wg), Melsbroek
CE-01
CE-02
CE-03* $
CE-04* $

Lockheed
C-130H Hercules
20 Smaldeel (15 Wg), Melsbroek
CH-01
CH-03
CH-04
CH-05
CH-07 $
CH-08

CH-09
CH-10
CH-11
CH-12
CH-13 $

Dassault
Falcon 20-5
21 Smaldeel (15 Wg), Melsbroek
CM-02

General Dynamics
F-16 MLU
1,350 Smaldeel (2 Wg),
    Florennes [FS];
31,349 Smaldeel, OCU (10 Wg),
    Kleine-Brogel [BL]

FA-56	F-16A	10 Wg
FA-57	F-16A	2 Wg
FA-67	F-16A	2 Wg
FA-68	F-16A	2 Wg $
FA-69	F-16A	2 Wg
FA-70	F-16A	10 Wg
FA-71	F-16A	2 Wg
FA-72	F-16A	2 Wg
FA-77	F-16A	10 Wg
FA-81	F-16A	10 Wg
FA-82	F-16A	10 Wg
FA-83	F-16A	2 Wg
FA-84	F-16A	2 Wg $
FA-86	F-16A	10 Wg
FA-87	F-16A	10 Wg
FA-89	F-16A	2 Wg
FA-91	F-16A	2 Wg
FA-92	F-16A	2 Wg
FA-94	F-16A	2 Wg $
FA-95	F-16A	10 Wg
FA-97	F-16A	10 Wg
FA-98	F-16A	2 Wg
FA-101	F-16A	10 Wg
FA-102	F-16A	10 Wg
FA-103	F-16A	10 Wg
FA-104	F-16A	10 Wg $
FA-106	F-16A	10 Wg $
FA-107	F-16A	10 Wg
FA-109	F-16A	10 Wg
FA-110	F-16A	10 Wg $
FA-114	F-16A	2 Wg
FA-116	F-16A	2 Wg
FA-117	F-16A	2 Wg
FA-118	F-16A	10 Wg
FA-119	F-16A	10 Wg
FA-121	F-16A	2 Wg
FA-123	F-16A	10 Wg
FA-124	F-16A	10 Wg
FA-126	F-16A	2 Wg
FA-127	F-16A	2 Wg
FA-128	F-16A	2 Wg
FA-129	F-16A	2 Wg
FA-130	F-16A	2 Wg
FA-131	F-16A	2 Wg
FA-132	F-16A	10 Wg
FA-133	F-16A	2 Wg
FA-134	F-16A	10 Wg
FA-135	F-16A	2 Wg
FA-136	F-16A	10 Wg
FB-12	F-16B	2 Wg
FB-14	F-16B	10 Wg
FB-15	F-16B	10 Wg
FB-17	F-16B	10 Wg
FB-18	F-16B	10 Wg
FB-20	F-16B	10 Wg
FB-21	F-16B	2 Wg
FB-22	F-16B	10 Wg
FB-23	F-16B	10 Wg
FB-24	F-16B	10 Wg $

Agusta A109HA
17 Smaldeel MRH (1 Wg),
    Beauvechain;
18 Smaldeel MRH (1 Wg),
    Beauvechain;
SLV (1 Wg), Beauvechain

H-20	17 Sm MRH
H-21	18 Sm MRH
H-22	17 Sm MRH
H-23	17 Sm MRH
H-24	17 Sm MRH $
H-25	18 Sm MRH
H-26	18 Sm MRH
H-27	18 Sm MRH
H-28	17 Sm MRH
H-29	18 Sm MRH
H-30	17 Sm MRH
H-31	18 Sm MRH
H-33	18 Sm MRH
H-35	18 Sm MRH
H-36	17 Sm MRH
H-38	17 Sm MRH
H-39	
H-40	18 Sm MRH
H-41	17 Sm MRH
H-42	SLV
H-44	17 Sm MRH
H-45	17 Sm MRH
H-46	17 Sm MRH

Piper L-21B Super Cub
Centre Militaire de Vol à Voile
    Bases: Florennes, Goetsenhoeven
    & Zoersel
LB-01
LB-02
LB-03
LB-05

Sud
SA.316B Alouette III
40 Smaldeel, Koksijde
M-1
M-2
M-3

NH Industries
NH.90-NFH/NH.90-TTH*
1 Wg, Beauvechain;
40 Smaldeel, Koksijde

RN-01	40 Sm
RN-02	40 Sm
RN-03	40 Sm
RN-04	40 Sm
RN-05*	1 Wg
RN-06*	1 Wg
RN-07*	1 Wg
RN-08*	1 Wg

Westland Sea
King Mk48
40 Smaldeel, Koksijde
RS-02
RS-04
RS-05 $

SIAI Marchetti
SF260D/SF260M+
Centre of Competence Air,
  Bevekom (5 Smaldeel & 9 Smaldeel);
  Red Devils *

ST-02	SF-260M+ *
ST-03	SF-260M+ *
ST-04	SF-260M+ *
ST-06	SF-260M+
ST-12	SF-260M+
ST-15	SF-260M+ *
ST-16	SF-260M+
ST-17	SF-260M+
ST-18	SF-260M+
ST-19	SF-260M+
ST-20	SF-260M+
ST-22	SF-260M+
ST-23	SF-260M+ *
ST-24	SF-260M+
ST-25	SF-260M+
ST-26	SF-260M+
ST-27	SF-260M+ *
ST-30	SF-260M+ $
ST-31	SF-260M+ *
ST-32	SF-260M+ *
ST-34	SF-260M+ *
ST-35	SF-260M+
ST-36	SF-260M+
ST-40	SF-260D
ST-41	SF-260D *
ST-42	SF-260D
ST-43	SF-260D
ST-44	SF-260D
ST-45	SF-260D
ST-46	SF-260D
ST-47	SF-260D
ST-48	SF-260D $

Police Fédérale/Federal Politie
Cessna 182 Skylane
Luchtsteundetachment, Melsbroek
| G-01 | C.182Q |
| G-04 | C.182R |

MDH
MD.520N
Luchtsteundetachment, Melsbroek
G-14
G-15

MDH
MD.900/MD.902* Explorer
Luchtsteundetachment, Melsbroek
G-10*
G-11*
G-12
G-16*

BOTSWANA
Botswana Defence Force
Bombardier Global Express
VIP Sqn, Sir Seretse Kharma IAP,
  Gaborone
OK1

Lockheed
C-130B Hercules
Z10 Sqn, Thebephatshwa
OM-1
OM-2
OM-3

BRAZIL
Força Aérea Brasileira
Airbus
A.319-133CJ (VC-1A)
1° GT, 1° Esq, Galeão;
2101

Embraer
EMB.190-190IGW (VC-2)
1° Grupo de Transport Especial,
  1° Esq, Brasilia
2590
2591

Embraer E-99/R-99
2° Esq, 6° GAv, Anapolis
6700	E-99
6701	R-99A
6702	R-99A
6703	E-99
6704	R-99A
6750	R-99B
6751	R-99B
6752	R-99B

Lockheed
C-130 Hercules
1° GT, 1° Esq, Galeão;
1° GTT, 1° Esq, Afonsos
2451	C-130E	1° GTT
2454	C-130E	1° GTT
2456	C-130E	1° GTT
2459	SC-130E	1° GTT
2461	KC-130H	1° GT
2462	KC-130H	1° GT
2463	C-130H	1° GT
2464	C-130H	1° GT
2465	C-130M	1° GT
2466	C-130M	1° GT
2467	C-130H	1° GT
2470	C-130M	1° GT
2472	C-130H	1° GT
2473	C-130H	1° GT
2474	C-130H	1° GT
2475	C-130H	1° GT
2476	C-130M	1° GTT
2479	C-130M	1° GT

BRUNEI
Airbus A.340-212
Brunei Govt, Bandar Seri Bergawan
V8-BKH

Boeing 747-430
Brunei Govt, Bandar Seri Bergawan
V8-ALI

Boeing 767-27GER
Brunei Govt, Bandar Seri Bergawan
V8-MHB

BULGARIA
Bulgarsky Voenno-Vazdushni Sily
Aeritalia C-27J Spartan
16 TAB, Sofia/Vrazhdebna
071
072
073

Antonov An-30
16 TAP, Sofia/Dobroslavtzi
055

Pilatus PC.XII/45
16 TAP, Sofia/Dobroslavtzi
020

Bulgarian Govt
Airbus A.319-112
Bulgarian Govt/BH Air, Sofia
LZ-AOA
LZ-AOB

Dassault Falcon 2000
Bulgarian Govt, Sofia
LZ-OOI

133

BURKINA FASO
Boeing 727-282
Govt of Burkina Faso,
   Ouagadougou
XT-BFA

CAMEROON
Grumman
G.1159A Gulfstream III
Govt of Cameroon, Yaounde
TJ-AAW

CANADA
Royal Canadian Air Force
Lockheed
CC-130 Hercules
413 Sqn, Greenwood (SAR) (14 Wing);
424 Sqn, Trenton (SAR) (8 Wing);
426 Sqn, Trenton (8 Wing);
435 Sqn, Winnipeg (17 Wing);
436 Sqn, Trenton (8 Wing)
CC-130E/CC-130E(SAR)*

130307*	8 Wing
130319	8 Wing
130327	8 Wing

CC-130H/CC-130H(SAR)*

130332*	14 Wing
130334*	14 Wing
130335	8 Wing
130336	8 Wing
130337	8 Wing

CC-130H(T)

130338	17 Wing
130339	17 Wing
130340	17 Wing
130341	17 Wing

CC-130H-30

130343	8 Wing
130344	8 Wing

CC-130J Hercules II

130601	8 Wing
130602	8 Wing
130603	8 Wing
130604	8 Wing
130605	8 Wing
130606	8 Wing
130607	8 Wing
130608	8 Wing
130609	8 Wing
130610	8 Wing
130611	8 Wing
130612	8 Wing
130613	8 Wing
130614	8 Wing
130615	8 Wing
130616	8 Wing
130617	8 Wing

Lockheed
CP-140/CP-140M* Aurora
404 Sqn, Greenwood (14 Wing);
405 Sqn, Greenwood (14 Wing);
407 Sqn, Comox (19 Wing)

140101*	14 Wing
140102	14 Wing
140103	14 Wing
140104	14 Wing
140105*	14 Wing
140106	14 Wing
140107	407 Sqn
140108*	14 Wing
140109	407 Sqn
140110*	14 Wing
140111*	14 Wing
140112	407 Sqn
140113	407 Sqn
140114*	14 Wing
140115*	407 Sqn
140116	14 Wing
140117	14 Wing
140118	14 Wing

De Havilland Canada
CT-142
402 Sqn, Winnipeg
   (17 Wing)

142803	CT-142
142805	CT-142
142805	CT-142
142806	CT-142

Canadair
CC-144 Challenger
412 Sqn, Ottawa (8 Wing)

144614	CC-144B
144615	CC-144B
144616	CC-144B
144617	CC-144C
144618	CC-144C

Airbus
CC-150 Polaris
(A310-304/A310-304F*)
437 Sqn, Trenton (8 Wing)

15001	[991]
15002*	[992]
15003*	[993]
15004*	[994]
15005*	[995]

Boeing CC-177
(C-17A Globemaster III)
429 Sqn, Trenton (8 Wing)

177701	
177702	
177703	
177704	
177705 (on order)	

CHAD
Boeing 737-74Q
Chad Government, N'djamena
TT-ABD

CHILE
Fuerza Aérea de Chile
Boeing 707
Grupo 10, Santiago

902	707-351C
904	707-358C

Boeing 737
Grupo 10, Santiago

921	737-58N
922	737-330

Boeing 767-3Y0ER
Grupo 10, Santiago
985

Boeing
KC-135E Stratotanker
Grupo 10, Santiago
982
983

Extra EA-300L
Los Halcones

39	
145	[2]
149	[1]
1304	[4]

Grumman
G.1159C Gulfstream IV
Grupo 10, Santiago
911

Lockheed
C-130H Hercules
Grupo 10, Santiago
995
996

CROATIA
Hrvatske Zračne Snage
Pilatus PC-9*/PC-9M
92 ZB, Pula;
93 ZB, Zadar

051*	
052*	
053*	
054	93 ZB
055	93 ZB
056	93 ZB
057	93 ZB
058	
059	93 ZB
060	
061	93 ZB
062	93 ZB

063	93 ZB
064	93 ZB
065	
066	93 ZB
067	92 ZB
068	93 ZB
069	93 ZB
070	93 ZB

Canadair
CL.601 Challenger
Croatian Govt, Zagreb
9A-CRO
9A-CRT

CZECH REPUBLIC
Ceske Vojenske Letectvo
Aero L-39/L-59 Albatros
213.vlt/21.zTL, Cáslav;
CLV, Pardubice

0103	L-39C	CLV
0106	L-39C	CLV
0107	L-39C	CLV
0108	L-39C	CLV
0113	L-39C	CLV
0115	L-39C	CLV
0441	L-39C	CLV
0444	L-39C	CLV
0445	L-39C	CLV
2344	L-39ZA	213.vlt/21.zTL
2415	L-39ZA	213.vlt/21.zTL
2421	L-39ZA	213.vlt/21.zTL
2433	L-39ZA	213.vlt/21.zTL $
2436	L-39ZA	213.vlt/21.zTL
3903	L-39ZA	213.vlt/21.zTL
5015	L-39ZA	213.vlt/21.zTL
5017	L-39ZA	213.vlt/21.zTL
5019	L-39ZA	213.vlt/21.zTL

Aero
L-159A ALCA/L-159B/L-159T-1
212.tl/21.zTL, Cáslav;
213.vlt/21.zTL, Cáslav;
LZO, Praha/Kbely
L-159A

6048	212.tl/21.zTL
6049	212.tl/21.zTL
6050	212.tl/21.zTL
6051	212.tl/21.zTL
6052	212.tl/21.zTL
6053	212.tl/21.zTL
6054	212.tl/21.zTL
6055	212.tl/21.zTL
6057	212.tl/21.zTL
6058	212.tl/21.zTL
6059	212.tl/21.zTL
6060	212.tl/21.zTL
6062	212.tl/21.zTL
6063	212.tl/21.zTL
6064	212.tl/21.zTL
6065	212.tl/21.zTL
6066	212.tl/21.zTL $

6068	212.tl/21.zTL
6070	212.tl/21.zTL
L-159B	
5831	LZO
5832	LZO
6069	212.tl/21.zTL
6073	212.tl/21.zTL
L-159T-1	
6046	213.vlt/21.zTL
6047	213.vlt/21.zTL
6067	213.vlt/21.zTL $
6075	213.vlt/21.zTL
6078	213.vlt/21.zTL
6079	213.vlt/21.zTL

Airbus A.319CJ-115X
241.dlt/24.zDL, Praha/Kbely
2801
3085

Canadair
CL.601-3A Challenger
241.dlt/24.zDL, Praha/Kbely
5105

CASA C-295M
242.tsl/24.zDL, Praha/Kbely
0452
0453
0454
0455

Evektor
EV-55M Outback
Evektor, Kunovice
0458

LET 410 Turbolet
242.tsl/24.zDL, Praha/Kbely;
CLV, Pardubice

0731	L-410UVP-E	CLV
0928	L-410UVP-T	CLV
1504	L-410UVP	242.dl
1526	L-410FG	242.dl
2601	L-410UVP-E	242.dl
2602	L-410UVP-E	242.dl

Mil Mi-17/
Mi-171Sh*
221.lbvr/22.zVrL, Náměšt;
222.vrtl/22.zVrL, Náměšt;
243.vrl/24.zDL, Praha/Kbely
CLV, Pardubice

0803	243.vrl/24.zDL
0811	243.vrl/24.zDL
0828	243.vrl/24.zDL
0832	243.vrl/24.zDL
0834	243.vrl/24.zDL
0835	243.vrl/24.zDL
0836	CLV
0837	CLV
0839	243.vrl/24.zDL

0840	243.vrl/24.zDL
0848	243.vrl/24.zDL
0849	243.vrl/24.zDL
0850	243.vrl/24.zDL
9767*	22.zVrL
9774*	22.zVrL
9781*	22.zVrL
9799*	22.zVrL
9806*	22.zVrL
9813*	22.zVrL
9825*	22.zVrL
9837*	22.zVrL
9844*	22.zVrL
9868*	22.zVrL
9873*	22.zVrL
9887*	22.zVrL
9892*	22.zVrL
9904*	22.zVrL
9915*	22.zVrL
9926*	22.zVrL

Mil Mi-24/Mi-35
221.lbvr/22.zL, Náměšt

0981	Mi-24V2
3361	Mi-35 $
3362	Mi-35
3365	Mi-35
3366	Mi-35
3367	Mi-35
3368	Mi-35
3369	Mi-35
3370	Mi-35
3371	Mi-35
7353	Mi-24V $
7354	Mi-24V
7355	Mi-24V
7356	Mi-24V
7357	Mi-24V
7360	Mi-24V

SAAB Gripen
211.tl/21.zTL, Cáslav
JAS 39C
9234
9235 $
9236
9237
9238 $
9239
9240 $
9241
9242
9243
9244
9245
JAS 39D
9819 $
9820 $

Yakovlev Yak-40
241.dlt/24.zDL, Praha/Kbely
0260	Yak-40
1257	Yak-40K

DENMARK
Flyvevåbnet
Lockheed
C-130J-30 Hercules II
Eskadrille 721, Aalborg
B-536
B-537
B-538
B-583

Canadair
CL.604 Challenger
Eskadrille 721, Aalborg
C-080
C-168 $
C-172
C-215

General Dynamics
F-16 MLU
Eskadrille 727, Skrydstrup;
Eskadrille 730, Skrydstrup;
416th FTS/412th TW, Edwards AFB, USA
E-004	F-16A	Esk 727
E-005	F-16A	Esk 727
E-006	F-16A	Esk 730
E-007	F-16A	Esk 727
E-008	F-16A	Esk 727
E-011	F-16A	Esk 727
E-016	F-16A	Esk 727
E-017	F-16A	Esk 727
E-018	F-16A	Esk 730
E-024	F-16A	Esk 727
E-070	F-16A	Esk 727
E-074	F-16A	Esk 730
E-075	F-16A	Esk 727
E-107	F-16A	Esk 727
E-189	F-16A	Esk 730
E-190	F-16A	Esk 730
E-191	F-16A	Esk 730
E-194	F-16A	Esk 730 $
E-596	F-16A	Esk 730
E-597	F-16A	Esk 727
E-598	F-16A	Esk 727
E-599	F-16A	Esk 727
E-600	F-16A	Esk 730
E-601	F-16A	Esk 730
E-602	F-16A	Esk 730
E-603	F-16A	Esk 730
E-604	F-16A	Esk 727
E-605	F-16A	Esk 730
E-606	F-16A	Esk 730
E-607	F-16A	Esk 727
E-608	F-16A	Esk 727
E-609	F-16A	Esk 727
E-610	F-16A	Esk 730
E-611	F-16A	Esk 727

ET-022	F-16B	Esk 730
ET-197	F-16B	Esk 727
ET-198	F-16B	Esk 730
ET-199	F-16B	Esk 730
ET-207	F-16B	Esk 727
ET-208	F-16B	Esk 730
ET-210	F-16B	412th TW
ET-612	F-16B	Esk 727
ET-613	F-16B	Esk 727
ET-614	F-16B	Esk 727
ET-615	F-16B	Esk 727

AgustaWestland
EH.101 Mk.512
Eskadrille 722, Karup
  Detachments at:
  Aalborg, Roskilde, Ronne, Skrydstrup
M-502
M-504
M-507
M-508
M-510
M-512
M-513
M-514
M-515
M-516
M-517
M-518
M-519
M-520

Sikorsky MH-60R
Sea Hawk
Eskadrille 723, Karup
N-971	(on order)
N-972	(on order)
N-973	(on order)
N-974	(on order)
N-975	(on order)
N-976	(on order)
N-977	(on order)
N-978	(on order)
N-979	(on order)

Aérospatiale
AS.550C-2 Fennec
Eskadrille 724, Karup
P-090
P-234
P-254
P-275
P-276
P-287
P-288
P-319
P-320
P-339
P-352
P-369

Westland Lynx
Mk 90B
Eskadrille 723, Karup
S-134
S-142
S-170
S-175
S-181
S-191
S-256

SAAB
T-17 Supporter
Aalborg Stn Flt;
Eskadrille 721, Aalborg;
Flyveskolen, Karup (FLSK);
Skrydstrup Stn Flt
T-401	FLSK
T-402	FLSK
T-403	FLSK
T-404	FLSK
T-405	Skrydstrup
T-407	Esk 721
T-409	Aalborg
T-410	FLSK
T-411	Skrydstrup
T-412	FLSK
T-413	FLSK
T-414	Esk 721
T-415	FLSK
T-417	FLSK
T-418	Skrydstrup
T-419	FLSK
T-420	FLSK
T-421	FLSK
T-423	FLSK
T-425	FLSK
T-426	FLSK
T-427	FLSK
T-428	FLSK
T-429	FLSK
T-430	FLSK
T-431	Esk 721
T-432	Skrydstrup

ECUADOR
Fuerza Aérea Ecuatoriana
Embraer
ERJ.135 Legacy 600
Escuadrón de Transporte 1114,
  Quito
FAE-051

EGYPT
**Al Quwwat al-Jawwiya
il Misriya**
Lockheed
C-130H/C-130H-30* Hercules
16 Sqn, Cairo West
1271/SU-BAB
1273/SU-BAD
1274/SU-BAE

OVERSEAS MILITARY AIRCRAFT MARKINGS

EGYPT-FINLAND

1275/SU-BAF	**Antonov An-2/**	HN-432
1277/SU-BAI	**WSK-PZL An-2***	HN-433 HavLLv 31
1278/SU-BAJ	Fixed Wing Squadron, Ämari	HN-434
1279/SU-BAK	20 y An-2T	HN-435
1280/SU-BAL	40 y An-2T*	HN-436 HavLLv 31
1281/SU-BAM	41 y An-2	HN-437 HavLLv 31
1282/SU-BAN	42 y An-2T*	HN-438
1283/SU-BAP		HN-439
1284/SU-BAQ	**FINLAND**	HN-440
1285/SU-BAR	**Suomen Ilmavoimat**	HN-441
1286/SU-BAS	**CASA C-295M**	HN-442
1287/SU-BAT	Tukilentolaivue,	HN-443
1288/SU-BAU	Jyväskylä/Tikkakoski	HN-444 HavLLv 31
1289/SU-BAV	CC-1	HN-445 HavLLv 31
1290/SU-BEW	CC-2	HN-446
1291/SU-BEX	CC-3	HN-447
1292/SU-BEY		HN-448 HavLLv 31
1293/SU-BKS*	**Fokker**	HN-449
1294/SU-BKT*	**F.27-100 Friendship**	HN-450 HavLLv 31
1295/SU-BKU*	Tukilentolaivue,	HN-451
1296/SU-BPJ	Jyväskylä/Tikkakoski	HN-452
1297/SU-BKW	FF-1	HN-453 HavLLv 31
1298/SU-BKX		HN-454 HavLLv 31
	McDonnell Douglas	HN-455 HavLLv 31
Egyptian Govt	**F-18 Hornet**	HN-456 HavLLv 31
Airbus A.340-211	Hävittäjälentolaivue 11,	HN-457
Egyptian Govt, Cairo	Roveniemi;	**F-18D Hornet**
SU-GGG	Hävittäjälentolaivue 31,	HN-461
	Kuopio/Rissala;	HN-462 KoeLntk
Cessna 680	Koelentokeskus,	HN-463
Citation Sovereign	Halli	HN-464 HavLLv 31
Egyptian Govt, Cairo	**F-18C Hornet**	HN-465
SU-BRF	HN-401 HavLLv 31	HN-466 KoeLntk
SU-BRG	HN-402 HavLLv 11	HN-467
	HN-403 KoeLntk	
Grumman	HN-404 HavLLv 31	**BAe Hawk 51/51A/66**
G.1159A Gulfstream III/	HN-405 HavLLv 31	Tukilentolaivue,
G.1159C Gulfstream IV/	HN-406 HavLLv 11	Jyväskylä/Tikkakoski
G.1159C Gulfstream IV-SP/	HN-407	**Hawk 51**
Gulfstream 400	HN-408	HW-307
Egyptian Air Force/Govt,	HN-409 HavLLv 11	HW-327
Cairo	HN-410	HW-330
SU-BGM Gulfstream IV	HN-411 HavLLv 11	HW-333
SU-BGU Gulfstream III	HN-412	HW-334
SU-BGV Gulfstream III	HN-413	HW-338
SU-BNC Gulfstream IV	HN-414 KoeLntk	HW-340
SU-BND Gulfstream IV	HN-415 HavLLv 11	HW-341
SU-BNO Gulfstream IV-SP	HN-416	HW-343
SU-BNP Gulfstream IV-SP	HN-417	HW-345
SU-BPE Gulfstream 400	HN-418 HavLLv 11	HW-350
SU-BPF Gulfstream 400	HN-419 HavLLv 11	**Hawk 51A**
	HN-420 HavLLv 11	HW-351
ESTONIA	HN-421	HW-352
Estonian Air Force	HN-422	HW-353
Aero L-39C Albatros	HN-423	HW-354
Fixed Wing Squadron, Ämari	HN-424 HavLLv 31	HW-355
10	HN-425	HW-356
11	HN-426 HavLLv 31	HW-357
	HN-427 HavLLv 11	
	HN-428 HavLLv 11	
	HN-429 HavLLv 31	
	HN-431 HavLLv 31	

Hawk 66
HW-360
HW-361
HW-362
HW-363
HW-364
HW-365
HW-366
HW-367
HW-368
HW-370
HW-371
HW-373
HW-374
HW-375
HW-376
HW-377

Gates
Learjet 35A
Tukilentolaivue,
  Jyväskylä/Tikkakoski;
Tukilentolaivue (Det.),
  Kuopio/Rissala*
LJ-1*
LJ-2
LJ-3

Pilatus PC-12/47E
Tukilentolaivue,
  Jyväskylä/Tikkakoski
PI-01
PI-02
PI-03
PI-04
PI-05
PI-06

Suomen Maavoimat
NH Industries
NH.90-TTH
1.HK/HekoP, Utti
NH-201
NH-202
NH-203
NH-204
NH-205
NH-206
NH-207
NH-208
NH-209
NH-210
NH-211
NH-212
NH-213
NH-214
NH-215
NH-216
NH-217
NH-218
NH-219
NH-220
NH-221

FRANCE
Armée de l'Air
Airbus A.310-304
ET 03.060 *Esterel*,
  Paris/Charles de Gaulle

418	F-RADC
421	F-RADA
422	F-RADB

Airbus A.330-223
ET 00.060, Evreux

240	(F-RARF)

Airbus A.330-243 MRTT
GRV 02.093 *Bretagne*, Istres

1...	(on order)
1...	(on order)
1...	(on order)
1...	(on order)
1...	(on order)
1...	(on order)
1...	(on order)
1...	(on order)
1...	(on order)
1...	(on order)
1...	(on order)
1...	(on order)

Airbus A.340-212
ET 03.060 *Esterel*,
  Paris/Charles de Gaulle

075	F-RAJA
081	F-RAJB

Airbus Military A.400M
CEAM (EC 05.330),
  Mont-de-Marsan (BA 118);
ET 01.061 *Touraine*
  Orléans

0007	F-RBAA	01.061
0008	F-RBAB	01.061
0010	F-RBAC	01.061
0011	F-RBAD	01.061
0012	F-RBAE	01.061
0014	F-RBAF	01.061
0019	F-RBAG	01.061
0031	F-RBAH	(on order)

Airtech CN-235M-200/-300*
ET 01.062 *Vercours*, Creil;
ET 03.062 *Ventoux*, Creil;
ET 00.052 *La Tontouta*, Noumea;
ET 00.058 *Antilles*, Fort de France;
ET 00.082 *Maine*, Faaa-Tahiti

045	62-IB	01.062
065	52-IC	00.052
066	82-ID	00.082
071	62-IE	01.062
072	62-IF	01.062
105	52-IG	00.052
107	52-IH	00.052
111	62-II	01.062
114	62-IJ	01.062
123	62-IM	01.062
128	62-IK	01.062
129	62-IL	01.062
137	62-IN	01.062
141	62-IO	00.058
152	62-IP	01.062
156	62-IQ	01.062
158	62-IR	01.062
160	62-IS	01.062
165	62-IT	01.062
193*	62-HA	03.062
194*	62-HB	03.062
195*	62-HC	03.062
196*	62-HD	03.062
197*	62-HE	03.062
198*	62-HF	03.062
199*	62-HG	03.062
200*	62-HH	03.062

Boeing C-135 Stratotanker
GRV 02.093 *Bretagne*, Istres

470	C-135FR	93-CA
471	C-135FR	93-CB
472	C-135FR	93-CC
474	C-135FR	93-CE
475	C-135FR	93-CF
497	KC-135R	93-CM$
525	KC-135R	93-CN
574	KC-135RG	93-CP
735	C-135FR	93-CG
736	C-135FR	93-CH
737	C-135FR	93-CI
738	C-135FR	93-CJ
739	C-135FR	93-CK
740	C-135FR	93-CL

Boeing E-3F Sentry
EDCA 00.036, Avord (BA 702)

201	702-CA
202	702-CB
203	702-CC
204	702-CD

CASA 212-300 Aviocar
DGA EV, Cazaux & Istres

378	MP
386	MQ

Cessna 310
DGA EV, Cazaux & Istres

190	310N	BL
193	310N	BG
194	310N	BH
513	310N	BE
569	310R	CS
820	310Q	CL
981	310Q	BF

D-BD Alpha Jet
AMD-BA, Istres;
CEAM (EC 05.330),
Mont-de-Marsan (BA 118);
DGA EV, Cazaux (BA 120)
& Istres (BA 125);
EAC 00.314, Tours (BA 705);
EE 02.002 *Côte d'Or*,
Dijon/Longvic (BA 102);
EPNER, Istres (BA 125);
ETO 01.008 *Saintonge*
& ETO 02.008 *Nice*,
Cazaux (BA 120);
GE 00.312, Salon de
Provence (BA 701);
Patrouille de France (PDF)
(EPAA 20.300),
Salon de Provence (BA 701)

01	F-ZJTS	DGA EV
E4		DGA EV
E7	705-TU	00.314
E8		DGA EV
E11	102-UB	02.002
E12		DGA EV
E13	102-MM	02.002
E17	705-AA	00.314
E18	118-AK	CEAM
E20	705-MS	00.314
E22	705-LS	00.314
E25	705-TJ	00.314
E26	102-ND	02.008
E28	705-AB	00.314
E29	102-NB	02.008
E30	705-MD	00.314
E31	705-RK	00.314
E32	102-FI	02.002
E33	120-FJ	01.008
E38	$	01.008
E41	705-RA	00.314
E44	F-UHRE	PDF
E45	705-TF	00.314
E46	F-UHRF	PDF [5]
E47	102-AC	02.002
E48	705-MH	00.314
E51	705-AD	00.314
E53	102-LI	02.002
E58	705-TK	00.314
E60		EPNER
E67	705-TB$	00.314
E68	705-MO	00.314
E72	705-LA	00.314
E73	F-TENE	PDF [6]

E74		
E76	102-RJ	02.002
E79	F-TENA	PDF [8]
E80		DGA EV
E81	705-FO	00.314
E82	120-LW	01.008
E83	102-TZ	02.002
E85	F-UGFF	PDF
E86	102-FB	02.002
E87	705-LC	00.314
E88	F-TELL	PDF [2]
E89	118-LX	CEAM
E90	120-TH	01.008
E93	120-TX	01.008
E94	F-TERH	PDF [1]
E95	F-TERQ	PDF [0]
E97	102-MB	02.002
E98	705-MF	00.314
E99	120-AH	01.008
E100		
E101	120-TT	01.008
E102	120-LM	01.008
E104	705-TG	00.314
E105	102-FM	02.002
E106		
E108	120-AF	01.008
E109	120-AG	01.008
E110	705-AH	00.314
E112	705-AO	00.314
E113	118-TD	CEAM
E114	F-TERR	PDF
E116	120-FN	01.008
E117	102-AI	02.002
E118	705-LN	00.314
E119	F-UGFE	PDF [7]
E120	705-LG	00.314
E121	705-LE	00.314
E124	120-RN	01.008
E125	705-LK	00.314
E127	705-FK	00.314
E128	705-TM	00.314
E129	705-LP	00.314
E130	F-TERP	PDF
E131	120-RO	01.008
E134	705-RM	00.314
E135		
E136	120-RP	01.008
E138	705-RQ	00.314
E139	705-FC	00.314
E140	102-FA	02.002
E141	705-NF	00.314
E142	120-LO	01.008
E143		
E144	120-AK	01.008
E145		
E146	705-RR	00.314
E147	118-LT	CEAM
E148	705-LU	00.314
E149	705-RS	00.314
E151	118-FD	CEAM
E152	F-UHRT	PDF
E153	705-RU	00.314

E154	120-AL	01.008
E156	118-TI	CEAM
E157	120-UC	02.008
E158	F-TERF	PDF [4]
E160	120-UH	01.008
E162	F-TERJ	PDF
E163	F-TERB	PDF [9]
E164	120-RV	01.008
E165	102-RE	02.002
E166	F-UHRW	PDF [3]
E167	705-MN	00.314
E168	102-FP	02.008
E169	102-RX	02.002
E170	705-RY	00.314
E171	705-RZ	00.314
E173	705-MA	00.314
E176	120-MB	01.008

Dassault
Falcon 7X
ET 00.060, Villacoublay

68	F-RAFA
86	F-RAFB

Dassault
Falcon 20
DGA EV, Cazaux & Istres

Falcon 20C

79	CT
104	CW
138	CR

Falcon 20E

252	CA
288	CV

Falcon 20F

342	CU
375	CZ

Dassault
Falcon 900
ET 00.060, Villacoublay

02	F-RAFP
004	(F-RAFQ)
...	(on order)
...	(on order)
...	(on order)

Dassault
Falcon 2000LX
ET 00.060, Villacoublay

231	(F-RAFC)
237	F-RAFD

FRANCE

Dassault
Mirage 2000B
AMD-BA, Istres;
CEAM (ECE 05.330) *Côte d'Argent*,
  Mont-de-Marsan (BA 118);
DGA EV, Cazaux (BA 120) & Istres
  (BA 125);
EC 02.005 *Ile de France*,
  Orange (BA 115)

501	(BX1)	DGA EV
523	115-KJ	02.005
524	115-OA	02.005
525	115-AM	02.005
527	115-OR	02.005
528	115-KS	02.005
529	115-OC	02.005

Dassault
Mirage 2000C/2000-5F*
CEAM (ECE 05.330) *Côte d'Argent*,
  Mont-de-Marsan (BA 118);
DGA EV, Istres (BA 125);
GC 01.002 *Cigognes*,
  Luxeuil (BA 116);
EC 02.005 *Ile de France*,
  Orange (BA 115);
EC 03.011 *Corse*, Djibouti (BA 188);
EC 03.030 *Lorraine*,
  Al Dhafra, UAE (BA 104)

01*		DGA EV
2		DGA EV
38*	116-EI	01.002
40*	116-EX	01.002
41*	118-FZ	CEAM
42*	116-EY	01.002
43*	116-EJ	01.002
44*	116-EQ	01.002
45*	116-EF	01.002
46*	116-EN	01.002
47*	116-EP	01.002
48*	116-EW	01.002
49*	116-EA	01.002
51*	118-AS$	CEAM
52*	116-EH	01.002
54*	116-EZ	01.002
55*	116-EU	01.002
56*	116-EG	01.002
57*	188-ET$	03.011
58*	116-EL$	01.002
59*	116-EV	01.002
61*	116-ME	01.002
62*	116-ED	01.002
63*	116-EM	01.002
64		
65*	116-MG	01.002
66*	116-EO	01.002
67*	116-MH	01.002
71*	116-EE	01.002
74*	116-MK	01.002
77*	118-AX	CEAM
78*	116-EC	01.002
82	115-YL	02.005
83	115-YC	02.005
85	115-LK	02.005
88	115-KV	02.005
93	115-YA	02.005
94	115-KB	02.005
96	115-KI	02.005
99	115-YB	02.005
100	115-YF	02.005
101	115-KE	02.005
102	115-KR	02.005
104	115-KG	02.005
105	115-LJ	02.005
106	118-KL	CEAM
107	115-YD	02.005
108	115-LC	02.005
109	115-YH	02.005
111	115-KF	02.005
113	115-YO	02.005
115	115-YM	02.005
117	115-LD	02.005
120	115-KC	02.005
121	115-KN	02.005
122	115-YE	02.005
123	115-KD	02.005
124	115-YT	02.005

Dassault
Mirage 2000D
AMD-BA, Istres;
CEAM (ECE 05.330) *Côte d'Argent*,
  Mont-de-Marsan (BA 118);
DGA EV, Istres (BA 125);
EC 01.003 *Navarre*,
  EC 02.003 *Champagne* &
  ETED 02.007 *Argonne*,
  Nancy (BA 133);
EC 03.003 *Ardennes*,
  Istres (BA 125);
EC 03.011 *Corse*, Djibouti (BA 188)

601	133-JG	01.003
602	133-XJ	03.003
603	133-XL	03.003
604	133-IP	01.003
605	133-LF	02.003
606	133-JC	02.003
607		DGA EV
609	118-IF	CEAM
610	133-XX	03.003
611	133-JP	02.003
613	133-MO	03.003
614	133-JU	02.003
615	133-JY	02.003
616	118-XH	CEAM
617	133-IS	01.003
618	133-XC	03.003
620	133-IU	02.007
622	133-IL	01.003
623	133-MP	02.003
624	133-IT	03.003
625	133-XG	03.003
626	133-IC	01.003
627	133-JO	02.003
628	133-JL	02.003
629	133-XO	03.003
630	133-XD	03.003
631	188-IH	03.011
632	133-XE	01.003
634	133-JE	01.003
635	133-AS	01.003
636	133-JV	02.003
637	133-XQ	03.003
638	133-IJ	03.003
639	133-JJ	02.003
640	118-IN	CEAM
641	133-JW	02.003
642	133-IE	01.003
643	133-JD	02.003
644		DGA EV
645	133-XP	03.003
646	118-MQ	CEAM
647	133-IO	01.003
648	133-XT$	01.003
649	133-XY	03.003
650	133-IA	02.003
651	133-LG	01.003
652	133-XN	03.003
653	133-AU	02.007
654	133-ID	01.003
655	133-LH	01.003
657	133-JM	03.003
658	133-JN	02.003
659	133-XR	03.003
660	118-JF	CEAM
661	133-XI	03.003
662	133-XA	03.003
664	133-IW	01.003
666	133-IQ	01.003
667	133-JZ	02.003
668	118-IG	CEAM
669	133-AL	02.007
670	133-XF	03.003
671	133-XK	03.003
672	133-XV	03.003
673		DGA EV
674	133-IR	01.003
675	133-JI	02.003
676	$	DGA EV
677	133-JT	01.003
678	133-JB	02.003
679	133-JX	01.003
680	133-XM	03.003
681	133-AG	01.003
682	133-JR	02.003
683	133-IV	01.003
684		
685	133-XZ	01.003
686	133-JH	02.003
D02		DGA EV

Dassault
Mirage 2000N
CEAM (ECE 05.330) *Côte d'Argent*,
  Mont-de-Marsan (BA 118);
DGA EV, Istres (BA 125);
EC 02.004 *Lafayette*,
  Istres (BA 125)

301		DGA EV
306	116-BL	02.004
316	116-AU$	02.004
319		
324		
333		
334		DGA EV
335	125-CI	02.004
338	125-CG	02.004
340	125-AA	02.004
342	125-BA	02.004
345	125-BU	02.004
348	125-AL	02.004
350	125-AJ	02.004
351	125-AQ	02.004
353	125-AM	02.004
354	125-BJ	02.004
355	125-AE	02.004
356	125-BX	02.004
357	125-CO	02.004
358	125-BQ	02.004
359	125-AK$	02.004
361	125-CK	02.004
362	125-CH	02.004
364	125-BB	02.004
365	125-AI	02.004
366	125-BC	02.004
367	125-AW	02.004
368	125-AR	02.004
369	125-AG	02.004
370	125-CQ	02.004
371	125-BD	02.004
372	125-CM	02.004
373	125-CF	02.004
374	125-BS	02.004
375	125-CL	02.004

Dassault
Rafale B/Rafale C
AMD-BA, Istres;
CEAM (ECE 05.330) *Côte d'Argent*,
  Mont-de-Marsan (BA 118);
DGA EV, Istres (BA 125);
EC 01.007 *Provence*,
  EC 01.091 *Gascogne*,
  ETR 02.092 *Aquitaine*,
  St Dizier (BA 113);
EC 02.030 *Normandie-Niémen*,
  Mont-de-Marsan (BA 118);
EC 03.030 *Lorraine*,
  Al Dhafra, UAE (BA 104)

Rafale F.2B

301		DGA EV
302		DGA EV
303	113-EA	02.092

304	118-EB	CEAM
305	118-EC$	CEAM
306	113-IB	01.091
307	113-IA	01.091
308	113-HA	01.007
309	113-HB	01.007
310	113-HC	02.092
311	113-HD	01.091
312	113-HF	01.007
313	113-HI	01.007
314	113-HP	01.091
315	113-HK	01.091
317	113-HO	01.091
318	113-HM	01.091
319	113-HN	01.091
320	113-HV	01.091
321	113-HQ	01.007
322	113-HU	01.091
323	118-HT$	CEAM
324	113-HW	02.092
325	113-HX	01.091
326	113-HY	01.091
327	113-HZ	02.092

Rafale F.3B

328	104-IC	03.030
329	104-ID	03.030
330	113-IE	01.091
331	113-IF	01.091
332	113-IG	01.091
333	113-IH	01.091
334	113-II	02.092
335	113-IJ	01.091
336	113-IK	01.091
337	113-IL	01.091
338	113-IO	01.091

Rafale F.3B

339	118-FF	CEAM
340	113-FG	01.091
341	113-FH	01.091
342	113-FI	01.091
343	113-FJ	01.091
344	113-FK	01.091
345	113-FL	01.091
346	113-FM	02.092
347	113-FN	01.091
348	113-FO	02.092
349	113-FP	01.091
350		
351		
352		
353		
354		
355		

Rafale F.2C

101		DGA EV
102	118-EF	CEAM
103	113-HR	01.007
104	113-HH$	01.007
105	113-HE	01.007
106	113-HG	01.007
107	113-HJ$	01.007
108	113-HS	01.007

Rafale F.3C

109	118-IM	02.030
110	118-IN	02.030
111	118-IP	02.030
112	118-IQ	02.030
113	104-IR	03.030
114	118-IS	02.030
115	113-IT	01.091
116	118-IU	02.030
117	118-IV	02.030
118	118-IW	02.030
119	118-IX$	02.030
120	113-IY	01.091
121	113-IZ	01.091
122	118-GA	02.030
123	118-GB	02.030
124	104-GC	03.030
125	118-GD	02.030
126	118-GE	02.030
127	118-GF	02.030
128	118-GG	02.030
129	118-GH	02.030
130	113-GI	02.092
131	104-GJ	03.030
132	118-GK	02.030
133	118-GL	02.030
134	118-GM	02.030
135	113-GN	02.092
136	104-GO	03.030
137	118-GP	02.030
138	118-GQ	02.030
139	113-GR	01.007
140	104-GS	03.030
141	113-GT	01.007
142	113-GU$	01.007
143	118-GV	02.030
144	113-GW	01.007

Rafale F.4C

145	113-GX	01.091
146		
147		
148		
149		
150		

DHC-6 Twin Otter 200/300*
GAM 00.056 *Vaucluse*, Evreux;
ET 03.061 *Poitou*, Orléans

292	F-RACC	00.056
298	F-RACD	00.056
300	F-RACE	00.056
730*	CA	03.061
745*	CV	03.061

FRANCE

Embraer
EMB.121AA/AN* Xingu
EAT 00.319,
  Capitaine Dartigues,
    Avord (BA 702)

054	YX
064	YY
066*	ZA
072	YA
075	YC
078	YE
082	YG
083*	ZE
084	YH
086	YI
089	YJ
090*	ZF
091	YK
092	YL
096	YN
098	YO
099	YP
102	YS
103	YT
105	YU
107	YV
108	YW

Eurocopter
AS.332 Super Puma/
EC.725 Cougar
EH 01.044 *Solenzara*, Solenzara;
EH 01.067 *Pyrénées*, Cazaux;
EH 03.067 *Parisis*, Villacoublay;
EH 05.067 *Alpilles*, Aix-en-Provence;
GAM 00.056 *Vaucluse*, Evreux;
CEAM, Mont-de-Marsan

2014	AS.332C	PN	05.067
2093	AS.332L	PP	05.067
2233	AS.332L-1	FY	00.056
2235	AS.332L-1	67-FZ	03.067
2244	AS.332C	PM	01.044
2377	AS.332L-1	FU	03.067
2461	EC.725AP	SA	01.067
2549	EC.725AP	SB	01.067
2552	EC.725AP	SE	01.067
2619	EC.725AP	SC	05.067
2626	EC.725AP	SD	01.067
2770	EC.725R2	SG	01.067
2772	EC.725R2	SH	01.067
2778	EC.725R2	SI	01.067
2789	EC.725R2	SJ	01.067
2802	EC.725R2	SK	01.067

Eurocopter AS.555AN Fennec
CIEH 00.341, Orange;
EH 03.067 *Parisis*, Villacoublay;
EH 05.067 *Alpilles*, Orange;
EH 06.067 *Solenzara*, Solenzara;
ET 00.068 *Antilles-Guyane*, Cayenne

5361	UT	00.068
5382	UV	
5386	UX	03.067
5387	UY	05.067
5390	UZ	
5391	VA	
5392	VB	05.067
5393	VC	
5396	VD	05.067
5397	VE	00.068
5398	VF	03.067
5399	VG$	
5400	VH	05.067
5412	VI	05.067
5427	VJ	03.067
5430	VL	06.067
5431	VM$	03.067
5441	VO	00.068
5444	VP	
5445	VQ	05.067
5448	VR	03.067
5452	VS	05.067
5455	VT	03.067
5457	VU	00.068
5458	VV	05.067
5466	VW	00.068
5468	VX	03.067
5490	VY	05.067
5506	WA	05.067
5509	WB	05.067
5511	WC	05.067
5516	WD	03.067
5520	WE	03.067
5523	WF	05.067
5526	WG	03.067
5530	WH	05.067
5532	WI	05.067
5534	WJ	05.067
5536	WK	05.067
5559	WL	05.067

Extra EA-330LC*/EA-330SC
EVAA, Salon de Provence

03*	F-TGCH
04	F-TGCI
05	F-TGCJ

Fokker 100
DGA EV, Istres
290
… (on order)

Lockheed
C-130H/C-130H-30* Hercules
ET 02.061 *Franche-Comté*,
  Orléans

4588	61-PM
4589	61-PN
5114	61-PA
5116	61-PB
5119	61-PC
5140	61-PD
5142*	61-PE
5144*	61-PF$
5150*	61-PG
5151*	61-PH
5152*	61-PI
5153*	61-PJ
5226*	61-PK
5227*	61-PL

SOCATA
TB-30 Epsilon
*Cartouche Dorée, (EPAA 00.315)
  Cognac (BA 709);
EPAA 00.315, Cognac (BA 709

7	315-UF
26	315-UY
27	315-UZ
30	315-VC
64	315-WG
65	315-WH
66	315-WI
67	315-WJ
69	F-SEWL*
73	315-WP
74	315-WQ
78	315-WU
82	315-WY
83	315-WZ
84	315-XA
90	F-SEXG*
91	315-XH
92	315-XI
95	315-XL
96	315-XM
97	315-XN
99	F-SEXP*
101	315-XR
102	F-SEXS [2]*
103	315-XT
104	F-SEXU [3]*
113	F-SEYD*
116	F-SEYG*
117	315-YH
118	315-YI
121	315-YL
127	315-YR
131	315-YV
133	315-YX
136	315-ZA
141	F-SEZF*
142	315-ZG
144	315-ZI

146	315-ZK
149	315-ZM
150	315-ZN

SOCATA TBM 700A

CEAM (EC 02.330), Mont-de-Marsan;
DGA EV, Cazaux & Istres;
ET 00.041 *Moselle*, Metz;
ET 00.043 *Médoc*, Bordeaux (BA 106);
ET 00.060, Villacoublay;
EdC 00.070, Chateaudun

33	XA	00.043
35	BW	DGA EV
77	XD	00.043
78	XE	CEAM
80	BY	DGA EV
93	XL	00.060
94	BZ	DGA EV
95	XH	CEAM
103	XI	00.060
104	XJ	00.070
105	XK	00.043
106	BX	DGA EV
110	XP	00.041
111	XM	01.040
115	BQ	DGA EV
117	XN	00.060
125	XO	00.070
131	XQ	00.060
146	XR	00.060
147	XS	00.060

Transall
C-160NG GABRIEL*/
C-160R

CEAM (EET 06.330), Mont-de-Marsan;
DGA EV, Cazaux & Istres;
EEA 00.054 *Dunkerque*, Metz;
ET 00.050 *Réunion*, St Denis;
ET 00.055 *Ouessant*, Dakar;
ET 00.058 *Guadeloupe*, Pointe-à-Pitre;
ET 01.064 *Bearn* & ET 02.064 *Anjou*, Evreux;
ET 00.088 *Larzac*, Djibouti

R51	C-160R	61-MW	00.064
R55	C-160R	61-ZC	00.064
R87	C-160R	61-ZE	00.050
R89	C-160R	61-ZG$	00.064
R90	C-160R	61-ZH	00.064
R91	C-160R	61-ZI	00.064
R93	C-160R	61-ZK	00.064
R94	C-160R	61-ZL	00.064
R96	C-160R	61-ZN	00.064
R97	C-160R	61-ZA	00.064

R158	C-160R	61-ZX	00.064
R159	C-160R	61-ZY	00.064
R160	C-160R	61-ZZ	00.064
R201	C-160R	64-GA	01.064
R202	C-160R	64-GB	02.064
R203	C-160R	64-GC$	01.064
R204	C-160R	64-GD	02.064
R206	C-160R	64-GF	02.064
R208	C-160R	64-GH$	02.064
R210	C-160R	64-GJ	02.064
R212	C-160R	64-GL	02.064
R213	C-160R	64-GM	01.064
R214	C-160R	64-GN	02.064
F216	C-160NG*	54-GT	00.054
R217	C-160R	64-GQ	01.064
R218	C-160R	64-GR$	02.064
F221	C-160NG*	GS$	00.054
R223	C-160R	64-GW	01.064
R224	C-160R	64-GX	02.064
R225	C-160R	64-GY	01.064
R226	C-160R	64-GZ	02.064

Aéronavale/Marine
Dassault-Breguet
Atlantique 2

21 Flottille & 23 Flottille, Lorient/Lann Bihoué

2	21F
3	21F
4	23F
5	23F
9	23F
11	21F
12	23F
13	21F
14	23F
16	21F
17	21F$
18	21F
19	23F
20	23F
21	21F
25	21F
26	21F
28	21F

Dassault
Falcon 10(MER)

ES 57, Landivisiau

32
101
129
133
143
185

Dassault
Falcon 20G Guardian

25 Flottille, Papeete & Tontouta

48
65
72
77
80

Dassault
Falcon 50 SURMAR

24 Flottille, Lorient/Lann Bihoué

5
7
27
30
34
36
78
132

Dassault
Rafale

11 Flottille, Landivisiau;
12 Flottille, Landivisiau;
17 Flottille, Landivisiau;
AMD-BA, Istres;
CEPA, Istres;
DGA EV, Istres;
ETR 02.092 *Aquitaine*, St Dizier (BA 113)

Rafale F.2

11	12F
12	12F
13	12F
14	12F
15	11F
16	11F
17	12F
19	12F
20	12F
21	12F
23	12F
26	12F

Rafale F.3

10	12F
27	11F $
28	11F
29	02.092
30	11F
31	12F
32	11F
33	11F
34	11F
35	12F
36	11F
37	17F
38	11F
39	12F

Rafale F.4

40	11F
41	
42	
43	
44	
45	
46	

Rafale M

M01	DGA EV
M02	DGA EV

Dassault
Super Etendard
17 Flottille, Landivisiau

1
2
8
10
12
17
19
31
32
33
35
41
43
44
46
51
52
65
69

Embraer
EMB.121AN Xingu
24 Flottille,
  Lorient/Lann Bihoué;
28 Flottille, Hyères

65	28F
67	28F $
68	28F
69	28F
71	28F
74	28F
77	28F
81	24F
85	24F
87	28F

Eurocopter
AS.365/AS.565 Panther
35 Flottille, Hyères,
  (with detachments at La Rochelle,
  Le Touquet & Tahiti*)
36 Flottille, Hyères

17	AS.365N	35F
19	AS.365N	35F
24	AS.365N	35F
57	AS.365N	35F
81	AS.365N	35F
91	AS.365N	35F
157	AS.365N	35F
313	AS.365F1	35F
318	AS.365F1	35F
322	AS.365F1	35F
355	AS.565MA	36F
362	AS.565MA	35F
403	AS.565UA	36F
436	AS.565MA	36F
452	AS.565MA	36F
453	AS.565MA	36F
466	AS.565MA	36F
482	AS.565MA	36F
486	AS.565MA	36F
488	AS.565MA	36F
503	AS.565MA	35F
505	AS.565MA	36F
506	AS.565MA	36F
507	AS.565MA	36F
511	AS.565MA	36F
519	AS.565MA	36F
522	AS.565MA	36F$
524	AS.565MA	36F
542	AS.565MA	36F
6872	AS.365N3	35F*
6928	AS.365N3	35F*

Eurocopter
EC.225LP Super Puma 2+
32 Flottille,
  Lanvéoc/Poulmic & Cherbourg

2741
2752
2851

NH Industries
NH.90-NFH Caiman
31 Flottille, Hyères;
33 Flottille,
  Lanvéoc/Poulmic
CEPA, Hyères

1	(F-ZWTO)
2	CEPA
3	31F
4	33F
5	33F
6	33F
7	33F
8	31F
9	31F
10	31F

11	33F
12	33F
13	31F
14	(on order)
15	(on order)
16	(on order)
17	(on order)
18	(on order)
19	(on order)
20	(on order)
21	(on order)
22	(on order)
23	(on order)
24	(on order)
25	(on order)
26	(on order)
27	(on order)

Northrop Grumman
E-2C Hawkeye
4 Flottille,
  Lorient/Lann Bihoué

1	(165455)
2	(165456)
3	(166417)
4	(on order)

Sud
SA.316B/SA.319B/SE.3160 Alouette III
35 Flottille, Hyères;
ES 22/ESHE, Lanvéoc/Poulmic

13	SE.3160	35F
14	SE.3160	22S
18	SE.3160	22S
41	SE.3160	35F
100	SA.319B	22S
114	SA.319B	22S
161	SA.319B	22S
237	SA.319B	35F
244	SE.3160	22S
245	SE.3160	22S
262	SA.319B	22S
268	SA.319B	22S
279	SE.3160	22S
302	SA.319B	22S
303	SA.319B	22S
309	SA.319B	22S
314	SA.319B	22S
347	SA.319B	22S
358	SA.319B	22S
731	SA.316B	22S
806	SA.316B	22S
809	SA.316B	22S
997	SA.319B	22S

Westland
Lynx HAS2(FN)/HAS4(FN)*
34 Flottille, Lanvéoc/Poulmic
  (with a detachment at Hyères)
CEPA, Hyères

260		
263		
264		
265		
267		
270		
271		
272 $		
273		
621		
623		
624		
625		
627		
801*		
802*		
804*		
806*		
807*		
808*		
810*		
811*		
813*		
814*		

**Aviation Legére de l'Armée
de Terre (ALAT)**
Aérospatiale
SA.330Ba Puma
1 RHC, Phalsbourg;
3 RHC, Etain;
5 RHC, Pau;
4 RHFS, Pau;
EALAT, Dax & Le Luc;
ESAM, Bourges;
GAM/STAT, Valence;
GIH, Cazaux

1005	DCA	ESAM
1006	DAA	
1020	DAB	5 RHC
1036	DAC	1 RHC
1037	DAD	5 RHC
1049	DAE	5 RHC
1052	DCB	EALAT
1055	DAF	GIH
1056	DCC	GAM/STAT
1057	DCD	EALAT
1069	DAG	1 RHC
1071	DCE	EALAT
1073	DCF	EALAT
1078	DAH	3 RHC
1092	DAI	1 RHC
1093	DCG	3 RHC
1100	DAJ	3 RHC
1102	DAK	
1107	DAL	1 RHC
1109	DAM	5 RHC

1114	DCH	3 RHC
1122	DCI	5 RHC
1123	DCJ	5 RHC
1128	DAN	1 RHC
1130	DCK	5 RHC
1135	DCL	1 RHC
1136	DCM	1 RHC
1142	DCN	EALAT
1143	DAO	1 RHC
1145	DCO	3 RHC
1149	DAP	5 RHC
1150	DCP	3 RHC
1155	DCQ	4 RHFS
1156	DAQ	1 RHC
1163	DCR	EALAT
1164	DCS	1 RHC
1165	DCT	1 RHC
1171	DCU	1 RHC
1172	DCV	EALAT
1173	DAR	
1176	DAS	3 RHC
1177	DCW	3 RHC
1179	DCX	EMB
1182	DCY	1 RHC
1186	DCZ	5 RHC
1189	DAT	3 RHC
1190	DDA	1 RHC
1192	DDB	1 RHC
1196	DDC	EALAT
1197	DAU	5 RHC
1198	DDD	1 RHC
1204	DAV	1 RHC
1206	DDE	3 RHC
1211	DAW	
1213	DDF	1 RHC
1214	DAX	3 RHC
1217	DAY	
1219	DAZ	3 RHC
1222	DDG	1 RHC
1223	DDH	EALAT
1228	DDI	
1229	DDJ	EALAT
1231	DDK	5 RHC
1232	DBA	5 RHC
1235	DDL	5 RHC
1236	DDM	EALAT
1239	DDN	EALAT
1243	DBB	5 RHC
1244	DDO	5 RHC
1248	DBC	3 RHC
1252	DDP	EALAT
1255	DDQ	1 RHC
1256	DDR	GAM/STAT
1260	DDS	EALAT
1262	DBD	5 RHC
1269	DDT	5 RHC
1277	DBE	1 RHC
1411	DDU	1 RHC
1417	DBF	
1419	DDV	1 RHC
1438	DBG	GAM/STAT
1447	DDW	1 RHC

1451	DBH	1 RHC
1507	DBI	EALAT
1510	DBJ	3 RHC
1512	DBK	
1519	DBL	3 RHC
1617	DBM	1 RHC
1632	DBN	5 RHC
1634	DBO	3 RHC
1654	DBP	5 RHC
1662	DDX	EALAT
1663	DBQ	4 RHFS
5682	DBR	4 RHFS

Aérospatiale
SA.342 Gazelle
1 RHC, Phalsbourg;
3 RHC, Etain;
5 RHC, Pau;
4 RHFS, Pau;
COMALAT, Villacoublay;
EALAT, Dax & Le Luc;
EFA, Le Luc;
EHADT, Etain;
ESAM, Bourges;
GAM/STAT, Valence
SA.342L1 Gazelle

4205	GEA	EALAT
4206	GEB	EALAT
4207	GEC	3 RHC
4208	GED	GAM/STAT
4209	GEE	EALAT
4210	GEF	1 RHC
4211	GEG	5 RHC
4212	GEH	ESAM
4214	GEI	EALAT
4215	GEJ	5 RHC
4216	GEK	GAM/STAT
4217	GEL	1 RHC
4218	GEM	1 RHC
4219	GEN	4 RHFS
4220	GEO	1 RHC
4221	GEP	EALAT
4222	GEQ	EALAT
4223	GER	3 RHC
4224	GES	1 RHC
4225	GET	GAM/STAT
4226	GEU	GAM/STAT
4227	GEV	1 RHC
4228	GEW	1 RHC
4229	GEX	4 RHFS
4230	GEY	GAM/STAT
4231	GEZ	5 RHC
4232	GFA	EALAT
4233	GFB	3 RHC
4234	GFC	

FRANCE

SA.342M Gazelle

1732	GJA	EALAT	4038	GJR	EALAT	4185	GKD	3 RHC
3458	GNA	EALAT	4039	GBC	1 RHC	4186	GCM	1 RHC
3459	GAA	1 RHC	4042	GMB	EALAT	4187	GKE	EALAT
3476	GAB	4 RHFS	4047	GJS	EALAT	4189	GCN	4 RHFS
3477	GJB	1 RHC	4048	GBD	EALAT	4190	GMJ	EALAT
3511	GJC	EALAT	4049	GNP	EALAT	4191	GCO	4 RHFS
3512	GAC	1 RHC	4053	GBE	1 RHC	4192	GOD	
3513	GNB	1 RHC	4059	GBF	1 RHC	4194	GMD	EALAT
3529	GJD	EALAT	4060	GNQ	EALAT	4195	GCP	3 RHC
3530	GAD	4 RHFS	4061	GBG	GAM/STAT	4198	GCQ	3 RHC
3546	GJF	EALAT	4065	GNR	EALAT	4201	GMK	EALAT
3548	GAE	1 RHC	4066	GBH	3 RHC			
3549	GNC	EALAT	4067	GJU	EALAT	**Aérospatiale**		
3567	GMA	EALAT	4071	GNS	EHADT	**AS.532UL Cougar**		
3615	GJH	1 RHC	4072	GBI	1 RHC	1 RHC, Phalsbourg;		
3617	GND	GAM/STAT	4078	GJV	EALAT	4 RHFS, Pau;		
3664	GAF	1 RHC	4079	GMC	EALAT	GAM/STAT, Valence		
3848	GAG	5 RHC	4083	GNT	EALAT	2252	CGA	GAM/STAT
3849	GAH	5 RHC	4084	GBJ	3 RHC	2266	CGB	5 RHC
3850	GAI	1 RHC	4095	GBL	5 RHC	2267	CGC	4 RHFS
3851	GJI	EALAT	4096	GNU	EALAT	2271	CGD	4 RHFS
3852	GNE	1 RHC	4102	GJW	EALAT	2272	CGE	GAM/STAT
3853	GNF	EALAT	4103	GJX	EALAT	2273	CGF	1 RHC
3855	GJJ	EALAT	4108	GBM	GAM/STAT	2282	CGG	
3856	GAJ	5 RHC	4109	GBN	5 RHC	2285	CGH	
3857	GJK	EALAT	4114	GBO	EALAT	2290	CGI	4 RHFS
3858	GNG	EMB	4115	GBP	5 RHC	2293	CGJ	4 RHFS
3859	GAK	5 RHC	4118	GNV	EALAT	2299	CGK	4 RHFS
3862	GAL	1 RHC	4119	GBQ	5 RHC	2300	CGL	1 RHC
3863	GAM	3 RHC	4120	GBR	EALAT	2301	CGM	1 RHC
3864	GJL	EMB	4123	GJY	EALAT	2303	CGN	1 RHC
3865	GAN	EALAT	4124	GBS	1 RHC	2316	CGO	4 RHFS
3866	GNI	1 RHC	4135	GNW		2323	CGQ	4 RHFS
3867	GJM	EALAT	4136	GBT	1 RHC	2324	CGR	4 RHFS
3868	GAO	1 RHC	4140	GBU	1 RHC	2325	CGS	
3870	GME		4141	GBV		2327	CGT	
3896	GNJ		4142	GBW	5 RHC	2331	CGU	
3911	GAP	1 RHC	4143	GNX	EALAT	2336	CGV	4 RHFS
3921	GAQ	EALAT	4144	GBX	1 RHC	2342	CHA	
3929	GJN	EALAT	4145	GBY	1 RHC	2369	CHB	
3930	GMF		4146	GNY	EALAT	2375	CHC	
3938	GAR	EALAT	4151	GBZ	3 RHC	2443	CGW	
3939	GMG		4155	GCA	3 RHC	2446	CGX	1 RHC
3947	GAS	5 RHC	4159	GNZ	EALAT			
3948	GAT	1 RHC	4160	GCC	1 RHC	**Eurocopter**		
3956	GNK		4161	GCD	1 RHC	**EC.665 Tigre HAP/Tigre HAD**		
3957	GAU	5 RHC	4162	GCE	5 RHC	1 RHC, Phalsbourg;		
3964	GAV	3 RHC	4164	GCF	5 RHC	5 RHC, Pau;		
3965	GJO	EALAT	4166	GJZ	EALAT	4 RHFS, Pau;		
3992	GNM	EALAT	4168	GCG	5 RHC	Airbus Helcioptiers, Marseille;		
3996	GAW	5 RHC	4171	GMI	EALAT	EFA, Le Luc;		
4008	GNN		4172	GCH	5 RHC	GAM/STAT, Valence		
4014	GJP	GAM/STAT	4175	GCI	4 RHFS	**Tigre HAP**		
4018	GAX	3 RHC	4176	GKA	EALAT	2001	BHH	EFA
4019	GAY	3 RHC	4177	GKB	1 RHC	2002	BHI	EFA
4020	GAZ	1 RHC	4178	GOA	EALAT	2003	BHJ	5 RHC
4022	GJQ	EALAT	4179	GCJ	1 RHC	2004	BHK	EFA
4023	GMH	ESAM	4180	GCK	GAM/STAT	2006	BHL	EFA
4026	GBA	1 RHC	4181	GCL	3 RHC	2009	BHB	EFA
4032	GNO	EALAT	4182	GKC	EALAT	2010	BHA	EFA
4034	GBB	1 RHC	4183	GOB	EALAT	2011	BHM	5 RHC
			4184	GOC	1 RHC	2012	BHT	GAM/STAT

2013	BHC	GAM/STAT
2015	BHD	5 RHC
2016	BIA	Airbus
2018	BHE	5 RHC
2019	BHF	5 RHC
2021	BHN$	EFA
2022	BHG	5 RHC
2023	BHP	5 RHC
2024	BHO	5 RHC
2025	BHQ	5 RHC
2026	BHR	GAM/STAT
2027	BHS	5 RHC
2028	BHU	5 RHC
2029	BHV	5 RHC
2031	BHX	EFA
2032	BHW	5 RHC
2033	BHZ	5 RHC
2034	BIB	EFA
2035	BIC	4 RHFS
2036	BID	EFA
2037	BIE	5 RHC
2038		Airbus
2039	BIG	
2040	BIH	5 RHC
2041	BII	
2042	BIJ	5 RHC
2043	BIK	5 RHC
2044	BIL	GAM/STAT
2045	BIM	EFA
2046	BIN	EFA
2047	BIO	
2048	BIP	
2049	BIQ	
2050	BIR	
2051	BIS	
2052	BIT	

Tigre HAD

6001	BJA	GAM/STAT
6002	BJB	GAM/STAT
6003	BJC	EFA
6004	BJD	EFA
6005	BJE	EFA
6006	BJF	1 RHC
6007	BJG	1 RHC
6008	BJH	
6009	BJI	
6010	BJK	
6011	BJL	

Eurocopter
EC.725AP Cougar
GIH, Cazaux

2611	CAA
2628	CAB
2630	CAC
2631	CAD
2633	CAE
2638	CAF
2640	CAG
2642	CAH

NH Industries
NH.90-TTH
1 RHC, Phalsbourg;
CFIA NH.90, Le Luc;
GAM/STAT, Valence

1239	EAA	CFIA
1256	EAB	GAM/STAT
1271	EAC	CFIA
1273	EAD	1 RHC
1290	EAE	CFIA
1291	EAF	CFIA
1292	EAG	1 RHC
1293	EAH	CFIA
1294	EAI	1 RHC
1295	EAJ	1 RHC
1306	EAK	
1307	EAL	1 RHC
1308	EAM	

Pilatus
PC-6B/B2-H4 Turbo Porter
1 GSALAT, Montauban

887	MCA
888	MCB
889	MCC
890	MCD
891	MCE

SOCATA
TBM 700A/TBM 700B*
EAAT, Rennes

99	ABO
100	ABP
136	ABR
139	ABS
156*	ABT
159*	ABU
160*	ABV

French Govt
Aérospatiale
AS.350B Ecureuil
Gendarmerie;
Sécurité Civile*

JCA	(1028)	AS.350BA
JCB	(1574)	AS.350B
JCD	(1576)	AS.350B
JCE	(1812)	AS.350B
JCF	(1691)	AS.350BA
JCG	(1753)	AS.350B
JCH	(1756)	AS.350B
JCI	(2222)	AS.350B-1
JCL	(1811)	AS.350B
JCN	(2044)	AS.350BA
JCP	(2045)	AS.350B
JCQ	(2057)	AS.350BA
JCR	(2088)	AS.350B
JCS	(1575)	AS.350B
JCT	(2104)	AS.350B
JCU	(2117)	AS.350B
JCV	(2118)	AS.350B
JCW	(2218)	AS.350BA
JCX	(2219)	AS.350B
JCY	(2221)	AS.350B
JCZ	(1467)	AS.350BA$
JEB	(1692)	AS.350B
JEC	(1810)	AS.350B
JEF	(2225)	AS.350B-1
F-ZBBN		AS.350B*
F-ZBEA		AS.350B
F-ZBFC		AS.350B-1*
F-ZBFD		AS.350B-1*

Aérospatiale
AS.355 Twin Ecureuil
Douanes Francaises

F-ZBAD	AS.355F-2
F-ZBEF	AS.355F-1
F-ZBEK	AS.355F-1
F-ZBEL	AS.355F-1

Beech
Super King Air B200
Sécurité Civile

F-ZBFJ	98
F-ZBFK	96
F-ZBMB	97

Canadair CL-415
Sécurité Civile

F-ZBEG	39
F-ZBEU	42
F-ZBFN	33
F-ZBFP	31
F-ZBFS	32 $
F-ZBFV	37
F-ZBFW	38
F-ZBFX	34
F-ZBFY	35
F-ZBME	44
F-ZBMF	45
F-ZBMG	48

Cessna F.406 Caravan II
Douanes Francaises

F-ZBAB	(0025)
F-ZBBB	(0039)
F-ZBCE	(0042)
F-ZBCF	(0077)
F-ZBCG	(0066)
F-ZBCH	(0075)
F-ZBCI	(0070)
F-ZBCJ	(0074)
F-ZBEP	(0006)
F-ZBES	(0017)
F-ZBGA	(0086)
F-ZBGD	(0090)
F-ZBGE	(0061)

Conair
Turbo Firecat
Sécurité Civile

F-ZBAA	22
F-ZBAP	12
F-ZBAZ	01
F-ZBCZ	23
F-ZBEH	20
F-ZBET	15
F-ZBEW	11 $
F-ZBEY	07
F-ZBMA	24

De Havilland Canada
DHC-8Q-402MR
Sécurité Civile

F-ZBMC	73
F-ZBMD	74

Eurocopter
EC.135T-2
Douanes Francaises*;
Gendarmerie

JDA	(0642)
JDB	(0654)
JDC	(0717)
JDD	(0727)
JDE	(0747)
JDF	(0757)
JDG	(0772)
JDH	(0787)
JDI	(0797)
JDJ	(0806)
JDK	(0857)
JDL	(0867)
JDM	(1055)
JDN	(1058)
JDO	(1086)
F-ZBGF*	
F-ZBGG*	
F-ZBGH*	
F-ZBGI*	
F-ZBGJ*	

Eurocopter
EC.145
Gendarmerie;
Sécurité Civile*

JBA	(9008)	EC.145C-1
JBB	(9014)	EC.145C-1
JBC	(9018)	EC.145C-1
JBD	(9019)	EC.145C-1
JBE	(9025)	EC.145C-1
JBF	(9035)	EC.145C-2
JBG	(9036)	EC.145C-2
JBH	(9037)	EC.145C-2
JBI	(9127)	EC.145C-2
JBJ	(9140)	EC.145C-2
JBK	(9162)	EC.145C-2
JBM	(9113)	EC.145C-2
JBN	(9173)	EC.145C-2
JBO	(9124)	EC.145C-2

JBR	(9169)	EC.145C-2
JBT	(9173)	EC.145C-2
F-ZBPA		EC.145C-1*
F-ZBPD		EC.145C-1*
F-ZBPE		EC.145C-1*
F-ZBPF		EC.145C-1*
F-ZBPG		EC.145C-1*
F-ZBPH		EC.145C-1*
F-ZBPI		EC.145C-1*
F-ZBPJ		EC.145C-1*
F-ZBPK		EC.145C-1*
F-ZBPL		EC.145C-1*
F-ZBPM		EC.145C-1*
F-ZBPN		EC.145C-1*
F-ZBPO		EC.145C-1*
F-ZBPP		EC.145C-2*
F-ZBPQ		EC.145C-2*
F-ZBPS		EC.145C-2*
F-ZBPT		EC.145C-2*
F-ZBPU		EC.145C-2*
F-ZBPV		EC.145C-2*
F-ZBPW		EC.145C-2*
F-ZBPX		EC.145C-2*
F-ZBPY		EC.145C-2*
F-ZBPZ		EC.145C-2*
F-ZBQA		EC.145C-2*
F-ZBQB		EC.145C-2*
F-ZBQC		EC.145C-2*
F-ZBQD		EC.145C-2*
F-ZBQE		EC.145C-2*
F-ZBQF		EC.145C-2*
F-ZBQG		EC.145C-2*
F-ZBQH		EC.145C-2*
F-ZBQI		EC.145C-2*
F-ZBQJ		EC.145C-2*
F-ZBQK		EC.145C-2*
F-ZBQL		EC.145C-2*

Hawker Beechcraft King Air B350ER
Douanes Francaises

F-ZBGK	(FL-682)
F-ZBGL	(FL-746)
F-ZBGM	(FL-752)
F-ZBGN	(FL-781)
F-ZBGO	(FL-800)
F-ZBGP	(FL-802)
F-ZBGQ	(on order)
F-ZBGR	(on order)

Civil operated aircraft in military use
BAE Jetstream 41
AVDEF, Nimes/Garons
F-HAVD
F-HAVF

Cirrus SR.20/SR.22
EADS Cognac Aviation Training
    Services/CFAIM 05.312,
    Salon de Provence

F-HGDU	SR.20
F-HKCA	SR.22
F-HKCB	SR.20

F-HKCD	SR.20
F-HKCE	SR.20
F-HKCF	SR.22
F-HKCG	SR.20
F-HKCH	SR.20
F-HKCI	SR.22
F-HKCJ	SR.20
F-HKCK	SR.20
F-HKCL	SR.22
F-HKCN	SR.20
F-HKCO	SR.22
F-HKCP	SR.20
F-HKCQ	SR.20
F-HKCR	SR.22
F-HKCS	SR.22
F-HKCT	SR.20
F-HKCU	SR.20
F-HKCV	SR.20
F-HKCX	SR.20

Dassault Falcon 20
AVDEF, Nimes/Garons

F-GJDB	Falcon 20C
F-GPAA	Falcon 20ECM
F-GPAD	Falcon 20E

Grob G120A-F
EADS Cognac Aviation Training
    Services/EPAA 00.315, Cognac
F-GUKA
F-GUKB
F-GUKC
F-GUKD
F-GUKE
F-GUKF
F-GUKG
F-GUKH
F-GUKI
F-GUKJ
F-GUKK
F-GUKL
F-GUKM
F-GUKN
F-GUKO
F-GUKP
F-GUKR
F-GUKS

GABON
Boeing 777-236
Gabonese Government, Libreville
TR-KPR

Dassault
Falcon 900EX
Gabonese Government, Libreville
TR-LEX

Grumman
G.1159C Gulfstream IVSP
Gabonese Government, Libreville
TR-KSP

GERMANY

Please note that German serials do not officially include the '+' part in them but aircraft wearing German markings are often painted with a cross in the middle, which is why it is included here.

Luftwaffe

Airbus A.310-304/MRTT*

1/FBS, Köln-Bonn

10+23
10+24*
10+25*
10+26*
10+27*

Bombardier Global 5000

3/FBS, Köln-Bonn

14+01
14+02
14+03
14+04

Airbus A.319CJ-115X

3/FBS, Köln-Bonn

15+01
15+02

Airbus A.340-313X

3/FBS, Köln-Bonn

16+01
16+02

Eurofighter
EF.2000GS/EF.2000GT*

EADS, Manching;
TLG 31 *Boelcke*,
   Nörvenich;
TLG 73 *Steinhoff*, Laage;
TLG 74 *Molders*, Neuburg/Donau;
TAubZLwSüd (TAZLS),
   Kaufbeuren;
WTD 61, Ingolstadt

Serial	Unit
30+02*	TLG 31
30+03*	TLG 73
30+04*	TLG 31
30+05*	TLG 73
30+06	TLG 74
30+07	TLG 74
30+09	TAZLS $
30+10*	TLG 73
30+11	TLG 74
30+12	TLG 31
30+14*	TLG 74
30+15	TLG 31
30+17*	TLG 73
30+20*	TLG 73
30+22	TLG 74
30+23	TLG 74
30+24*	TLG 73
30+25	TLG 74
30+26	TAZLS
30+27*	TLG 73
30+28	TLG 74
30+29	TLG 74 $
30+30	TLG 74
30+31*	TLG 31
30+32	TLG 74
30+33	TLG 74
30+35*	TLG 73
30+38*	TLG 73
30+39	TLG 74
30+40	TLG 74
30+42*	TLG 74
30+45	TLG 73 $
30+46	TLG 73
30+47	TLG 73
30+48	TLG 73 $
30+49	TLG 73
30+50	TLG 73
30+51	TLG 73
30+52	TLG 73
30+53	TLG 31
30+54*	TLG 73
30+55	TLG 73
30+56	TLG 73
30+57	TLG 73
30+58	TLG 74
30+59*	TLG 73
30+60	TLG 73
30+61	TLG 74
30+62	TLG 74
30+63	TLG 74
30+64	TLG 31
30+65	TLG 31
30+66	
30+67*	TLG 73
30+68	TLG 74
30+69	TLG 74
30+70	TLG 74
30+71*	TLG 73
30+72	TLG 31
30+73	TLG 73
30+74	TLG 74
30+75	TLG 31
30+76	TLG 31
30+77*	TLG 73
30+78	TLG 31
30+79	TLG 31
30+80	TLG 73
30+81	TLG 73
30+82	TLG 31
30+83	TLG 31
30+84*	TLG 31
30+85	TLG 31
30+86	TLG 74
30+87	TLG 74
30+88	TLG 74
30+89	TLG 31
30+90	TLG 31
30+91	TLG 31
30+92	TLG 31
30+93	TLG 31
30+94	TLG 31
30+95*	TLG 74
30+96	TLG 31
30+97	TLG 31
30+98	TLG 31
30+99*	TLG 73
31+00	TLG 31 $
31+01	TLG 31
31+02	TLG 31
31+03*	EADS
31+04	TLG 31
31+05	TLG 31
31+06	TLG 31 $
31+07	TLG 31
31+08	TLG 31
31+09	TLG 74
31+10	EADS
31+11	TLG 74
31+12	
31+13*	
31+14	TLG 31
31+15	TAZLS
31+16	TLG 73
31+17	TLG 73
31+18	TLG 73
31+19	TLG 73
31+20	TLG 73
31+21	TLG 73
31+22	TLG 73
31+24*	
31+25*	
31+26*	
31+27*	
31+28*	
31+29	
31+30	
31+31	
31+32	
31+33	
31+34	
31+35	
31+36	
31+37	
31+38	
31+39	
31+40	
31+41	
31+42	
31+43	
31+44	
31+45	
31+46	
31+47	
31+48	
31+49	
31+50	
31+51	
31+52	
31+53	
98+03*	WTD 61
98+04	EADS
98+07	WTD 61
98+08*	
98+31*	EADS

149

GERMANY

Panavia
Tornado Strike/
Trainer[1]/ECR[2]
EADS, Manching;
GAFFTC, Holloman AFB, USA;
TLG 33, Büchel;
TLG 51 *Immelmann*, Schleswig/Jagel;
TAubZLwSüd (TAZLS), Kaufbeuren;
WTD 61, Ingolstadt

43+01[1]	TLG 33 $	45+28	TLG 33	46+48[2]	TLG 51
43+07[1]	TLG 51	45+35	TLG 33	46+49[2]	TLG 51
43+10[1]	TLG 33	45+39	TLG 33	46+50[2]	TLG 51
43+25	TLG 33	45+50	TLG 51	46+51[2]	TLG 51
43+29[1]	GAFFTC	45+53	TAZLS	46+52[2]	TLG 51
43+32		45+54	GAFFTC	46+53[2]	
43+34	TAZLS	45+57	TLG 33	46+54[2]	TLG 51
43+37[1]		45+59	TLG 33	46+55[2]	TLG 51
43+38	GAFFTC	45+60[1]	TLG 51	46+56[2]	TLG 51
43+42[1]	GAFFTC	45+61[1]	TLG 51	46+57[2]	TLG 51 $
43+45[1]	GAFFTC	45+64	TLG 33	98+59	WTD 61
43+46	TLG 33	45+66	TLG 33	98+60	WTD 61
43+48	TLG 33	45+67	TLG 33	98+77	WTD 61 $
43+50	TLG 33	45+68	TLG 33	98+79[2]	WTD 61
43+52	TLG 33	45+69	GAFFTC		
43+54	TAZLS	45+70[1]	TLG 51	**Transall C-160D**	
43+59	TAZLS	45+71	TLG 33	LTG 61 (1.LwDiv), Landsberg;	
43+71		45+72		LTG 62 (1.LwDiv), Wunstorf;	
43+92[1]	GAFFTC	45+74	TAZLS	LTG 63 (4.LwDiv), Hohn;	
43+94[1]	GAFFTC	45+76	TLG 33	WTD 61, Ingolstadt	
43+97[1]	TLG 51	45+77[1]	TLG 51	50+08	LTG 61
43+98	TLG 33	45+84	TLG 33	50+09	LTG 61
44+06	GAFFTC	45+85	TLG 33	50+10	LTG 62
44+16[1]	GAFFTC	45+88	TLG 33	50+17	LTG 63
44+17	TLG 51	45+90	TLG 33	50+29	LTG 62
44+21	WTD 61	45+92	TLG 33	50+33	LTG 63
44+23	TLG 33	45+93	TLG 33	50+34	LTG 63
44+29	TLG 33	45+94	TLG 33	50+36	LTG 63
44+30	GAFFTC	46+00[1]	GAFFTC	50+38	LTG 61
44+33	GAFFTC	46+02	TLG 33	50+40	LTG 63
44+34	GAFFTC	46+05[1]	GAFFTC	50+41	LTG 61
44+58	TLG 33	46+07[1]	TLG 33	50+42	LTG 62
44+61	TLG 33	46+08[1]		50+44	LTG 61
44+64	TLG 33	46+09[1]	GAFFTC	50+45	LTG 63
44+65	TLG 33	46+10	TLG 33	50+46	LTG 62
44+69	TLG 33	46+11	TLG 33	50+47	LTG 61
44+70	TLG 33	46+15	TLG 33	50+48	LTG 61 $
44+72[1]	TLG 33	46+18	TLG 33	50+49	LTG 61
44+73	WTD 61	46+20	TLG 51	50+51	LTG 61
44+75[1]	TLG 51	46+21	TLG 33	50+53	LTG 61
44+78	TLG 33	46+22	TLG 33	50+54	LTG 63
44+79	TLG 33	46+23[2]	TLG 51	50+55	LTG 62
44+90	TLG 51	46+24[2]	TLG 51	50+57	LTG 62
45+00	GAFFTC	46+25[2]	TLG 51	50+58	LTG 63
45+08	GAFFTC	46+26[2]		50+59	LTG 63
45+09	TLG 33	46+28[2]	TLG 51 $	50+61	LTG 62
45+13[1]	GAFFTC	46+29[2]		50+64	LTG 63
45+14[1]	TLG 33	46+30[2]		50+65	LTG 62
45+16[1]	TLG 33	46+32[2]	TLG 51	50+66	LTG 61
45+19	TLG 33	46+33[2]		50+67	LTG 63
45+20	TLG 33	46+34[2]		50+69	LTG 63
45+22	TLG 51	46+35[2]	TLG 51	50+70	LTG 62
45+23	TLG 33	46+36[2]	TLG 51	50+71	LTG 63
		46+37[2]		50+72	LTG 61
		46+38[2]	WTD 61	50+73	LTG 61
		46+39[2]		50+74	LTG 61
		46+40[2]	TLG 51	50+75	LTG 63
		46+41[2]		50+76	LTG 63
		46+43[2]		50+77	LTG 62
		46+44[2]	TLG 51	50+78	LTG 63
		46+45[2]	TLG 51	50+79	LTG 63
		46+46[2]	TLG 51	50+81	LTG 62

50+82	LTG 61	
50+83	LTG 62	
50+84	LTG 61	
50+86	LTG 61	
50+87	LTG 63	
50+88	LTG 61	
50+89	LTG 62	
50+90	LTG 62	
50+91	LTG 62	
50+92	LTG 61	
50+93	LTG 63	
50+95	LTG 63	
50+96	LTG 61	
50+97	LTG 63	
51+01	LTG 62	
51+02	LTG 63	
51+03	LTG 63	
51+04	WTD 61	
51+05	LTG 62	
51+06	LTG 63	
51+08	WTD 61	
51+09	LTG 63	
51+10	LTG 61	
51+12	LTG 61	
51+13	LTG 61	
51+14	LTG 62	
51+15	LTG 61	

Airbus Military A.400M
LTG 62, Wunstorf

54+01		
54+02	(on order)	
54+03	(on order)	
54+04	(on order)	
54+05	(on order)	
54+06	(on order)	
54+07	(on order)	
54+08	(on order)	
54+09	(on order)	
54+10	(on order)	

Eurocopter
AS.532U-2 Cougar
3/FBS, Berlin-Tegel

82+01	
82+02	
82+03	

Sikorsky/VFW
CH-53G/CH-53GA/CH-53GE/CH-53GS
HFWS, Bückeburg;
HSG 64, Holzdorf, Laupheim
    & Rheine-Bentlage;
TsLw 3, Fassberg;
WTD 61, Ingolstadt

84+05	CH-53G	HFWS
84+06	CH-53G	HSG 64 $
84+09	CH-53G	TsLw 3
84+10	CH-53G	HFWS
84+12	CH-53G	HSG 64
84+13	CH-53GS	HFWS
84+14	CH-53GE	HSG 64
84+15	CH-53GS	HSG 64
84+16	CH-53G	HFWS
84+18	CH-53G	HFWS
84+19	CH-53G	TsLw 3
84+24	CH-53GS	HSG 64
84+25	CH-53GS	HSG 64
84+26	CH-53GE	HSG 64
84+27	CH-53G	HFWS
84+28	CH-53G	HSG 64
84+29	CH-53G	HSG 64
84+30	CH-53GS	HSG 64
84+31	CH-53GA	HSG 64
84+32	CH-53G	HSG 64
84+33	CH-53GA	HSG 64
84+34	CH-53G	HSG 64
84+35	CH-53G	HSG 64
84+37	CH-53GA	HSG 64
84+38	CH-53G	HSG 64
84+39	CH-53GA	HSG 64
84+40	CH-53G	HSG 64
84+41	CH-53G	HFWS
84+42	CH-53GS	HSG 64
84+43	CH-53G	HSG 64
84+44	CH-53G	HSG 64
84+45	CH-53GS	HSG 64
84+46	CH-53G	HSG 64
84+47	CH-53GA	HSG 64
84+48	CH-53G	HSG 64
84+49	CH-53GA	HSG 64
84+50	CH-53G	HFWS
84+51	CH-53GS	HSG 64
84+52	CH-53GS	HSG 64
84+53	CH-53GE	HSG 64
84+54	CH-53G	HSG 64
84+55	CH-53G	WTD 61
84+57	CH-53G	HSG 64
84+58	CH-53G	HSG 64
84+59	CH-53G	HFWS
84+60	CH-53G	HSG 64
84+62	CH-53GS	HSG 64
84+63	CH-53G	HSG 64
84+64	CH-53GS	HSG 64
84+65	CH-53G	HSG 64
84+66	CH-53GS	HSG 64
84+67	CH-53GS	HSG 64
84+68	CH-53G	HSG 64
84+70	CH-53GA	TsLw 3
84+71	CH-53G	HSG 64

84+72	CH-53G	HSG 64
84+73	CH-53GS	HSG 64
84+74	CH-53G	HFWS
84+75	CH-53G	HSG 64
84+76	CH-53G	HSG 64
84+77	CH-53G	HSG 64
84+78	CH-53GS	TsLw 3
84+79	CH-53GS	HSG 64
84+80	CH-53G	HSG 64
84+82	CH-53GE	HSG 64
84+83	CH-53G	TsLw 3
84+84	CH-53G	HSG 64
84+85	CH-53GS	HSG 64
84+86	CH-53GA	Airbus
84+87	CH-53G	HSG 64
84+88	CH-53G	HSG 64
84+89	CH-53GA	HSG 64
84+90	CH-53G	HSG 64
84+91	CH-53GS	HSG 64
84+92	CH-53GS	HSG 64
84+94	CH-53G	WTD 61
84+95	CH-53GA	HSG 64
84+96	CH-53GE	HSG 64
84+97	CH-53G	HSG 64
84+98	CH-53GS	HSG 64
84+99	CH-53GA	HFWS
85+00	CH-53GS	HSG 64
85+01	CH-53GS	HSG 64
85+02	CH-53G	HSG 64
85+03	CH-53G	HSG 64
85+04	CH-53GA	HSG 64
85+05	CH-53GS	HSG 64
85+06	CH-53GA	HSG 64
85+07	CH-53GS	HSG 64
85+08	CH-53G	HSG 64
85+10	CH-53GS	HSG 64
85+11	CH-53G	HSG 64
85+12	CH-53GS	HSG 64

Northrop Grumman
RQ-4E Euro Hawk
WTD 61, Manching

99+01	WTD 61

Marineflieger
Dornier Do.228LM/Do.228NG*
MFG 3, Nordholz

57+04	
57+05*	

Lockheed
P-3C Orion
MFG 3, Nordholz

60+01 $	
60+02	
60+03	
60+04	
60+05 $	
60+06	
60+07	
60+08	

Westland
Super Lynx Mk88A
MFG 5, Nordholz;
WTD 61, Ingolstadt

83+02	MFG 3
83+03	MFG 3
83+04	MFG 3
83+05	MFG 3
83+06	MFG 3
83+07	WTD 61
83+09	MFG 3
83+10	MFG 3
83+11	MFG 3 $
83+12	MFG 3
83+13	MFG 3
83+15	MFG 3
83+17	MFG 3
83+18	MFG 3
83+19	MFG 3
83+20	MFG 3 $
83+21	MFG 3
83+22	MFG 3
83+23	MFG 3
83+24	MFG 3
83+25	MFG 3

Westland
Sea King HAS41
MFG 5, Nordholz

89+50
89+51
89+52
89+53
89+54
89+55 $
89+56
89+57
89+58 $
89+60
89+61
89+62
89+63
89+64 $
89+65
89+66
89+67
89+68
89+69
89+70
89+71

Heeresfliegertruppe
Eurocopter
EC.665 Tiger UHT
Airbus Helicopters, Donauwörth;
HFAZT, Le Luc, France;
KHR 36, Fritzlar;
TsLw 3, Fassberg;
WTD 61, Ingolstadt

74+01	KHR 36
74+02	Airbus
74+03	TsLw 3
74+05	HFAZT
74+06	HFAZT
74+07	HFAZT
74+08	TsLw 3
74+09	HFAZT
74+10	KHR 36
74+11	HFAZT
74+14	HFAZT
74+15	HFAZT
74+16	KHR 36
74+17	KHR 36
74+18	KHR 36
74+19	KHR 36
74+20	KHR 36
74+21	KHR 36
74+22	KHR 36
74+23	KHR 36
74+24	
74+26	KHR 36
74+28	KHR 36
74+29	
74+30	KHR 36
74+31	
74+32	
74+34	
74+35	KHR 36
74+36	KHR 36
74+37	
74+38	HFAZT
74+39	
74+40	
74+41	
74+42	
74+43	
74+44	
74+45	
74+46	
74+47	
74+48	
74+49	
74+50	
74+51	
74+52	
74+53	
74+54	
74+55	
74+56	
74+57	
74+58	
74+59	
98+16	TsLw 3
98+17	WTD 61
98+18	Airbus

Airbus Helicopters EC.645T-2

76+01
76+02

NH Industries
NH.90-TTH
Airbus Helicopters, Donauwörth;
HFWS, Bückeburg;
THR 10, Fassberg;
TsLw 3, Fassberg;
WTD 61, Ingolstadt

78+01	HFWS
78+02	TsLw 3
78+03	HFWS
78+04	TsLw 3
78+05	TsLw 3
78+06	HFWS
78+07	HFWS
78+08	HFWS
78+09	HFWS
78+10	HFWS
78+11	THR 10
78+12	THR 10
78+13	HFWS
78+14	THR 10
78+15	THR 10
78+16	
78+17	
78+18	
78+20	HFWS
78+21	HFWS
78+22	HFWS
78+23	
78+24	Airbus
78+25	
78+26	HWFS
78+27	
78+28	
78+29	
78+30	
78+31	
78+32	
78+33	
78+34	
78+35	
78+36	
78+37	
78+38	
78+39	
78+40	
78+41	
78+42	
78+43	
78+44	
78+45	
78+46	
78+47	
78+48	
78+49	
78+50	
79+01	HSG 64
79+02	HSG 64

79+03	THR 10	86+41		87+88	
79+04	THR 10	86+44	HFWS	87+99	
79+05	THR 10	86+45		88+04	
79+06	HSG 64	86+47	HFWS	88+06	
79+07	HSG 64	86+48	HFWS	88+09	
79+09		86+49	HFWS $	88+10	HFWS
79+10		86+50			
79+24	THR 10	86+51	HFWS	**GHANA**	
79+25		86+55	HFWS	**Ghana Air Force**	
79+26	TsLw 3	86+56	HFWS	**Dassault**	
79+27		86+58		**Falcon 900EASy**	
79+28		86+59	HFWS	VIP Flight, Accra	
79+29		86+60	HFWS	9G-EXE	
79+30		86+62	HFWS		
98+49	Airbus	86+66	HFWS	**GREECE**	
98+50	Airbus	86+68		**Elliniki Polemiki Aeroporía**	
98+56	Airbus	86+70	HFWS $	**Aeritalia C-27J Spartan**	
98+90		86+77	HFWS	354 Mira, Elefsís	
98+99		86+78	HFWS	4117	
		86+79		4118	
Eurocopter EC.135P-1		86+80		4120	
HFWS, Bückeburg		86+83	TsLw 3	4121	
82+51		86+85		4122	
82+52		86+87	HFWS	4123	
82+53		86+90		4124	
82+54		86+92		4125	
82+55		86+95	HFWS		
82+56		86+99	HFWS	**Dassault Mirage 2000**	
82+57		87+01	HFAZT	331 MAPK/114 PM, Tanagra;	
82+59		87+02		332 MAPK/114 PM, Tanagra	
82+60		87+04		**Mirage 2000BG**	
82+61		87+06	HFWS	201	332 MAPK
82+62		87+09		202	332 MAPK
82+63		87+11		**Mirage 2000EG**	
82+64		87+12	TsLw 3	210	332 MAPK
82+65		87+15	TsLw 3	212	332 MAPK
		87+16	HFWS	213	332 MAPK
MBB Bo.105P		87+23		215	332 MAPK
HFAZT, Le Luc, France;		87+25		216	332 MAPK
HFWS, Bückeburg;		87+26		217	332 MAPK
THR 30, Niederstetten;		87+28	HFWS $	218	332 MAPK
TsLw 3, Fassberg;		87+35		219	332 MAPK
WTD 61, Ingolstadt		87+39		220	332 MAPK
86+07	HFWS	87+41	HFWS	221	332 MAPK
86+10		87+43		226	332 MAPK
86+11		87+50		231	332 MAPK
86+15	HFAZT	87+51		232	332 MAPK
86+16		87+52	HFWS	233	332 MAPK
86+18	TsLw 3	87+55		237	332 MAPK
86+19	HFWS	87+56	TsLw 3	239	332 MAPK $
86+21		87+60		241	332 MAPK
86+24	HFWS	87+61	HFWS	242	332 MAPK
86+25	HFWS	87+62	HFWS	**Mirage 2000-5BG**	
86+27		87+66	HFWS	505	331 MAPK
86+28	HFWS	87+71	HFWS	506	331 MAPK
86+29	HFWS	87+73	HFWS	507	331 MAPK
86+30		87+76	HFWS	508	331 MAPK
86+33		87+77		509	331 MAPK
86+34	HFWS	87+78	HFWS		
86+36	HFWS	87+79			
86+39	HFWS	87+81	TsLw 3		
86+40	HFWS	87+87	HFWS		

Mirage 2000-5EG

511	331 MAPK
514	331 MAPK
527	331 MAPK
530	331 MAPK
534	331 MAPK
535	331 MAPK
536	331 MAPK
540	331 MAPK
543	331 MAPK
545	331 MAPK
546	331 MAPK
547	331 MAPK
548	331 MAPK
549	331 MAPK
550	331 MAPK
551	331 MAPK
552	331 MAPK
553	331 MAPK
554	331 MAPK
555	331 MAPK

Embraer ERJ.135BJ Legacy/ERJ.145H/ERJ.135LR

352 MMYP/112 PM, Elefsís;
380 Mira/112 PM, Elefsís

374	ERJ.145H	380 Mira
671	ERJ.145H	380 Mira
729	ERJ.145H	380 Mira
757	ERJ.145H	380 Mira
135L-484	ERJ.135BJ	352 MMYP
145-209	ERJ.135LR	352 MMYP

Gulfstream Aerospace Gulfstream V

352 MMYP/112 PM, Elefsís

678

Lockheed C-130H Hercules

356 MTM/112 PM, Elefsís
*ECM

741*
742
743
744
745 $
746
747*
749
751
752 $

Lockheed F-16C/F-16D*
Fighting Falcon

330 Mira/111 PM,
  Nea Ankhialos;
335 Mira/116 PM,
  Áraxos;
336 Mira/116 PM,
  Áraxos;
337 Mira/110 PM, Lárissa;
340 Mira/115PM, Soúda;
341 Mira/111 PM,
  Nea Ankhialos;
343 Mira/115PM, Soúda;
347 Mira/111 PM,
  Nea Ankhialos

001	335 Mira
002	335 Mira
003	335 Mira
004	335 Mira
005	335 Mira
006	335 Mira
007	335 Mira
008	335 Mira
009	335 Mira
010	335 Mira
011	335 Mira
012	335 Mira
013	335 Mira
014	335 Mira
015	335 Mira
016	335 Mira
017	335 Mira
018	335 Mira
019	335 Mira
020	335 Mira
021*	335 Mira
022*	335 Mira
023*	335 Mira
024*	335 Mira
025*	335 Mira
026*	335 Mira
027*	335 Mira
028*	335 Mira
029*	335 Mira
030*	335 Mira
045	347 Mira
046	341 Mira
047	347 Mira
048	341 Mira
049	347 Mira
050	341 Mira
051	347 Mira
052	341 Mira
053	347 Mira
054	341 Mira
055	347 Mira
056	341 Mira
057	347 Mira
058	341 Mira
060	341 Mira
061	347 Mira
062	341 Mira $
063	347 Mira
064	341 Mira
065	347 Mira
066	341 Mira
067	347 Mira
068	341 Mira
069	347 Mira
070	341 Mira
071	347 Mira
072	341 Mira
073	347 Mira
074	341 Mira
075	347 Mira
076	341 Mira $
077*	341 Mira
078*	347 Mira
079*	347 Mira
080*	341 Mira
081*	347 Mira
082*	341 Mira
083*	347 Mira
110	330 Mira
111	330 Mira
112	
113	330 Mira
114	330 Mira
115	330 Mira
116	330 Mira
117	330 Mira
118	
119	330 Mira
120	
121	341 Mira
122	330 Mira
124	
125	330 Mira
126	330 Mira
127	330 Mira
128	330 Mira
129	330 Mira
130	330 Mira
132	
133	330 Mira
134	330 Mira
136	330 Mira
138	
139	330 Mira
140	330 Mira
141	330 Mira
143	330 Mira
144*	330 Mira
145*	330 Mira
146*	
147*	330 Mira
148*	330 Mira
149*	330 Mira
500	343 Mira
501	337 Mira
502	337 Mira
503	343 Mira
504	343 Mira

505	343 Mira $
506	340 Mira
507	337 Mira
508	337 Mira
509	343 Mira
510	343 Mira
511	343 Mira
512	343 Mira
513	343 Mira
515	343 Mira
517	337 Mira
518	340 Mira
519	340 Mira
520	343 Mira
521	340 Mira
523	340 Mira
524	337 Mira
525	340 Mira
526	343 Mira
527	343 Mira
528	337 Mira
529	343 Mira
530	337 Mira
531	337 Mira
532	337 Mira
533	340 Mira
534	340 Mira
535	340 Mira
536	340 Mira $
537	340 Mira
538	340 Mira
539	337 Mira
600*	337 Mira
601*	340 Mira
602*	340 Mira
603*	337 Mira
605*	340 Mira
606*	337 Mira
607*	343 Mira
608*	340 Mira
609*	337 Mira
610*	340 Mira
611*	340 Mira
612*	337 Mira
613*	343 Mira
615*	343 Mira
616*	343 Mira
617*	343 Mira
618*	343 Mira
619*	337 Mira $

HUNGARY
Hungarian Defence Forces
Antonov An-26
59 Sz.D.REB, Kecskemét

110
405
406
407
603

Boeing
C-17A Globemaster III
Strategic Airlift Capability (SAC)
Heavy Airlift Wing,
Pápa, Hungary

01	(08-0001)
02	(08-0002)
03	(08-0003)

SAAB 39C/39D* Gripen
59 Sz.D.REB, Kecskemét

30
31
32
33
34
35
36
37
38
39
40
41
42*
43*

INDIA
Bharatiya Vayu Sena
Boeing
C-17A Globemaster III
81 Sqn, Hindon AB

CB-8001
CB-8002
CB-8003
CB-8004
CB-8005
CB-8006
CB-8007
CB-8008
CB-8009
CB-8010

ISRAEL
Heyl ha'Avir
Boeing 707
120 Sqn, Nevatim

120	RC-707
128	RC-707
137	RC-707
140	KC-707
248	KC-707
250	KC-707
255	EC-707
260	KC-707
264	KC-707
272	VC-707
275	KC-707
290	KC-707
295	VC-707

Lockheed
C-130E/H Karnaf/
C-130J-30 Shimshon
103 Sqn & 131 Sqn, Nevatim

102	C-130H	131 Sqn
208	C-130E	131 Sqn
305	C-130E	131 Sqn
309	C-130E	131 Sqn
310	C-130E	131 Sqn
314	C-130E	131 Sqn
316	C-130E	131 Sqn
318	C-130E	131 Sqn
420	KC-130H	131 Sqn
427	C-130H	131 Sqn
428	C-130H	131 Sqn
435	C-130H	131 Sqn
436	C-130H	131 Sqn
522	KC-130H	131 Sqn
545	KC-130H	131 Sqn
661	C-130J-30	103 Sqn
662	C-130J-30	103 Sqn
663	C-130J-30	(on order)
664	C-130J-30	(on order)

ITALY
Aeronautica Militare Italiana
Aeritalia G222/
C-27J Spartan/MC-27J Pretorian
9ª Brigata Aerea, Pratica di Mare:
14° Stormo/8° Gruppo;
46ª Brigata Aerea, Pisa:
98° Gruppo;
RSV, Pratica di Mare
G222RM (RC-222)

MM62139	14-20	8

G222TCM (C-222)

MM62124	RS-46	RSV
MM62146	14-11	8

C-27J Spartan

MM62214	46-84	98
MM62215	46-80	98
MM62217	46-81	98
MM62218	46-82	98
CSX62219	RS-50	RSV
MM62221	46-85	98
MM62222	46-86	98
MM62223	46-88	98
MM62224	46-89	98
MM62225	46-90	98
MM62250	46-91	98

MC-27J Pretorian

CSX62127	Alenia	
MM62220	46-83	98

Aeritalia-EMB
AMX (A-11A)/
AMX-ACOL (A-11B)/
AMX-T (TA-11A)/
AMX-T-ACOL (TA-11B)
51° Stormo, Istrana:
  101° Gruppo, 103° Gruppo
  & 132° Gruppo;
RSV, Pratica di Mare

MM7126	A-11B	32-12	101
MM7129	A-11B	32-15	101
MM7147	A-11A	32-01$	101
MM7148	A-11B		
MM7149	A-11B	32-24	101
MM7151	A-11B	51-51	132
MM7159	A-11A	51-10$	103
MM7160	A-11B		
MM7161	A-11B	51-37	132
MM7162	A-11B	51-33	132
MM7163	A-11B	51-72	132
MM7164	A-11B	51-40	132
MM7165	A-11B	32-16	101
MM7166	A-11B	32-13	101
MM7167	A-11B	51-56	132
MM7168	A-11B	51-55	132
MM7169	A-11B	51-66	132
MM7170	A-11B	51-30	132
MM7171	A-11B	51-52	132
MM7172	A-11B		
MM7173	A-11B	51-63	132
MM7174	A-11B	51-60	132
MM7175	A-11A	51-45	132
MM7176	A-11B		
MM7177	A-11B	51-42	132
MM7178	A-11B	51-43	132
MM7179	A-11B	51-64	132
MM7180	A-11B	51-53	132
MM7182	A-11B	51-62	132
MM7183	A-11B	51-41	132
MM7184	A-11B	51-65	132
MM7185	A-11B	51-36	132
MM7186	A-11B	51-50	132
MM7189	A-11B		
MM7190	A-11B	51-57	132
MM7191	A-11B	51-34	132
MM7192	A-11B	51-70	132
MM7193	A-11B	51-54	132
MM7194	A-11B	51-67	132
CSX7195	A-11A		Alenia
MM7196	A-11B	51-35	103
MM7197	A-11B	51-46	132
MM7198	A-11B	51-44$	132
MM55029	TA-11A	32-50$	101
MM55030	TA-11A	32-41	101
CSX55034	TA-11B	RS-18	RSV
MM55036	TA-11A	32-51	101
MM55037	TA-11A	51-80	132
MM55042	TA-11A	32-56	101
MM55043	TA-11A	51-21	101
MM55044	TA-11A	51-82	132
MM55046	TA-11A	32-47	101
MM55047	TA-11A	32-53	101
MM55049	TA-11A	51-23	101
MM55051	TA-11A	32-42	101

Aermacchi
MB339A (T-339A)/
MB339PAN (AT-339A)/
MB339CD-1 (FT-339B)*/
MB339CD-2 (FT-339C)*/
32° Stormo, Amendola:
  632ª SC;
36° Stormo, Gioia del Colle:
  12° Gruppo;
51° Stormo, Istrana:
  651ª SC;
61° Stormo, Lecce:
  212° Gruppo & 213° Gruppo;
Aermacchi, Venegono;
Frecce Tricolori [FT]
  (313° Gruppo), Rivolto
  (MB339A/PAN);
RSV, Pratica di Mare

MM54443	T-339A	61-50	
MM54446	T-339A	61-01	
MM54452	T-339A		
CSX54453	T-339A	RS-11	RSV
MM54457	T-339A	61-11	
MM54458	T-339A	61-12	
MM54465	T-339A	61-21	
MM54467	T-339A	61-23	
MM54468	T-339A	61-24	
MM54473	AT-339A	5	[FT]
MM54475	AT-339A		[FT]
MM54477	AT-339A	9	[FT]
MM54479	AT-339A		[FT]
MM54480	AT-339A	2	[FT]
MM54482	AT-339A		[FT]
MM54487	AT-339A		[FT]
MM54488	T-339A	61-32	
MM54492	T-339A		
MM54496	T-339A	61-42	
MM54499	T-339A		
MM54500	AT-339A	12	[FT]
MM54504	T-339A	61-52	
MM54505	AT-339A	4	[FT]
MM54507	T-339A	61-55	
MM54509	T-339A		
MM54510	AT-339A		[FT]
MM54511	T-339A	61-61	
MM54512	T-339A	61-62	
MM54514	AT-339A	10	[FT]
MM54515	T-339A	61-65	
MM54516	T-339A	61-66	
MM54517	AT-339A		[FT]
MM54518	AT-339A	8	[FT]
MM54532	T-339A		
MM54533	T-339A	61-72	
MM54534	AT-339A		[FT]
MM54536	T-339A		
MM54537	T-339A		
MM54538	AT-339A	1	[FT]
MM54539	AT-339A		[FT]
MM54542	AT-339A		[FT]

CSX54544	FT-339		RSV
MM54547	AT-339A		[FT]
MM54548	T-339A	61-106	
MM54549	T-339A	61-107	
MM54551	AT-339A		[FT]
MM55052	AT-339A		[FT]
MM55053	AT-339A	0	[FT]
MM55054	AT-339A	6	[FT]
MM55055	AT-339A	11	[FT]
MM55058	AT-339A	7	[FT]
MM55059	AT-339A	3	[FT]
MM55062	FT-339	61-126	
MM55063	FT-339	61-127	
MM55064	FT-339	61-130	
MM55065	FT-339	61-131	
MM55066	FT-339	61-132	
MM55067	FT-339	61-133	
MM55068	FT-339	RS-33	RSV
MM55069	FT-339	61-135	
MM55070	FT-339	61-136	
MM55072	FT-339	61-140	
MM55073	FT-339	61-141	
MM55074	FT-339	36-06	
MM55075	FT-339	61-143	
MM55076	FT-339	36-04	
MM55077	FT-339	61-145	
MM55078	FT-339	61-146	
MM55079	FT-339	61-147	
MM55080	FT-339	61-150	
MM55081	FT-339	61-151	
MM55082	FT-339	61-152	
MM55084	FT-339	61-154	
MM55085	FT-339	61-155	
MM55086	FT-339	61-156	
MM55087	FT-339	61-167	
MM55088	FT-339	61-160	
MM55089	FT-339	32-161	
MM55090	FT-339	61-162	
MM55091	FT-339	RS-32	RSV

Aermacchi
M345
Aermacchi, Venegono
| CPX619 | | | |

Aermacchi
M-346/T-346A Master
61° Stormo, Lecce;
Aermacchi, Venegono;
RSV, Pratica di Mare

CPX616	M-346		
CPX622	M-346A		
MT55144	T-346A		RSV
MT55145	T-346A		RSV
CSX55152	T-346A	(on order)	
MM55153	T-346A	(on order)	
CSX55154	T-346A	61-01	
MM55...	T-346A	(on order)	
MM55...	T-346A	(on order)	

Aérospatiale
ATR.72-600MP (P-72A)
CSX62279	(on order)
CSX62280	Alenia
CSX62281	(on order)
CSX62282	(on order)

AgustaWestland
HH-101A Caesar
15° Stormo:
81° Centro, Cervia;
82° Centro SAR, Trapani/Birgi;
83° Gruppo SAR, Cervia;
84° Centro SAR, Gioia del Colle;
85° Centro SAR, Pratica di Mare
MM8....	15-01	(on order)
MM8....	15-02	(on order)
MM8....	15-03	(on order)

AgustaWestland
AW139
15° Stormo:
81° Centro, Cervia;
82° Centro SAR, Trapani/Birgi;
83° Gruppo SAR, Cervia;
84° Centro SAR, Gioia del Colle;
85° Centro SAR, Pratica di Mare;
31° Stormo, Roma-Ciampino:
93° Gruppo
HH-139A
MM81796	15-40	83
MM81797	15-41	84
MM81798	15-42	81
MM81799	15-43	81
MM81800	15-44	
MM81801	15-45	84
MM81802	15-46	81
MM81803	15-47	81
MM81804	15-48	85
MM81805	15-49	82
MM81822	15-50	85
MM81823	15-51	
MM81824	15-52	

VH-139A
MM81806	93
MM81807	93
MM81811	93
MM81812	93

Airbus A.319CJ-115X (VC-319A)
31° Stormo, Roma-Ciampino:
306° Gruppo
MM62174
MM62209
MM62243

Boeing 767-2EYER (KC-767A)
14° Stormo, Pratica di Mare:
8° Gruppo
MM62226	14-01
MM62227	14-02
MM62228	14-03
MM62229	14-04

Breguet
Br.1150 Atlantic (P-1150A)
41° Stormo, Catania:
88° Gruppo
MM40115	41-77 $
MM40118	41-80
MM40122	41-84
MM40125	41-87

Dassault Falcon 50 (VC-50A)
31° Stormo, Roma-Ciampino:
93° Gruppo
MM62026
MM62029

Dassault
Falcon 900EX (VC-900A)/
900EX EASy (VC-900B)*
31° Stormo, Roma-Ciampino:
93° Gruppo
MM62171 $
MM62172
MM62210
MM62244*
MM62245*

Eurofighter
F-2000A/TF-2000A* Typhoon
4° Stormo, Grosseto:
9° Gruppo & 20° Gruppo;
36° Stormo, Gioia del Colle:
12° Gruppo;
37° Stormo, Trapani/Birgi:
18° Gruppo;
Alenia, Torino/Caselle;
RSV, Pratica di Mare
MMX602	RS-01	Alenia
MMX603	RMV-01	Alenia
MMX614*		RSV
MM7235	4-13	9
MM7270	4-1	9
MM7271	4-15	9
MM7272	36-14	12
MM7273	4-10	20
MM7274	4-4	9
MM7275	36-11	12
MM7276	36-05	12
MM7277	4-40	20
MM7278	RS-23	RSV
MM7279	4-20	9
MM7280	36-30	12
MM7281	4-14	9
MM7282	36-15	12
MM7284	36-10	12

MM7285	4-16	9
MM7286	36-02	12
MM7287	4-3	9
MM7288	36-42	12
MM7289	4-5	9
MM7290	4-7	9
MM7291	4-11	9
MM7292	36-21	12
MM7293	36-33	12
MM7294	4-52	20
MM7295		9
MM7296	36-22	12
MM7297	36-23	12
MM7298	36-24	12
MM7299	4-41	20
MM7300	36-12	12
MM7301	37-11	18
MM7302	36-25	12
MM7303	4-2	9
MM7304	4-21	9
CSX7305		RSV
MM7306	4-50	20
MM7307	37-01	18
MM7308	36-31	12
MM7309	37-03	18
MM7310	36-32	12
MM7311	37-10	18
MM7312	36-34	12
MM7313	36-35	12
MM7314	36-37	12
MM7315	37-04	18
MM7316	37-07	18
MM7317	4-43	20
MM7318		
MM7319	37-05	18
MM7320	4-45	20
MM7321	37-12	18
MM7322	36-40	12
MM7323	4-6	9
MM7324	36-41	12
MM7325	4-42	20
MM7326	4-46$	20
MM7327	4-47	20
MM7328		
MM7329	37-15	18
MM7330	36-43	12
MM7331	37-16	18
MM7332		
MM7333		
MM7334		
MM7335		
MM7336		
MM7337		
CSX7338		Alenia
MM7339		
MM7340		
MM7341		
MM7342		
MM7343		
MM7344		
MM7345		
MM7346		

ITALY

MM7347		
MM7348		
MM7349		
MM7350		
MM55092*	4-25	20
MM55093*	4-31	20
MM55094*	4-27	20
MM55095*	4-23	20
MM55096*	4-30	20
MM55097*	4-24	20
MM55128*	4-26	20
MM55129*	4-32	20
MM55130*	4-33	20
MM55131*	4-34	20
MM55132*	4-35	20
MM55133*	4-36	20

Lockheed
C-130J/C-130J-30/KC-130J*
Hercules II
46ª Brigata Aerea, Pisa:
  2° Gruppo & 50° Gruppo
C-130J

MM62175	46-40	
MM62177	46-42	
MM62178	46-43	
MM62179	46-44	
MM62180	46-45	
MM62181	46-46	
MM62182	46-47	
MM62183*	46-48	
MM62184*	46-49	
MM62185	46-50	
MM62186	46-51	

C-130J-30

MM62187	46-53	
MM62188	46-54	
MM62189	46-55	
MM62190	46-56	
MM62191	46-57	
MM62192	46-58	
MM62193	46-59	
MM62194	46-60	
MM62195	46-61	
MM62196	46-62	

Panavia
Tornado IDS (A-200A)/
Tornado IDS[MLU](A-200C)/
Tornado IDS Trainer(TA-200A)/
Tornado IDS Trainer[MLU](TA-200C)/
Tornado ECR(EA-200B)/
Tornado ECR [MLU](EA-200C)
6° Stormo, Ghedi:
  102° Gruppo, 154° Gruppo &
  156° Gruppo;
50° Stormo, Piacenza:
  155° Gruppo;
RSV, Pratica di Mare

MM7003	A-200C		
MM7004	A-200C	6-55	102
MM7006	A-200C	6-31$	102
MM7007	A-200A	6-01	154
MM7008	A-200C	50-53	155
MM7013	A-200C	6-75	102
MM7015	A-200C	6-32	102
MM7019	EA-200B	50-05	155
MM7020	EA-200B	50-41	155
MM7021	EA-200B	50-01$	155
MM7023	A-200A	36-31	156
MM7024	A-200C	6-50	102
MM7025	A-200A	6-05	154
MM7026	A-200C	6-35	102
MM7028	A-200A	6-67	102
MM7029	A-200A	6-22	154
MM7030	EA-200B	50-04$	155
MM7034	EA-200B		
MM7035	A-200C	6-27	154
MM7036	EA-200B	50-55	155
MM7037	A-200A	6-16	102
MM7038	A-200C	6-37	102
MM7039	EA-200B	50-50	155
MM7040	A-200A	6-21	102
CSX7041	A-200C	RS-06	RSV
MM7042	A-200A		
MM7043	A-200C	6-25	154
MM7044	A-200C	6-76	156
MM7047	EA-200C	50-43	155
MM7048	A-200C		
MM7051	EA-200B	50-45	155
MM7052	EA-200C	50-02	155
MM7053	EA-200B	50-07	155
MM7054	EA-200C	50-40$	155
MM7055	A-200A	50-42	155
MM7056	A-200A	6-66	102
MM7057	A-200A	6-04	154
MM7058	A-200C	6-11	154
MM7059	A-200A	50-47	155
MM7062	EA-200C	50-44	155
MM7063	A-200C	6-26	154
MM7065	A-200C	50-56	155
MM7066	A-200A	50-03	155
MM7067	EA-200B	50-57	155
MM7068	EA-200B	50-46	155
MM7070	EA-200C	50-06	155
MM7071	A-200C	6-12	154
MM7072	A-200C	6-30	154
MM7073	EA-200B	6-34	102

MM7075	A-200A	6-07	154
CMX7079	EA-200B		Alenia
MM7081	EA-200C	50-51	155
MM7082	EA-200B	50-54	155
MM7083	A-200A	6-72	156
MM7084	EA-200C	50-52	155
CMX7085	A-200A	36-50	Alenia
MM7088	A-200A	6-10	154
MM55002	TA-200A	6-52$	102
MM55003	TA-200A		
MM55004	TA-200A	6-53	102
MM55006	TA-200C	6-15	154
MM55007	TA-200A	6-51	102
MM55008	TA-200A	6-45	102
MM55009	TA-200A	6-44	102
MM55010	TA-200A	6-42	102

Piaggio
P-180AM Avanti (VC-180A)/
P-180RM Avanti (RC-180A)
14° Stormo, Pratica di Mare:
  71° Gruppo;
36° Stormo, Gioia del Colle:
  636ª SC;
RSV, Pratica di Mare:
  311° Gruppo

MM62159	VC-180A	71
MM62160	RC-180A	71
MM62161	VC-180A	RSV
MM62162	RC-180A	71
MM62163	VC-180A	71
MM62164	VC-180A	RSV
MM62199	VC-180A	636
MM62201	VC-180A	71
MM62202	VC-180A	71
MM62203	VC-180A	71
MM62204	VC-180A	71
MM62205	VC-180A	71
MM62206	VC-180A	71
MM62207	VC-180A	71
MM62286	VC-180A	
MM62287	VC-180A	

Aviazione dell'Esercito
Dornier Do.228-212 (UC-228)
28° Gruppo Squadrone
  Cavalleria dell'Aria,
  Viterbo

MM62156	E.I.101
MM62157	E.I.102
MM62158	E.I.103

Piaggio
P-180AM Avanti (VC-180A)
28° Gruppo Squadrone Det,
  Cavalleria dell'Aria,
  Roma/Ciampino
MM62167
MM62168
MM62169

Guardia Costiera
Aérospatiale
ATR.42-400MP (P-42A)/
ATR.42-500MP (P-42B)*
2° Nucleo, Catania;
3° Nucleo, Pescara

MM62170	10-01	3
MM62208	10-02	2
MM62270*	10-03	2

Guardia di Finanza
Aérospatiale
ATR.42-400MP (P-42A)
Gruppo Esplorazione Aeromarittima,
  Pratica di Mare

MM62165	GF-13
MM62166	GF-14
MM62230	GF-15
MM62251	GF-16

Piaggio
P-180AM Avanti (VC-180A)/
P-180AM Avanti II (VC-180B)*
Gruppo Esplorazione Aeromarittima,
  Pratica di Mare

| MM62248 | GF-18 |
| MM62249* | GF-19 |

Marina Militare Italiana
AgustaWestland
EH-101
1° Grupelicot, La Spezia/Luni;
3° Grupelicot, Catania
Mk110 ASW (SH-101A)

MM81480	2-01	3
MM81481	2-02	3
MM81482	2-03	3
MM81483	2-04	3
MM81484	2-05	3
MM81485	2-06	3
MM81486	2-07	3
MM81487	2-08	3
MM81719	2-22	1
MM81726	2-23	1

Mk112 AEW (EH-101A)

MM81488	2-09	1
MM81489	2-10	3
MM81490	2-11	3
MM81491	2-12	1

Mk410 UTY (UH-101A)

MM81492	2-13	1
MM81493	2-14	1
MM81494	2-15	1
MM81495	2-16	1

Mk413 ASH

MM81633	2-18	1
MM81634	2-19	1
MM81635	2-20	1
MM81636	2-21	1

McDonnell Douglas
AV-8B/TAV-8B Harrier II+
Gruppo Aerei Imbarcati,
  Taranto/Grottaglie
AV-8B

MM7199	1-03
MM7200	1-04
MM7201	1-05
MM7212	1-06
MM7213	1-07
MM7214	1-08
MM7215	1-09
MM7217	1-11
MM7218	1-12
MM7219	1-13
MM7220	1-14
MM7222	1-16
MM7223	1-18
MM7224	1-19 $

TAV-8B

| MM55032 | 1-01 |
| MM55033 | 1-02 |

Piaggio
P-180AM Avanti (VC-180A)
9ª Brigata Aerea, AMI,
  Pratica di Mare:
  71° Gruppo

MM62200	9-01
MM62211	9-02
MM62212	9-03
MM62213	

Italian Govt
Dassault Falcon 900
Italian Govt/Soc. CAI,
  Roma/Ciampino
I-CAEX
I-DIES

IVORY COAST
Airbus A.319CJ-133
Ivory Coast Govt, Abidjan
TU-VAS

Boeing 727-2Y4
Ivory Coast Govt, Abidjan
TU-VAO

Grumman
G.1159C Gulfstream IV
Ivory Coast Govt, Abidjan
TU-VAD

JAPAN
Japan Air Self Defence Force
Boeing 747-47C
701st Flight Sqn, Chitose
20-1101
20-1102

Boeing KC-767J
404th Hikotai, Nagoya
07-3604
87-3601
87-3602
97-3603

JORDAN
Al Quwwat al Jawwiya
al Malakiya al Urduniya
Extra EA-300LP
Royal Jordanian Falcons,
  Amman
JY-RFA
JY-RFB
JY-RFC
JY-RFD
JY-RFE

Lockheed C-130H Hercules
3 Sqn, Amman/Marka
344
345 $
346
347

Jordanian Govt
Airbus A.318-112
Jordanian Govt, Amman
VQ-BDD

Airbus A.340-642
Jordanian Govt, Amman
VQ-CDD

Gulfstream Aerospace G.450
Jordanian Govt, Amman
VQ-BCE

Gulfstream Aerospace G.650
Jordanian Govt, Amman
VQ-BNZ

KAZAKHSTAN
Airbus A.330-223
Govt of Kazakhstan, Almaty
UP-A3001

Boeing 757-2M6
Govt of Kazakhstan, Almaty
UP-B5701

Tupolev Tu-154M
Govt of Kazakhstan, Almaty
UP-T5401

KENYA
Kenyan Air Force
Fokker 70ER
Kenyan Govt, Nairobi
KAF 308

KUWAIT
Al Quwwat al Jawwiya al Kuwaitiya
Boeing
C-17A Globemaster III
41 Sqn Kuwait International
KAF 342 (13-0001)
KAF 343 (13-0002)

Lockheed
KC-130J Hercules II
41 Sqn, Kuwait International
KAF 326
KAF 327
KAF 328

Lockheed
L100-30 Hercules
41 Sqn, Kuwait International
KAF 323
KAF 324
KAF 325

Kuwaiti Govt
Airbus A.300C4-620
Kuwaiti Govt, Safat
9K-AHI

Airbus A.310-308
Kuwaiti Govt, Safat
9K-ALD

Airbus A.319CJ-115X
Kuwaiti Govt, Safat
9K-GEA

Airbus A.320-212
Kuwaiti Govt, Safat
9K-AKD

Airbus A.340-542
Kuwaiti Govt, Safat
9K-GBA
9K-GBB

Boeing 737-9BQER
Kuwaiti Govt, Safat
9K-GCC

Boeing 747-469
Kuwaiti Govt, Safat
9K-ADE

Boeing 747-8JK
Kuwaiti Govt, Safat
9K-GAA

Gulfstream Aerospace G.550
Kuwaiti Govt, Safat
9K-GFA

Gulfstream Aerospace
Gulfstream V
Kuwaiti Govt/Kuwait Airways, Safat
9K-AJD
9K-AJE
9K-AJF

KYRGYZSTAN
Tupolev Tu-154M
Govt of Kyrgyzstan, Bishkek
EX-00001

LITHUANIA
Karines Oro Pajegos
Aeritalia C-27J Spartan
Transporto Eskadrile, Siauliai-Zokniai
06
07
08

Antonov An-26RV
Transporto Eskadrile, Siauliai-Zokniai
03
04
05

LET 410UVP Turbolet
Transporto Eskadrile, Siauliai-Zokniai
01

Mil Mi-8
Sraigtasparniu Eskadrile, Panevezys/Pajuostis
02	Mi-8T
09	Mi-8T
10	Mi-8T
11	Mi-8PS
21	Mi-8MTV-1
22	Mi-8MTV-1
28	Mi-8T

LUXEMBOURG
NATO
Boeing 757-28A
NATO, Geilenkirchen, Germany
OO-TFA

Boeing E-3A
NAEW&CF, Geilenkirchen, Germany
LX-N90442
LX-N90443 $
LX-N90444
LX-N90445
LX-N90446
LX-N90447
LX-N90448
LX-N90449
LX-N90450
LX-N90451
LX-N90452
LX-N90453
LX-N90454
LX-N90455
LX-N90456
LX-N90458
LX-N90459

MACEDONIA
Macedonian Govt
Bombardier Lear 60
Macedonian Govt, Skopje
Z3-MKD

MALAYSIA
Royal Malaysian Air Force/ Tentera Udara Diraja Malaysia
Boeing 737-7H6
2 Sqn, Simpang
M53–01

Bombardier
BD.700-1A10 Global Express
2 Sqn, Simpang
M48-02

Dassault Falcon 900
2 Sqn, Simpang
M37-01

Lockheed
C-130 Hercules
14 Sqn, Labuan;
20 Sqn, Subang
M30-01	C-130T	20 Sqn
M30-02	C-130H	20 Sqn
M30-03	C-130H	14 Sqn
M30-04	C-130H-30	20 Sqn
M30-05	C-130H-30	14 Sqn
M30-06	C-130H-30	14 Sqn
M30-07	C-130T	20 Sqn
M30-08	C-130H(MP)	20 Sqn
M30-09	C-130H(MP)	20 Sqn
M30-10	C-130H-30	20 Sqn
M30-11	C-130H-30	20 Sqn
M30-12	C-130H-30	20 Sqn
M30-14	C-130H-30	14 Sqn
M30-15	C-130H-30	20 Sqn
M30-16	C-130H-30	20 Sqn

Malaysian Govt
Airbus A.319CJ-115X
9M-NAA

MALTA
Bombardier
Learjet 60
Govt of Malta, Luqa
9H-AFK

MEXICO
Fuerza Aérea Mexicana
Boeing 737-33A
8° Grupo Aéreo, Mexico City
TP-02   (XC-UJB)

Boeing 757-225
8° Grupo Aéreo, Mexico City
TP-..   (XC-UJM)

Boeing 787-8
8° Grupo Aéreo, Mexico City
TP-01

Armada de Mexico
Gulfstream Aerospace G.450
PRIMESCTRANS, Mexico City
XC-LMF

MOROCCO
Force Aérienne Royaume
Marocaine/ Al Quwwat al
Jawwiya al Malakiya
Marakishiya
Aeritalia
C-27J Spartan
Escadrille de Transport 3,
Kenitra
CN-AMN
CN-AMO
CN-AMP
CN-AMQ

Airtech
CN.235M-100
Escadrille de Transport 3,
Kenitra
023	CN-AMA
024	CNA-MB
025	CNA-MC
026	CN-AMD
027	CN-AME
028	CN-AMF
031	CNA-MG

CAP-232
Marche Verte
28	CN-ABP [2]
29	CN-ABQ [5]
31	CN-ABR [3]
36	CN-ABS [4]
37	CN-ABT [1]

41	CN-ABU [2]
42	CN-ABV [6]
43	CN-ABW [7]
44	CN-ABX

Lockheed
C-130H Hercules
Escadrille de Transport 3,
Kenitra
4535	CN-AOA	C-130H
4551	CNA-OC	C-130H
4575	CNA-OD	C-130H
4581	CNA-OE	C-130H
4583	CNA-OF	C-130H
4713	CN-AOG	C-130H
4733	CN-AOI	C-130H
4738	CNA-OJ	C-130H
4739	CNA-OK	C-130H
4742	CNA-OL	C-130H
4875	CN-AOM	C-130H
4876	CN-AON	C-130H
4877	CNA-OO	C-130H
4888	CNA-OP	C-130H
4907	CN-AOR	KC-130H
4909	CNA-OS	KC-130H

Govt of Morocco
Boeing 737-8KB
Govt of Morocco, Rabat
CN-MVI

Cessna 560 Citation V
Govt of Morocco, Rabat
CNA-NV
CNA-NW

Cessna
560XLS Citation Excel
Govt of Morocco, Rabat
CN-AMJ
CN-AMK

Dassault Falcon 50
Govt of Morocco, Rabat
CN-ANO

Grumman
G.1159 Gulfstream IITT/
G.1159A Gulfstream III
Govt of Morocco, Rabat
CN-ANL   Gulfstream IITT
CN-ANU   Gulfstream III

Gulfstream Aerospace G.550
Govt of Morocco, Rabat
CN-AMS

NAMIBIA
Dassault Falcon 7X
Namibian Govt, Eros
V5-GON

Dassault Falcon 900B
Namibian Govt, Eros
V5-NAM

NETHERLANDS
Koninklijke Luchtmacht
Boeing-Vertol
CH-47 Chinook
298 Sqn, Defence Helicopter Command,
Gilze-Rijen;
302 Sqn, Fort Hood, Texas, USA
CH-47D Chinook
D-101	298 Sqn
D-102	298 Sqn
D-103	298 Sqn
D-106	298 Sqn
D-661	298 Sqn
D-662	298 Sqn
D-663	298 Sqn
D-664	298 Sqn
D-665	298 Sqn
D-666	298 Sqn
D-667	298 Sqn
CH-47F Chinook
D-890	298 Sqn
D-891	298 Sqn
D-892	298 Sqn
D-893	302 Sqn
D-894	302 Sqn
D-895	302 Sqn

Eurocopter
AS.532U-2 Cougar
300 Sqn, Defence Helicopter Command,
Gilze-Rijen
S-419
S-440
S-441
S-442
S-445
S-447
S-453
S-454
S-456
S-457
S-459

NETHERLANDS

General Dynamics
F-16
Now operated on a pool basis. Squadron markings carried do not necessarily reflect the squadron operating.
312/313 Sqns, Volkel;
322 Sqn & Test Flt, Leeuwarden;
148th FS/162th FW, Tucson International Airport, Arizona, USA

J-001	F-16AM	312 Sqn
J-002	F-16AM	322 Sqn $
J-003	F-16AM	313 Sqn $
J-004	F-16AM	148th FS
J-005	F-16AM	313 Sqn
J-006	F-16AM	322 Sqn $
J-008	F-16AM	313 Sqn $
J-009	F-16AM	322 Sqn
J-010	F-16AM	148th FS
J-011	F-16AM	312/313 Sqn
J-013	F-16AM	322 Sqn
J-014	F-16AM	313 Sqn
J-015	F-16AM	322 Sqn
J-016	F-16AM	312 Sqn
J-017	F-16AM	322 Sqn
J-018	F-16AM	148th FS
J-019	F-16AM	148th FS
J-020	F-16AM	312 Sqn
J-021	F-16AM	322 Sqn
J-055	F-16AM	313 Sqn
J-057	F-16AM	312/313 Sqn
J-060	F-16AM	322 Sqn
J-061	F-16AM	322 Sqn
J-062	F-16AM	313 Sqn
J-063	F-16AM	313 Sqn
J-064	F-16BM	148th FS
J-065	F-16BM	322 Sqn
J-066	F-16BM	Test Flt
J-067	F-16BM	148th FS
J-135	F-16AM	322 Sqn
J-136	F-16AM	312 Sqn
J-142	F-16AM	322 Sqn
J-144	F-16AM	312/313 Sqn
J-145	F-16AM	
J-146	F-16AM	322 Sqn
J-193	F-16AM	322 Sqn
J-196	F-16AM	313 Sqn $
J-197	F-16AM	312/313 Sqn
J-199	F-16AM	311 Sqn
J-201	F-16AM	322 Sqn
J-202	F-16AM	322 Sqn
J-208	F-16BM	
J-209	F-16BM	148th FS
J-210	F-16BM	148th FS
J-362	F-16AM	322 Sqn
J-366	F-16AM	148th FS
J-367	F-16AM	312 Sqn
J-368	F-16BM	322 Sqn
J-369	F-16BM	148th FS
J-508	F-16AM	313 Sqn
J-509	F-16AM	322 Sqn
J-510	F-16AM	313 Sqn
J-511	F-16AM	313 Sqn

J-512	F-16AM	322 Sqn
J-513	F-16AM	322 Sqn
J-514	F-16AM	313 Sqn
J-515	F-16AM	322 Sqn
J-516	F-16AM	322 Sqn
J-616	F-16AM	312 Sqn
J-623	F-16AM	148th FS
J-624	F-16AM	322 Sqn
J-628	F-16AM	322 Sqn
J-630	F-16AM	312/313 Sqn
J-631	F-16AM	322 Sqn $
J-632	F-16AM	322 Sqn
J-635	F-16AM	312 Sqn
J-637	F-16AM	312 Sqn
J-638	F-16AM	313 Sqn
J-641	F-16AM	312 Sqn
J-642	F-16AM	322 Sqn
J-643	F-16AM	313 Sqn
J-644	F-16AM	322 Sqn
J-646	F-16AM	312 Sqn
J-647	F-16AM	312/313 Sqn
J-866	F-16AM	312 Sqn
J-868	F-16AM	322 Sqn
J-870	F-16AM	312/313 Sqn
J-871	F-16AM	322 Sqn
J-872	F-16AM	322 Sqn
J-873	F-16AM	322 Sqn
J-877	F-16AM	322 Sqn
J-879	F-16AM	322 Sqn
J-881	F-16AM	322 Sqn
J-882	F-16BM	322 Sqn

Grumman
G.1159C Gulfstream IV
334 Sqn, Eindhoven
V-11

Lockheed
C-130H/C-130H-30* Hercules
336 Sqn, Eindhoven
G-273*
G-275*
G-781
G-988

Lockheed Martin
F-35A Lightning II
323 Sqn, Edward AFB, California, USA;
LMTAS, USA

F-001	323 Sqn
F-002	323 Sqn

MDH
AH-64D Apache Longbow
301 Sqn, Defence Helicopter Command, Gilze-Rijen;
302 Sqn, Fort Hood, Texas, USA

Q-01	301 Sqn
Q-02	302 Sqn
Q-03	302 Sqn
Q-04	301 Sqn
Q-05	301 Sqn

Q-06	302 Sqn
Q-07	302 Sqn
Q-08	301 Sqn
Q-09	301 Sqn
Q-10	301 Sqn
Q-11	302 Sqn
Q-12	302 Sqn
Q-13	301 Sqn
Q-14	301 Sqn
Q-15	301 Sqn
Q-16	301 Sqn
Q-17	301 Sqn
Q-18	301 Sqn
Q-19	301 Sqn
Q-21	301 Sqn
Q-22	301 Sqn
Q-23	301 Sqn
Q-24	301 Sqn
Q-25	301 Sqn
Q-26	301 Sqn
Q-27	302 Sqn
Q-28	302 Sqn
Q-29	301 Sqn
Q-30	301 Sqn

NH Industries
NH.90-NFH
860 Sqn, De Kooij

N-088	
N-102	
N-110	
N-164	
N-175	
N-195	
N-227	
N-228	
N-233	
N-234	
N-258	
N-277	
N-316	
N-317	
N-318	
N-319	
N-324	(on order)
N-325	(on order)
N-326	(on order)
N-327	(on order)

McDonnell Douglas
KDC-10
334 Sqn, Eindhoven
T-235
T-264

Pilatus
PC-7 Turbo Trainer
131 EMVO Sqn,
  Woensdrecht
L-01
L-02 $
L-03
L-04
L-05
L-06
L-07
L-08
L-09
L-10
L-11
L-12
L-13

Sud Alouette III
300 Sqn, Defence Helicopter Command,
  Gilze-Rijen
A-247
A-275
A-292
A-301 $

Kustwacht
Dornier Do.228-212
Base: Schiphol
PH-CGC
PH-CGN

Netherlands Govt
Fokker 70
Dutch Royal Flight, Schiphol
PH-KBX

NEW ZEALAND
Royal New Zealand Air Force
Boeing 757-2K2
40 Sqn, Whenuapai
NZ7571
NZ7572

Lockheed
C-130H/C-130H(NZ)* Hercules
40 Sqn, Whenuapai
NZ7001*
NZ7002
NZ7003*
NZ7004*
NZ7005

Lockheed
P-3K2 Orion
5 Sqn, Whenuapai
NZ4201
NZ4202
NZ4203
NZ4204
NZ4205
NZ4206

NIGER
Government of Niger
Boeing 737-2N9C
Government of Niger, Niamey
5U-BAG

Boeing 737-75U
Government of Niger, Niamey
5U-GRN

NIGERIA
Federal Nigerian Air Force
Lockheed
C-130H/C-130H-30* Hercules
88 MAG, Abuja
NAF-910
NAF-912
NAF-913
NAF-917*
NAF-918*

Nigerian Govt
Boeing 737-7N6
Federal Govt of Nigeria,
  Abuja
5N-FGT   [001]

Dassault Falcon 7X
Federal Govt of Nigeria,
  Abuja
5N-FGU
5N-FGV

Dassault Falcon 900
Federal Govt of Nigeria,
  Abuja
NAF-961

Gulfstream Aerospace G.550
Federal Govt of Nigeria,
  Abuja
5N-FGW

Gulfstream Aerospace
Gulfstream V
Federal Govt of Nigeria,
  Abuja
5N-FGS

Hawker 4000 Horizon
Federal Govt of Nigeria,
  Abuja
5N-FGX

NORWAY
Luftforsvaret
Bell 412HP*/412SP
339 Skv, Bardufoss
139
140
141
142

143
144
145*
146
147
148
149
161
162
163
164*
165*
166*
167
194

Dassault
Falcon 20 ECM
717 Skv, Gardermoen
041
053
0125

General Dynamics
F-16 MLU
331 Skv, Bodø;
332 Skv, Bodø;
338 Skv, Ørland
[All operate from a pool
and belong to the FLO];
412th TW, Edwards AFB, USA

272	F-16A	FLO $
273	F-16A	FLO
275	F-16A	FLO
276	F-16A	FLO
277	F-16A	FLO
279	F-16A	FLO
281	F-16A	FLO
282	F-16A	FLO
284	F-16A	FLO
285	F-16A	FLO
286	F-16A	FLO
288	F-16A	FLO
289	F-16A	FLO
291	F-16A	FLO
292	F-16A	FLO
293	F-16A	FLO
295	F-16A	FLO
297	F-16A	FLO
298	F-16A	FLO $
299	F-16A	FLO
302	F-16B	FLO
304	F-16B	FLO
305	F-16B	FLO
306	F-16B	FLO
658	F-16A	FLO
659	F-16A	FLO
660	F-16A	FLO
661	F-16A	FLO
662	F-16A	FLO
663	F-16A	FLO
664	F-16A	FLO $

NORWAY-POLAND

665	F-16A	FLO
666	F-16A	412th TW
667	F-16A	FLO
668	F-16A	FLO
669	F-16A	FLO
670	F-16A	FLO
671	F-16A	FLO $
672	F-16A	FLO
673	F-16A	FLO
674	F-16A	FLO
675	F-16A	FLO
677	F-16A	FLO
678	F-16A	FLO
680	F-16A	FLO
681	F-16A	FLO
682	F-16A	FLO
683	F-16A	FLO
686	F-16A	FLO $
687	F-16A	FLO
688	F-16A	FLO
689	F-16B	FLO
690	F-16B	FLO
691	F-16B	FLO
692	F-16B	FLO $
693	F-16B	FLO
711	F-16B	FLO

Lockheed
C-130J-30 Hercules II
335 Skv, Gardermoen
5601
5607
5629
5699

Lockheed P-3C Orion
333 Skv, Andøya
3296
3297
3298
3299

Lockheed P-3N Orion
333 Skv, Andøya
4576
6603

NH Industries
NH.90-NFH
337 Skv, Bardufoss
013
027
049
057
058
087   (on order)
171
212   (on order)
217   (on order)

Westland
Sea King Mk 43/
Mk 43A/Mk 43B
330 Skv:
  A Flt, Bodø;
  B Flt, Banak;
  C Flt, Ørland;
  D Flt, Sola

060	Mk 43
062	Mk 43
066	Mk 43
069	Mk 43
070	Mk 43
071	Mk 43B
072	Mk 43
073	Mk 43
074	Mk 43
189	Mk 43A
322	Mk 43B
329	Mk 43B
330	Mk 43B

OMAN
Royal Air Force of Oman
Airbus A.320-214X
4 Sqn, Seeb
554
555
556

Grumman
G.1159C Gulfstream IV
4 Sqn, Seeb
557
558

Lockheed
C-130H Hercules/
C-130J-30 Hercules II
16 Sqn, Seeb;
Royal Flt, Seeb

501	C-130H	16 Sqn
502	C-130H	16 Sqn
503	C-130H	16 Sqn
505	C-130J	16 Sqn
506	C-130J	16 Sqn
507	C-130J-30	(on order)
508	C-130J-30	(on order)
525	C-130J-30	Royal Flt

Omani Govt
Airbus A.319-115CJ
Govt of Oman, Seeb
A4O-AJ

Airbus A.320-233
Govt of Oman, Seeb
A4O-AA

Boeing 747-430/8H0*
Govt of Oman, Seeb
A4O-HMS*
A4O-OMN

Boeing 747-430
Govt of Oman, Seeb
A4O-OMN

Boeing 747SP-27
Govt of Oman, Seeb
A4O-SO

Gulfstream Aerospace G.550
Govt of Oman, Seeb
A4O-AD
A4O-AE

PAKISTAN
Pakistani Govt
Airbus A.310-304
12 VIP Communications Sqn, Islamabad
J-757

Cessna 560 Citation VI
12 VIP Communications Sqn, Islamabad
J-754

Gulfstream Aerospace
G.1159C Gulfstream IV-SP
12 VIP Communications Sqn, Islamabad
4270
J-755

Gulfstream Aerospace
G.450
12 VIP Communications Sqn, Islamabad
J-756

POLAND
Sily Powietrzne RP
CASA C-295M
13.eltr/8.BLTr, Kraków/Balice
011
012 $
013
014
015
016
017
018
020
021
022
023
024
025
026
027

Lockheed
C-130E Hercules
14.eltr/33.BLTr, Powidz
1501
1502
1503
1504
1505

Lockheed Martin (GD)
F-16C/F-16D* Fighting Falcon
3.elt/31.BLT, Poznan/Krzesiny;
6.elt/31.BLT, Poznan/Krzesiny;
10.elt/32.BLT, Lask

4040	6.elt
4041	6.elt
4042	3.elt
4043	3.elt
4044	3.elt
4045	3.elt
4046	3.elt
4047	3.elt
4048	3.elt
4049	3.elt
4050	3.elt
4051	3.elt
4052	6.elt
4053	6.elt
4054	6.elt
4055	6.elt $
4056	6.elt
4057	6.elt
4058	6.elt
4059	6.elt
4060	6.elt
4061	6.elt
4062	6.elt $
4063	10.elt
4064	10.elt
4065	10.elt
4066	10.elt
4067	10.elt
4068	10.elt
4069	10.elt
4070	10.elt
4071	10.elt
4072	10.elt
4073	10.elt
4074	10.elt
4075	10.elt
4076*	3.elt
4077*	3.elt
4078*	3.elt
4079*	3.elt
4080*	3.elt
4081*	3.elt
4082*	6.elt $
4083*	6.elt
4084*	6.elt
4085*	10.elt
4086*	10.elt
4087*	10.elt

Mikoyan MiG-29A/UB
1.elt/23.BLT, Minsk/Mazowiecki;
41.elt/22.BLT, Malbork

15*	1.elt $
28*	1.elt
38	1.elt
40	1.elt
42*	1.elt
54	1.elt
56	1.elt
59	1.elt
64*	
65	41.elt
66	41.elt
67	41.elt
70	41.elt
83	1.elt
89	1.elt
92	41.elt
105	1.elt
108	1.elt
111	1.elt
114	1.elt
115	1.elt
4101	41.elt
4103	41.elt
4104	41.elt
4105	1.elt
4110*	41.elt
4113	41.elt
4116	41.elt
4120	41.elt
4121	41.elt
4122	41.elt
4123*	41.elt

PZL 130TC-I/TC-II* Orlik
2.OSzL/42.BLSz, Radom

012
014
015
016
019
020
022
023
024
025
026
029
030
031*
032
033
035
036
037*
038*
040*
041*
042*
043*

044
045
046
047*
048*
049
050*
051*
052*

PZL M28B Bryza
1.OSzL/41.BLSz, Deblin;
13.eltr/8.BLTr, Kraków/Balice;
14.eltr/33.BLTr, Powidz

0204	M28B-TD	1.OSzL
0205	M28B-TD	1.OSzL
0206	M28B-TD	14.eltr
0207	M28B-TD	13.eltr
0208	M28B-TD	14.eltr
0209	M28B-TD	14.eltr
0210	M28B-TD	14.eltr
0211	M28B-TD	14.eltr
0212	M28B-TD	13.eltr
0213	M28B/PT	13.eltr
0214	M28B/PT	13.eltr
0215	M28B/PT	13.eltr
0216	M28B/PT	13.eltr
0217	M28B/PT	13.eltr
0218	M28B/PT	13.eltr
0219	M28B/PT	13.eltr
0220	M28B/PT	13.eltr
0221	M28B/PT	13.eltr
0222	M28B/PT	13.eltr
0223	M28B/PT	13.eltr
0224	M28B/PT	13.eltr
0225	M28B/PT	13.eltr

Sukhoi
Su-22UM-3K*/Su-22M-4
8.elt/21.BLT, Swidwin;
40.elt/21.BLT, Swidwin

305*	8.elt
308*	40.elt
310*	40.elt
508*	8.elt
509*	40.elt
707*	8.elt
3201	8.elt
3304	8.elt
3612	40.elt
3713	40.elt $
3715	40.elt
3816	40.elt
3817	40.elt
3819	8.elt
3920	40.elt
7411	40.elt
7412	8.elt
8101	40.elt
8102	8.elt
8103	8.elt
8205	40.elt

8308	8.elt $	
8309	40.elt	
8310	40.elt	
8715	40.elt	
8816	40.elt	
8818	8.elt $	
8919	8.elt $	
8920	40.elt	
9102	40.elt $	
9615	8.elt	
9616	40.elt	

Lotnictwo Marynarki Wojennej
Mil Mi-14
29.el/44.BLMW, Darlowo;

1001	Mi-14PL	29.el
1002	Mi-14PL	29.el
1003	Mi-14PL	29.el
1005	Mi-14PL	29.el
1008	Mi-14PL	29.el
1009	Mi-14PL	29.el
1010	Mi-14PL	29.el
1011	Mi-14PL	29.el
1012	Mi-14PL/R	29.el

PZL M28B Bryza
28.el/43.BLMW, Gdynia/Babie Doly;
30.el/44.BLMW, Cewice/Siemirowice

0404	M28B-1E	30.el
0405	M28B-1E	30.el
0723	M28B-TD	28.el
0810	M28B-1RM	30.el
1003	M28B-TD	28.el
1006	M28B-1R	30.el
1008	M28B-1R	30.el
1017	M28B-1R	30.el
1022	M28B-1R	30.el
1114	M28B-1R	30.el
1115	M28B-1R	30.el
1116	M28B-1R	30.el
1117	M28B-1	28.el
1118	M28B-1	28.el

Straz Graniczna (Polish Border Guard)
PZL M28-05 Skytruck
Base: Gdansk/Rebiechowo
SN-50YG
SN-60YG

Polish Government
Embraer
EMB.175-200LR
Polish Government, Warszawa
SP-LIG
SP-LIH

PORTUGAL
Força Aérea Portuguesa
Aérospatiale
TB-30 Epsilon
Esq 101, Sintra
11401
11402
11403
11404
11405 $
11406
11407
11409
11410 $
11411
11413
11414
11415
11416 $
11417
11418

CASA C-295M/
CASA C-295MPA*
Esq 502, Lisbon/Montijo
 (with a detachment at Lajes)
16701
16702
16703
16704
16705
16706
16707
16708*
16709*
16710*
16711*
16712*

D-BD Alpha Jet
Esq 103, Beja;
Asas de Portugal, Beja*
15202*
15206*
15208*
15211 $
15220*
15226
15227*
15230
15250

Dassault
Falcon 50
Esq 504, Lisbon/Montijo
17401
17402
17403

EHI EH-101
Mk514/Mk515/Mk516
Esq 751, Lisbon/Montijo
Mk514
19601
19602
19603
19604
19605*
19606*
Mk515
19607
19608
Mk516
19609
19610
19611
19612

Lockheed
C-130H/C-130H-30* Hercules
Esq 501, Lisbon/Montijo
16801*
16802*
16803 $
16804
16805
16806*

Lockheed Martin (GD)
F-16(MLU) Fighting Falcon
Esq 201, Monte Real;
Esq 301, Monte Real

15101	F-16A
15102	F-16A
15103	F-16A
15104	F-16A
15105	F-16A
15106	F-16A $
15107	F-16A
15108	F-16A
15109	F-16A
15110	F-16A
15112	F-16A
15113	F-16A
15114	F-16A
15115	F-16A
15116	F-16A
15117	F-16A
15118	F-16B
15119	F-16B
15120	F-16B
15121	F-16A
15122	F-16A
15123	F-16A
15124	F-16A
15125	F-16A
15126	F-16A
15127	F-16A
15128	F-16A
15129	F-16A
15130	F-16A

15131	F-16A
15132	F-16A
15133	F-16A
15134	F-16A
15135	F-16A
15136	F-16A $
15137	F-16B
15138	F-16A
15139	F-16B
15141	F-16A

Lockheed
P-3C Orion
Esq 601, Beja
14807
14808 $
14809
14810
14811

Marinha
Westland
Super Lynx Mk 95
Esq de Helicopteros,
  Lisbon/Montijo
19201
19202
19203
19204
19205

QATAR
Qatar Emiri Air Force
Boeing
C-17A Globemaster III
12 Transport Sqn, Al-Udeid

MAA	(08-0201)
MAB	(08-0202)
MAC	(10-0203)
MAE	(10-0204)

Lockheed
C-130J-30 Hercules II
12 Transport Sqn, Al-Udeid

MAH	(08-0211)
MAI	(08-0212)
MAJ	(08-0213)
MAK	(08-0214)

Qatar Government
Airbus A.310-308
Qatari Govt, Doha
A7-AFE

Airbus A.319-115X/
A.319CJ-133*
Qatari Govt, Doha
A7-HHJ*
A7-MED*
A7-MHH

Airbus A.320-232
Qatari Govt, Doha
A7-AAG
A7-MBK

Airbus A.330-202/-203*
Qatari Govt, Doha
A7-HHM*
A7-HJJ

Airbus A.340-211/-313X/-541
Qatari Govt, Doha

A7-AAH	A.340-313X
A7-HHH	A.340-541
A7-HHK	A.340-211

Boeing 747-8KB/-8KJ*
Qatari Govt, Doha
A7-HHE*
A7-HJA

Bombardier
Global Express
Qatari Govt/Qatar Airways, Doha
A7-AAM

ROMANIA
Fortele Aeriene Romania
Alenia C-27J Spartan
Escadrilla 902,
  Bucharest/Otopeni
2701
2702
2703
2704
2705
2706
2707

Lockheed
C-130B/C-130H* Hercules
Escadrilla 901,
  Bucharest/Otopeni
5927
5930
6150
6166*
6191*

RUSSIA
Voenno-Vozdushniye Sily Rossioki
Federatsii (Russian Air Force)
Antonov An-30/An-30B
(Open Skies)
Bases: Chkalovskiy & Kubinka

01 bl	An-30B	Chkalovskiy
RA-30078	An-30	Kubinka

Antonov An-124
224th Transport Regiment,
  Seshcha/Bryansk
RA-82011
RA-82021
RA-82023
RA-82025
RA-82028
RA-82030
RA-82032
RA-82035
RA-82036
RA-82038
RA-82039
RA-82040
RA-82041

Sukhoi Su-27
TsAGI, Gromov Flight
  Research Institute, Zhukhovsky

595 w	Su-27P
597 w	Su-30
598 w	Su-27P

Russian Govt
Ilyushin Il-62M
Rossiya Special Flight Detachment,
  Moscow/Vnukovo
RA-86466
RA-86467
RA-86468
RA-86536
RA-86540
RA-86559
RA-86561
RA-86710
RA-86712

Ilyushin
Il-96-300/
Il-96-300PU/
Il-96-300S/
Il-96-400T
Rossiya Special Flight Detachment,
  Moscow/Vnukovo

RA-96012	Il-96-300PU
RA-96016	Il-96-300PU
RA-96017	Il-96-300S
RA-96018	Il-96-300
RA-96019	Il-96-300
RA-96020	Il-96-300PU
RA-96021	Il-96-300PU
RA-96102	Il-96-400T
RA-96104	Il-96-400T

Tupolev Tu-134AK-3
Rossiya Special Flight Detachment,
  Moscow/Vnukovo
RA-65904

Tupolev Tu-154M
8 oae, Chalovskiy;
6991 AvB, Chalovskiy;
Rossiya Special Flight Detachment,
  Moscow/Vnukovo;
Yuri Gagarin Cosmonaut Training
  Centre (Open Skies), Chalovskiy

RA-85041	6991 AvB
RA-85155	8 oae
RA-85629	Rossiya
RA-85631	Rossiya
RA-85655	Open Skies
RA-85659	Rossiya
RA-85686	Rossiya
RA-85843	Rossiya

Tupolev Tu-204-300
Rossiya Special Flight Detachment,
  Moscow/Vnukovo
RA-64057
RA-64058

Tupolev Tu-214/Tu-214ON/Tu-214PU/
Tu-214SR/Tu-214SUS
(Open Skies), Chalovskiy;
Rossiya Special Flight Detachment,
  Ulyanovsk*;
Rossiya Special Flight Detachment,
  Moscow/Vnukovo

RA-64504	Tu-214	Rossiya
RA-64505	Tu-214	Rossiya
RA-64506	Tu-214	Rossiya
RA-64515	Tu-214SR	Rossiya
RA-64516	Tu-214SR	Rossiya
RA-64517	Tu-214SR	Rossiya
RA-64520	Tu-214PU	Rossiya
RA-64521	Tu-214	Rossiya*
RA-64522	Tu-214SUS	Rossiya
RA-64524	Tu-214SUS	Rossiya
RF-64519	Tu-214ON	Open Skies

SAUDI ARABIA
Al Quwwat al Jawwiya
as Sa'udiya
Airbus A.330-203 MRTT
24 Sqn, Al Kharj
2401
2402
2403
2404   (on order)
2405   (on order)
2406   (on order)

Airbus A.340-213X
RSAF, Riyadh
HZ-HMS2

BAe 125-800/-800B*
1 Sqn, Riyadh
HZ-105
HZ-109*
HZ-110*
HZ-130*

BAe Hawk 65/65A
88 Sqn, *Saudi Hawks*, Tabuk

8805	Hawk 65A
8806	Hawk 65A
8807	Hawk 65
8808	Hawk 65
8810	Hawk 65
8811	Hawk 65A
8812	Hawk 65A
8813	Hawk 65
8814	Hawk 65
8816	Hawk 65
8818	Hawk 65A
8820	Hawk 65

Boeing 737-7DP/-8DP*
1 Sqn, Riyadh
HZ-101
HZ-102*

Boeing
E-3A/KE-3A/RE-3A/RE-3B
Sentry
18 Sqn, Al Kharj;
19 Sqn, Al Kharj;
23 Sqn, Al Kharj

1801	E-3A	18 Sqn
1802	E-3A	18 Sqn
1803	E-3A	18 Sqn
1804	E-3A	18 Sqn
1805	E-3A	18 Sqn
1901	RE-3A	19 Sqn
1902	RE-3B	19 Sqn
2301	KE-3A	23 Sqn
2302	KE-3A	23 Sqn
2303	KE-3A	23 Sqn
2304	KE-3A	23 Sqn
2305	KE-3A	23 Sqn
2306	KE-3A	23 Sqn
2307	KE-3A	23 Sqn

Cessna 550 Citation II
1 Sqn, Riyadh
HZ-133
HZ-134
HZ-135
HZ-136

Grumman
G.1159C Gulfstream IV
1 Sqn, Riyadh
HZ-103

Lockheed
C-130/L.100 Hercules
1 Sqn, Riyadh;
4 Sqn, Jeddah;
16 Sqn, Jeddah;
32 Sqn, Al Kharj

111	VC-130H	1 Sqn
112	VC-130H	1 Sqn
464	C-130H	4 Sqn
465	C-130H	4 Sqn
466	C-130H	4 Sqn
467	C-130H	4 Sqn
468	C-130H	4 Sqn
472	C-130H	4 Sqn
473	C-130H	4 Sqn
474	C-130H	4 Sqn
475	C-130H	4 Sqn
477	C-130H	4 Sqn
478	C-130H	4 Sqn
482	C-130H	4 Sqn
483	C-130H	4 Sqn
484	C-130H	4 Sqn
485	C-130H	4 Sqn
486	C-130H	4 Sqn
1601	C-130H	16 Sqn
1602	C-130H	16 Sqn
1604	C-130H	16 Sqn
1605	C-130H	16 Sqn
1615	C-130H	16 Sqn
1622	C-130H-30	16 Sqn
1623	C-130H-30	16 Sqn
1624	C-130H	16 Sqn
1625	C-130H	16 Sqn
1626	C-130H	16 Sqn
1627	C-130H	16 Sqn
1628	C-130H	16 Sqn
1629	C-130H	16 Sqn
1630	C-130H-30	16 Sqn
1631	C-130H-30	16 Sqn
1632	L.100-30	16 Sqn
3201	KC-130H	32 Sqn
3202	KC-130H	32 Sqn
3203	KC-130H	32 Sqn
3204	KC-130H	32 Sqn
3205	KC-130H	32 Sqn
3206	KC-130H	32 Sqn
3207	KC-130H	32 Sqn
HZ-117	L.100-30	1 Sqn
HZ-128	L.100-30	1 Sqn
HZ-129	L.100-30	1 Sqn
HZ-132	L.100-30	1 Sqn

Saudi Govt
Boeing 747-3G1/468*
Saudi Royal Flight, Jeddah
HZ-HM1A
HZ-HM1*

Boeing 747SP-68
Saudi Govt, Jeddah;
Saudi Royal Flight, Jeddah
HZ-AIF   Govt
HZ-AIJ   Royal Flight
HZ-HM1B Royal Flight

Boeing 757-23A
Saudi Govt, Jeddah
HZ-HMED

Boeing MD-11
Saudi Royal Flight, Jeddah
HZ-AFAS
HZ-HM7

Bombardier Lear 60
Armed Forces Medical
  Services, Riyadh
HZ-MS1A
HZ-MS1B

Dassault Falcon 900
Saudi Govt, Jeddah
HZ-AFT
HZ-AFZ

Grumman
G.1159A Gulfstream III
Saudi Govt, Jeddah
HZ-AFN
HZ-AFR

Grumman
G.1159C Gulfstream IV/
Gulfstream IV-SP*
Armed Forces Medical
  Services, Riyadh;
Saudi Govt, Jeddah
HZ-AFU Govt
HZ-AFV Govt
HZ-AFW Govt
HZ-AFX Govt
HZ-AFY Govt
HZ-MFL Govt
HZ-MS4* AFMS

Gulfstream Aerospace
Gulfstream V
Armed Forces Medical
  Services, Riyadh
HZ-MS5A
HZ-MS5B

Lockheed
C-130H/L.100 Hercules
Armed Forces Medical
  Services, Riyadh
HZ-MS02 C-130H
HZ-MS06 L.100-30
HZ-MS07 C-130H
HZ-MS09 L.100-30

SERBIA
Dassault Falcon 50
Govt of Serbia, Belgrade
YU-BNA

SINGAPORE
Republic of Singapore Air Force
Boeing
KC-135R Stratotanker
112 Sqn, Changi
750
751
752
753

Lockheed
C-130 Hercules
122 Sqn, Paya Labar
720 KC-130B
721 KC-130B
724 KC-130B
725 KC-130B
730 C-130H
731 KC-130H
732 C-130H
733 C-130H
734 KC-130H
735 C-130H

SLOVAKIA
Slovenské Vojenske Letectvo
Aero L-39 Albatros
Zmiešané Letecké Kridlo
  'Otta Smika', Sliač:
  2 Stihacia Letka [SL]
1701 L-39ZAM
1730 L-39ZAM
4703 L-39ZAM
5251 L-39CM
5252 L-39CM
5253 L-39CM
5301 L-39CM $
5302 L-39CM

Antonov An-26
Dopravního Kridlo 'Generála
  Milana Ratislava Štefánika',
  Malacky:
  1 Dopravná Roj
2506
3208

LET 410 Turbolet
Dopravního Kridlo 'Generála
  Milana Ratislava Štefánika',
  Malacky:
  2 Dopravná Roj
1133 L-410T
1521 L-410FG
2311 L-410UVP
2421 L-410UVP
2718 L-401UVP-E
2721 L-401UVP-E
2818 L-401UVP-20
2901 L-401UVP-20

Mikoyan MiG-29AS/UBS
Zmiešané Letecké Kridlo
  'Otta Smika', Sliač:
  1 Stihacia Letka [SL]
0619 MiG-29AS
0921 MiG-29AS
1303 MiG-29UBS $
2123 MiG-29AS
3709 MiG-29AS
3911 MiG-29AS
5304 MiG-29UBS $
6124 MiG-29AS
6526 MiG-29AS
6627 MiG-29AS
6728 MiG-29AS

Mil M-17/M-17M*
Vrtulnikové Letecské Kridlo
  'Generálplukovníka Jána
  Ambrusa', Prešov:
  2 Bitevná Vrtul'nikova
  Letka
0807
0808
0812
0820
0821
0823*
0824
0826
0841
0842
0844
0845
0846
0847

Slovak Govt
Yakovlev Yak-40
Slovak Govt,
  Bratislava/Ivanka
OM-BYE
OM-BYL

SLOVENIA
Slovene Army
LET 410UVP-E
LTO, Brnik
L4-01

Pilatus PC-9/PC-9M*
1/2/3 OSBL, Cerklje
L9-51
L9-53
L9-61*
L9-62*
L9-63*
L9-64*
L9-65*
L9-66*
L9-67*
L9-68*
L9-69*

Slovenian Govt
Dassault Falcon 2000EX
Govt of Slovenia, Ljubjana
L1-01

SOUTH AFRICA
South African Air Force/
Suid Afrikaanse Lugmag
Boeing 737-7ED
21 Sqn, Waterkloof
ZS-RSA

Dassault Falcon 900
21 Sqn, Waterkloof
ZS-NAN

Lockheed
C-130B/C-130BZ* Hercules
28 Sqn, Waterkloof
401
402*
403
404
405*
406*

SOUTH KOREA
Republic of Korea Air Force
Boeing 747-4B5
296 Sqn, 35 Combined Group,
Seoul AB
10001

SPAIN
Ejército del Aire
Airbus A.310-304
Grupo 45, Torrejón
T.22-1   45-50
T.22-2   45-51

Airtech
CN.235M-10 (T.19A)/
CN.235M-100 (T.19B)/
CN.235M-100(MPA) (D.4)/
CN.235M VIGMA (T.19B)*
Grupo Esc, Matacán (74);
801 Esc, Palma/
   Son San Juan;
802 Esc, Gando/Las Palmas;
803 Esc, Cuatro Vientos;
Guardia Civil (09)

D.4-01	[T.19B-12]	(801 Esc)
D.4-02	[T.19B-09]	(801 Esc)
D.4-03	[T.19B-10]	(801 Esc)
D.4-04	[T.19B-08]	(802 Esc)
D.4-05	[T.19B-06]	(801 Esc)
D.4-06	[T.19B-05]	(801 Esc)
D.4-07	[T.19B-15]	(803 Esc)
D.4-08	[T.19B-14]	(802 Esc)
T.19A-01		403-01
T.19A-02		403-02
T.19B-07		74-25
T.19B-11		74-29
T.19B-13		74-31
T.19B-16		74-34
T.19B-17		74-17
T.19B-18		74-36
T.19B-19		74-19
T.19B-20		74-38
T.19B-21*		09-501
T.19B-22*		09-502

Boeing 707
47 Grupo Mixto, Torrejón

TK.17-1	331B	47-01
T.17-2	331B	47-02
T.17-3	368C	47-03
TM.17-4	351C	47-04

CASA 101EB Aviojet
Grupo 54, Torrejón;
Grupo de Escuelas de
   Matacán (74);
AGA, San Javier (79);
Patrulla Aguila, San Javier*

E.25-05	79-05
E.25-06	79-06 [6]*
E.25-08	79-08
E.25-09	79-09
E.25-10	79-20
E.25-11	79-11
E.25-12	79-12
E.25-13	79-13
E.25-14	79-14 [7]*
E.25-15	79-15
E.25-16	79-16
E.25-17	74-40
E.25-19	79-19
E.25-20	79-20
E.25-21	79-21
E.25-22	79-22 [2]*
E.25-23	79-23

E.25-24	
E.25-25	79-25
E.25-26	79-26
E.25-27	79-27
E.25-29	74-45
E.25-31	79-31 [3]*
E.25-34	74-44
E.25-35	54-20
E.25-37	79-37
E.25-38	79-38
E.25-40	79-40
E.25-41	74-41
E.25-43	74-43
E.25-44	79-44
E.25-45	79-45
E.25-46	79-46
E.25-50	79-33
E.25-51	74-07
E.25-52	79-34 [4]*
E.25-53	74-09
E.25-54	79-35
E.25-55	54-21
E.25-56	74-11
E.25-57	74-12
E.25-59	74-13
E.25-61	54-22
E.25-62	79-17
E.25-63	74-17 [1]*
E.25-65	79-95
E.25-66	74-20
E.25-68	74-22
E.25-69	79-97
E.25-72	74-26
E.25-73	79-98
E.25-74	74-28
E.25-76	74-30
E.25-78	79-02 [8]*
E.25-80	79-03
E.25-81	74-34
E.25-83	74-35
E.25-84	79-04
E.25-86	79-32
E.25-87	79-29$
E.25-88	74-39

CASA 212 Aviocar
212A (T.12B)/
212ECM (TM.12D)/
212-200 (T.12D)/
212-200 (TR.12D)
Ala 72, Alcantarilla;
47 Grupo Mixto, Torrejón

T.12B-13	72-01
T.12B-49	72-07
T.12B-55	72-08
T.12B-63	72-14
T.12B-65	72-11
T.12B-66	72-09
T.12B-67	72-12
T.12B-69	72-15
T.12B-70	72-17
T.12B-71	72-10
TM.12D-72	47-12
T.12D-75	47-14
TR.12D-76	72-21
TR.12D-77	72-22
TR.12D-79	72-23
TR.12D-81	72-24

CASA C-295M
Ala 35, Getafe

T.21-01	35-39
T.21-02	35-40
T.21-03	35-41
T.21-04	35-42
T.21-05	35-43
T.21-06	35-44
T.21-07	35-45
T.21-08	35-46
T.21-09	35-47
T.21-10	35-48
T.21-11	35-49
T.21-12	35-50
T.21-13	35-51

Cessna 560 Citation VI
403 Esc, Getafe

TR.20-01	403-11
TR.20-02	403-12
TR.20-03	403-21

Dassault Falcon 20D/E
47 Grupo Mixto, Torrejón

TM.11-1	20E	47-21
TM.11-2	20D	47-22
TM.11-3	20D	47-23
TM.11-4	20E	47-24

Dassault Falcon 900/900B*
Grupo 45, Torrejón

T.18-1	45-40
T.18-2	45-41
T.18-3*	45-42
T.18-4*	45-43
T.18-5*	45-44

Eurocopter
EC.120B Colibri
Ala 78, Granada;
*Patrulla Aspa, Granada**

HE.25-1	78-20*
HE.25-2	78-21*
HE.25-3	78-22*
HE.25-4	78-23*
HE.25-5	78-24*
HE.25-6	78-25*
HE.25-7	78-26*
HE.25-8	78-27*
HE.25-9	78-28*
HE.25-10	78-29
HE.25-11	78-30
HE.25-12	78-31*
HE.25-13	78-32
HE.25-14	78-33*
HE.25-15	78-34*

Eurofighter
EF.2000/EF.2000(T)* Tifón
NB: Where serials have been allocated using the new system, the old 'last two' have been added in brackets e.g. C.16-10001 would have been C.16-52.
Ala 11, Morón;
Ala 14, Albacete;
CASA, Getafe

CE.16-01*	11-70
CE.16-02*	11-71
CE.16-03*	11-72
CE.16-04*	11-73
CE.16-05*	11-74
CE.16-06*	11-75
CE.16-07*	11-76
CE.16-09*	11-78
CE.16-10*	11-79
CE.16-11*	14-70
C.16-20	11-91
C.16-21	11-01
C.16-22	11-02
C.16-23	11-03
C.16-24	11-04
C.16-25	11-05
C.16-26	11-06
C.16-27	11-07
C.16-28	11-08
C.16-29	11-09
C.16-30	11-10
C.16-31	14-01
C.16-32	11-12
C.16-33	11-13
C.16-35	14-02
C.16-36	14-03
C.16-37	14-04
C.16-38	14-05
C.16-39	14-06
C.16-40	11-14
C.16-41	14-07
C.16-42	14-08
C.16-43	11-15
C.16-44	14-09
C.16-45	14-10
C.16-46	11-16
C.16-47	14-11
C.16-48	14-12
C.16-49	14-13
C.16-50	14-14
C.16-51	11-17
CE.16-10000*(12)	11-81
C.16-10001(52)	14-
C.16-10002(53)	14-
C.16-10003(54)	14-
C.16-10004(55)	14-
CE.16-10005*(13)	
C.16-10007(56)	
C.16-10012(57)	14-
C.16-10015*(14)	
C.16-100..(58)	
C.16-100..(59)	
C.16-10040(60)	

Lockheed
C-130H/C-130H-30/
KC-130H Hercules
311 Esc/312 Esc (Ala 31),
Zaragoza

TL.10-01	C-130H-30	31-01
T.10-03	C-130H	31-03
T.10-04	C-130H	31-04
TK.10-05	KC-130H	31-50
TK.10-06	KC-130H	31-51
TK.10-07	KC-130H	31-52
T.10-8	C-130H	31-05
T.10-9	C-130H	31-06
T.10-10	C-130H	31-07
TK.10-11	KC-130H	31-53
TK.10-12	KC-130H	31-54

Lockheed
P-3A/P-3M Orion
Grupo 22, Morón

P.3A-01	P-3A	22-21
P.3M-08	P-3M	22-31
P.3M-09	P-3M	22-32
P.3M-12	P-3M	22-35

McDonnell Douglas
F-18 Hornet
Ala 12, Torrejón;
Ala 15, Zaragoza;
Esc 462, Gran Canaria
EF-18BM Hornet

CE.15-01	15-70$
CE.15-2	15-71
CE.15-03	15-72
CE.15-04	15-73
CE.15-5	15-74
CE.15-06	15-75
CE.15-07	15-76
CE.15-08	12-71
CE.15-9	15-77
CE.15-10	12-73$
CE.15-11	12-74
CE.15-12	12-75

EF-18A+/EF-18AM* Hornet

C.15-13	12-01
C.15-14	15-01
C.15-15	15-02$
C.15-16	15-03
C.15-18	15-05
C.15-20	15-07
C.15-21	15-08
C.15-22	15-09
C.15-23	15-10
C.15-24	15-11
C.15-25	15-12
C.15-26	15-13$
C.15-27	15-14
C.15-28	15-15
C.15-29	15-16
C.15-30	15-17
C.15-31	15-18
C.15-32	15-19
C.15-33	15-20
C.15-34	12-50$
C.15-35	15-22
C.15-36	15-23
C.15-37	15-24
C.15-38	15-25
C.15-39	15-26
C.15-40	15-27
C.15-41	15-28$
C.15-43	15-30
C.15-44	12-02
C.15-45	12-03
C.15-46	12-04
C.15-47	15-31
C.15-48	12-06
C.15-49	12-07
C.15-50	12-08
C.15-51	12-09
C.15-52	12-10
C.15-53	12-11
C.15-54	12-12
C.15-55	12-13
C.15-56	12-14
C.15-57	12-15
C.15-59	12-17
C.15-60	12-18
C.15-61	12-19
C.15-62	12-20
C.15-64	15-34
C.15-65	12-23
C.15-66	12-24
C.15-67	15-33
C.15-68	12-26
C.15-69	12-27
C.15-70	12-28
C.15-72	12-30

F/A-18A Hornet

C.15-73	46-01$
C.15-75	46-03
C.15-77	46-05
C.15-79	46-07
C.15-80	46-08
C.15-81	46-09

C.15-82	46-10
C.15-83	46-11
C.15-84	46-12
C.15-85	46-13
C.15-86	46-14
C.15-87	46-15
C.15-88	46-16
C.15-89	46-17
C.15-90	46-18
C.15-92	46-20
C.15-93	46-21
C.15-94	46-22
C.15-95	46-23
C.15-96	46-24

Arma Aérea de l'Armada Española
BAe/McDonnell Douglas
EAV-8B/EAV-8B+/
TAV-8B Harrier II
Esc 009, Rota

EAV-8B

VA.1A-15	01-903
VA.1A-19	01-907
VA.1A-20	01-909
VA.1A-22	01-911

EAV-8B+

VA.1B-24	01-914
VA.1B-25	01-915
VA.1B-26	01-916
VA.1B-27	01-917
VA.1B-28	01-918
VA.1B-29	01-919
VA.1B-30	01-920
VA.1B-35	01-923
VA.1B-36	01-924
VA.1B-37	01-925
VA.1B-38	01-926
VA.1B-39	01-927

TAV-8B

VA.1B-33	01-922

Cessna 550 Citation II
Esc 004, Rota

U.20-1	01-405
U.20-2	01-406
U.20-3	01-407

Cessna 650 Citation VII
Esc 004, Rota

U.21-01	01-408

SUDAN
Dassault Falcon 50
Sudanese Govt, Khartoum
ST-PSR

Dassault Falcon 900B
Sudanese Govt, Khartoum
ST-PSA

SWEDEN
Svenska Flygvapnet
Grumman
G.1159C Gulfstream 4
(Tp.102A/S.102B Korpen/
Tp.102C)
Flottiljer 17M,
Stockholm/Bromma &
Malmslätt

Tp.102A

102001	021

S.102B Korpen

102002	022
102003	023

Tp.102C

102004	024

Gulfstream Aerospace G.550
(Tp.102D)
Flottiljer 17M,
Stockholm/Bromma &
Malmslätt

102005	025

Lockheed
C-130H Hercules (Tp.84)
Flottiljer 7, Såtenäs

84002	842
84004	844
84005	845
84006	846
84007	847
84008	848

Rockwell
Sabreliner-40 (Tp.86)
FMV, Malmslätt

86001	861

SAAB JAS 39 Gripen
Flottiljer 4, Östersund/
Frösön;
Flottiljer 7, Såtenäs [G];
Flottiljer 17, Ronneby/
Kallinge;
Flottiljer 21, Luleå/
Kallax
FMV, Malmslätt

JAS 39A

39131	131$	
39132	132	
39133	133	FMV
39135	135	F21
39136	136	
39138	138	
39144	44	SAAB
39146	146	
39150	150	
39151	151	
39159	159	
39167	167	
39168	168	

39170	170	F17
39171	171	F17
39172	172	
39174	174	
39176	176	F17
39179	179	F17
39180	180	F17
39181	181	
39182	182	
39183	183	F17
39185	185	
39188	188	F17
39189	189	F17
39190	190	F17
39191	191	F17
39192	192	
39193	193	
39194	194	
39195	195	
39196	196	
39198	198	
39199	199	
39200	200	
39201	201	
39203	203	
39205	205	
39206	206	
JAS 39B		
39800	58	FMV
39801	801	
39802	802	SAAB
39804	804	
39806	806	
39807	807	
39808	808	
39809	809	
39810	810	
39811	811	
39812	812	
39813	813	
39814	814	
JAS 39C		
39-6	6	SAAB
39208	208	F17
39209	209	F7
39210	210	F17
39211	211	F7
39212	212	F17
39213	213	F7
39214	214	F7
39215	215	F21
39216	216	F17
39217	217	F7
39218	218	F21
39219	219	F21
39220	220	F7
39221	221	F17
39222	222	F17
39223	223	F21
39224	224	F17
39225	225	F7
39226	226	F17
39227	227	F7
39228	228	F17
39229	229	F21
39230	230	F17
39231	231	F7
39232	232	F21
39233	233	F17
39246	246	F21
39247	247	F17
39248	248	F21
39249	249	F17
39250	250	FMV
39251	251	FMV
39252	252	F17
39253	253	F21
39254	254	F17
39255	255	F17
39256	256	F17
39257	257	F21
39258	258	F21
39260	260	F21
39261	261	F17
39262	262	F17
39263	263	F21
39264	264	F17
39265	265	F21
39266	266	F17
39267	267	F21
39268	268	F7
39269	269	F17
39270	270	F17
39271	271	F21
39272	272	F7
39273	273	SAAB
39274	274	FMV
39275	275	F7
39276	276	F21
39277	277	F7
39278	278	F17
39279	279	F21
39280	280	F21
39281	281	F17
39282	282	F17
39283	283	F21
39284	284	F17
39285	285	F21
39286	286	F21
39287	287	F17
39288	288	F21
39289	289	F17
39290		
39291		
39292		
39293		
39294		
JAS 39D		
39815	815	F21
39816	816	F17
39817	817	F21
39818	818	
39821	821	F21
39822	822	SAAB
39823	823	F21
39824	824	F17
39825	825	F7
39826	826	F17
39827	827	SAAB
39829	829	FMV
39830	830	F7
39831	831	F17
39832	832	FMV
39833	833	F21
39834	834	F7
39835	835	F7
39836	836	F17
39837	837	F21
39838	838	
39839	839	
39840	840	F17
39841	841	F7
39842		

JAS 39E

39-8	SAAB	

JAS 39NG

39-7	SAAB	

SAAB
SF.340 (OS.100 & Tp.100C)/
SF.340AEW&C (S.100D)
Argus
Flottiljer 17M, Malmslätt;
TSFE, Malmslätt
OS.100

100001	001	F17M

Tp.100C

100008	008	TSFE

S.100D

100003	003	TSFE
100004	004	TSFE

Förvarsmaktens Helikopterflottilj
Aérospatiale
AS.332M-1 Super Puma
(Hkp.10/Hkp.10B[1]/Hkp.10D[2])
1.HkpSkv, Luleå/Kallax;
3.HkpSkv, Berga, Goteborg/Säve,
  & Ronneby/Kallinge

10401	91	3.HkpSkv
10402[2]	92	3.HkpSkv
10403[1]	93	1.HkpSkv
10405	95	3.HkpSkv
10406[1]	96	1.HkpSkv
10407[2]	97	3.HkpSkv
10408[1]	98	1.HkpSkv
10410[1]	90	3.HkpSkv
10411	88	3.HkpSkv
10412	89	3.HkpSkv

Agusta
A109LUH Power (Hkp.15)
2.HkpSkv, Linkoping/Malmen;
3.HkpSkv, Berga & Ronneby/Kallinge
Hkp.15A

15021	21	2.HkpSkv
15022	22	2.HkpSkv
15023	23	2.HkpSkv
15024	24	2.HkpSkv
15025	25	2.HkpSkv
15026	26	2.HkpSkv
15027	27	2.HkpSkv
15028	28	2.HkpSkv
15029	29	2.HkpSkv
15030	30	2.HkpSkv
15031	31	2.HkpSkv
15032	32	2.HkpSkv
15034	34	2.HkpSkv

Hkp.15B

15033	33	3.HkpSkv
15035	35	3.HkpSkv
15036	36	3.HkpSkv
15037	37	3.HkpSkv
15038	38	3.HkpSkv
15039	39	3.HkpSkv
15040	40	3.HkpSkv

NH Industries
NH.90
1.HkpSkv, Luleå/Kallax;
2.HkpSkv, Linkoping/Malmen;
3.HkpSkv, Berga & Ronneby/Kallinge;
FMV, Malmslätt
NH.90-TTH (Hkp.14A/Hkp.14B)/
NH.90-TTT (Hkp.14D)

141041	41	Hkp.14A	Eurocopter
142042	42	Hkp.14B	2.HkpSkv
142043	43	Hkp.14B	2.HkpSkv
142044	44	Hkp.14B	1.HkpSkv
142045	45	Hkp.14B	2.HkpSkv
141046	46	Hkp.14A	FMV
141047	47	Hkp.14A	2.HkpSkv
144048	48	Hkp.14D	(on order)
144049	49	Hkp.14D	(on order)
144050	50	Hkp.14D	Eurocopter
144051	51	Hkp.14D	3.HkpSkv
141052	52	Hkp.14D	(on order)
141053	53	Hkp.14D	(on order)
144054	54	Hkp.14D	3.HkpSkv
144055	55	Hkp.14D	(on order)
144056	56	Hkp.14D	(on order)
144057	57	Hkp.14D	(on order)
144058	58	Hkp.14D	(on order)

Sikorsky UH-60M
Black Hawk (Hkp.16A)
2.HkpSkv, Linkoping/Malmen;
FMV, Malmslätt

161226	01	FMV
161227	02	2.HkpSkv
161228	03	2.HkpSkv
161229	04	2.HkpSkv
161230	05	2.HkpSkv
161231	06	2.HkpSkv
161232	07	2.HkpSkv
161233	08	2.HkpSkv
161234	09	2.HkpSkv
161235	10	2.HkpSkv
161236	11	2.HkpSkv
161237	12	2.HkpSkv
161238	13	2.HkpSkv
161239	14	2.HkpSkv
161240	15	2.HkpSkv

Swedish Coast Guard
Bombardier DHC-8Q-311
Base: Nykoping

SE-MAA	[501]
SE-MAB	[502]
SE-MAC	[503]

SWITZERLAND
Schweizerische Flugwaffe
(Most aircraft are pooled centrally. Some
carry unit badges but these rarely indicate
actual operators.)
Aérospatiale
AS.332M-1/AS.532UL
Super Puma
Lufttransport Staffel 3
 (LtSt 3), Dübendorf;
Lufttransport Staffel 5
 (LtSt 5), Payerne;
Lufttransport Staffel 6
 (LtSt 6), Alpnach;
Lufttransport Staffel 8
 (LtSt 8), Alpnach
Detachments at Emmen,
 Meiringen & Sion
AS.332M-1

T-311
T-312
T-313
T-314
T-315
T-316 $
T-317
T-318
T-319
T-320
T-321
T-322
T-323
T-324
T-325

AS.532UL

T-331
T-332
T-333
T-334
T-335
T-336
T-337
T-338
T-339
T-340
T-342

Aurora Flight Services
Centaur OPA
Armasuisse, Emmen
R-711

Beech 1900D
Lufttransportdienst des
 Bundes, Dübendorf
T-729

Beech
King Air 350C
Lufttransportdienst des
 Bundes, Dübendorf
T-721

Cessna
560XL Citation Excel
Lufttransportdienst des
 Bundes, Dübendorf
T-784

Dassault Falcon 900EX
Lufttransportdienst des
 Bundes, Dübendorf
T-785

Eurocopter
EC.135P-2*/EC.635P-2
Lufttransportdienst des
   Bundes, Dübendorf (LTDB);
Lufttransport Geschwader 2
   (LTG 2), Alpnach;
Lufttransport Staffel 3
   (LtSt 3), Dübendorf;
Lufttransport Staffel 5
   (LtSt 5), Payerne;
Lufttransport Staffel 6
   (LtSt 6), Alpnach;
Lufttransport Staffel 8
   (LtSt 8), Alpnach
Detachments at Emmen,
   Meiringen & Sion

T-351*	LTDB
T-352*	LTDB
T-353	LTG 2
T-354	LTG 2
T-355	LTG 2
T-356	LTG 2
T-357	LTG 2
T-358	LTG 2
T-359	LTG 2
T-360	LTG 2
T-361	LTG 2
T-362	LTG 2
T-363	LTG 2
T-364	LTG 2
T-365	LTG 2
T-366	LTG 2
T-367	LTG 2
T-368	LTG 2
T-369	LTG 2
T-370	LTG 2

McDonnell Douglas
F/A-18 Hornet
Flieger Staffel 11 (FlSt 11),
   Meiringen;
Escadrille d'Aviation 17 (EdAv 17),
   Payerne;
Flieger Staffel 18 (FlSt 18),
   Payerne
F/A-18C
J-5001
J-5002
J-5003
J-5004
J-5005
J-5006
J-5007
J-5008
J-5009
J-5010
J-5011 $
J-5012
J-5013
J-5014 $
J-5015
J-5016

J-5017 $
J-5018 $
J-5019
J-5020
J-5021
J-5022
J-5023
J-5024
J-5025
J-5026
F/A-18D
J-5232
J-5233
J-5234
J-5235
J-5236
J-5238

Northrop F-5 Tiger II
Armasuisse, Emmen;
Escadrille d'Aviation 6 (EdAv 6),
   Sion;
Flieger Staffel 8 (FlSt 8),
   Meiringen;
Flieger Staffel 19 (FlSt 19),
   Sion;
Patrouille Suisse, Emmen (*P. Suisse*)
F-5E
J-3005
J-3014
J-3015
J-3030
J-3033
J-3036
J-3038 $
J-3041
J-3044
J-3056
J-3057
J-3062
J-3063
J-3065
J-3067
J-3068
J-3069
J-3070
J-3072
J-3073
J-3074
J-3076
J-3077
J-3079
J-3080
J-3081   *P. Suisse*
J-3082   *P. Suisse*
J-3083   *P. Suisse*
J-3084   *P. Suisse*
J-3085   *P. Suisse*
J-3086   *P. Suisse*
J-3087   *P. Suisse*
J-3088   *P. Suisse*
J-3089

J-3090   *P. Suisse*
J-3091   *P. Suisse*
J-3092
J-3093
J-3094
J-3095
J-3097
F-5F
J-3201
J-3202
J-3203
J-3204
J-3206
J-3207
J-3210
J-3211
J-3212

Pilatus
PC.6B/B2-H2 Turbo Porter
Lufttransport Staffel 7
   (LtSt 7), Emmen
V-612
V-613
V-614
V-616
V-617
V-618
V-619
V-620
V-622 $
V-623
V-631
V-632
V-633
V-634
V-635

Pilatus NCPC-7
Turbo Trainer
Instrumentation Flieger
   Staffel 14 (InstruFlSt 14),
   Dübendorf;
Pilotenschule, Emmen
A-912
A-913
A-914
A-915
A-916
A-917
A-918
A-919
A-922
A-923
A-924
A-925
A-926
A-927
A-928
A-929
A-930
A-931

A-932
A-933
A-934
A-935
A-936
A-937
A-938
A-939
A-940
A-941

Pilatus PC-9
Zielfliegerstaffel 12,
   Sion
C-401
C-402
C-403
C-405
C-406
C-407
C-408
C-409
C-410
C-411
C-412

Pilatus PC-21
Pilotenschule, Emmen
A-101
A-102
A-103
A-104
A-105
A-106
A-107
A-108

Swiss Govt
Pilatus PC-12/45
Swiss Govt, Emmen
HB-FOG

SYRIA
Dassault Falcon 900
Govt of Syria, Damascus
YK-ASC

TANZANIA
Airbus A.340-542
Tanzanian Govt, Dar-es-Salaam
5H-...   (on order)

Gulfstream Aerospace G.550
Tanzanian Govt, Dar-es-Salaam
5H-ONE

THAILAND
Airbus A.319CJ-133
Royal Flight, Bangkok
B.L15-1/47 (HS-TYR) [60221]

Airbus A.320CJ-214
6 Wing, Don Muang
HS-TYT   60203

Boeing 737-448/-4Z6/-8Z6
Royal Flight, Bangkok
HS-CMV   4Z6   [11-111, 90401]
HS-HRH   448   [99-999, 90409]
HS-TYS   8Z6   [55-555]

TUNISIA
Boeing 737-7HJ
Govt of Tunisia, Tunis
TS-IOO

TURKEY
Türk Hava Kuvvetleri
Airbus Military A.400M
221 Filo, Erkilet
13-0009
14-0013
1.-....   (on order)
1.-....   (on order)
1.-....   (on order)
1.-....   (on order)
1.-....   (on order)
1.-....   (on order)
1.-....   (on order)
1.-....   (on order)

Boeing 737-7FS AEW&C
131 Filo, Konya
13-001
13-002
13-003
..-004   (on order)

Boeing
KC-135R Stratotanker
101 Filo, Incirlik
57-2609
58-0110
60-0325
60-0326
62-3539
62-3563
62-3567

Canadair
NF-5A-2000/NF-5B-2000
Freedom Fighter
134 Filo, *Turkish Stars*, Konya
NF-5A-2000
3004   [3]
3023   [6]
3025   [5]
3027
3032
3036
3039
3046
3048

3049   [8]
3052
3058
3066
3072   [7]
NF-5B-2000
4005
4009
4020   [1]
4021   [0]

Cessna 650 Citation VII
212 Filo, Ankara/Etimesgut
004
93-005

Grumman
G.1159C Gulfstream IV
211 Filo, Ankara/Etimesgut
91-003
TC-ATA/91-002
TC-GAP/001

Gulfstream Aerospace G.550
212 Filo, Ankara/Etimesgut
09-001
TC-CBK

Lockheed
C-130E Hercules
222 Filo, Erkilet
63-3186
63-3187
63-3188
63-3189
65-0451
67-0455
68-1606
68-1608
68-1609
70-1610
70-1947
71-1468
73-0991 $

Transall C-160D
221 Filo, Erkilet
68-020
68-023
69-019
69-021
69-024
69-026
69-027
69-029
69-031
69-032
69-033 $
69-034
69-035
69-036
69-040 $

TUSAS-GD F-16C/F-16D*		87-0014	Öncel Filo	90-0018	161 Filo
Fighting Falcon		87-0015	Öncel Filo	90-0019	161 Filo
3 AJEÜ, Konya:		87-0016	Öncel Filo	90-0020	162 Filo
132 Filo;		87-0017	152 Filo	90-0021	161 Filo
4 AJÜ, Akinci:		87-0018	152 Filo	90-0022*	161 Filo
141 Filo, 142 Filo		87-0019	Öncel Filo	90-0023*	161 Filo
& 143/Öncel Filo;		87-0020	152 Filo	90-0024*	161 Filo
5 AJÜ, Merzifon:		87-0021	Öncel Filo	91-0001	161 Filo
151 Filo & 152 Filo;		88-0013*	152 Filo	91-0002	161 Filo
6 AJÜ, Bandirma:		88-0014*	141 Filo	91-0003	161 Filo
161 Filo & 162 Filo;		88-0015*	141 Filo	91-0004	161 Filo
8 AJÜ, Diyarbakir:		88-0019	Öncel Filo	91-0005	161 Filo
181 Filo & 182 Filo;		88-0020	Öncel Filo	91-0006	162 Filo
9 AJÜ, Balikesir:		88-0021	142 Filo	91-0007	161 Filo
191 Filo & 192 Filo		88-0024	152 Filo	91-0008	141 Filo
07-1001	142 Filo	88-0025	142 Filo	91-0010	161 Filo
07-1002	142 Filo	88-0026	142 Filo	91-0011	141 Filo $
07-1003	141 Filo	88-0027	142 Filo	91-0012	141 Filo
07-1004	142 Filo	88-0028	191 Filo	91-0013	192 Filo
07-1005		88-0029	142 Filo	91-0014	141 Filo
07-1006	142 Filo	88-0030	142 Filo	91-0015	182 Filo
07-1007	142 Filo	88-0031	142 Filo	91-0016	182 Filo
07-1008		88-0032	142 Filo	91-0017	182 Filo
07-1009	142 Filo	88-0033	141 Filo	91-0018	182 Filo
07-1010	142 Filo	88-0034	141 Filo	91-0020	162 Filo
07-1011		88-0035	141 Filo	91-0022*	161 Filo
07-1012		88-0036	141 Filo	91-0024*	161 Filo
07-1013		88-0037	141 Filo	92-0001	182 Filo
07-1014		89-0022	141 Filo	92-0002	182 Filo
07-1015*		89-0023	141 Filo	92-0003	162 Filo
07-1016*		89-0024	141 Filo	92-0004	182 Filo
07-1017*		89-0025	141 Filo	92-0005	191 Filo
07-1018*	141 Filo	89-0026	141 Filo	92-0006	182 Filo
07-1019*		89-0027	141 Filo	92-0007	182 Filo
07-1020*	142 Filo	89-0028	141 Filo	92-0008	152 Filo
07-1021*		89-0030	141 Filo	92-0009	191 Filo
07-1022*	142 Filo	89-0031	141 Filo	92-0010	182 Filo
07-1023*	142 Filo	89-0034	162 Filo	92-0011	182 Filo
07-1024*	142 Filo	89-0035	162 Filo	92-0012	182 Filo
07-1025*		89-0036	141 Filo	92-0013	161 Filo
07-1026*	142 Filo	89-0037	162 Filo	92-0014	182 Filo
07-1027*	142 Filo	89-0038	162 Filo	92-0015	161 Filo
07-1028*	142 Filo	89-0039	152 Filo	92-0016	161 Filo
07-1029*	142 Filo	89-0040	162 Filo	92-0017	152 Filo
07-1030*		89-0041	161 Filo	92-0018	161 Filo
86-0066	Öncel Filo	89-0042*	162 Filo	92-0019	181 Filo
86-0068	Öncel Filo	89-0043*	162 Filo	92-0020	181 Filo
86-0069	Öncel Filo	89-0044*	162 Filo	92-0021	181 Filo
86-0070	Öncel Filo $	89-0045*	182 Filo	92-0022*	181 Filo
86-0071	152 Filo	90-0004	152 Filo	92-0023*	181 Filo
86-0072	Öncel Filo	90-0005	162 Filo	92-0024*	141 Filo
86-0192*	Öncel Filo	90-0006	182 Filo	93-0001	181 Filo
86-0193*	Öncel Filo	90-0007	162 Filo	93-0003	181 Filo
86-0194*	Öncel Filo	90-0008	162 Filo	93-0004	181 Filo
86-0195*	Öncel Filo	90-0009	162 Filo	93-0005	181 Filo
86-0196*	Öncel Filo	90-0010	161 Filo	93-0006	181 Filo
87-0002*	152 Filo	90-0011	141 Filo $	93-0007	181 Filo
87-0003*	Öncel Filo	90-0012	161 Filo	93-0008	181 Filo
87-0009	Öncel Filo	90-0013	162 Filo	93-0009	181 Filo
87-0010	Öncel Filo	90-0014	134 Filo $	93-0010	132 Filo
87-0011	Öncel Filo	90-0016	161 Filo	93-0011	181 Filo
87-0013	Öncel Filo $	90-0017	161 Filo	93-0012	181 Filo

93-0013	181 Filo
93-0658	192 Filo
93-0659	191 Filo
93-0660	152 Filo
93-0661	151 Filo
93-0663	151 Filo
93-0664	191 Filo
93-0665	191 Filo
93-0667	191 Filo
93-0668	141 Filo
93-0669	151 Filo
93-0670	191 Filo
93-0671	191 Filo
93-0672	132 Filo
93-0673	132 Filo
93-0674	191 Filo
93-0675	191 Filo
93-0676	192 Filo
93-0677	192 Filo $
93-0678	152 Filo
93-0679	192 Filo
93-0680	192 Filo $
93-0681	192 Filo
93-0682	192 Filo $
93-0683	192 Filo
93-0684	141 Filo
93-0685	192 Filo
93-0687	192 Filo
93-0688	151 Filo
93-0689	192 Filo
93-0690	192 Filo
93-0691*	192 Filo $
93-0692*	151 Filo $
93-0693*	152 Filo
93-0694*	151 Filo
93-0695*	141 Filo
93-0696*	151 Filo $
94-0071	192 Filo
94-0072	191 Filo
94-0073	151 Filo
94-0074	191 Filo
94-0075	191 Filo
94-0076	191 Filo
94-0077	191 Filo
94-0078	191 Filo
94-0079	191 Filo
94-0080	191 Filo
94-0082	151 Filo
94-0083	151 Filo
94-0084	191 Filo
94-0085	191 Filo
94-0086	191 Filo
94-0088	152 Filo
94-0089	152 Filo
94-0090	192 Filo $
94-0091	132 Filo
94-0092	132 Filo
94-0093	192 Filo
94-0094	151 Filo
94-0095	192 Filo
94-0096	192 Filo
94-0105*	141 Filo

94-0106*	191 Filo
94-0108*	191 Filo
94-0109*	191 Filo
94-0110*	151 Filo
94-1557*	152 Filo
94-1558*	132 Filo
94-1559*	152 Filo
94-1560*	192 Filo
94-1561*	192 Filo
94-1562*	192 Filo
94-1563*	192 Filo
94-1564*	192 Filo

Turkish Govt
Airbus A.319CJ-115X
Turkish Govt, Ankara
TC-ANA

Airbus A.330-243
Turkish Govt, Ankara
TC-TUR

Gulfstream Aerospace G.550
Turkish Govt, Ankara
TC-DAP

TURKMENISTAN
BAe 1000B
Govt of Turkmenistan,
  Ashkhabad
EZ-B021

Boeing 757-23A
Govt of Turkmenistan,
  Ashkhabad
EZ-A010

Boeing 777-22KLR
Govt of Turkmenistan,
  Ashkhabad
EZ-A777

UGANDA
Gulfstream Aerospace G.550
Govt of Uganda,
  Entebbe
5X-UGF

UKRAINE
Ukrainian Air Force
Antonov An-30
Open Skies
81 y

Ilyushin Il-76MD
25 TABR, Melitopol
  & Zaporozhye

76413	25 TABR
76423	25 TABR
76531	25 TABR
76557	25 TABR
76559	
76564	25 TABR
76566	
76580	25 TABR
76585	25 TABR
76596	25 TABR
76601	25 TABR
76621	25 TABR
76622	25 TABR
76633	25 TABR
76637	25 TABR
76645	25 TABR
76647	25 TABR
76660	25 TABR
76661	25 TABR
76683	25 TABR
76697	25 TABR
76698	25 TABR
76699	25 TABR
76700	25 TABR
76732	25 TABR
78772	25 TABR
78820	25 TABR
86915	

Airbus A.319CJ-115X
Govt of Ukraine, Kiev
UR-ABA

Ilyushin Il-62M
Govt of Ukraine, Kiev
UR-86527
UR-86528

UNITED ARAB EMIRATES
United Arab Emirates Air
Force
Aermacchi MB339A*/MB339NAT
Al Fursan, Al Ain
430
431
432
433*
434
435
436*
437
438
439
440

AgustaWestland AW.139
Dubai Air Wing
DU-140
DU-142

Airbus A.330-243 MRTT
MRTT Sqn, Al Ain
1300
1301
1302
1303 (on order)

Boeing C-17A
Globemaster III
Heavy Transport Sqn, Dubai/Minhad
1223 (10-0401)
1224 (10-0402)
1225 (10-0403)
1226 (10-0404)
1227 (10-0405)
1228 (10-0406)

Lockheed
C-130H/L.100-30*
Hercules
Transport Wing, Abu Dhabi/Bateen
1211
1212
1213
1214
1215*
1216*
1217*
Dubai
311*
312*

UAE Govt
Airbus A.319CJ-113X
Dubai Air Wing
A6-ESH

Airbus A.320-232
Govt of Abu Dhabi;
Govt of Dubai
A6-DLM Abu Dhabi
A6-HMS Dubai

BAE RJ.85/RJ.100*
Govt of Abu Dhabi;
Govt of Dubai
A6-AAB* Abu Dhabi
A6-RJ1 Dubai
A6-RJ2 Dubai

Boeing
737-7BC/7F0/8AJ/8EC/
8EO/8EX
Govt of Abu Dhabi;
Govt of Dubai

A6-AUH	8EX	Dubai
A6-DFR	7BC	Abu Dhabi
A6-HEH	8AJ	Dubai
A6-HRS	7F0	Dubai
A6-MRM	8EC	Dubai
A6-MRS	8EO	Dubai

Boeing
747-412F/422/433/48E/4F6/8Z5
Dubai Air Wing;
Govt of Abu Dhabi

A6-COM	433	Dubai
A6-GGP	412F	Dubai
A6-HRM	422	Dubai
A6-MMM	422	Dubai
A6-PFA	8Z5	(on order)
A6-UAE	48E	Dubai
A6-YAS	4F6	Abu Dhabi

Boeing 747SP-31
Govt of Dubai
A6-SMR

Boeing 777-2ANER/-35RER*
Govt of Abu Dhabi
A6-ALN
A6-SIL*

Boeing 787-8
Govt of Abu Dhabi
A6-PFC

Grumman
G.1159C Gulfstream IV
Dubai Air Wing
A6-HHH

YEMEN
Boeing 747SP-27
Govt of Yemen, Sana'a
7O-YMN

This section lists the codes worn by some overseas air forces and, alongside, the serial of the aircraft currently wearing this code. This list will be updated occasionally and those with Internet access can download the latest version via the 'Military Aircraft Markings' Web Site, www.militaryaircraftmarkings.co.uk and via the MAM2009 Yahoo! Group.

FRANCE		120-UH	E160	F-UHRF	E48	116-EW	48
FRENCH AIR FORCE		705-AA	E17	F-UHRT	E152	116-EX	40
D-BD Alpha Jet		705-AB	E28	F-UHRW	E166	116-EY	42
0 [PDF]	E95	705-AD	E51			116-EZ	54
1 [PDF]	E94	705-AH	E110	**Dassault Mirage 2000**		116-MG	65
2 [PDF]	E88	705-AO	E112	115-AM	525	116-MH	67
3 [PDF]	E166	705-FC	E139	115-KB	94	116-MK	74
4 [PDF]	E158	705-FK	E127	115-KC	120	118-AS	51
5 [PDF]	E46	705-FO	E81	115-KD	123	118-AX	77
6 [PDF]	E73	705-LA	E72	115-KE	101	118-EB	76
7 [PDF]	E119	705-LC	E87	115-KF	111	118-FZ	41
8 [PDF]	E79	705-LE	E121	115-KG	104	118-IF	609
9 [PDF]	E163	705-LG	E120	115-KI	96	118-IG	668
102-AC	E47	705-LK	E125	115-KJ	523	118-IN	640
102-AI	E117	705-LN	E118	115-KN	121	118-JF	660
102-FA	E140	705-LP	E129	115-KR	102	118-JK	612
102-FB	E86	705-LS	E22	115-KS	528	118-KL	106
102-FI	E32	705-LU	E148	115-KV	88	118-KQ	112
102-FM	E105	705-MA	E173	115-LB	87	118-MQ	646
102-FP	E168	705-MD	E30	115-LC	108	118-XH	616
102-LI	E53	705-MF	E98	115-LD	117	125-AA	340
102-MB	E97	705-MH	E48	115-LE	79	125-AE	355
102-MM	E13	705-MN	E167	115-LJ	105	125-AI	365
102-NA	E79	705-MO	E68	115-LK	85	125-AL	348
102-NB	E29	705-MS	E20	115-OA	524	125-AK	359
102-ND	E26	705-NF	E141	115-OC	529	125-AM	353
102-RE	E165	705-RA	E41	115-OR	527	125-AQ	351
102-RJ	E76	705-RK	E31	115-YA	93	125-AR	368
102-RX	E169	705-RM	E134	115-YB	99	125-BA	342
102-TZ	E83	705-RQ	E138	115-YC	85	125-BB	364
102-UB	E11	705-RR	E146	115-YD	107	125-BC	366
118-AK	E18	705-RS	E149	115-YE	122	125-BD	371
118-FD	E151	705-RT	E152	115-YF	100	125-BJ	354
118-FE	E119	705-RU	E153	115-YH	109	125-BQ	358
118-LT	E147	705-RW	E166	115-YL	82	125-BS	374
118-LX	E89	705-RY	E170	115-YM	115	125-BU	345
118-TD	E113	705-RZ	E171	115-YO	113	125-BX	356
118-TI	E156	705-TB	E67	115-YT	124	125-CF	373
120-AF	E108	705-TF	E42	116-AU	316	125-CG	338
120-AG	E109	705-TG	E104	116-BL	306	125-CI	335
120-AH	E99	705-TJ	E25	116-EA	49	125-CL	375
120-AK	E144	705-TK	E58	116-EC	78	125-CM	372
120-AL	E154	705-TM	E128	116-ED	62	125-CO	357
120-FJ	E33	705-TU	E7	116-EE	71	125-CQ	370
120-FN	E116	F-TELL	E88	116-EF	45	125-CU	362
120-LM	E102	F-TENE	E73	116-EG	56	133-AG	681
120-LO	E142	F-TERB	E163	116-EH	52	133-AL	669
120-LW	E82	F-TERD	E121	116-EI	38	133-AS	635
120-MB	E176	F-TERF	E158	116-EJ	43	133-AU	653
120-RE	E44	F-TERH	E94	116-EL	58	133-IA	650
120-RN	E124	F-TERI	E117	116-EM	63	133-IC	626
120-RO	E131	F-TERJ	E162	116-EN	46	133-ID	654
120-RP	E136	F-TERK	E31	116-EO	66	133-IE	642
120-RV	E164	F-TERN	E46	116-EP	47	133-IJ	638
120-TH	E90	F-TERP	E130	116-EQ	44	133-IL	622
120-TT	E101	F-TERQ	E95	116-ER	68	133-IO	647
120-TX	E93	F-TERR	E114	116-EU	55	133-IP	604
120-UC	E157	F-UHRE	E44	116-EV	59	133-IQ	666

Code	No.	Code	No.	Code	No.	Code	No.
133-IR	674	104-GS	140	118-EC	305	DBH	1451
133-IS	617	104-IC	328	118-EF	102	DBI	1507
133-IT	624	104-ID	329	118-FF	339	DBJ	1510
133-IU	620	104-IF	331	118-GA	122	DBK	1512
133-IV	683	104-IR	113	118-GB	123	DBL	1519
133-IW	664	113-EA	303	118-GD	125	DBM	1617
133-JB	678	113-FG	340	118-GE	126	DBN	1632
133-JC	606	113-FH	341	118-GF	127	DBO	1634
133-JD	643	113-FI	342	118-GH	129	DBP	1654
133-JE	634	113-FJ	343	118-GK	132	DBQ	1663
133-JG	601	113-FK	344	118-GL	133	DBR	5682
133-JH	686	113-FL	345	118-GM	134	DCA	1005
133-JI	675	113-FM	346	118-GP	137	DCB	1052
133-JJ	639	113-FN	347	118-GQ	138	DCC	1056
133-JL	628	113-FO	348	118-GV	143	DCD	1057
133-JM	657	113-FP	349	118-HT	323	DCE	1071
133-JN	658	113-GG	128	118-IM	109	DCF	1073
133-JO	627	113-GI	130	118-IN	110	DCG	1093
133-JP	611	113-GN	135	118-IP	111	DCH	1114
133-JR	682	113-GR	139	118-IQ	112	DCI	1122
133-JT	677	113-GT	141	118-IR	113	DCJ	1122
133-JU	614	113-GU	142	118-IS	114	DCK	1130
133-JV	636	113-GW	144	118-IU	116	DCL	1135
133-JW	641	113-GX	145	118-IV	117	DCM	1136
133-JX	679	113-HA	308	118-IW	118	DCN	1142
133-JY	615	113-HB	309	118-IX	119	DCO	1145
133-JZ	667	113-HC	310			DCP	1150
133-LF	605	113-HD	311	**FRENCH ARMY**		DCQ	1155
133-LG	651	113-HE	105	**SA.330 Puma**		DCR	1163
133-LH	655	113-HF	312	DAA	1006	DCS	1164
133-MO	613	113-HG	106	DAB	1020	DCT	1165
133-MP	623	113-HH	104	DAC	1036	DCU	1171
133-XA	662	113-HI	313	DAD	1037	DCV	1172
133-XC	618	113-HJ	107	DAE	1049	DCW	1177
133-XD	630	113-HK	315	DAF	1055	DCX	1179
133-XE	632	113-HM	318	DAG	1069	DCY	1182
133-XF	670	113-HN	319	DAH	1078	DCZ	1186
133-XG	625	113-HO	317	DAI	1092	DDA	1190
133-XI	661	113-HP	314	DAJ	1100	DDB	1192
133-XK	671	113-HQ	321	DAK	1102	DDC	1196
133-XL	603	113-HR	103	DAL	1107	DDD	1198
133-XM	680	113-HS	108	DAM	1109	DDE	1206
133-XN	652	113-HU	322	DAN	1128	DDF	1213
133-XO	629	113-HV	320	DAO	1143	DDG	1222
133-XP	645	113-HX	325	DAP	1149	DDH	1223
133-XQ	637	113-HY	326	DAQ	1156	DDI	1228
133-XR	659	113-HZ	327	DAR	1173	DDJ	1229
133-XT	648	113-IA	307	DAS	1176	DDK	1231
133-XV	672	113-IB	306	DAT	1189	DDL	1235
133-XX	610	113-IE	330	DAU	1197	DDM	1236
133-XY	649	113-IF	331	DAV	1204	DDN	1239
133-XZ	685	113-IG	332	DAW	1211	DDO	1244
188-ET	57	113-IH	333	DAX	1214	DDP	1252
188-IH	631	113-II	334	DAY	1217	DDQ	1255
188-ME	61	113-IJ	335	DAZ	1219	DDR	1256
188-XJ	602	113-IK	336	DBA	1232	DDS	1260
188-YR	91	113-IL	337	DBB	1243	DDT	1269
		113-IO	338	DBC	1248	DDU	1411
Dassault Rafale		113-IT	115	DBD	1262	DDV	1419
104-GC	124	113-IY	120	DBE	1277	DDW	1447
104-GJ	131	113-IZ	121	DBF	1417	DDX	1662
104-GO	136	118-EB	304	DBG	1438		

SA.342 Gazelle

Code	No.	Code	No.	Code	No.
GAA	3459	GCL	4181	GKE	4187
GAB	3476	GCM	4186	GMA	3567
GAC	3512	GCN	4189	GMB	4042
GAD	3530	GCO	4191	GMC	4079
GAE	3548	GCP	4195	GMD	4194
GAF	3564	GCQ	4198	GME	3870
GAG	3848	GEA	4204	GMF	3930
GAH	3849	GEB	4206	GMG	3939
GAI	3850	GEC	4207	GMH	4023
GAJ	3856	GED	4208	GMI	4171
GAK	3859	GEE	4209	GMJ	4190
GAL	3862	GEF	4210	GMK	4201
GAM	3863	GEG	4211	GNA	3458
GAN	3865	GEH	4212	GNB	3513
GAO	3868	GEI	4214	GNC	3549
GAP	3911	GEJ	4215	GND	3617
GAQ	3921	GEK	4216	GNE	3852
GAR	3938	GEL	4217	GNF	3853
GAS	3947	GEM	4218	GNG	3858
GAT	3948	GEN	4219	GNI	3866
GAU	3957	GEO	4220	GNJ	3896
GAV	3964	GEP	4221	GNK	3956
GAW	3996	GEQ	4222	GNM	3992
GAX	4018	GER	4223	GNN	4008
GAY	4019	GES	4224	GNO	4032
GAZ	4020	GET	4225	GNP	4049
GBA	4026	GEU	4226	GNQ	4060
GBB	4034	GEV	4227	GNR	4065
GBC	4039	GEW	4228	GNS	4071
GBD	4048	GEX	4229	GNT	4083
GBE	4053	GEY	4230	GNU	4096
GBF	4059	GEZ	4231	GNV	4118
GBG	4061	GFA	4231	GNW	4135
GBH	4066	GFB	4233	GNX	4143
GBI	4072	GFC	4234	GNY	4146
GBJ	4084	GJA	1732	GNZ	4159
GBL	4095	GJB	3477	GOA	4178
GBM	4108	GJC	3511	GOB	4183
GBN	4109	GJD	3529	GOC	4184
GBO	4114	GJF	3546	GOD	4192
GBP	4115	GJH	3615		
GBQ	4119	GJI	3851		
GBR	4120	GJJ	3855		

AS.532UL/EC.725AP Cougar

Code	No.
CAA	2611
CAB	2628
CAC	2630
CAD	2631
CAE	2633
CAF	2638
CAG	2640
CAH	2642
CGA	2252
CGB	2266
CGC	2267
CGD	2271
CGE	2272
CGF	2273
CGG	2282
CGH	2285
CGI	2290
CGJ	2292
CGK	2299
CGL	2300
CGM	2301
CGN	2303
CGO	2316
CGQ	2323
CGR	2324
CGS	2325
CGT	2327
CGU	2331
CGV	2336
CGW	2443
CGX	2446
CHA	2342
CHB	2369
CHC	2375

(continued from Gazelle column 2)

Code	No.
GBS	4124
GBT	4136
GBU	4140
GBV	4141
GBW	4142
GBX	4144
GBY	4145
GBZ	4151
GCA	4155
GCC	4160
GCD	4161
GCE	4162
GCF	4164
GCG	4168
GCH	4172
GCI	4175
GCJ	4179
GCK	4180
GJK	3857
GJL	3864
GJM	3867
GJN	3929
GJO	3965
GJP	4014
GJQ	4022
GJR	4038
GJS	4047
GJU	4067
GJV	4078
GJW	4102
GJX	4103
GJY	4123
GJZ	4166
GKA	4176
GKB	4177
GKC	4182
GKD	4185

EC.665 Tigre

Code	No.
BHA	2010
BHB	2009
BHC	2013
BHD	2015
BHE	2018
BHF	2019
BHG	2022
BHH	2001
BHI	2002
BHJ	2003
BHK	2004
BHL	2006
BHM	2011
BHN	2021
BHO	2024
BHP	2023
BHQ	2025
BHR	2026
BHS	2027
BHT	2012
BHU	2028
BHV	2029
BHW	2032
BHZ	2033
BIA	2016
BIB	2034
BIC	2035
BID	2036
BIE	2037
BIG	2039
BIH	2040
BII	2041
BIJ	2042
BIK	2043
BIL	2044
BIM	2045
BIN	2046
BIO	2047
BIP	2048
BIQ	2049
BIR	2050
BIS	2051
BIT	2052
BJA	6001
BJB	6002
BJC	6003

Code	Serial	Code	Serial	Code	Serial	Code	Serial
BJD	6004	51-60	MM7174	61-136	MM55070	36-12	MM7300
BJE	6005	51-62	MM7182	61-140	MM55072	36-14	MM7272
BJF	6006	51-63	MM7173	61-141	MM55073	36-15	MM7282
BJG	6007	51-64	MM7179	61-143	MM55075	36-21	MM7292
BJH	6008	51-65	MM7184	61-145	MM55077	36-22	MM7296
BJI	6009	51-66	MM7169	61-146	MM55078	36-23	MM7297
BJK	6010	51-67	MM7194	61-147	MM55079	36-24	MM7298
BJL	6011	51-70	MM7192	61-150	MM55080	36-25	MM7302
		51-72	MM7163	61-151	MM55081	36-26	MM7294
NH.90-TTH		51-80	MM55037	61-152	MM55082	36-30	MM7280
EAA	1239	51-82	MM55044	61-154	MM55084	36-31	MM7308
EAB	1256	RS-14	MM7177	61-155	MM55085	36-32	MM7310
EAC	1271	RS-18	MM55034	61-156	MM55086	36-33	MM7293
EAD	1273			61-157	MM55087	36-34	MM7312
EAF	1291	**Aermacchi MB339**		61-160	MM55088	36-35	MM7313
EAG	1292	0 [FT]	MM55053	61-162	MM55090	36-37	MM7314
EAH	1293	1 [FT]	MM54538	RS-11	CSX54453	36-40	MM7322
EAI	1294	2 [FT]	MM54480	RS-32	MM55091	36-41	MM7324
EAJ	1295	3 [FT]	MM55059	RS-33	MM55068	36-42	MM7288
EAK	1306	4 [FT]	MM54505			36-43	MM7330
EAL	1307	5 [FT]	MM54473	**Aermacchi T-346A Master**		37-01	MM7307
EAM	1308	6 [FT]	MM55054	61-01	CSX55054	37-03	MM7309
		7 [FT]	MM55058			37-04	MM7315
ITALY		8 [FT]	MM54518	**Eurofighter Typhoon**		37-05	MM7319
Aeritalia-EMB AMX		9 [FT]	MM54477	4-1	MM7270	37-07	MM7316
32-01	MM7147	10 [FT]	MM54514	4-2	MM7303	37-10	MM7311
32-12	MM7126	11 [FT]	MM55055	4-3	MM7287	37-11	MM7301
32-13	MM7166	12 [FT]	MM54500	4-4	MM7274	37-12	MM7321
32-15	MM7129	32-161	MM55089	4-5	MM7289	37-15	MM7329
32-16	MM7165	36-04	MM55076	4-6	MM7323	37-16	MM7331
32-24	MM7149	36-06	MM55074	4-7	MM7290	RMV-01	MMX603
32-41	MM55030	61-01	MM54446	4-11	MM7291	RS-01	MMX602
32-42	MM55051	61-11	MM54457	4-13	MM7235	RS-23	MM7278
32-47	MM55046	61-12	MM54458	4-14	MM7281		
32-50	MM55029	61-15	MM54454	4-15	MM7271	**Panavia Tornado**	
32-51	MM55036	61-20	MM55055	4-16	MM7285	6-01	MM7007
32-53	MM55047	61-21	MM54465	4-20	MM7279	6-04	MM7057
32-56	MM55042	61-23	MM54467	4-21	MM7304	6-05	MM7025
51-10	MM7159	61-24	MM54468	4-23	MM55095	6-07	MM7075
51-21	MM55043	61-26	MM55059	4-24	MM55097	6-10	MM7088
51-23	MM55049	61-32	MM54488	4-25	MM55092	6-11	MM7058
51-30	MM7170	61-42	MM54496	4-26	MM55128	6-12	MM7071
51-33	MM7162	61-50	MM54443	4-27	MM55094	6-15	MM55006
51-34	MM7191	61-52	MM54504	4-30	MM55096	6-16	MM7037
51-35	MM7196	61-55	MM54507	4-31	MM55093	6-21	MM7040
51-36	MM7185	61-60	MM54510	4-32	MM55129	6-22	MM7029
51-37	MM7161	61-61	MM54511	4-33	MM55130	6-25	MM7043
51-40	MM7164	61-62	MM54512	4-34	MM55131	6-26	MM7063
51-41	MM7183	61-64	MM54514	4-35	MM55132	6-27	MM7035
51-42	MM7177	61-65	MM54515	4-36	MM55133	6-30	MM7072
51-43	MM7178	61-66	MM54516	4-40	MM7277	6-31	MM7006
51-44	MM7198	61-70	MM54518	4-41	MM7299	6-32	MM7015
51-45	MM7175	61-72	MM54533	4-42	MM7325	6-34	MM7073
51-46	MM7197	61-106	MM54548	4-45	MM7320	6-35	MM7026
51-50	MM7186	61-107	MM54549	4-46	MM7326	6-37	MM7038
51-51	MM7151	61-126	MM55062	4-47	MM7327	6-42	MM55010
51-52	MM7171	61-127	MM55063	4-50	MM7306	6-44	MM55009
51-53	MM7180	61-130	MM55064	4-52	MM7294	6-45	MM55008
51-54	MM7193	61-131	MM55065	36-02	MM7286	6-50	MM7024
51-55	MM7168	61-132	MM55066	36-05	MM7276	6-51	MM55007
51-56	MM7167	61-133	MM55067	36-10	MM7284	6-52	MM55001
51-57	MM7190	61-135	MM55069	36-11	MM7275	6-53	MM55004

Code	Serial	Code	Serial	Code	Serial	Code	Serial
6-55	MM7004	74-28	E.25-74	**CASA 212 Aviocar**		14-03	C.16-36
6-66	MM7056	74-30	E.25-76	47-12	TM.12D-72	14-04	C.16-37
6-67	MM7028	74-34	E.25-81	47-14	T.12D-75	14-05	C.16-38
6-72	MM7083	74-35	E.25-83	72-01	T.12B-13	14-06	C.16-39
6-75	MM7013	74-39	E.25-88	72-07	T.12B-49	14-07	C.16-41
6-76	MM7044	74-40	E.25-17	72-08	T.12B-55	14-08	C.16-42
36-31	MM7023	74-41	E.25-41	72-09	T.12B-66	14-09	C.16-44
36-50	CSX7085	74-42	E.25-18	72-11	T.12B-65	14-10	C.16-45
50-01	MM7021	74-43	E.25-43	72-12	T.12B-67	14-11	C.16-47
50-02	MM7052	74-44	E.25-34	72-14	T.12B-63	14-12	C.16-48
50-03	MM7066	74-45	E.25-29	72-15	T.12B-69	14-13	C.16-49
50-04	MM7030	79-02	E.25-78	72-17	T.12B-70	14-14	C.16-50
50-05	MM7019	79-03	E.25-80	72-21	TR.12D-76	14-70	CE.16-11
50-06	MM7070	79-04	E.25-84	72-22	TR.12D-77		
50-07	MM7053	79-05	E.25-05	72-23	TR.12D-79		
50-40	MM7054	79-06	E.25-06	72-24	TR.12D-81		
50-41	MM7020	79-08	E.25-08				
50-42	MM7055	79-09	E.25-09	**Eurofighter Tifón**			
50-43	MM7047	79-11	E.25-11	11-01	C.16-21		
50-44	MM7062	79-12	E.25-12	11-02	C.16-22		
50-45	MM7051	79-13	E.25-13	11-03	C.16-23		
50-46	MM7068	79-14	E.25-14	11-04	C.16-24		
50-47	MM7059	79-15	E.25-15	11-05	C.16-25		
50-50	MM7039	79-16	E.25-16	11-06	C.16-26		
50-51	MM7081	79-17	E.25-62	11-07	C.16-27		
50-52	MM7084	79-20	E.25-20	11-08	C.16-28		
50-53	MM7008	79-21	E.25-21	11-09	C.16-29		
50-54	MM7082	79-22	E.25-22	11-10	C.16-30		
50-55	MM7036	79-23	E.25-23	11-12	C.16-32		
50-56	MM7065	79-25	E.25-25	11-13	C.16-33		
50-57	MM7067	79-27	E.25-27	11-14	C.16-40		
RS-05	CSX7047	79-29	E.25-87	11-15	C.16-43		
RS-06	CSX7041	79-31	E.25-31	11-16	C.16-46		
		79-32	E.25-86	11-17	C.16-51		
SPAIN		79-33	E.25-50	11-70	CE.16-01		
CASA 101EB Aviojet		79-34	E.25-52	11-71	CE.16-02		
54-20	E.25-35	79-35	E.25-54	11-72	CE.16-03		
54-21	E.25-55	79-37	E.25-37	11-73	CE.16-04		
54-22	E.25-61	79-38	E.25-38	11-74	CE.16-05		
74-07	E.25-51	79-40	E.25-40	11-75	CE.16-06		
74-09	E.25-53	79-44	E.25-44	11-76	CE.16-07		
74-11	E.25-56	79-45	E.25-45	11-78	CE.16-09		
74-12	E.25-57	79-46	E.25-46	11-79	CE.16-10		
74-13	E.25-59	79-49	E.25-49	11-81	CE.16-10000		
74-17	E.25-63	79-95	E.25-65		(CE.16-12)		
74-20	E.25-66	79-97	E.25-69	11-91	C.16-20		
74-22	E.25-68	79-98	E.25-73	14-01	C.16-31		
74-26	E.25-72			14-02	C.16-35		

US MILITARY AIRCRAFT MARKINGS

All USAF and US Army aircraft have been allocated a fiscal year (FY) number since 1921. Individual aircraft are given a serial according to the fiscal year in which they are ordered. The numbers commence at 0001 and are prefixed with the year of allocation. For example F-15C Eagle 84-001 (84-0001) was the first aircraft ordered in 1984. The fiscal year (FY) serial is carried on the technical data block which is usually stencilled on the left-hand side of the aircraft just below the cockpit. The number displayed on the fin is a corruption of the FY serial. Most tactical aircraft carry the fiscal year in small figures followed by the last three or four digits of the serial in large figures. Large transport and tanker aircraft such as C-130s and KC-135s sometimes display a five-figure number commencing with the last digit of the appropriate fiscal year and four figures of the production number. An example of this is KC-135R 58-0128 which displays 80128 on its fin.

US Army serials have been allocated in a similar way to USAF serials although in recent years an additional zero has been added so that all US Army serials now have the two-figure fiscal year part followed by five digits. This means that, for example C-20E 70140 is officially 87-00140 although as yet this has not led to any alterations to serials painted on aircraft.

USN and USMC serials follow a straightforward numerical sequence which commenced, for the present series, with the allocation of 00001 to an SB2C Helldiver by the Bureau of Aeronautics in 1940. Numbers in the 168000 series are presently being issued. They are usually carried in full on the rear fuselage of the aircraft.

US Coast Guard serials began with the allocation of the serial 1 to a Loening OL-5 in 1927.

UK-BASED USAF AIRCRAFT

The following aircraft are normally based in the UK. They are listed in numerical order of type with individual aircraft in serial number order, as depicted on the aircraft. The number in brackets is either the alternative presentation of the five-figure number commencing with the last digit of the fiscal year, or the fiscal year where a five-figure serial is presented on the aircraft. Where it is possible to identify the allocation of aircraft to individual squadrons by means of colours carried on fin or cockpit edge, this is also provided.

Notes	Type			Type			Notes
	McDonnell Douglas			86-0178	F-15C	y	
	F-15C Eagle/F-15D Eagle/			91-0301	F-15E	bl	
	F-15E Strike Eagle			91-0302	F-15E	bl	
	LN: 48th FW, RAF Lakenheath:			91-0303	F-15E		
	492nd FS blue/white			91-0306	F-15E	bl	
	493rd FS black/yellow			91-0307	F-15E	bl	
	494th FS red/white			91-0308	F-15E	bl	
	00-3000	F-15E	r	91-0309	F-15E	m [48 OG]	
	00-3001	F-15E	r	91-0310	F-15E	r	
	00-3002	F-15E	r	91-0311	F-15E	r	
	00-3003	F-15E	r	91-0312	F-15E	bl	
	00-3004	F-15E	r	91-0313	F-15E	r	
	01-2000	F-15E	r	91-0314	F-15E	r	
	01-2001	F-15E	r	91-0315	F-15E	bl	
	01-2002	F-15E	r	91-0316	F-15E	bl	
	01-2003	F-15E	r	91-0317	F-15E	bl	
	01-2004	F-15E	r	91-0318	F-15E	m [48 OG]	
	84-0001	F-15C	y	91-0320	F-15E	r [494 FS]	
	84-0010	F-15C	y	91-0321	F-15E	bl	
	84-0015	F-15C	y	91-0324	F-15E	r	
	84-0019	F-15C	y	91-0326	F-15E	r	
	84-0027	F-15C	y	91-0327	F-15E	bl	
	84-0044	F-15D	y	91-0329	F-15E	r	
	86-0154	F-15C	y	91-0331	F-15E	bl	
	86-0156	F-15C	y	91-0332	F-15E	bl	
	86-0159	F-15C		91-0334	F-15E	r	
	86-0160	F-15C	y	91-0335	F-15E	r	
	86-0163	F-15C	y	91-0602	F-15E	r	
	86-0164	F-15C	y	91-0603	F-15E	r	
	86-0165	F-15C	y	91-0604	F-15E	r	
	86-0166	F-15C	y	91-0605	F-15E	bl	
	86-0171	F-15C		92-0364	F-15E	r	
	86-0172	F-15C	y	96-0201	F-15E	r	
	86-0174	F-15C	y	96-0202	F-15E	bl	
	86-0175	F-15C	y	96-0204	F-15E	r	
	86-0176	F-15C	y	96-0205	F-15E	bl	

Notes	Type			Type	Notes
	97-0218	F-15E	bl	**Lockheed**	
	97-0219	F-15E	bl	**MC-130J Commando II**	
	97-0220	F-15E	bl	352nd SOG, RAF Mildenhall:	
	97-0221	F-15E	bl [492 FS]	67th SOS	
	97-0222	F-15E	bl	05714 (FY10)	
	98-0131	F-15E	bl	15731 (FY11)	
	98-0132	F-15E	r	15733 (FY11)	
	98-0133	F-15E	bl	15737 (FY11)	
	98-0134	F-15E	bl	25757 (FY12)	
	98-0135	F-15E	bl	25759 (FY12)	
				25760 (FY12)	
	Bell-Boeing				
	CV-22B Osprey			**Boeing KC-135R/KC-135T**	
	352nd SOG, RAF Mildenhall:			**Stratotanker**	
	7th SOS			351st ARS/100th ARW,	
	FY09			RAF Mildenhall [D] (*r/w/bl*)	
	0046			10299 (FY61)	KC-135R
	FY11			10321 (FY61)	KC-135R
	0057			14837 (FY64)	KC-135R
	0058			23551 (FY62)	KC-135R
	0059			23565 (FY62)	KC-135R
	0060			38021 (FY63)	KC-135R
				38027 (FY63)	KC-135R
	Sikorsky HH-60G			38871 (FY63)	KC-135R
	Pave Hawk			38884 (FY63)	KC-135R
	56th RQS/48th FW,			80034 (FY58)	KC-135R
	RAF Lakenheath [LN]			80100 (FY58)	KC-135R
	26205 (FY89)			80118 (FY58)	KC-135R
	26206 (FY89)			91464 (FY59)	KC-135T
	26208 (FY89)			91492 (FY59)	KC-135R
	26212 (FY89)			91513 (FY59)	KC-135T
	26353 (FY91)				

F-15E Strike Eagle 91-0335 wears the red and white tail markings of the 494th Fighter Squadron and the LN code of the USAF's 48th Fighter Wing, based at Lakenheath.

These aircraft are normally based in Western Europe with the USAFE. They are shown in numerical order of type designation, with individual aircraft in serial number order as carried on the aircraft. Fiscal year (FY) details are also provided if necessary. The unit allocation and operating bases are given for most aircraft.

Notes	Type				Type				Notes
	Beech				89-2118	AV	gn		
	C-12 Super King Air				89-2137	AV	pr	[31 OG]	
	US Embassy Flight, Taszár, Hungary				89-2152	AV	gn		
	FY83				89-2178*	AV	gn		
	30497	C-12D			90-0709	AV	pr		
	FY76				90-0772	AV	gn		
	60168	C-12C			90-0773	AV	gn		
					90-0777*	AV	pr		
	Lockheed (GD)				90-0795*	AV	gn		
	F-16CM/F-16DM* Fighting Falcon				90-0796*	AV	gn		
	AV: 31st FW, Aviano, Italy:				90-0800*	AV	gn		
	510th FS *purple/white*				90-0813	SP	r		
	555th FS *green/yellow*				90-0818	SP	r		
	SP: 52nd FW, Spangdhlem,				90-0827	SP	r		
	Germany:				90-0829	SP	r	[52 OG]	
	480th FS *red*				90-0833	SP	r		
	87-0350	AV	gn		91-0338	SP	r		
	87-0351	AV	gn		91-0340	SP	r		
	87-0355	AV	pr		91-0342	SP	r		
	87-0359	AV	gn		91-0343	SP	r		
	88-0413	AV	pr		91-0344	SP	r		
	88-0425	AV	gn		91-0351	SP	r		
	88-0435	AV	gn		91-0352	SP	m	[52 FW]	
	88-0443	AV	pr		91-0358	SP	r		
	88-0444	AV	pr		91-0360	SP	r		
	88-0446	AV	gn		91-0361	SP	r		
	88-0491	AV	pr		91-0366	SP	r	[480 FS]	
	88-0516	AV	pr		91-0368	SP			
	88-0525	AV	pr		91-0402	SP	r		
	88-0526	AV	gn		91-0403	SP	r		
	88-0532	AV	gn		91-0407	SP	r		
	88-0535	AV	gn		91-0412	SP	r		
	88-0541	AV	pr		91-0416	SP	r		
	89-2001	AV	m	[31 FW]	91-0417	SP	r		
	89-2008	AV	pr		91-0472*	SP			
	89-2009	AV	pr		91-0481*	SP			
	89-2011	AV	pr		92-3918	SP	r		
	89-2016	AV	gn		96-0080	SP	r		
	89-2018	AV	gn		96-0083	SP	r		
	89-2023	AV	gn						
	89-2024	AV	gn		**Grumman**				
	89-2026	AV	pr		**C-20H Gulfstream IV**				
	89-2029	AV	pr		76th AS/86th AW, Ramstein,				
	89-2030	AV	pr	[510 FS]	Germany				
	89-2035	AV	gn	[555 FS]	*FY90*				
	89-2038	AV	pr		00300				
	89-2039	AV	gn		*FY92*				
	89-2041	AV	gn		20375				
	89-2044	AV	gn						
	89-2046	AV	pr						
	89-2047	AV	pr						
	89-2049	AV	pr	[USAFE]					
	89-2057	AV	pr						
	89-2068	AV	gn						
	89-2096	AV	pr						
	89-2102	AV	pr						

Notes	Type	Type	Notes
	Gates C-21A	**Lockheed C-130J-30**	
	Learjet	**Hercules II**	
	76th AS/86th AW, Ramstein,	37th AS/86th AW, Ramstein,	
	Germany	Germany [RS] (bl/w)	
	FY84	43142 (FY04)	
	40085	68610 (FY06)	
	40096	68611 (FY06)	
	40126	68612 (FY06)	
		78608 (FY07)	
		78609 (FY07)	
	Gulfstream Aerospace	78613 (FY07)	
	C-37A Gulfstream V	78614 (FY07)	
	309th AS/86th AW, Chievres,	88601 (FY08)	[86 AW]
	Belgium	88602 (FY08)	[86 OG]
	FY01	88603 (FY08)	[37 AS]
	10076	88604 (FY08)	
	FY99	88605 (FY08)	
	90402	88607 (FY08)	
	Boeing C-40B		
	76th AS/86th AW, Ramstein,		
	Germany		
	FY02		
	20042		

Notes	Type		Type		Notes
	Fairchild C-26D		900530	Sigonella	
	NAF Naples, Italy;		900531	Naples	
	NAF Sigonella, Italy		910502	Naples	
	900528	Sigonella			

Notes	Type			Type			Notes
	Beech			40157	C-12U	E/6-52nd AVN	
	C-12 Huron			40158	C-12U	F/6-52nd AVN	
	'E' Co, 6th Btn, 52nd Avn Reg't,			40160	C-12U	E/6-52nd AVN	
	Stuttgart;			40161	C-12U	F/6-52nd AVN	
	'F' Co, 6th Btn, 52nd Avn Reg't,			40162	C-12U	F/6-52nd AVN	
	Wiesbaden;			40163	C-12U	F/6-52nd AVN	
	'A' Co, 2nd Btn, 228th Avn Reg't,			40165	C-12U	F/6-52nd AVN	
	Heidelberg;			40173	C-12U	F/6-52nd AVN	
	1st Military Intelligence Btn,			40180	C-12U	E/6-52nd AVN	
	Wiesbaden;			FY86			
	SHAPE Flight Det, Chievres			60079	C-12J	SHAPE Flt Det	
	FY91			FY89			
	00516	RC-12X	1st MIB	00273	RC-12X	1st MIB	
	FY92						
	13123	RC-12X	1st MIB	**Cessna UC-35A**			
	13125	RC-12X	1st MIB	**Citation V**			
	FY93			'F' Co, 6th Btn, 52nd Avn Reg't,			
	30699	RC-12X	1st MIB	Wiesbaden			
	30701	RC-12X	1st MIB	FY95			
	FY84			50123			
	40156	C-12U	F/6-52nd AVN	50124			

Notes	Type			Type			Notes
	FY97			*FY84*			
	70101			23936	UH-60A	C/5-158th AVN	
	70102			23951	UH-60A		
	70105			*FY85*			
	FY99			24397	UH-60A	C/5-158th AVN	
	90102			24422	UH-60A		
				24437	UH-60A		
	Boeing-Vertol CH-47F			24446	UH-60A	C/5-158th AVN	
	Chinook			*FY86*			
	'B' Co, 5th Btn, 158th Avn Reg't,			24532	UH-60A	C/1-214th AVN	
	Ansbach			24538	UH-60A	C/1-214th AVN	
	FY04			24551	UH-60A	C/1-214th AVN	
	08713			24552	UH-60A		
	08716			*FY87*			
	FY06			24583	UH-60A	SHAPE Flt Det	
	08023			24584	UH-60A	SHAPE Flt Det	
	08027			24589	UH-60A	G/6-52nd AVN	
	08030			24614	UH-60A	C/5-158th AVN	
	08031			24642	UH-60A	G/6-52nd AVN	
	FY07			24644	UH-60A	C/1-214th AVN	
	08034			24645	UH-60A		
	08037			24647	UH-60A	C/1-214th AVN	
	08724			24656	UH-60A	C/5-158th AVN	
	FY08			26004	UH-60A	C/5-158th AVN	
	08045			26005	UH-60A	C/5-158th AVN	
	08047			*FY88*			
	08049			26019	UH-60A	C/1-214th AVN	
				26023	UH-60A	C/1-214th AVN	
	Sikorsky H-60 Black Hawk			26027	UH-60A	G/6-52nd AVN	
	'A' Co, 3rd Btn, 158th Avn Reg't,			26032	UH-60A	C/5-158th AVN	
	Ansbach;			26037	UH-60A	A/1-214th AVN	
	'B' Co, 3rd Btn, 158th Avn Reg't,			26039	UH-60A	C/1-214th AVN	
	Ansbach;			26045	UH-60A	C/1-214th AVN	
	'A' Co, 5th Btn, 158th Avn Reg't,			26054	UH-60A	C/1-214th AVN	
	Ansbach;			26071	UH-60A	G/6-52nd AVN	
	'C' Co, 5th Btn, 158th Avn Reg't,			26075	UH-60A	C/5-158th AVN	
	Ansbach;			26080	UH-60A	C/1-214th AVN	
	SHAPE Flight Det, Chievres;			26082	UH-60A	C/5-158th AVN	
	'G' Co, 6th Btn, 52nd Avn Reg't,			*FY89*			
	Coleman Barracks;			26132	UH-60A	C/5-158th AVN	
	'B' Co, 70th Transportation Reg't,			26138	UH-60A	C/1-214th AVN	
	Coleman Barracks;			26142	UH-60A	G/6-52nd AVN	
	'C' Co, 1st Btn, 214th Avn Reg't,			26157	UH-60A	C/5-158th AVN	
	Hohenfels;			26158	UH-60A	C/5-158th AVN	
	'A' Co, 1st Btn, 214th Avn Reg't,			26163	UH-60A	C/5-158th AVN	
	Landstuhl;			26165	UH-60A	G/6-52nd AVN	
	6th Avn Co, Vicenza, Italy			*FY92*			
	FY79			26452	UH-60L	A/5-158th AVN	
	23330	UH-60A		*FY94*			
	FY82			26551	UH-60L	G/6-52nd AVN	
	23745	UH-60A	C/1-214th AVN	26570	UH-60L	A/5-158th AVN	
	23750	UH-60A	C/1-214th AVN	26572	UH-60L	A/5-158th AVN	
	23754	UH-60A		26573	UH-60L	A/5-158th AVN	
	23755	UH-60A		26577	UH-60L	A/5-158th AVN	
	23756	UH-60A		*FY95*			
	23757	UH-60A	C/1-214th AVN	26637	UH-60L	A/3-158th AVN	
	FY83			26639	UH-60L	A/5-158th AVN	
	23855	UH-60A	C/1-214th AVN	26641	UH-60L	3-158th AVN	
	23869	UH-60A	C/1-214th AVN	26642	UH-60L	3-158th AVN	
	23875	UH-60A	C/1-214th AVN	26643	UH-60L	3-158th AVN	
	23888	UH-60A	C/5-158th AVN	26645	UH-60L	A/3-158th AVN	
				26646	UH-60L	3-158th AVN	

Notes	Type			Type		Notes
	26649	UH-60L	B/3-158th AVN	05426	2-159th AVN	
	26650	UH-60L	A/3-158th AVN	05428	2-159th AVN	
	26651	UH-60L	B/3-158th AVN	05430	3-159th AVN	
	26652	UH-60L	3-158th AVN	05431	2-159th AVN	
	26654	UH-60L	C/3-158th AVN	05433	3-159th AVN	
	26655	UH-60L	C/3-158th AVN	05437	3-159th AVN	
	FY96			05439	3-159th AVN	
	26674	UH-60L	C/3-158th AVN	05441	3-159th AVN	
	26675	UH-60L	A/3-158th AVN	05442	3-159th AVN	
	26676	UH-60L	A/3-158th AVN	05445	2-159th AVN	
	26677	UH-60L	A/3-158th AVN	05446	2-159th AVN	
	26678	UH-60L	A/3-158th AVN	05453	2-159th AVN	
	26680	UH-60L	3-158th AVN	05460	3-159th AVN	
	26682	UH-60L	3-158th AVN	05467	3-159th AVN	
	26683	UH-60L	3-158th AVN	*FY05*		
	26684	UH-60L	3-158th AVN	05485	2-159th AVN	
	26685	UH-60L	A/3-158th AVN	07001	2-159th AVN	
	26686	UH-60L	3-158th AVN	07010	2-159th AVN	
	26687	UH-60L	3-158th AVN	*FY06*		
	26688	UH-60L	3-158th AVN	07021	3-159th AVN	
	26689	UH-60L	A/3-158th AVN	*FY08*		
	26690	UH-60L	3-158th AVN	05547	3-159th AVN	
	26691	UH-60L	3-158th AVN	05551	2-159th AVN	
	26692	UH-60L	A/3-158th AVN	05557	3-159th AVN	
	26696	UH-60L	3-158th AVN	*FY09*		
	FY97			05580	2-159th AVN	
	26763	UH-60L	A/5-158th AVN	05581	2-159th AVN	
	26766	UH-60L	A/5-158th AVN	05587	2-159th AVN	
	26768	UH-60L	A/5-158th AVN	05589	2-159th AVN	
	FY05					
	27062	UH-60L	B/3-158th AVN	**Eurocopter UH-72A Lakota**		
	FY06			Joint Multinational Readiness		
	27110	UH-60L	A/5-158th AVN	Centre, Hohenfels		
				FY07		
	MDH AH-64D Apache			72029		
	2nd Btn, 159th Avn Reg't, Illesheim;			*FY09*		
	3rd Btn, 159th Avn Reg't, Illesheim			72095		
	FY02			72096		
	05302	3-159th AVN		72097		
	05304	2-159th AVN		72098		
	05311	3-159th AVN		72100		
	05312	3-159th AVN		72105		
	05316	2-159th AVN		72106		
	05321	2-159th AVN		72107		
	05322	2-159th AVN		72108		
	05323	3-159th AVN				
	05326	2-159th AVN				
	05343	3-159th AVN				
	05344	2-159th AVN				
	FY03					
	05352	2-159th AVN				
	05353	2-159th AVN				
	05361	3-159th AVN				
	05367	3-159th AVN				
	05371	2-159th AVN				
	05381	2-159th AVN				
	05384	2-159th AVN				
	05389	2-159th AVN				
	05403	3-159th AVN				
	05418	3-159th AVN				
	FY04					
	05421	3-159th AVN				

The following aircraft are normally based in the USA but are likely to be seen visiting the UK from time to time. The presentation is in numerical order of the type, commencing with the B-1B and concluding with the C-135. The aircraft are listed in numerical progression by the serial actually carried externally. Fiscal year information is provided, together with details of mark variations and in some cases operating units. Where base-code letter information is carried on the aircrafts' tails, this is detailed with the squadron/base data; for example the 7th Wing's B-1B 60105 carries the letters DY on its tail, thus identifying the Wing's home base as Dyess AFB, Texas.

Notes	Type			Type			Notes
	Rockwell B-1B Lancer			60119	7th BW	bl/w	
	7th BW, Dyess AFB, Texas [DY]:			60120	53rd Wg	bk/gy	
	9th BS (bk/w), 13th BS (r)			60121	28th BW	bk/y	
	& 28th BS (bl/w);			60122	53rd Wg	bk/gy	
	28th BW, Ellsworth AFB,			60123	7th BW	bl/w	
	South Dakota [EL]:			60124	7th BW	bl/w	
	34th BS (bk/r) & 37th BS (bk/y);			60126	7th BW	bl/w	
	337th TES/53rd Wg, Dyess AFB,			60129	28th BW	bk/r	
	Texas [OT] bk/gy;			60132	53rd Wg	bk/gy	
	77th WPS/57th Wg, Dyess AFB,			60133	7th BW	bk/w	
	Texas [WA] y/bk;			60134	28th BW	bk/r	
	419th FLTS/412th TW, Edwards AFB,			60135	7th BW	bl/w	
	California [ED]			60136	7th BW	bl/w	
	FY85			60138	7th BW	y/bk	
	50059	7th BW	bk/w $	60139	28th BW	bl/y $	
	50060	28th BW	bk/r	60140	7th BW	bk/w	
	50061	7th BW	bl/w				
	50064	7th BW	bl/w	**Northrop B-2 Spirit**			
	50066	28th BW		419th FLTS/412th TW, Edwards AFB,			
	50068	412th TW		California;			
	50069	28th BW	bk/r	509th BW, Whiteman AFB,			
	50072	28th BW	bk/r	Missouri [WM]:			
	50073	7th BW	bk/w $	13th BS, 393rd BS & 715th BS			
	50074	7th BW	bk/w	(Names are given where known.			
	50075	412th TW		Each begins Spirit of ...)			
	50077	57th Wg	y/bk	FY90			
	50079	28th BW	bk/r	00040	509th BW	Alaska	
	50080	7th BW	bl/w $	00041	509th BW	Hawaii	
	50081	28th BW	bk/r	FY92			
	50084	28th BW	bk/r	20700	509th BW	Florida	
	50085	28th BW	bk/y	FY82			
	50088	7th BW	bl/w	21066	509th BW	America	
	50089	7th BW	bl/w	21067	509th BW	Arizona	
	50090	7th BW	bl/w	21068	509th BW	New York	
	FY86			21069	509th BW	Indiana	
	60094	28th BW	bk/y $	21070	509th BW	Ohio	
	60095	28th BW	bk/r	21071	509th BW	Mississippi	
	60097	7th BW	bl/w	FY93			
	60098	7th BW	bk/w	31085	412th TW	Oklahoma	
	60099	28th BW	bk/y $	31086	509th BW	Kitty Hawk	
	60101	7th BW	bl/w	31087	509th BW	Pennsylvania	
	60102	28th BW	bk/y	31088	509th BW	Louisiana	
	60103	7th BW	bl/w	FY88			
	60104	28th BW	bk/y	80328	509th BW	Texas	
	60105	7th BW	bl/w	80329	509th BW	Missouri	
	60107	7th BW	bl/w	80330	509th BW	California	
	60108	28th BW	bk/r	80331	509th BW	South Carolina	
	60109	7th BW	bk/w	80332	509th BW	Washington	
	60110	7th BW	bl/w	FY89			
	60111	28th BW	bk/r	90128	509th BW	Nebraska	
	60112	7th BW	bk/w	90129	509th BW	Georgia	
	60113	28th BW	bk/y				
	60115	28th BW	bk/r				
	60117	7th BW	bk/w				
	60118	28th BW	bk/y				

Notes	Type			Type			Notes
	Lockheed U-2			11408	E-3B	w	
	9th RW, Beale AFB, California [BB]:			*FY82*			
	1st RS, 5th RS			20006	E-3C		
	& 99th RS (*bk/r*);			20007	E-3G	Boeing *r*	
	Lockheed, Palmdale;			*FY83*			
	Warner Robins			30009	E-3C		
	Air Logistics Centre [WR]			*FY73*			
	FY68			31675	E-3B	*or*	
	68-10329	U-2S	9th RW	*FY75*			
	68-10331	U-2S	9th RW	50556	E-3B	w	
	68-10336	U-2S	9th RW	50557	E-3B	r	
	68-10337	U-2S	9th RW	50558	E-3B		
	FY80			50559	E-3B	r	
	80-1064	TU-2S	9th RW	50560	E-3B	w	
	80-1065	TU-2S	9th RW	*FY76*			
	80-1066	U-2S	9th RW	61604	E-3B	r	
	80-1067	U-2S	Lockheed	61605	E-3B	w	
	80-1068	U-2S	9th RW	61606	E-3B	*or*	
	80-1069	U-2S	9th RW	61607	E-3B	w	
	80-1070	U-2S	9th RW	*FY77*			
	80-1071	U-2S	9th RW	70351	E-3G	w	
	80-1073	U-2S	9th RW	70352	E-3B	r	
	80-1074	U-2S	9th RW	70353	E-3B	w	
	80-1076	U-2S	9th RW	70355	E-3B	gn	
	80-1077	U-2S	9th RW	70356	E-3B	w	
	80-1078	TU-2S	9th RW	*FY78*			
	80-1079	U-2S	9th RW	80576	E-3G	w	
	80-1080	U-2S	9th RW	80577	E-3B	*or*	
	80-1081	U-2S	9th RW	80578	E-3B	w	
	80-1083	U-2S	9th RW	*FY79*			
	80-1084	U-2S	9th RW	90001	E-3G	w	
	80-1085	U-2S	9th RW	90002	E-3B	w	
	80-1086	U-2S	9th RW	90003	E-3B	m	
	80-1087	U-2S	9th RW				
	80-1089	U-2S	9th RW	**Boeing E-4B**			
	80-1090	U-2S	9th RW	1st ACCS/55th Wg, Offutt AFB,			
	80-1091	TU-2S	9th RW	Nebraska [OF]			
	80-1092	U-2S	9th RW	*FY73*			
	80-1093	U-2S	9th RW	31676			
	80-1094	U-2S	9th RW	31677			
	80-1096	U-2S	9th RW	*FY74*			
	80-1099	U-2S	9th RW	40787			
				FY75			
	Boeing E-3 Sentry			50125			
	552nd ACW, Tinker AFB,						
	Oklahoma [OK] (*w*):						
	960th ACS, 963rd ACS,						
	964th ACS, 965th ACS						
	& 966th AACS;						
	961st ACS/18th Wg, Kadena AB,						
	Japan [ZZ] (*or*);						
	962nd ACS/3rd Wg, Elmendorf AFB,						
	Alaska [AK] (*gn*)						
	FY80						
	00137	E-3C	gn				
	00138	E-3G	w				
	00139	E-3C	*or*				
	FY81						
	10004	E-3C	w				
	10005	E-3G	*or* $				
	FY71						
	11407	E-3B	w				

Notes	Type				Type				Notes
	Lockheed				70028	C-5B	60th AMW	w	
	C-5 Galaxy/C-5M Super Galaxy				70029	C-5M	LMTAS		
	60th AMW, Travis AFB, California:				70030	C-5B	60th AMW	bk/bl	
	21st AS (bk/gd) & 22nd AS (bk/bl);				70031	C-5B	439th AW	bl/r	
	105th AW, Stewart AFB, New York:				70032	C-5M	LMTAS		
	137th AS (bl);				70033	C-5B	439th AW	bl/r	
	164th AW, Memphis, Tennessee ANG:				70034	C-5M	60th AMW	bk/bl	
	155th AS (r);				70035	C-5M	436th AW	bl/y	
	167th AW, Martinsburg,				70036	C-5M	436th AW	bl/y	
	West Virginia ANG [WV]:				70037	C-5B	439th AW	bl/r	
	167th AS (r);				70038	C-5B	439th AW	bl/r	
	412th TW, Edwards AFB, California:				70039	C-5B	439th AW	bl/r	
	418th FLTS;				70040	C-5M	436th AW	bl/y	
	433rd AW AFRC, Kelly AFB, Texas:				70041	C-5B	439th AW	bl/r	
	68th AS;				70042	C-5M	60th AMW	bk/bl	
	436th AW, Dover AFB, Delaware:				70043	C-5B	439th AW	bl/r	
	9th AS (bl/y);				70044	C-5M	60th AMW	bk/bl	
	439th AW AFRC, Westover ARB,				70045	C-5M	436th AW	bl/y	
	Massachusetts:				*FY68*				
	337th AS (bl/r)				80213	C-5M	LMTAS		
	FY70				80216	C-5C	60th AMW	bk/gd	
	00448	C-5A	433rd AW		*FY69*				
	00451	C-5A	433rd AW		90006	C-5A	433rd AW		
	00456	C-5A	433rd AW		90009	C-5A	167th AW	r	
	00460	C-5A	167th AW	r	90012	C-5A	167th AW	r	
	00461	C-5A	433rd AW		90020	C-5A	433rd AW		
	FY83				90023	C-5A	167th AW	r	
	31285	C-5M	436th AW	bl/y	90024	C-5M	436th AW	bl/y	
	FY84				90025	C-5A	167th AW	r	
	40060	C-5B	60th AMW	bk/gd					
	40061	C-5M	436th AW	bl/y					
	40062	C-5B	60th AMW	bk/gd					
	FY85								
	50001	C-5M	436th AW	bl/y					
	50002	C-5M	436th AW	bl/y					
	50003	C-5M	436th AW	bl/y					
	50004	C-5M	436th AW	bl/y					
	50005	C-5M	436th AW	bl/y					
	50006	C-5B	439th AW	bl/r					
	50007	C-5M	436th AW	bl/y					
	50008	C-5M	436th AW	bl/y					
	50009	C-5B	439th AW	bl/r					
	50010	C-5M	60th AMW	bk/bl					
	FY86								
	60011	C-5M	60th AMW	bk/bl					
	60012	C-5B	439th AW	bl/r					
	60013	C-5M	436th AW	bl/y					
	60014	C-5B	439th AW	bl/r					
	60015	C-5M	LMTAS						
	60016	C-5B	60th AMW	bk/bl					
	60017	C-5M	436th AW	bl/y					
	60018	C-5B	439th AW	bl/r					
	60019	C-5B	439th AW	bl/r					
	60020	C-5M	436th AW	bl/y					
	60021	C-5B	439th AW	bl/r					
	60022	C-5M	LMTAS						
	60023	C-5B	439th AW	bl/r					
	60024	C-5B	60th AMW	bk/bl					
	60025	C-5M	436th AW	bl/y					
	60026	C-5B	60th AMW	bk/gd					
	FY87								
	70027	C-5B	439th AW	bl/r					

Boeing E-8C J-STARS

116th ACW, Robins AFB, Georgia [WR]:
12th ACCS (gn), 16th ACCS (bk),
128th ACS/Georgia ANG (r)
& 330th CTS (y)

FY00		
02000		
FY90		
FY01		
12005		
FY02		
29111		
FY92		
23289	m	
23290	bk	
FY93		
31097	r	
FY94		
40284	r	
40285	bk	
FY95		
50121	r	
50122	r	
FY96		
60042	bk	
60043	bl	
FY97		
70100	r	
70200	bk	
70201	r	
FY99		
90006	gn	

McDonnell Douglas
KC-10A Extender

60th AMW, Travis AFB, California:
6th ARS & 9th ARS;
305th AMW, McGuire AFB,
New Jersey:
2nd ARS (bl/r) & 32nd ARS (bl)

FY82		
20191	60th AMW	
20192	60th AMW	
20193	60th AMW	
FY83		
30075	60th AMW	
30076	60th AMW	
30077	60th AMW	
30078	60th AMW	
30079	305th AMW	bl
30080	60th AMW	
30081	305th AMW	bl
30082	305th AMW	bl
FY84		
40185	60th AMW	
40186	305th AMW	bl/r
40187	60th AMW	
40188	305th AMW	bl/r
40189	305th AMW	bl/r
40190	305th AMW	bl/r
40191	60th AMW	
40192	305th AMW	bl/y

FY85		
50027	305th AMW	bl/r
50028	305th AMW	bl/r
50029	60th AMW	
50030	305th AMW	bl/r
50031	305th AMW	bl/r
50032	305th AMW	bl/y
50033	60th AMW	
50034	305th AMW	bl/y
FY86		
60027	305th AMW	bl
60028	305th AMW	bl/r
60029	60th AMW	
60030	305th AMW	bl/r
60031	60th AMW	
60032	305th AMW	
60033	60th AMW	
60034	60th AMW	
60035	305th AMW	bl
60036	305th AMW	bl
60037	60th AMW	
60038	60th AMW	
FY87		
70117	60th AMW	
70118	305th AMW	
70119	60th AMW	
70120	305th AMW	bl
70121	305th AMW	bl
70122	305th AMW	bl/r
70123	60th AMW	
70124	305th AMW	bl/y
FY79		
90433	305th AMW	bl
90434	305th AMW	bl/r
91710	305th AMW	bl/r
91711	305th AMW	bl
91712	305th AMW	bl/r
91713	60th AMW	
91946	60th AMW	
91947	305th AMW	bl
91948	60th AMW	
91949	305th AMW	bl/r
91950	60th AMW	
91951	60th AMW	

Bombardier
E-11A Global Express

653rd ELSG, Hanscom Field

FY11	
19001	
19355	
19358	
FY12	
29506	

Notes	Type			Type			Notes
	Boeing			00219	62nd AW	gn	
	C-17A Globemaster III			00220	62nd AW	gn	
	3rd Wg, Elmendorf AFB,			00221	437th AW	y/bl	
	Alaska: 517th AS (w/bk);			00222	437th AW	y/bl	
	15th Wg, Hickam AFB,			00223	437th AW	y/bl	
	Hawaii: 535th AS (r/y);			FY90			
	60th AMW, Travis AFB, California:			00532	62nd AW	gn	
	21st AS (bk/w);			00533	15th Wg	r/y	
	62nd AW, McChord AFB,			00534	437th AW	y/bl	
	Washington (gn):			00535	445th AW	r/w	
	4th AS, 7th AS,8th AS			FY01			
	& 10th AS;			10186	436th AW	bl/y	
	97th AMW, Altus AFB, Oklahoma:			10187	62nd AW	gn	
	58th AS(r/y);			10188	105th AW	bl	
	105th AW, Stewart AFB, New York:			10189	164th AW	r	
	137th AS (bl);			10190	97th AMW	r/y	
	164th AW, Memphis, Tennessee ANG:			10191	436th AW	bl/y	
	155th AS (r);			10192	105th AW	bl	
	167th AW, Martinsburg,			10193	437th AW	y/bl	
	West Virginia ANG [WV]:			10194	445th AW	r/w	
	167th AS (r);			10195	97th AMW	r/y	
	172nd AW, Jackson Int'l Airport,			10196	167th AW	r	
	Mississippi ANG:			10197	97th AMW	r/y	
	183rd AS (bl/gd);			FY02			
	305th AMW, McGuire AFB, New Jersey:			21098	305th AMW	bl	
	6th AS (bl);			21099	97th AMW	r/y	
	412th TW, Edwards AFB,			21100	164th AW	r	
	California [ED]:			21101	437th AW	y/bl	
	418th FLTS;			21102	97th AMW	r/y	
	436th AW, Dover AFB, Delaware:			21103	62nd AW	r/y	
	3rd AS (bl/y);			21104	62nd AW	gn	
	437th AW, Charleston AFB,			21105	62nd AW	gn	
	South Carolina (y/bl):			21106	62nd AW	gn	
	14th AS, 15th AS, & 16th AS			21107	97th AMW	r/y	
	445th AW AFRC,			21108	62nd AW	gn	
	Wright-Patterson AFB, Ohio:			21109	62nd AW	gn	
	89th AS (r/w);			21110	164th AW	r	
	452nd AMW AFRC, March ARB,			21111	62nd AW	gn	
	California:			21112	172nd AW	bl/gd	
	729th AS (or/y)			FY92			
	FY00			23291	164th AW	r	
	00171	3rd Wg	w/bk	23292	437th AW	y/bl	
	00172	97th AMW	r/y	23293	437th AW	y/bl	
	00174	3rd Wg	w/bk	23294	62nd AW	gn	
	00175	62nd AW	gn	FY93			
	00176	164th AW	r	30599	3rd Wg	w/bk	
	00177	105th AW	bl	30600	164th AW	r	
	00178	97th AMW	r/y	30601	62nd AW	gn	
	00179	97th AMW	r/y	30602	437th AW	y/bl	
	00180	62nd AW	gn	30603	445th AW	r/w	
	00181	62nd AW	gn	30604	445th AW	r/w	
	00182	62nd AW	gn	FY03			
	00183	62nd AW	gn	33113	172nd AW	bl/gd	
	00184	62nd AW	gn	33114	172nd AW	bl/gd	
	00185	3rd Wg	w/bk	33115	172nd AW	bl/gd	
	FY10			33116	172nd AW	bl/gd	
	00213	437th AW	y/bl	33117	172nd AW	bl/gd	
	00214	437th AW	y/bl	33118	172nd AW	bl/gd	
	00215	437th AW	y/bl	33119	172nd AW	bl/gd	
	00216	62nd AW	gn	33120	62nd AW	gn	
	00217	62nd AW	gn	33121	412th TW		
	00218	62nd AW	gn	33122	437th AW	y/bl	

C-17A

Notes	Type			Type			Notes
	33123	437th AW	y/bl	66159	60th AMW	bk/w	
	33124	437th AW	y/bl	66160	60th AMW	bk/w	
	33125	305th AMW	bl	66161	60th AMW	bk/w	
	33126	305th AMW	bl	66162	60th AMW	bk/w	
	33127	62nd AW	gn	66163	60th AMW	bk/w	
	FY94			66164	60th AMW	bk/w	
	40065	164th AW	r	66165	436th AW	bl/y	
	40066	62nd AW	gn	66166	436th AW	bl/y	
	40067	105th AW	bl	66167	436th AW	bl/y	
	40068	445th AW	r/w	66168	436th AW	bl/y	
	40069	437th AW	y/bl	FY97			
	40070	167th AW	r	70041	437th AW	y/bl	
	FY04			70042	62nd AW	gn	
	44128	305th AMW	bl	70043	452nd AMW	or/y	
	44129	437th AW	y/bl	70044	445th AW	r/w	
	44130	305th AMW	bl	70045	105th AW	bl	
	44131	305th AMW	bl	70046	164th AW	r	
	44132	305th AMW	bl	70047	437th AW	y/bl	
	44133	305th AMW	bl	70048	445th AW	r/w	
	44134	305th AMW	bl	FY07			
	44135	437th AW	y/bl	77169	436th AW	bl/y	
	44136	305th AMW	bl	77170	436th AW	bl/y	
	44137	305th AMW	bl	77171	305th AMW	bl	
	44138	452nd AMW	or/y	77172	60th AMW	bk/w	
	FY95			77173	436th AW	bl/y	
	50102	437th AW	y/bl	77174	436th AW	bl/y	
	50103	62nd AW	gn	77175	436th AW	bl/y	
	50104	164th AW	r	77176	436th AW	bl/y	
	50105	105th AW	bl	77177	436th AW	bl/y	
	50106	62nd AW	gn	77178	305th AMW	bl	
	50107	437th AW	y/bl	77179	60th AMW	bk/w	
	FY05			77180	437th AW	y/bl	
	55139	452nd AMW	or/y	77181	437th AW	y/bl	
	55140	452nd AMW	or/y	77182	437th AW	y/bl	
	55141	452nd AMW	or/y	77183	437th AW	y/bl	
	55142	452nd AMW	or/y	77184	437th AW	y/bl	
	55143	452nd AMW	or/y	77185	437th AW	y/bl	
	55144	452nd AMW	or/y	77186	437th AW	y/bl	
	55145	452nd AMW	or/y	77187	437th AW	y/bl	
	55146	15th Wg	r/y	77188	437th AW	y/bl	
	55147	15th Wg	r/y	77189	437th AW	y/bl	
	55148	15th Wg	r/y	FY98			
	55149	15th Wg	r/y	80049	97th AMW	r/y	
	55150	15th Wg	r/y	80050	97th AMW	r/y	
	55151	15th Wg	r/y	80051	3rd Wg	w/bk	
	55152	15th Wg	r/y	80052	62nd AW	gn	
	55153	15th Wg	r/y	80053	62nd AW	gn	
	FY96			80054	164th AW	r	
	60001	62nd AW	gn	80055	97th AMW	r/y $	
	60002	437th AW	y/bl	80056	3rd Wg	w/bk	
	60003	62nd AW	gn	80057	105th AW	bl	
	60004	62nd AW	gn	FY88			
	60005	105th AW	bl	80265	62nd AW	gn	
	60006	437th AW	y/bl	80266	437th AW	y/bl	
	60007	172nd AW	bl/gd	FY08			
	60008	97th AMW	r/y	88190	437th AW	y/bl	
	FY06			88191	437th AW	y/bl	
	66154	60th AMW	bk/w	88192	62nd AW	gn	
	66155	60th AMW	bk/w	88193	62nd AW	gn	
	66156	60th AMW	bk/w	88194	62nd AW	gn	
	66157	60th AMW	bk/w	88195	62nd AW	gn	
	66158	60th AMW	bk/w	88196	62nd AW	gn	

Notes	Type			Type	Notes
	88197	62nd AW	gn	**C-20K Gulfstream III**	
	88198	437th AW	y/bl	*FY87*	
	88199	62nd AW	gn	70139 412th TW	
	88200	62nd AW	gn		
	88201	62nd AW	gn	**Lockheed Martin**	
	88202	62nd AW	gn	**F-22A Raptor**	
	88203	62nd AW	gn	1st FW, Langley AFB, Virginia [FF]:	
	88204	437th AW	y/bl	27th FS & 94th FS;	
	FY99			3rd Wg, Elmendorf AFB, Alaska [AK]:	
	90058	62nd AW	gn	90th FS & 525th FS;	
	90059	62nd AW	gn	15th Wg, Hickam AFB, Hawaii [HH]:	
	90060	62nd AW	gn	19th FS;	
	90061	97th AMW	r/y	44th FG AFRES, Tyndall AFB,	
	90062	437th AW	y/bl	Florida [TY]:	
	90063	97th AMW	r/y	301st FS;	
	90064	97th AMW	r/y	53rd Wg, Nellis AFB, Nevada [OT]:	
	90165	445th AW	r/w	422nd TES;	
	90166	62nd AW	gn	57th Wg, Nellis AFB, Nevada [WA]:	
	90167	3rd Wg	w/bk	433rd WPS;	
	90168	3rd Wg	w/bk	154th Wg, Hickam AFB, Hawaii ANG [HH]:	
	90169	437th AW	y/bl	199th FS;	
	90170	3rd Wg	w/bk	192nd FW, Langley AFB, Virginia ANG [FF]:	
	FY89			149th FS;	
	91189	437th AW	y/bl	325th FW, Tyndall AFB, Florida [TY]:	
	91190	62nd AW	gn	43rd FS & 95th FS;	
	91191	105th AW	bl	412nd FW, Edwards AFB, California [ED]:	
	91192	437th AW	y/bl	411th FLTS;	
	FY09			477th FG AFRES, Elmendorf AFB,	
	99205	437th AW	y/bl	Alaska [AK]:	
	99206	437th AW	y/bl	302nd FS	
	99207	437th AW	y/bl	*FY00*	
	99208	437th AW	y/bl	00-4012 TY 43rd FS	
	99209	62nd AW	gn	00-4015 TY 43rd FS	
	99210	62nd AW	gn	00-4016 TY 43rd FS	
	99211	62nd AW	gn	00-4017 TY 43rd FS	
	99212	437th AW	y/bl	*FY01*	
				01-4018 TY 43rd FS	
	Grumman			01-4019 TY 43rd FS	
	C-20 Gulfstream III/IV			01-4020 TY 43rd FS	
	89th AW, Andrews AFB, Maryland:			01-4021 TY 43rd FS	
	99th AS;			01-4022 TY 43rd FS	
	412th FLTS/412th TW Edwards AFB,			01-4023 TY 43rd FS	
	California;			01-4024 TY 43rd FS	
	OSAC/PAT, US Army, Andrews AFB,			01-4025 TY 43rd FS	
	Maryland;			01-4026 TY 43rd FS	
	Pacific Flight Detachment,			01-4027 TY 43rd FS	
	Hickam AFB, Hawaii			*FY02*	
	C-20B Gulfstream III			02-4028 TY 43rd FS	
	FY86			02-4029 TY 43rd FS	
	60201	89th AW		02-4030 TY 43rd FS	
	60202	89th AW		02-4031 TY 43rd FS	
	60203	89th AW		02-4032 TY 43rd FS	
	60204	89th AW		02-4033 TY 43rd FS	
	60206	89th AW		02-4034 TY 43rd FS	
	C-20F Gulfstream IV			02-4035 TY 43rd FS [325 OG]	
	FY91			02-4036 TY 43rd FS	
	10108	OSAC/PAT		02-4038 TY 43rd FS	
				02-4039 TY 43rd FS	
				02-4040 TY 43rd FS [325 FW]	
				FY03	
				03-4041 TY 43rd FS	
				03-4042 TY 43rd FS	

Notes	Type			Type			Notes
	03-4043	TY	43rd FS [43 FS]	05-4103	AK	90th FS	[3 Wg]
	03-4044	TY	43rd FS	05-4104	TY	95th FS	
	03-4045	HH	199th FS	05-4105	TY	95th FS	[95 FS]
	03-4046	HH	199th FS	05-4106	TY	95th FS	
	03-4047	HH	199th FS	05-4107	TY	95th FS	
	03-4048	HH	199th FS	*FY06*			
	03-4049	HH	199th FS	06-4108	AK	525th FS	
	03-4050	HH	199th FS	06-4109	WA	433rd WPS	
	03-4051	HH	199th FS	06-4110	AK	525th FS	[11 AF]
	03-4052	HH	199th FS	06-4111	OT	422nd TES	
	03-4053	HH	199th FS	06-4112	AK	525th FS	
	03-4054	HH	199th FS	06-4113	AK	525th FS	[3 OG]
	03-4055	HH	199th FS	06-4114	AK	525th FS	
	03-4056	HH	199th FS	06-4115	AK	525th FS	[525 FS]
	03-4057	HH	199th FS	06-4116	WA	433rd WPS	
	03-4058	HH	199th FS	06-4117	AK	525th FS	
	03-4059	HH	199th FS	06-4118	AK	525th FS	
	03-4060	HH	199th FS	06-4119	AK	525th FS	
	03-4061	HH	199th FS	06-4120	OT	422nd TES	
	FY04			06-4121	AK	525th FS	
	04-4062	HH	199th FS	06-4122	AK	525th FS	
	04-4063	HH	199th FS	06-4123	AK	525th FS	
	04-4064	HH	199th FS	06-4124	OT	422nd TES	
	04-4065	HH	199th FS	06-4126	AK	525th FS	
	04-4066	OT	422nd TES	06-4127	AK	525th FS	
	04-4067	FF	94th FS	06-4128	OT	422nd TES	
	04-4068	OT	422nd TES	06-4129	AK	525th FS	
	04-4069	OT	422nd TES	06-4130	AK	525th FS	
	04-4070	FF	94th FS	*FY07*			
	04-4071	WA	433rd WPS	07-4131	AK	525th FS	
	04-4072	TY	95th FS	07-4132	ED	411th FLTS	[411 FLTS]
	04-4073	FF	94th FS	07-4133	AK	525th FS	
	04-4074	AK	90th FS	07-4134	AK	525th FS	
	04-4075	AK	90th FS	07-4135	AK	90th FS	
	04-4076	TY	95th FS	07-4136	AK	90th FS	
	04-4077	AK	525th FS	07-4137	AK	90th FS	
	04-4078	TY	95th FS	07-4138	AK	90th FS	
	04-4079	TY	95th FS	07-4139	AK	90th FS	
	04-4080	TY	95th FS	07-4140	AK	90th FS	
	04-4081	TY	95th FS	07-4141	AK	90th FS	
	04-4082	FF	94th FS [192 FW]	07-4142	AK	90th FS	
	04-4083	TY	301st FS [301 FS]	07-4143	AK	90th FS	
	FY05			07-4144	AK	90th FS	
	05-4084	TY	95th FS	07-4145	AK	90th FS	
	05-4085	FF	94th FS	07-4146	AK	90th FS	
	05-4086	TY	95th FS	07-4147	AK	90th FS	
	05-4087	TY	95th FS	07-4148	AK	90th FS	
	05-4088	TY	95th FS	07-4149	AK	90th FS	
	05-4089	TY	95th FS	07-4150	AK	90th FS	
	05-4090	AK	90th FS	07-4151	AK	90th FS	
	05-4091	TY	95th FS	*FY08*			
	05-4092	AK	525th ES	08-4152	FF	94th FS	
	05-4093	TY	95th FS	08-4153	FF	27th FS	
	05-4094	TY	43rd FS	08-4154	FF	94th FS	
	05-4095	TY	95th FS [95 FS]	08-4155	FF	27th FS	
	05-4096	WA	433rd WPS	08-4156	FF	94th FS	
	05-4097	TY	95th FS	08-4157	FF	27th FS	
	05-4098	TY	95th FS	08-4158	FF	94th FS	
	05-4099	TY	95th FS	08-4159	FF	27th FS	
	05-4100	TY	95th FS	08-4160	FF	94th FS	
	05-4101	TY	95th FS	08-4161	FF	27th FS	
	05-4102	AK	90th FS [302 FS]	08-4162	FF	94th FS	

Notes	Type				Type		Notes
	08-4163	FF	27th FS		**Pilatus PC-12/U-28A**		
	08-4164	FF	94th FS		1st SOW, Hurlburt Field, Florida:		
	08-4165	FF	27th FS		34th SOS & 319th SOS;		
	08-4166	FF	94th FS		318th SOS/27th SOW, Cannon AFB,		
	08-4167	FF	27th FS		New Mexico		
	08-4168	FF	94th FS		*FY01*		
	08-4169	FF	94th FS		10415	318th SOS	
	08-4170	FF	27th FS		*FY04*		
	08-4171	FF	94th FS		40688	318th SOS	
	FY09				*FY05*		
	09-4172	FF	27th FS		50409	1st SOW	
	09-4173	FF	27th FS		50419	1st SOW	
	09-4174	FF	27th FS		50424	1st SOW	
	09-4175	FF	94th FS		50446	1st SOW	
	09-4176	FF	27th FS		50447	1st SOW	
	09-4177	FF	94th FS		50482	1st SOW	
	09-4178	FF	27th FS		50556	1st SOW	
	09-4179	FF	94th FS		50573	1st SOW	
	09-4180	FF	27th FS		50597	318th SOS	
	09-4181	FF	94th FS		*FY06*		
	09-4182	FF	27th FS		60740	318th SOS	
	09-4183	FF	94th FS		60692	1st SOW	
	09-4184	FF	27th FS		*FY07*		
	09-4185	FF	27th FS	[1 OG]	70488	1st SOW	
	09-4186	FF	27th FS		70691	318th SOS	
	09-4187	FF	94th FS		70711	1st SOW	
	09-4188	FF	27th FS		70712	1st SOW	
	09-4189	OT	422nd TES		70718	1st SOW	
	09-4190	AK	90th FS		70777	318th SOS	
	09-4191	FF	94th FS		70779	1st SOW	
	FY10				70790	1st SOW	
	10-4192	FF	94th FS	[192 FW]	70793	1st SOW	
	10-4193	AK	525th FS	[3 Wg]	70808	318th SOS	
	10-4194	FF	94th FS	[94 FS]	70809	1st SOW	
	10-4195	AK	525th FS	[525 FS]	70821	1st SOW	
	FY91				70829	1st SOW	
	91-4004	ED	411th FLTS		70838	1st SOW	
	91-4006	ED	411th FLTS		70840	1st SOW	
	91-4007	ED	411th FLTS	[412 TW]	*FY08*		
	91-4009	ED	411th FLTS		80519	1st SOW	
	FY99				80581	1st SOW	
	99-4010	OT	422nd TES	[422 TES]	80646	1st SOW	
	99-4011	WA	433rd WPS		80790	1st SOW	
					80809	1st SOW	
	Boeing VC-25A				80822	1st SOW	
	Presidential Airlift Sqn/				80835	1st SOW	
	89th AW, Andrews AFB,				80850	1st SOW	
	Maryland						
	FY82						
	28000						
	FY92						
	29000						

Boeing C-32
1st AS/89th AW, Andrews AFB,
   Maryland;
150th SOS/108th Wg, McGuire AFB,
   New Jersey;
486th FLTS/46th TW, Eglin AFB, Florida
FY00

09001	C-32B	150th SOS

FY02

24452	C-32B	150th SOS
25001	C-32B	486th FLTS

FY98

80001	C-32A	89th AW
80002	C-32A	89th AW
86006	C-32B	486th FLTS

FY99

90003	C-32A	89th AW
90004	C-32A	89th AW

FY09

90015	C-32A	89th AW
90016	C-32A	89th AW
90017	C-32A	89th AW

FY99

96143	C-32B	486th FLTS

Gulfstream Aerospace
C-37 Gulfstream V
6th AMW, MacDill AFB, Florida:
   310th AS;
15th Wg, Hickam AFB, Hawaii:
   65th AS;
89th AW, Andrews AFB, Maryland:
   99th AS;
OSAC/PAT, US Army, Andrews AFB,
   Maryland
C-37A Gulfstream V
FY01

10028	6th AMW
10029	6th AMW
10030	6th AMW
10065	15th Wg

FY02

21863	OSAC/PAT

FY04

41778	OSAC/PAT

FY97

71944	OSAC/PAT
70400	89th AW
70401	89th AW

FY99

90404	89th AW

C-37B Gulfstream V
FY11

10550	89th AW

FY06

60500	89th AW

FY09

90525	89th AW

IAI C-38A Astra
201st AS/113th FW, DC ANG,
   Andrews AFB, Maryland
FY94
41569
41570

Boeing C-40
15th Wg, Hickam AFB, Hawaii:
   65th AS;
86th AW, Ramstein, Germany:
   76th AS;
89th AW, Andrews AFB, Maryland:
   1st AS;
113th FW, DC ANG,
   Andrews AFB, Maryland:
   201st AS;
932nd AW AFRC, Scott AFB,
   Illinois:
   73rd AS
FY01

10015	C-40B	15th Wg
10040	C-40B	89th AW
10041	C-40B	89th AW

FY02

20042	C-40B	86th AW
20201	C-40C	201st AS
20202	C-40C	201st AS
20203	C-40C	201st AS
20204	C-40C	201st AS

FY05

50730	C-40C	932nd AW
50932	C-40C	932nd AW
54613	C-40C	932nd AW

FY09

90540	C-40C	932nd AW

Boeing KC-46A Pegasus
FY11

46001	(on order)
46002	(on order)
46003	(on order)
46004	(on order)
46005	(on order)

Notes	Type			Type			Notes
	Boeing B-52H Stratofortress			00062	2nd BW	r	
	2nd BW, Barksdale AFB, Louisiana [LA]:			*FY61*			
	20th BS (*bl*), 96th BS (*r*)			10001	5th BW	bk/y	
	& 343rd BS (*bk*);			10002	2nd BW	gn $	
	5th BW, Minot AFB, North Dakota [MT]:			10003	93rd BS	or/bl	
	23rd BS (*r/y*) & 69th BS (*bk/y*);			10004	2nd BW	bl	
	53rd TEG, Barksdale AFB,			10006	2nd BW	r	
	Louisiana [OT]:			10008	93rd BS	or/bl	
	49th TES;			10010	2nd BW	bl	
	93rd BS/307th BW AFRC,			10011	2nd BW	bk $	
	Barksdale AFB,			10012	2nd BW	r	
	Louisiana [BD] (*or/bl*);			10013	2nd BW	bl	
	419th FLTS/412th TW Edwards AFB,			10014	5th BW	bk/y	
	California [ED]			10015	2nd BW	r	
	FY60			10016	2nd BW	r	
	00001	2nd BW	r	10017	93rd BS	or/bl	
	00002	2nd BW	gn $	10018	5th BW	bk/y	
	00003	93rd BS	or/bl	10019	2nd BW	r	
	00004	5th BW	r/y	10020	2nd BW	bl $	
	00005	5th BW	r/y	10021	93rd BS	or/bl	
	00007	5th BW	r/y	10028	5th BW	bk/y	
	00008	2nd BW	bl $	10029	93rd BS	or/bl	
	00009	5th BW	bk/y	10031	93rd BS	or/bl	
	00011	93rd BS	r/y $	10032	5th BW	r/y	
	00012	5th BW	bk/y	10034	5th BW	r/y	
	00013	2nd BW	bl	10035	5th BW	r/y	
	00015	93rd BS	or/bl	10036	2nd BW	r	
	00017	5th BW	bk/y	10038	93rd BS	or/bl	
	00018	5th BW	bk/y	10039	5th BW	bk/y	
	00021	2nd BW	r	10040	5th BW	r/y	
	00022	2nd BW	r				
	00023	93rd BS	or/bl	**Lockheed**			
	00024	2nd BW	bl	**C-130H Hercules/**			
	00025	2nd BW	bl	**C-130J Hercules II*/**			
	00026	5th BW	r/y	**C-130J-30 Hercules II/**			
	00028	2nd BW	r	**AC-130H Spectre/**			
	00029	5th BW	r/y	**AC-130J Ghostrider/**			
	00031	53rd TEG		**AC-130U Spooky/**			
	00032	2nd BW	r	**AC-130W Stinger II/**			
	00033	5th BW	r/y	**EC-130H Compass Call/**			
	00035	93rd BS	y/bl	**EC-130J Commando Solo III/**			
	00036	412th TW		**HC-130J Combat King II/**			
	00037	5th BW	r/y	**HC-130N Combat King/**			
	00038	93rd BS	or/bl	**HC-130P Combat King/**			
	00041	93rd BS	or/bl	**MC-130H Combat Talon II/**			
	00042	93rd BS	or/bl	**MC-130J Commando II/**			
	00044	5th BW	bk/y	**MC-130P Combat Shadow/**			
	00045	93rd BS	or/bl	**MC-130W Combat/Dragon Spear**			
	00047	5th BW		1st SOS/353rd SOG, Kadena AB,			
	00048	2nd BW	bl	Japan;			
	00049	53rd TEG		4th SOS/1st SOW, Hurlburt Field,			
	00050	412th TW		Florida;			
	00051	93rd BS	or/bl	7th SOS/352nd SOG,			
	00052	2nd BW	r	RAF Mildenhall, UK;			
	00054	2nd BW	bl	8th SOS/1st SOW, Duke Field,			
	00055	5th BW	or/bk $	Florida;			
	00056	5th BW	r/y	9th SOS/27th SOW, Cannon AFB,			
	00057	93rd BS	or/bl	New Mexico;			
	00058	2nd BW	bl	15th SOS/1st SOW, Hurlburt Field,			
	00059	2nd BW	r $	Florida;			
	00060	5th BW	bk/y	16th SOS/27th SOW, Cannon AFB,			
	00061	93rd BS	or/bl $	New Mexico;			

Notes	Type	Type	Notes
	17th SOS/353rd SOG, Kadena AB, Japan;	Rhode Island ANG [RI] (r);	
	19th AW Little Rock AFB, Arkansas [LK]:	144th AS/176th CW, Elmendorf AFB, Alaska ANG (bk/y);	
	41st AS (gn), 50th AS (r),	154th TS/189th AW, Little Rock,	
	53rd AS (bk) & 61st AS (w);	Arkansas ANG (r);	
	37th AS/86th AW, Ramstein AB,	156th AS/145th AW, Charlotte,	
	Germany [RS] (bl/w);	North Carolina ANG [NC] (bl);	
	39th RQS/920th RQW AFRC,	157th FS/169th FW, McEntire ANGS,	
	Patrick AFB, Florida [FL];	South Carolina ANG [SC];	
	40th FTS/46th TW, Eglin AFB, Florida;	158th AS/165th AW, Savannah,	
	41st RQS/347th Wg, Moody AFB,	Georgia ANG (r);	
	Georgia [FT];	159th FS/125th FW, Jacksonville,	
	53rd WRS/403rd AW AFRC,	Florida ANG;	
	Keesler AFB, Missouri;	164th AS/179th AW, Mansfield,	
	55th ECG, Davis-Monthan AFB,	Ohio ANG [OH] (bl);	
	Arizona [DM]:	165th AS/123rd AW, Standiford Field,	
	41st ECS (bl) & 43rd ECS (r);	Kentucky ANG [KY];	
	58th SOW, Kirtland AFB,	167th AS/167th AW, Martinsburg,	
	New Mexico:	West Virginia ANG [WV] (r);	
	415th SOS & 550th SOS;	169th AS/182nd AW, Peoria,	
	67th SOS/352nd SOG,	Illinois ANG [IL] (or);	
	RAF Mildenhall, UK;	180th AS/139th AW,	
	71st RQS/23rd Wg, Moody AFB,	Rosencrans Memorial Airport,	
	Georgia [FT] (bl);	Missouri ANG [XP] (y);	
	73rd SOS/27th SOW, Cannon AFB,	181st AS/136th AW, NAS Dallas,	
	New Mexico;	Texas ANG (bl/w);	
	79th RQS/563rd RQG, Davis-Monthan	186th AS/120th AW, Great Falls,	
	AFB, Arizona [FT];	Montana ANG;	
	85th TES/53rd Wg, Eglin AFB,	187th AS/153rd AW, Cheyenne,	
	Florida [OT];	Wyoming ANG [WY] (y/bk);	
	95th AS/440th AW AFRC, Pope AFB,	192nd AS/152nd AW, Reno,	
	North Carolina (w/r);	Nevada ANG [NV] (bl/y);	
	96th AS/934th AW AFRC,	193rd SOS/193rd SOW, Harrisburg,	
	Minneapolis/St Paul,	Pennsylvania ANG [PA];	
	Minnesota (pr);	198th AS/156th AW, San Juan,	
	102nd RQS/106th RQW, Suffolk Field,	Puerto Rico ANG;	
	New York ANG [LI];	204th AS/15th Wg, Hickam AFB,	
	105th AS/118th AW, Nashville,	Hawaii ANG [HH];	
	Tennessee ANG (r);	210th RQS/176th CW, Elmendorf AFB,	
	109th AS/133rd AW,	Alaska ANG [AK];	
	Minneapolis/St Paul,	314th AW, Little Rock AFB,	
	Minnesota [MN] (pr/bk);	Arkansas:	
	115th AS/146th AW,	48th AS (y) & 62nd AS (bl);	
	Channel Island ANGS,	317th AG, Dyess AFB, Texas:	
	California ANG [CI] (gn);	39th AS (r) & 40th AS (bl);	
	118th AS/103rd AW,	327th AS/913th AW AFRC,	
	Bradley ANGB,	NAS Willow Grove,	
	Connecticut ANG (y/bk);	Pennsylvania (bk);	
	122nd FS/159th FW,	328th AS/914th AW AFRC,	
	NAS New Orleans,	Niagara Falls,	
	Louisiana ANG [JZ];	New York [NF] (bl);	
	130th AS/130th AW, Yeager Int'l	357th AS/908th AW AFRC, Maxwell AFB,	
	Airport, Charleston	Alabama (bl);	
	West Virginia ANG [WV] (pr/y);	374th AW, Yokota AB, Japan [YJ]:	
	130th RQS/129th RQW, Moffet Field,	36th AS (r);	
	California ANG [CA] (bl);	412th TW Edwards AFB, California:	
	139th AS/109th AW, Schenectady,	413rd FLTS & 452nd FLTS [ED];	
	New York ANG [NY];	645th Materiel Sqn, Palmdale,	
	142nd AS/166th AW,	California [D4];	
	New Castle County Airport,	700th AS/94th AW AFRC, Dobbins ARB,	
	Delaware ANG [DE] (bl);	Georgia [DB] (bl);	
	143rd AS/143rd AW, Quonset,	731st AS/302nd AW AFRC, Peterson AFB,	

Notes	Type				Type			Notes
	Colorado (pr/w);				11233	C-130H	165th AS	
	757th AS/910th AW AFRC,				11234	C-130H	165th AS	
	Youngstown ARS,				11235	C-130H	165th AS	
	Ohio [YO] (bl);				11236	C-130H	165th AS	
	758th AS/911th AW AFRC,				11237	C-130H	165th AS	
	Pittsburgh ARS,				11238	C-130H	165th AS	
	Pennsylvania (bk/y);				11239	C-130H	154th TS	r
	815th AS/403rd AW AFRC, Keesler AFB,				FY01			
	Missouri [KT] (r);				11461	C-130J*	115th AS	gn
	LMTAS, Marietta, Georgia				11462	C-130J*	115th AS	gn
	FY90				FY91			
	00162	MC-130H	1st SOS		11651	C-130H	154th TS	r
	00163	AC-130U	4th SOS		11652	C-130H	180th AS	y
	00164	AC-130U	4th SOS		11653	C-130H	187th AS	y/bk
	00165	AC-130U	4th SOS		FY01			
	00166	AC-130U	4th SOS		11935	EC-130J	193rd SOS	
	00167	AC-130U	4th SOS		FY64			
	FY80				14852	HC-130P	71st RQS	bl
	00320	C-130H	158th AS	r	14853	HC-130P	39th RQS	
	00321	C-130H	158th AS	r	14855	HC-130P	39th RQS	
	00322	C-130H	158th AS	r	14859	C-130E		
	00323	C-130H	158th AS	r	14860	HC-130P	71st RQS	bl
	00324	C-130H	158th AS	r	14861	C-130H	198th AS	
	00325	C-130H	158th AS	r	14862	EC-130H	55th ECG	
	00326	C-130H	158th AS	r	14864	HC-130P	39th RQS	
	00332	C-130H	158th AS	r	14866	C-130H	198th AS	
	FY90				FY11			
	01057	C-130H	142nd AS	bl	15719	HC-130J	79th RQS	
	01058	MC-130W	73rd SOS		15725	HC-130J	71st RQS	bl
	FY90				15727	HC-130J	71st RQS	bl
	01791	C-130H	180th AS	y	15729	MC-130J	LMTAS	
	01792	C-130H	180th AS	y	15731	MC-130J	67th SOS	
	01793	C-130H	180th AS	y	15732	C-130J	314th AW	y
	01794	C-130H	180th AS	y	15733	MC-130J	67th SOS	
	01795	C-130H	180th AS	y	15734	C-130J	19th AW	w
	01796	C-130H	180th AS	y	15735	MC-130J	9th SOS	
	01797	C-130H	180th AS	y	15736	C-130J	19th AW	w
	01798	C-130H	180th AS	y	15737	MC-130J	67th SOS	
	FY00				15738	C-130J	19th AW	w
	01934	EC-130J	193rd SOS		15740	C-130J	19th AW	w
	FY90				15745	C-130J	19th AW	w
	02103	HC-130N	210th RQS		15748	C-130J	19th AW	w
	FY10				15752	C-130J	19th AW	w
	05700	C-130J	317th AG	r	FY91			
	05701	C-130J	317th AG	r	19141	C-130H	328th AS	bl
	05714	MC-130J	67th SOS		19142	C-130H	328th AS	bl
	05716	HC-130J	79th RQS		19143	C-130H	328th AS	bl
	05717	HC-130J	79th RQS		19144	C-130H	328th AS	bl
	05728	C-130J	314th AW	y	FY82			
	FY90				20054	C-130H	164th AS	
	09107	C-130H	757th AS	bl	20055	C-130H	154th TS	r
	09108	C-130H	757th AS	bl	20056	C-130H	164th AS	
	FY81				20057	C-130H	144th AS	bk/y
	10626	C-130H	164th AS	bl	20058	C-130H	144th AS	bk/y
	10627	C-130H	154th TS	r	20059	C-130H	144th AS	bk/y
	10628	C-130H	154th TS	r	20060	C-130H	144th AS	bk/y
	10629	C-130H	164th AS	bl	20061	C-130H	144th AS	bk/y
	10630	C-130H	164th AS	bl	FY92			
	10631	C-130H	154th TS	r	20253	AC-130U	4th SOS	
	FY91				FY02			
	11231	C-130H	165th AS		20314	C-130J*	314th AW	y $
	11232	C-130H	165th AS					

Notes	Type				Type				Notes
	FY92				30487	C-130H	139th AS		
	20547	C-130H	130th AS	pr/y	30488	C-130H	164th AS	bl	
	20548	C-130H	169th AS	or	30489	C-130H	139th AS		
	20549	C-130H	187th AS	y/bk	30490	LC-130H	139th AS		
	20550	C-130H	700th AS	bl	30491	LC-130H	139th AS		
	20551	C-130H	700th AS	bl	30492	LC-130H	139th AS		
	20552	C-130H	700th AS	bl	30493	LC-130H	139th AS		
	20553	C-130H	156th AS	bl	FY93				
	20554	C-130H	192nd AS	bl/y	31036	C-130H	700th AS	bl	
	21094	LC-130H	139th AS		31037	C-130H	700th AS	bl	
	21095	LC-130H	139th AS		31038	C-130H	700th AS	bl	
	FY02				31039	C-130H	700th AS	bl	
	21434	C-130J	143rd AS	r	31040	C-130H	700th AS	bl	
	FY92				31041	C-130H	700th AS	bl	
	21451	C-130H	169th AS	or	31096	LC-130H	139th AS		
	21452	C-130H	169th AS	or	FY83				
	21453	C-130H	156th AS	bl	31212	MC-130H	1st SOS		
	21454	C-130H	156th AS	bl	FY93				
	FY02				31455	C-130H	156th AS	bl	
	21463	C-130J*	115th AS	gn	31456	C-130H	156th AS	bl	
	21464	C-130J*	115th AS	gn	31457	C-130H	156th AS	bl	
	FY92				31458	C-130H	156th AS	bl	
	21531	C-130H	187th AS	y/bk	31459	C-130H	156th AS	bl	
	21532	C-130H	187th AS	y/bk	31561	C-130H	156th AS	bl	
	21533	C-130H	187th AS	y/bk	31562	C-130H	156th AS	bl	
	21534	C-130H	187th AS	y/bk	31563	C-130H	156th AS	bl	
	21535	C-130H	187th AS	y/bk	FY73				
	21536	C-130H	187th AS	y/bk	31580	EC-130H	55th ECG		
	21537	C-130H	187th AS	y/bk	31581	EC-130H	55th ECG	$	
	21538	C-130H	187th AS	y/bk	31582	C-130H	19th AW	bk	
	FY62				31583	EC-130H	55th ECG	r	
	21863	HC-130P	71st RQS	bl	31584	EC-130H	55th ECG		
	FY92				31585	EC-130H	55th ECG	bl	
	23021	C-130H	327th AS	bk	31586	EC-130H	55th ECG	bl	
	23022	C-130H	757th AS	bl	31587	EC-130H	55th ECG		
	23023	C-130H	757th AS	bl	31588	EC-130H	55th ECG		
	23024	C-130H	757th AS	bl	31590	EC-130H	55th ECG		
	23281	C-130H	96th AS	pr	31592	EC-130H	55th ECG	bl	
	23282	C-130H	96th AS	pr	31594	EC-130H	55th ECG		
	23283	C-130H	96th AS	pr	31595	EC-130H	55th ECG		
	23284	C-130H	96th AS	pr	31597	C-130H	186th AS		
	23285	C-130H	96th AS	pr	31598	C-130H	186th AS		
	23286	C-130H	96th AS	pr	FY93				
	23287	C-130H	96th AS	pr	32041	C-130H	169th AS	or	
	23288	C-130H	96th AS	pr	32042	C-130H	169th AS	or	
	FY12				32104	HC-130N	210th RQS		
	25754	AC-130J	LMTAS		32105	HC-130N	210th RQS		
	25755	HC-130J	58th SOW		32106	HC-130N	210th RQS		
	25756	C-130J	19th AW	w	FY73				
	25757	MC-130J	67th SOS		33300	LC-130H	139th AS		
	25759	MC-130J	67th SOS		FY93				
	25760	MC-130J	67th SOS		37311	C-130H	187th AS	y/bk	
	25761	MC-130J	17th SOS		37312	C-130H	169th AS	or	
	25762	MC-130J			37313	C-130H	187th AS	y/bk	
	25763	MC-130J			37314	C-130H	187th AS	y/bk	
	25765	HC-130J			FY03				
	25768	HC-130J			38154	C-130J	815th AS	r	
	25769	HC-130J			FY84				
	FY02				40204	C-130H	327th AS	bk	
	28155	C-130J	815th AS	r	40205	C-130H	327th AS	bk	
	FY83				40206	C-130H	142nd AS	bl	
	30486	C-130H	154th TS	r	40207	C-130H	142nd AS	bl	

Notes	Type				Type				Notes
	40208	C-130H	142nd AS	bl	FY04				
	40209	C-130H	142nd AS	bl	48153	C-130J	815th AS	r	
	40210	C-130H	142nd AS	bl	FY85				
	40212	C-130H	142nd AS	bl	50011	MC-130H	15th SOS		
	40213	C-130H	142nd AS	bl	50035	C-130H	357th AS	bl	
	40476	MC-130H	15th SOS		50036	C-130H	357th AS	bl	
	FY74				50037	C-130H	95th AS	w/r	
	41658	C-130H	19th AW	bk	50038	C-130H	357th AS	bl	
	41659	C-130H	374th AW	r	50039	C-130H	357th AS	bl	
	41660	C-130H	374th AW	r	50040	C-130H	357th AS	bl	
	41661	C-130H	374th AW	r	50041	C-130H	357th AS	bl	
	41663	C-130H	374th AW	r $	50042	C-130H	357th AS	bl	
	41664	C-130H	118th AS	y/bk	FY65				
	41666	C-130H	374th AW	r $	50962	TC-130H	55th ECG		
	41667	C-130H	118th AS	y/bk	50963	C-130H	198th AS		
	41668	C-130H	19th AW	bk	50966	C-130H	198th AS		
	41669	C-130H	374th AW	r	50967	HC-130H	122nd FS		
	41670	C-130H	186th AS		50968	C-130H	198th AS		
	41671	C-130H	186th AS		50970	HC-130P	39th RQS		
	41674	C-130H	374th AW	r	50974	HC-130P	102nd RQS		
	41679	C-130H	186th AS		50976	HC-130P	39th RQS		
	41680	C-130H	118th AS	y/bk	50978	HC-130P	102nd RQS		
	41682	C-130H	374th AW	r	50980	WC-130H	198th AS		
	41685	C-130H	374th AW	r	50981	HC-130P	71st RQS	bl	
	41687	C-130H	118th AS	y/bk	50982	HC-130P	71st RQS	bl	
	41688	C-130H	118th AS	y/bk	50984	WC-130H	198th AS		
	41690	C-130H	186th AS		50985	WC-130H	198th AS		
	41691	C-130H	186th AS		50989	EC-130H	55th ECG	r	
	41692	C-130H	374th AW	r	50991	MC-130P	1st SOG		
	42061	C-130H	374th AW	r	FY95				
	42063	C-130H	118th AS	y/bk	51001	C-130H	109th AS	pr/bk	
	42065	C-130H	374th AW	r $	51002	C-130H	109th AS	pr/bk	
	42067	C-130H	374th AW	r	FY85				
	42069	C-130H	118th AS	y/bk	51361	C-130H	181st AS	bl/w	
	42070	C-130H	19th AW	bk	51362	C-130H	181st AS	bl/w	
	42131	C-130H	19th AW	bk	51363	C-130H	181st AS	bl/w	
	42132	C-130H	374th AW	r	51364	C-130H	181st AS	bl/w	
	42133	C-130H	118th AS	y/bk	51365	C-130H	181st AS	bl/w	
	42134	C-130H	118th AS	y/bk	51366	C-130H	181st AS	bl/w	
	FY04				51367	C-130H	181st AS	bl/w	
	43142	C-130J	37th AS	bl/w	51368	C-130H	181st AS	bl/w	
	43143	C-130J	19th AW	w	FY05				
	FY94				51435	C-130J	143rd AS	r	
	46701	C-130H	169th AS	or	51436	C-130J	143rd AS	r	
	46702	C-130H	169th AS	or	51465	C-130J	115th AS	gn	
	46703	C-130H	169th AS	or	51466	C-130J	115th AS	gn	
	46704	C-130H	154th TS	r	53145	C-130J	19th AW	bk	
	46705	C-130H	167th AS	r	53146	C-130J	314th AW	y	
	46706	C-130H	167th AS	r	53147	C-130J	19th AW	w	
	46707	C-130H	130th AS	pr/y	FY95				
	46708	C-130H	130th AS	pr/y	56709	C-130H	156th AS	bl	
	47310	C-130H	731st AS	pr/w	56710	C-130H	130th AS	pr/y	
	47315	C-130H	731st AS	pr/w	56711	C-130H	156th AS	bl	
	47316	C-130H	731st AS	pr/w	56712	C-130H	130th AS	pr/y	
	47317	C-130H	731st AS	pr/w	FY05				
	47318	C-130H	731st AS	pr/w	58152	C-130J	815th AS	r	
	47319	C-130H	731st AS	pr/w	58156	C-130J	815th AS	r	
	47320	C-130H	731st AS	pr/w	58157	C-130J	815th AS	r	
	47321	C-130H	731st AS	pr/w	58158	C-130J	815th AS	r	
	FY94								
	48151	C-130J*	314th AW	y					
	48152	C-130J*	815th AS	r					

Notes	Type				Type				Notes
	FY66					68611	C-130J	37th AS	bl/w
	60212	HC-130P	130th RQS	bl		68612	C-130J	37th AS	bl/w
	60216	HC-130P	130th RQS	bl		*FY87*			
	60217	MC-130P	1st SOG			70023	MC-130H		
	60219	HC-130P	130th RQS	bl		70024	MC-130H		
	60221	HC-130P	58th SOW			70125	MC-130H	15th SOS	
	60222	HC-130P	102nd RQS			70126	MC-130H	15th SOS	
	60223	MC-130P	130th RQS	bl		70128	AC-130U	4th SOS	
	60225	MC-130P				*FY97*			
	FY86					71351	C-130J*	314th AW	y
	60410	C-130H	95th AS	w/r		71352	C-130J*	314th AW	y
	60411	C-130H	95th AS	w/r		71353	C-130J*	314th AW	y
	60413	C-130H	327th AS	bk		71354	C-130J*	314th AW	y
	60414	C-130H	95th AS	w/r		*FY07*			
	60415	C-130H	327th AS	bk		71468	C-130J	115th AS	gn
	60418	C-130H	95th AS	w/r		*FY97*			
	60419	C-130H	327th AS	bk		71931	EC-130J	193rd SOS	
	FY96					*FY07*			
	61003	C-130H	109th AS	pr/bk		73170	C-130J	317th AG	bl
	61004	C-130H	109th AS	pr/bk		74635	C-130J	19th AW	w
	61005	C-130H	109th AS	pr/bk		74636	C-130J	19th AW	w
	61006	C-130H	109th AS	pr/bk		74637	C-130J	19th AW	w
	61007	C-130H	109th AS	pr/bk		74638	C-130J	19th AW	w
	61008	C-130H	109th AS	pr/bk		74639	C-130J	19th AW	w
	FY86					*FY97*			
	61391	C-130H	154th TS	r		75303	WC-130J	53rd WRS	
	61392	C-130H	154th TS	r		75304	WC-130J	53rd WRS	
	61393	C-130H	154th TS	r		75305	WC-130J	53rd WRS	
	61394	C-130H	154th TS	r		75306	WC-130J	53rd WRS	
	61395	C-130H	154th TS	r		*FY07*			
	61396	C-130H	154th TS	r		76310	C-130J	19th AW	w
	61397	C-130H	154th TS	r		76311	C-130J	19th AW	w
	61398	C-130H	154th TS	r		76312	C-130J	19th AW	w
	FY06					78608	C-130J	37th AS	bl/w
	61437	C-130J	143rd AS	r		78609	C-130J	37th AS	bl/w
	61438	C-130J	143rd AS	r		78613	C-130J	37th AS	bl/w
	61467	C-130J	115th AS	gn		78614	C-130J	37th AS	bl/w
	FY86					*FY87*			
	61699	MC-130H	58th SOW			79281	C-130H	328th AS	bl
	FY06					79282	C-130H	95th AS	w/r
	63171	C-130J	317th AG	bl		79283	C-130H	96th AS	pr
	FY76					79284	C-130H	95th AS	w/r
	63301	LC-130H	139th AS			79285	C-130H	328th AS	bl
	63302	LC-130H	139th AS			79286	MC-130W	73rd SOS	
	FY06					79287	C-130H	758th AS	bl
	64631	C-130J	19th AW	w		79288	MC-130W	73rd SOS	
	64632	C-130J	19th AW	w		*FY88*			
	64633	C-130J	19th AW	w		80191	MC-130H	1st SOS	
	64634	C-130J	19th AW	w		80192	MC-130H	15th SOS	
	FY96					80193	MC-130H	16th SOS	
	65300	WC-130J	53rd WRS			80194	MC-130H	58th SOW	
	65301	WC-130J	53rd WRS			80195	MC-130H		
	65302	WC-130J	53rd WRS			80264	MC-130H	15th SOS	
	67322	C-130H	731st AS	pr/w		*FY78*			
	67323	C-130H	731st AS	pr/w		80806	C-130H	327th AS	bk
	67324	C-130H	731st AS	pr/w		80807	C-130H	758th AS	bk/y
	67325	C-130H	731st AS	pr/w		80808	C-130H	758th AS	bk/y
	68153	EC-130J	193rd SOS			80809	C-130H	758th AS	bk/y
	68154	EC-130J	193rd SOS			80810	C-130H	758th AS	bk/y
	FY06					80811	C-130H	758th AS	bk/y
	68159	C-130J	815th AS	r		80812	C-130H	758th AS	bk/y
	68610	C-130J	37th AS	bl/w		80813	C-130H	327th AS	bk

Notes	Type				Type				Notes
	FY88				86205	MC-130J	9th SOS		
	81301	MC-130W	73rd SOS		86206	MC-130J	9th SOS		
	81302	MC-130W	73rd SOS		88601	C-130J	37th AS	bl/w $	
	81303	MC-130W	73rd SOS		88602	C-130J	37th AS	bl/w $	
	81304	MC-130W	73rd SOS		88603	C-130J	37th AS	bl/w $	
	81305	MC-130W	73rd SOS		88604	C-130J	37th AS	bl/w	
	81306	MC-130W	73rd SOS		88605	C-130J	37th AS	bl/w	
	81307	MC-130W	73rd SOS		88606	C-130J	314th AW	y	
	81308	MC-130W	73rd SOS		88607	C-130J	37th AS	bl/w	
	FY98				*FY09*				
	81355	C-130J*	314th AW	y	90108	HC-130J	85th TES		
	81356	C-130J*	314th AW	y	90109	HC-130J	58th SOW		
	81357	C-130J*	314th AW	y	*FY89*				
	81358	C-130J*	314th AW	y	90280	MC-130H	1st SOS		
	FY88				90281	MC-130H	15th SOS		
	81803	MC-130H			90282	MC-130H	15th SOS		
	FY98				90283	MC-130H	1st SOS		
	81932	EC-130J	193rd SOS		*FY79*				
	FY88				90473		C-130H	192nd AS	bl/y
	82101	HC-130N	102nd RQS		90474		C-130H	154th TS	r
	82102	HC-130N	102nd RQS		90475		C-130H	192nd AS	bl/y
	FY08				90476		C-130H	192nd AS	bl/y
	83172	C-130J	317th AG	bl	90477		C-130H	192nd AS	bl/y
	83173	C-130J	317th AG	bl	90478		C-130H	192nd AS	bl/y
	83174	C-130J	317th AG	bl	90479		C-130H	192nd AS	bl/y
	83175	C-130J	317th AG	bl	90480		C-130H	192nd AS	bl/y
	83176	C-130J	317th AG	bl	*FY89*				
	83177	C-130J	317th AG	bl	90509	AC-130U	4th SOS		
	83178	C-130J	317th AG	bl	90510	AC-130U	4th SOS		
	83179	C-130J	317th AG	bl	90511	AC-130U	4th SOS		
	FY88				90512	AC-130U	4th SOS		
	84401	C-130H	95th AS	w/r $	90513	AC-130U	4th SOS		
	84402	C-130H	95th AS	w/r	90514	AC-130U	4th SOS		
	84403	C-130H	95th AS	w/r	91051	MC-130W	73rd SOS		
	84404	C-130H	95th AS	w/r	91052	AC-130U	4th SOS		
	84405	C-130H	95th AS	w/r	91053	AC-130U	4th SOS		
	84406	C-130H	327th AS	bk	91054	AC-130U	4th SOS		
	84407	C-130H	327th AS	bk	91055		C-130H	758th AS	bl
	FY98				91056	AC-130U	4th SOS		
	85307	WC-130J	53rd WRS		91181		C-130H	154th TS	r
	85308	WC-130J	53rd WRS		91182		C-130H	144th AS	bk/y
	FY08				91183		C-130H	144th AS	bk/y
	85675	C-130J	317th AG	bl	91184		C-130H	144th AS	bk/y
	85678	C-130J	317th AG	bl	91185		C-130H	144th AS	bk/y
	85679	C-130J	317th AG	bl	91186		C-130H	327th AS	bl
	85683	C-130J	317th AG	r	91187		C-130H	327th AS	bl
	85684	C-130J	317th AG	r	91188		C-130H	327th AS	bl
	85685	C-130J	317th AG	r	*FY99*				
	85686	C-130J	317th AG	r	91431		C-130J	143rd AS	r
	85691	C-130J	317th AG	r	91432		C-130J	143rd AS	r
	85692	C-130J	317th AG	r	91433		C-130J	143rd AS	r
	85693	C-130J	317th AG	r	91933	EC-130J	193rd SOS		
	85697	MC-130J	9th SOS		95309	WC-130J	53rd WRS		
	85705	C-130J	317th AG	r	*FY09*				
	85712	C-130J	317th AG	r	95706	HC-130J	58th SOW		
	85715	C-130J	317th AG	r	95707	HC-130J	79th RQS		
	85724	C-130J	317th AG	r	95708	HC-130J	79th RQS		
	85726	C-130J	317th AG	r	95709	HC-130J	79th RQS		
	86201	MC-130J	19th SOS		95710	AC-130J	412th TW		
	86202	MC-130J	58th SOW		95711	MC-130J	522nd SOS		
	86203	MC-130J	58th SOW		95713	MC-130J	522nd SOS		
	86204	MC-130J	9th SOS						

Notes	Type			

FY69

95819	MC-130P	1st SOG	
95820	MC-130P	1st SOG	
95824	HC-130N	39th RQS	
95826	MC-130P		
95828	MC-130P	1st SOG	
95829	HC-130N	58th SOW	
95830	HC-130N	58th SOW	
95832	MC-130P		
95833	HC-130N	58th SOW	

FY09

96207	MC-130J	9th SOS	
96208	MC-130J	58th SOW	
96209	MC-130J	58th SOW	
96210	MC-130J	9th SOS	

FY89

99101	C-130H	327th AS	bk
99102	C-130H	327th AS	bk
99103	C-130H	757th AS	bl
99104	C-130H	757th AS	bl/r
99105	C-130H	757th AS	bl
99106	C-130H	757th AS	bl

Boeing C-135

6th AMW, MacDill AFB, Florida:
  91st ARS (*y/bl*);
15th Wg, Hickam AFB, Hawaii:
  65th AS;
18th Wg, Kadena AB, Japan [ZZ]:
  909th ARS (*or/bk*);
22nd ARW, McConnell AFB, Kansas:
  344th ARS (*y/bk*), 349th ARS (*y/bl*)
  350th ARS (*y/r*) & 384th ARS (*y/pr*);
55th Wg, Offutt AFB, Nebraska [OF]:
  38th RS (*gn*), 45th RS (*bk*)
  & 343rd RS;
88th ABW, Wright-Patterson AFB, Ohio:
92nd ARW, Fairchild AFB, Washington:
  92nd ARS (*bk*), 93rd ARS (*bl*),
  & 97th ARS (*y*);
97th AMW, Altus AFB, Oklahoma:
  54th ARS (*y/r*);
100th ARW, RAF Mildenhall, UK [D]:
  351st ARS (*r/w/bl*);
106th ARS/117th ARW, Birmingham,
  Alabama ANG (*w/r*);
108th ARS/126th ARW,
  Scott AFB, Illinois ANG (*w/bl*);
117th ARS/190th ARW, Forbes Field,
  Kansas ANG (*bl/y*);
121st ARW, Rickenbacker ANGB,
  Ohio ANG:
  145th ARS & 166th ARS (*bl*);
126th ARS/128th ARW, Mitchell Field,
  Wisconsin ANG (*w/bl*);
132nd ARS/101st ARW, Bangor,
  Maine ANG (*w/gn*);
133rd ARS/157th ARW, Pease ANGB,
  New Hampshire ANG (*bl*);
141st ARS/108th ARW, McGuire AFB,
  New Jersey ANG (*r*);
151st ARS/134th ARW, Knoxville,

Tennessee ANG (*w/or*);
153rd ARS/186th AW, Meridian,
  Mississippi ANG;
168th ARS/168th ARW, Eielson AFB,
  Alaska ANG [AK] (*bl/y*);
171st ARS/127th Wg, Selfridge ANGB,
  Michigan ANG (*bk/y*);
171st ARW, Greater Pittsburgh,
  Pennsylvania ANG:
  146th ARS (*y/bk*) &
  147th ARS (*bk/y*);
173rd ARS/155th ARW, Lincoln,
  Nebraska ANG (*r/w*);
174th ARS/185th ARW, Sioux City,
  Iowa ANG (*y/bk*);
191st ARS/151st ARW, Salt Lake City,
  Utah ANG (*bl/bk*);
196th ARS/163rd ARW, March ARB,
  California ANG (*bl/w*);
197th ARS/161st ARW, Phoenix,
  Arizona ANG;
203rd ARS/15th Wg, Hickam AFB,
  Hawaii ANG [HH] (*y/bk*);
366th Wg, Mountain Home AFB,
  Idaho [MO]:
  22nd ARS (*y/gn*);
412th TW, Edwards AFB,
  California [ED]:
  418th FLTS (*or*);
434th ARW AFRC, Grissom AFB,
  Indiana:
  72nd ARS (*bl*) & 74th ARS (*r/w*);
452nd AMW AFRC, March ARB,
  California:
  336th ARS (*or/y*);
459th ARW AFRC, Andrews AFB,
  Maryland:
  756th ARS (*y/bk*);
507th ARW AFRC, Tinker AFB,
  Oklahoma:
  465th ARS (*bl/y*);
645th Materiel Sqn, Greenville, Texas;
916th ARW AFRC, Seymour Johnson AFB,
  North Carolina:
  77th ARS (*gn/y*);
927th ARW AFRC, MacDill AFB, Florida:
  63rd ARS (*pr/w*);
L3 Systems, Greenville, Texas

FY60

00313	KC-135R	22nd ARW	
00314	KC-135R	434th ARW	r/w
00315	KC-135R	126th ARS	w/bl
00316	KC-135R	191st ARS	bl/bk
00318	KC-135R	203rd ARS	y/bk
00320	KC-135R	6th AMW	y/bl
00322	KC-135R	434th ARW	bl
00323	KC-135R	203rd ARS	y/bk
00324	KC-135R	6th AMW	y/bl
00328	KC-135R	18th Wg	or/bk
00329	KC-135R	203rd ARS	y/bk
00331	KC-135R	22nd ARW	
00332	KC-135R	168th ARS	bl/y

Notes	Type				Type				Notes
	00333	KC-135R	92nd ARW	bk	10323	KC-135R	6th AMW	y/bl	
	00334	KC-135R	168th ARS	bl/y	10324	KC-135R	452nd AMW	or/y	
	00335	KC-135T	6th AMW	y/bl	12662	RC-135S	55th Wg	bk	
	00336	KC-135T	6th AMW	y/bl	12663	RC-135S	55th Wg	bk	
	00337	KC-135T	22nd ARW		12666	NC-135W	645th MS		
	00339	KC-135T	6th AMW	y/bl	12667	WC-135W	55th Wg	bk	
	00341	KC-135R	153rd ARS		12670	OC-135B	55th Wg		
	00342	KC-135T	92nd ARW		12672	OC-135B	55th Wg		
	00343	KC-135T	22nd ARW		*FY64*				
	00344	KC-135T	22nd ARW		14828	KC-135R	191st ARS	bl/bk	
	00345	KC-135T	171st ARS	bk/y	14829	KC-135R	197th ARS		
	00346	KC-135T	171st ARS	bk/y	14830	KC-135R	L3 Systems		
	00347	KC-135R	121st ARW	bl	14831	KC-135R	197th ARS		
	00348	KC-135R	18th Wg	or/bk	14832	KC-135R	151st ARS	w/or	
	00349	KC-135R	916th ARW	gn/y	14834	KC-135R	434th ARW	r/w	
	00350	KC-135R	18th Wg	or/bk	14835	KC-135R	452nd AMW	or/y	
	00351	KC-135R	97th AMW	y/r	14836	KC-135R	133rd ARS	bl	
	00353	KC-135R	92nd ARW		14837	KC-135R	100th ARW	r/w/bl	
	00355	KC-135R	22nd ARW		14838	KC-135R	6th AMW	y/bl	
	00356	KC-135R	22nd ARW		14839	KC-135R	108th ARS	w/bl	
	00357	KC-135R	22nd ARW		14840	KC-135R	121st ARW	bl	
	00358	KC-135R	108th ARS	w/bl	14841	RC-135V	55th Wg	bl	
	00359	KC-135R	434th ARW	r/w	14842	RC-135V	55th Wg	gn	
	00360	KC-135R	97th AMW	y/r	14843	RC-135V	55th Wg	gn	
	00362	KC-135R	22nd ARW		14844	RC-135V	55th Wg	gn	
	00363	KC-135R	434th ARW	bl	14845	RC-135V	55th Wg	bl	
	00364	KC-135R	434th ARW	r/w	14846	RC-135V	55th Wg	gn	
	00365	KC-135R	117th ARS	bl/y	14847	RC-135U	55th Wg	bk	
	00366	KC-135R	141st ARS	r	14848	RC-135V	55th Wg	gn	
	00367	KC-135R	121st ARW	bl	14849	RC-135U	55th Wg	bk	
	FY61				*FY62*				
	10264	KC-135R	121st ARW	bl	23498	KC-135R	18th Wg	or/bk	
	10266	KC-135R	117th ARS	bl/y	23499	KC-135R	92nd ARW		
	10267	KC-135R	22nd ARW		23500	KC-135R	126th ARS	w/bl	
	10272	KC-135R	434th ARW	r/w	23502	KC-135R	97th AMW	y/r	
	10275	KC-135R	191st ARS	bl/bk	23503	KC-135R	507th ARW	bl/y	
	10276	KC-135R	173rd ARS	r/w	23505	KC-135R	97th AMW	y/r	
	10277	KC-135R	117th ARS	bl/y	23506	KC-135R	133rd ARS	bl	
	10280	KC-135R	452nd AMW	or/y	23507	KC-135R	22nd ARW		
	10284	KC-135R	197th ARS		23508	KC-135R	141st ARS	r	
	10288	KC-135R	92nd ARW		23509	KC-135R	916th ARW	gn/y	
	10290	KC-135R	203rd ARS	y/bk	23510	KC-135R	434th ARW	r/w	
	10292	KC-135R	97th AMW	y/r	23511	KC-135R	121st ARW	bl	
	10293	KC-135R	22nd ARW		23512	KC-135R	126th ARS	w/bl	
	10294	KC-135R	916th ARW	gn/y	23513	KC-135R	132nd ARS	w/gn	
	10295	KC-135R	22nd ARW		23514	KC-135R	141st ARS	r	
	10298	KC-135R	126th ARS	w/bl	23515	KC-135R	133rd ARS	bl	
	10299	KC-135R	100th ARW	r/w/bl	23516	KC-135R	197th ARS $		
	10300	KC-135R	108th ARS	w/bl	23517	KC-135R	97th AMW	y/r	
	10305	KC-135R	6th AMW	y/bl	23518	KC-135R	434th ARW	bl	
	10307	KC-135R	459th ARW	y/bk	23519	KC-135R	92nd ARW		
	10308	KC-135R	97th AMW	y/r	23521	KC-135R	434th ARW	bl	
	10309	KC-135R	126th ARS	w/bl	23523	KC-135R	22nd ARW		
	10310	KC-135R	133rd ARS	bl	23524	KC-135R	168th ARS	bl/y	
	10311	KC-135R	22nd ARW	y	23526	KC-135R	173rd ARS	r/w	
	10313	KC-135R	916th ARW	gn/y	23528	KC-135R	916th ARW	gn/y	
	10314	KC-135R	22nd ARW		23529	KC-135R	6th AMW	y/bl	
	10315	KC-135R	92nd ARW		23530	KC-135R	434th ARW	bl	
	10317	KC-135R	117th ARS	bl/y	23531	KC-135R	121st ARW	bl	
	10318	KC-135R	106th ARS	w/r	23533	KC-135R	452nd AMW	or/y	
	10320	KC-135R	412th TW	or	23534	KC-135R	22nd ARW		
	10321	KC-135R	100th ARW	r/w/bl	23537	KC-135R	507th ARW	bl/y	

Notes	Type				Type				Notes
	23538	KC-135R	22nd ARW		37995	KC-135R	22nd ARW	gy	
	23540	KC-135R	22nd ARW		37996	KC-135R	434th ARW	bl	
	23541	KC-135R	22nd ARW		37997	KC-135R	22nd ARW		
	23542	KC-135R	916th ARW	gn/y	37999	KC-135R	92nd ARW		
	23543	KC-135R	459th ARW	y/bk	38000	KC-135R	92nd ARW		
	23544	KC-135R	141st ARS	r	38002	KC-135R	22nd ARW	gy/si	
	23545	KC-135R	22nd ARW		38003	KC-135R	141st ARS	r	
	23547	KC-135R	133rd ARS	bl	38004	KC-135R	117th ARS	bl/y	
	23549	KC-135R	97th AMW	y/r	38006	KC-135R	97th AMW	y/r	
	23550	KC-135R	197th ARS		38007	KC-135R	106th ARS	w/r	
	23551	KC-135R	100th ARW	r/w/bl	38008	KC-135R			
	23552	KC-135R	22nd ARW		38011	KC-135R	6th AMW	y/bl	
	23553	KC-135R	18th Wg	or/bk	38012	KC-135R	412th TW		
	23554	KC-135R	22nd ARW		38013	KC-135R	121st ARW	bl	
	23556	KC-135R	459th ARW	y/bk	38014	KC-135R	916th ARW	gn/y	
	23557	KC-135R	916th ARW	gn/y	38015	KC-135R	168th ARS	bl/y	
	23558	KC-135R	452nd AMW	or/y	38017	KC-135R	92nd ARW		
	23559	KC-135R	92nd ARW		38018	KC-135R	173rd ARS	r/w	
	23561	KC-135R	97th AMW	y/r	38019	KC-135R	22nd ARW		
	23562	KC-135R	18th Wg	or/bk	38020	KC-135R	22nd ARW		
	23564	KC-135R	22nd ARW		38021	KC-135R	100th ARW	r/w/bl	
	23565	KC-135R	100th ARW	r/w/bl	38022	KC-135R	22nd ARW	m	
	23566	KC-135R	174th ARS	y/bk	38023	KC-135R	126th ARS	w/bl	
	23568	KC-135R	6th AMW	y/bl	38024	KC-135R	452nd AMW	or/y	
	23569	KC-135R	22nd ARW		38025	KC-135R			
	23571	KC-135R	168th ARS	bl/y	38026	KC-135R	126th ARS	w/bl	
	23572	KC-135R	117th ARS	bl/y	38027	KC-135R	100th ARW	r/w/bl	
	23573	KC-135R	6th AMW	y/bl	38028	KC-135R	168th ARS	bl/y	
	23575	KC-135R	22nd ARW		38029	KC-135R	141st ARS	r	
	23576	KC-135R	133rd ARS	bl	38030	KC-135R	203rd ARS	y/bk	
	23577	KC-135R	916th ARW	gn/y	38031	KC-135R	92nd ARW		
	23578	KC-135R	141st ARS	r	38032	KC-135R	434th ARW	bl	
	23580	KC-135R	916th ARW	gn/y	38033	KC-135R	22nd ARW		
	23582	WC-135C	55th Wg	bk	38034	KC-135R	18th Wg	or/bk	
	24125	RC-135W	55th Wg	bk	38035	KC-135R	106th ARS	w/r	
	24126	RC-135W	55th Wg	bl	38036	KC-135R	197th ARS		
	24127	TC-135W	55th Wg	bl	38038	KC-135R	203rd ARS	y/bk	
	24128	RC-135S	55th Wg		38039	KC-135R	507th ARW	bl/y	
	24129	TC-135W	55th Wg	gn	38040	KC-135R	141st ARS	r	
	24130	RC-135W	55th Wg	gn	38041	KC-135R	434th ARW	bl	
	24131	RC-135W	55th Wg	gn	38043	KC-135R	168th ARS	bl/y	
	24132	RC-135W	55th Wg	gn	38044	KC-135R	916th ARW	gn/y	
	24133	TC-135W	55th Wg	bk	38045	KC-135R	92nd ARW		
	24134	RC-135W	55th Wg	bl	38871	KC-135R	100th ARW	r/w/bl	
	24135	RC-135W	55th Wg	bk	38872	KC-135R	132nd ARS	w/gn	
	24138	RC-135W	55th Wg	gn	38873	KC-135R	132nd ARS	w/gn	
	24139	RC-135W	55th Wg	gn	38874	KC-135R	22nd ARW		
	FY63				38875	KC-135R	117th ARS	bl/y	
	37976	KC-135R	97th AMW	y/r	38876	KC-135R	168th ARS	bl/y	
	37977	KC-135R	18th Wg	or/bk	38878	KC-135R	18th Wg	or/bk	
	37978	KC-135R	22nd ARW		38879	KC-135R	92nd ARW		
	37979	KC-135R	22nd ARW		38880	KC-135R	203rd ARS	y/bk	
	37980	KC-135R	412th TW	or	38881	KC-135R	191st ARS	bl/bk	
	37981	KC-135R	108th ARS	w/bl	38883	KC-135R	18th Wg	or/bk	
	37982	KC-135R	92nd ARW	bk	38884	KC-135R	100th ARW	r/w/bl	
	37984	KC-135R	106th ARS	w/r	38885	KC-135R	92nd ARW		
	37985	KC-135R	507th ARW	bl/y	38887	KC-135R	92nd ARW		
	37987	KC-135R	22nd ARW		38888	KC-135R	92nd ARW		
	37988	KC-135R	173rd ARS	r/w	39792	RC-135V	55th Wg	gn $	
	37991	KC-135R	173rd ARS	r/w	FY57				
	37992	KC-135R	153rd ARS		71419	KC-135R	117th ARS	bl/y	
	37993	KC-135R	121st ARW	m	71427	KC-135R	117th ARS	bl/y	

Notes	Type				Type				Notes
	71428	KC-135R	151st ARS	w/or	80051	KC-135R	507th ARW	bl/y	
	71430	KC-135R	133rd ARS	bl	80052	KC-135R	452nd AMW	or/y	
	71432	KC-135R	191st ARS	bl/bk	80054	KC-135T	171st ARW	y/bk	
	71435	KC-135R	191st ARS	bl/bk	80055	KC-135T	6th AMW	y/bl	
	71436	KC-135R	151st ARS	w/or	80056	KC-135R	15th Wg		
	71437	KC-135R	916th ARW	gn/y	80057	KC-135R	174th ARS	y/bk	
	71438	KC-135R	452nd AMW	or/y	80058	KC-135R	507th ARW	bl/y	
	71439	KC-135R	22nd ARW		80059	KC-135R	117th ARS	bl/y	
	71440	KC-135R	22nd ARW		80060	KC-135T	171st ARW	y/bk	
	71441	KC-135R	174th ARS	y/bk	80061	KC-135T	173rd ARS	r/w	
	71451	KC-135R	151st ARS	w/or	80062	KC-135T	171st ARS	bk/y	
	71453	KC-135R	106th ARS	w/r	80063	KC-135R	133rd ARS	bl	
	71454	KC-135R	22nd ARW		80065	KC-135T	22nd ARW		
	71456	KC-135R	916th ARW	gn/y	80066	KC-135R	133rd ARS	bl	
	71459	KC-135R	18th Wg	or/bk	80067	KC-135R	174th ARS	y/bk	
	71461	KC-135R	173rd ARS	r/w	80069	KC-135T	22nd ARW		
	71462	KC-135R	153rd ARS		80071	KC-135T	22nd ARW		
	71468	KC-135R	916th ARW	gn/y	80072	KC-135T	171st ARW	y/bk	
	71469	KC-135R	197th ARS		80073	KC-135R	106th ARS	w/r	
	71472	KC-135R	434th ARW	bl	80074	KC-135T	171st ARW	y/bk	
	71473	KC-135R	106th ARS	w/r	80075	KC-135R	459th ARW	y/bk	
	71474	KC-135R	97th AMW	y/r	80076	KC-135R	434th ARW	r/w	
	71483	KC-135R	92nd ARW		80077	KC-135T	171st ARW	y/bk	
	71486	KC-135R	153rd ARS		80079	KC-135R	153rd ARS		
	71487	KC-135R	459th ARW	y/bk	80083	KC-135R	121st ARW	bl	
	71488	KC-135R	6th AMW	y/bl	80084	KC-135T	171st ARW	y/bk	
	71493	KC-135R	18th Wg	or/bk	80085	KC-135R	452nd AMW	or/y	
	71499	KC-135R	191st ARS	bl/bk	80086	KC-135T	6th AMW	y/bl	
	71502	KC-135R	22nd ARW		80088	KC-135T	171st ARS	bk/y	
	71506	KC-135R	18th Wg	or/bk	80089	KC-135T	22nd ARW		
	71508	KC-135R	203rd ARS	y/bk	80092	KC-135R	92nd ARW		
	71512	KC-135R	459th ARW	y/bk	80093	KC-135R	18th Wg	or/bk	
	71514	KC-135R	126th ARS	w/bl	80094	KC-135T	92nd ARW		
	72597	KC-135R	151st ARS	w/or	80095	KC-135T	92nd ARW		
	72598	KC-135R	452nd AMW	or/y	80098	KC-135R	132nd ARS	w/gn	
	72599	KC-135R	916th ARW	gn/y	80099	KC-135T	171st ARW	y/bk	
	72603	KC-135R	452nd AMW	or/y	80100	KC-135R	100th ARW	r/w/bl	
	72605	KC-135R	18th Wg	or/bk	80102	KC-135R	507th ARW	bl/y	
	72606	KC-135R	174th ARS	y/bk	80103	KC-135T	92nd ARW	bk	
	FY58				80104	KC-135R	108th ARS	w/bl	
	80001	KC-135R			80106	KC-135R	106th ARS	w/r	
	80004	KC-135R	106th ARS	w/r	80107	KC-135R	132nd ARS	w/gn	
	80009	KC-135R	126th ARS	w/bl	80109	KC-135R	174th ARS	y/bk	
	80010	KC-135R	141st ARS	r	80112	KC-135T	171st ARW	y/bk	
	80011	KC-135R	22nd ARW		80113	KC-135R	97th AMW	y/r	
	80015	KC-135R	507th ARW	bl/y	80117	KC-135T	171st ARW	y/bk	
	80016	KC-135R	92nd ARW		80118	KC-135R	100th ARW	r/w/bl	
	80018	KC-135R	22nd ARW		80119	KC-135R	151st ARS	w/or	
	80021	KC-135R	132nd ARS	w/gn	80120	KC-135R	153rd ARS		
	80023	KC-135R	108th ARS	w/bl	80121	KC-135R	507th ARW	bl/y	
	80027	KC-135R	191st ARS	bl/bk	80122	KC-135R	117th ARS	bl/y	
	80030	KC-135R	132nd ARS	w/gn	80123	KC-135R	18th Wg	or/bk	
	80034	KC-135R	100th ARW	r/w/bl	80124	KC-135R	22nd ARW		
	80035	KC-135R	92nd ARW		80125	KC-135T	6th AMW	y/bl	
	80036	KC-135R	97th AMW	y/r	80126	KC-135R	22nd ARW	y	
	80038	KC-135R	916th ARW	gn/y	80128	KC-135R	97th AMW	y/r	
	80042	KC-135T	22nd ARW		80129	KC-135T	171st ARS	bk/y	
	80045	KC-135T	171st ARW		FY59				
	80046	KC-135T	92nd ARW		91444	KC-135R	121st ARW	bl	
	80047	KC-135T	22nd ARW		91446	KC-135R	132nd ARS	w/gn	
	80049	KC-135T	22nd ARW		91448	KC-135R	153rd ARS		
	80050	KC-135T	6th AMW	y/bl	91450	KC-135R	197th ARS		

Notes	Type				Type	Notes
	91453	KC-135R	153rd ARS		10329	
	91455	KC-135R	203rd ARS	y/bk	FY12	
	91458	KC-135R	121st ARW	bl	20331	
	91459	KC-135R	18th Wg	or/bk	20335	
	91460	KC-135T	171st ARW	y/bk	20336	
	91461	KC-135R	126th ARS	w/bl	20337	
	91462	KC-135T	92nd ARW		20338	
	91463	KC-135R	173rd ARS	r/w	FY13	
	91464	KC-135T	100th ARW	r/w/bl	30341	
	91466	KC-135R	108th ARS	w/bl	30342	
	91467	KC-135T	171st ARW	y/bk	FY08	
	91468	KC-135T	171st ARW	y/bk	80310	
	91469	KC-135R	459th ARW	y/bk	FY09	
	91470	KC-135T	92nd ARW	gn	90317	
	91471	KC-135T	6th AMW	y/bl	90320	
	91472	KC-135R	203rd ARS	y/bk		
	91474	KC-135T	171st ARS	bk/y	**Dornier**	
	91475	KC-135R	18th Wg	or/bk	**C-146A Wolfhound**	
	91476	KC-135R	92nd ARW		524th SOS/27th SOW, Cannon AFB,	
	91478	KC-135R	151st ARS	w/or	New Mexico	
	91480	KC-135T	92nd ARW	bk	FY10	
	91482	KC-135R	452nd AMW	or/y	03026	
	91483	KC-135R	121st ARW	bl	03068	
	91486	KC-135R	92nd ARW		03077	
	91488	KC-135R	132nd ARS	w/gn	03097	
	91490	KC-135T	171st ARW	y/bk	FY11	
	91492	KC-135R	100th ARW	r/w/bl	13016	
	91495	KC-135R	173rd ARS	r/w	13031	
	91498	KC-135R	132nd ARS	w/gn	13075	
	91499	KC-135R	203rd ARS	y/bk	13104	
	91500	KC-135R	108th ARS	w/bl	FY12	
	91501	KC-135R	97th AMW	y/r	23040	
	91502	KC-135R	92nd ARW		23047	
	91504	KC-135T	171st ARW	y/bk	23050	
	91505	KC-135R	151st ARS	w/or	23060	
	91506	KC-135R	174th ARS	y/bk	23085	
	91507	KC-135R	117th ARS	bl/y	FY95	
	91508	KC-135R	22nd ARW		53058	
	91509	KC-135R	141st ARS	r	FY97	
	91510	KC-135T	92nd ARW		73091	
	91511	KC-135R	22nd ARW		73093	
	91512	KC-135T	171st ARS	bk/y	FY99	
	91513	KC-135T	100th ARW	r/w/bl	93106	
	91515	KC-135R	92nd ARW			
	91516	KC-135R	126th ARS	w/bl		
	91517	KC-135R	151st ARS	w/or		
	91519	KC-135R	174th ARS	y/bk		
	91520	KC-135T	92nd ARW			
	91521	KC-135R	168th ARS	bl/y		
	91522	KC-135R	108th ARS	w/bl		
	91523	KC-135T	171st ARW	y/bk		

PZL-Mielec
C-145A Skytruck
6th SOS/1st SOW, Duke Field, Florida
FY10
00321
00322
00323
00324
FY11
10326

Notes	Type			Type			Notes
	Lockheed P-3 Orion			158571	[YB-571]	P-3C	VP-1
	CinCLANT/VP-30, NAS Jacksonville,			158573	[YD-573]	P-3C	VP-4
	Florida;			158912	[RL-912]	NP-3C	VXS-1
	CNO/VP-30, NAS Jacksonville, Florida;			158914	[914]	P-3C	VP-45
	NASC-FS, Point Mugu, California;			158916	[916]	P-3C	VP-47
	USNTPS, NAS Point Mugu, California;			158917	[917]	P-3C	VQ-1
	VP-1, NAS Whidbey Island, Washington			158919	[919]	P-3C	VP-10
	[YB];			158922	[922]	P-3C	VP-9
	VP-4, MCBH Kaneohe Bay, Hawaii [YD];			158923	[RD-923]	P-3C	VP-47
	VP-8, NAS Jacksonville, Florida [LC];			158926	[926]	P-3C	VP-62
	VP-9, MCBH Kaneohe Bay, Hawaii [PD];			158934	[LK-934]	P-3C	VP-26
	VP-10, NAS Jacksonville, Florida [LD];			158935	[935]	P-3C	VQ-1
	VP-26, NAS Jacksonville, Florida [LK];			159318	[318]	P-3C	VP-10
	VP-30, NAS Jacksonville, Florida [LL];			159323	[323]	P-3C	VP-9
	VP-40, NAS Whidbey Island, Washington			159326	[326]	P-3C	VP-26
	[QE];			159504		P-3C	VPU-2
	VP-45, NAS Jacksonville, Florida [LN];			159887	[887]	EP-3E	VQ-1
	VP-46, NAS Whidbey Island, Washington			159893	[893]	EP-3E	VQ-1
	[RC];			159894	[894]	P-3C	VP-40
	VP-47, MCBH Kaneohe Bay, Hawaii			160285		P-3C	VPU-2
	[RD];			160287	[287]	P-3C	VP-9
	VP-62, NAS Jacksonville, Florida [LT];			160290	[290]	P-3C	VX-20
	VP-69, NAS Whidbey Island, Washington			160291	[291]	EP-3E	VQ-1
	[PJ];			160292	[292]	P-3C	VPU-2
	VPU-2, MCBH Kaneohe Bay, Hawaii;			160293		P-3C	NASC-FS
	VQ-1, NAS Whidbey Island, Washington			160610	[LK-610]	P-3C	VP-26
	[PR];			160761	[761]	P-3C	VP-40
	VX-1, NAS Patuxent River, Maryland			160764	[764]	EP-3E	VQ-1
	[JA];			160999	[999]	P-3C	VP-9
	VX-20, Patuxent River, Maryland;			161001	[001]	P-3C	VP-69
	VX-30, NAS Point Mugu, California;			161011	[011]	P-3C	VP-10
	VXS-1, Patuxent River, Maryland [RL]			161012	[RD-012]	P-3C	VP-47
	150521	[341]	NP-3D	VX-30	161121	[RC-121]	P-3C	VP-46
	150522	[340]	NP-3D	VX-30	161122	[226]	P-3C	VPU-2
	153442	[357]	NP-3D	VX-30	161124	[JA-124]	P-3C	VX-1
	154589	[RL-589]	NP-3D	VXS-1	161126	[126]	P-3C	VP-8
	156507	[507]	EP-3E	VQ-1	161127	[127]	P-3C	VP-26
	156510	[LL-510]	P-3C	VP-30	161129	[129]	P-3C	VP-69
	156511	[511]	EP-3E	VQ-1	161132	[132]	P-3C	VP-46
	156514	[514]	EP-3E	VQ-1	161333	[333]	P-3C	VP-30
	156515	[LD-515]	P-3C	VP-10	161337	[337]	P-3C	VP-47
	156517	[517]	EP-3E	VQ-1	161339	[339]	P-3C	VP-1
	156528	[528]	EP-3E	VQ-1	161404	[404]	P-3C	VP-1
	156529	[529]	EP-3E	VQ-1	161405	[405]	P-3C	VP-10
	157316	[316]	EP-3E	VQ-1	161406	[406]	P-3C	VP-30
	157318	[318]	EP-3E	VQ-1	161407	[YD-407]	P-3C	VP-4
	157319	[319]	P-3C	VP-10	161408	[408]	P-3C	VP-40
	157325	[325]	EP-3E	VQ-1	161409	[409]	P-3C	VP-62
	157326	[326]	EP-3E	VQ-1	161410	[410]	EP-3E	VQ-1
	157329	[LL-329]	P-3C	VP-30	161411	[LK-411]	P-3C	VP-26
	157331	[331]	P-3C	VP-4	161412	[PJ-412]	P-3C	VP-69
	158204	[204]	NP-3C	VX-20	161413	[LD-413]	P-3C	VP-10
	158210	[210]	P-3C	VP-40	161414	[YB-414]	P-3C	VP-1
	158215	[LC-215]	P-3C	VP-8	161415	[LK-415]	P-3C	VP-26
	158222	[YD-222]	P-3C	VP-4	161586	[QE-586]	P-3C	VP-40
	158224	[224]	P-3C	VP-47	161587	[RC-587]	P-3C	VP-46
	158225	[225]	P-3C	VP-26	161588	[588]	P-3C	VP-46
	158227	[300]	NP-3D	VX-30	161589	[LK-589]	P-3C	VP-26
	158563	[PD-563]	P-3C	VP-9	161590	[590]	P-3C	VP-1
	158564	[564]	P-3C	VP-26	161593	[593]	P-3C	VP-1
	158567	[PD-567]	P-3C	VP-9	161594	[YD-594]	P-3C	VP-4
	158570	[570]	P-3C	VXS-1	161596	[596]	P-3C	VP-40

Notes	Type			Type	Notes
	161763 [PJ-763]	P-3C	VP-69	**Boeing P-8A Poseidon**	
	161764 [764]	P-3C	VP-1	Boeing, Seattle;	
	161765 [765]	P-3C	VP-10	VP-5, NAS Jacksonville, Florida [LA];	
	161766 [766]	P-3C	VP-40	VP-8, NAS Jacksonville, Florida [LC];	
	161767 [PD-767]	P-3C	VP-9	VP-10, NAS Jacksonville, Florida [LD];	
	162314 [RC-314]	P-3C	VP-46	VP-16, NAS Jacksonville, Florida [LF];	
	162315 [RC-315]	P-3C	VP-46	VP-26, NAS Jacksonville, Florida [LK];	
	162316 [316]	P-3C	VP-62	VP-30, NAS Jacksonville, Florida [LL];	
	162317 [317]	P-3C	VP-30	VP-45, NAS Jacksonville, Florida [LN];	
	162318 [LL-318]	P-3C	VP-30	VX-1, NAS Patuxent River, Maryland;	
	162770 [770]	P-3C	VP-30	VX-20, Patuxent River, Maryland [JA]	
	162771 [771]	P-3C	VP-26	167951	VX-20	
	162772 [772]	P-3C	VP-40	167952 [JA-952]	VX-1	
	162773 [YD-773]	P-3C	VP-4	167953	VX-20	
	162774 [774]	P-3C	VX-20	167954	VX-20	
	162775 [775]	P-3C	VP-4	167955 [JA-955]	VX-1	
	162776 [RC-776]	P-3C	VP-46	167956 [JA-956]	VX-1	
	162777 [777]	P-3C	VP-69	168428 [LN-428]	VP-45	
	162778 [778]	P-3C	VP-47	168429 [LF-429]	VP-16	
	162998 [QE-998]	P-3C	VP-40	168430 [LF-430]	VP-16	
	162999 [RC-999]	P-3C	VP-46	168431 [LA-431]	VP-5	
	163000 [000]	P-3C	VP-8	168432 [LF-432]	VP-16	
	163001 [001]	P-3C		168433 [LA-433]	VP-5	
	163002 [002]	P-3C	VP-4	168434 [LN-434]	VP-45	
	163003 [003]	P-3C	VP-4	168435 [435]	VP-5	
	163004 [004]	P-3C	VP-46	168436 [LA-436]	VP-5	
	163006 [006]	P-3C	VP-40	168437 [LA-437]	VP-5	
	163289 [LD-289]	P-3C	VP-10	168438 [438]	VP-5	
	163290 [290]	P-3C	VP-30	168439 [LF-439]	VP-16	
	163291 [291]	P-3C	VP-69	168440 [440]	VP-30	
	163293 [293]	P-3C	VP-9	168754 [754]	VP-30	
	163294 [YD-294]	P-3C	VP-4	168755 [755]	VP-30	
	163295 [295]	P-3C	VP-40	168756 [756]	VP-45	
				168757 [757]	VP-30	
	Boeing E-6B Mercury			168758 [LN-758]	VP-45	
	Boeing, McConnell AFB, Kansas;			168759 [759]	VP-16	
	VQ-3 & VQ-4, SCW-1,			168760 [760]	VP-30	
	Tinker AFB, Oklahoma			168761 [761]	Boeing	
	162782 VQ-3			168762 [762]	Boeing	
	162783 VQ-3			168763 (on order)		
	162784 VQ-4			168764 (on order)		
	163918 VQ-3			168848 (on order)		
	163919 VQ-3			168849 (on order)		
	163920 VQ-3				
	164386 VQ-4			**Grumman C-20A/C-20D Gulfstream III/**	
	164387 VQ-3			**C-20G Gulfstream IV***	
	164388 VQ-4			VMR-Det, MCBH Kaneohe Bay, Hawaii;	
	164404 VQ-4			VR-1, NAF Washington, Maryland;	
	164405 VQ-4			VR-51, MCBH Kaneohe Bay, Hawaii [RG]	
	164406 VQ-3			**C-20A**	
	164407 VQ-4			830500 VR-1	
	164408 VQ-4			**C-20D**	
	164409 VQ-4			163691 VR-1	
	164410 VQ-4			163692 VR-1	
				C-20G	
				165093 VR-51	
				165094 VR-51	
				165151 VR-51	
				165152 VR-51	
				165153 VMR-Det	

Notes	Type				Type	Notes

Cessna C-35 Citation V
MAW-4, Miramar MCAS, California;
MWHS-1, Futenma MCAS, Japan;
MWHS-4, NAS New Orleans;
VMR-1, Cherry Point MCAS,
   North Carolina;
VMR-2, NAF Washington, Maryland

165740	[EZ]	UC-35C	MWHS-4
165741	[EZ]	UC-35C	MWHS-4
165939		UC-35D	MWHS-1
166374		UC-35D	VMR-2
166474		UC-35D	VMR-1
166500		UC-35D	MAW-4
166712		UC-35D	MWHS-1
166713		UC-35D	MWHS-1
166714	[VM]	UC-35D	VMR-2
166715		UC-35D	VMR-1
166766		UC-35D	MWHS-1
166767	[VM]	UC-35D	VMR-2

Gulfstream Aerospace
C-37B Gulfstream V
CFLSW Det, Hawaii;
VR-1, NAF Washington, Maryland

166375	CFLSW Det
166376	VR-1
166377	VR-1
166378	VR-1
166379	VR-1

Boeing C-40A Clipper
VR-56, NAS Oceana, Virginia;
VR-57, NAS North Island, California;
VR-58, NAS Jacksonville, Florida;
VR-59, NAS Fort Worth JRB, Texas
VR-61, NAS Fort Worth JRB, Texas

165829	VR-58
165830	VR-59
165831	VR-59
165832	VR-58
165833	VR-59
165834	VR-58
165835	VR-57
165836	VR-57
166693	VR-57
166694	VR-56
166695	VR-56
166696	VR-56
168980	VR-61
168981	(on order)
168...	(on order)
168...	(on order)
168...	(on order)

Lockheed C-130 Hercules
VR-53, NAF Washington, Maryland [AX];
VR-54, NAS New Orleans, Louisiana
   [CW];
VR-55, NAS Point Mugu, California [RU];
VR-62, NAS Jacksonville, Florida [JW];
VR-64, NAS Willow Grove, Pennsylvania
   [BD];
VMGR-152, Iwakuni MCAS, Japan [QD];
VMGR-234, NAS Fort Worth, Texas [QH];
VMGR-252, Cherry Point MCAS,
   North Carolina [BH];
VMGR-352, MCAS Miramar, California
   [QB];
VMGR-452, Stewart Field, New York
   [NY];
VX-20, Patuxent River, Maryland;
VX-30, NAS Point Mugu, California

148891	[403]	KC-130F	VX-30
148893	[402]	KC-130F	VX-30
148897	[400]	KC-130F	VX-30
160625		KC-130R	VX-20
160626		KC-130R	VX-20
160627		KC-130R	VX-20
162308	[QH]	KC-130T	VMGR-234
162309	[QH]	KC-130T	VMGR-234
162310	[QH]	KC-130T	VMGR-234
162311	[QH]	KC-130T	VMGR-234
162785	[QH]	KC-130T	VMGR-234
162786	[QH]	KC-130T	VMGR-234
163022	[QH]	KC-130T	VMGR-234
163023	[QH]	KC-130T	VMGR-234
163310	[QH]	KC-130T	VMGR-234
163311	[NY]	KC-130T	VMGR-452
163591	[RU]	KC-130T	VR-55
163592	[NY]	KC-130T	VMGR-452
164105	[NY]	KC-130T	VMGR-452
164106	[RU]	KC-130T	VR-55
164180	[NY]	KC-130T	VMGR-452
164181	[NY]	KC-130T	VMGR-452
164441	[NY]	KC-130T	VMGR-452
164442	[NY]	KC-130T	VMGR-452
164597	[RU]	KC-130T-30	VR-55
164598	[RU]	KC-130T-30	VR-55
164762	[CW]	C-130T	VR-54
164763		C-130T	*Blue Angels*
164993	[BD]	C-130T	VR-64
164994	[CW]	C-130T	VR-54
164995	[AX]	C-130T	VR-53
164996	[BD]	C-130T	VR-64
164997	[AX]	C-130T	VR-53
164998	[AX]	C-130T	VR-53
164999	[NY]	KC-130T	VMGR-452
165000	[QH]	KC-130T	VMGR-234
165158	[CW]	C-130T	VR-54
165159	[CW]	C-130T	VR-54
165160	[JW]	C-130T	VR-62
165161	[BD]	C-130T	VR-64
165162	[NY]	KC-130T	VMGR-452
165163	[NY]	KC-130T	VMGR-452
165313	[JW]	C-130T	VR-62
165314	[JW]	C-130T	VR-62

Notes	Type				Type				Notes
	165315	[NY]	KC-130T	VMGR-452	166764	[BH]	KC-130J	VMGR-252	
	165316	[NY]	KC-130T	VMGR-452	166765	[QB]	KC-130J	VMGR-352	
	165348	[JW]	C-130T	VR-62	167108	[BH]	KC-130J	VMGR-252	
	165349	[JW]	C-130T	VR-62	167109	[QB]	KC-130J	VMGR-352	
	165350	[AX]	C-130T	VR-53	167110	[QB]	KC-130J	VMGR-352	
	165351	[AX]	C-130T	VR-53	167111	[QB]	KC-130J	VMGR-352	
	165352	[NY]	KC-130T	VMGR-452	167112	[BH]	KC-130J	VMGR-252	
	165353	[NY]	KC-130T	VMGR-452	167923	[QD]	KC-130J	VMGR-152	
	165378	[RU]	C-130T	VR-55	167924	[QB]	KC-130J	VMGR-352	
	165379	[BD]	C-130T	VR-64	167925	[QD]	KC-130J	VMGR-152	
	165735	[QB]	KC-130J	VMGR-352	167926	[QD]	KC-130J	VMGR-152	
	165736	[QB]	KC-130J	VMGR-352	167927	[QD]	KC-130J	VMGR-152	
	165737	[BH]	KC-130J	VMGR-252	167981	[QD]	KC-130J	VMGR-152	
	165738	[BH]	KC-130J	VMGR-252	167982	[QD]	KC-130J	VMGR-152	
	165739	[QB]	KC-130J	VMGR-352	167983	[QD]	KC-130J	VMGR-152	
	165809	[BH]	KC-130J	VMGR-252	167984	[QB]	KC-130J	VMGR-352	
	165810	[BH]	KC-130J	VMGR-252	167985	[QB]	KC-130J	VMGR-352	
	165957	[QD]	KC-130J $	VMGR-152	168065	[QD]	KC-130J	VMGR-152	
	166380	[BH]	KC-130J	VMGR-252	168066	[QD]	KC-130J	VMGR-152	
	166381		KC-130J	VX-20	168067	[QB]	KC-130J	VMGR-352	
	166382	[QB]	KC-130J	VMGR-352	168068	[QB]	KC-130J	VMGR-352	
	166472	[BH]	KC-130J	VMGR-252	168069	[BH]	KC-130J	VMGR-252	
	166473	[QH]	KC-130J	VMGR-234	168070	[BH]	KC-130J	VMGR-252	
	166511	[BH]	KC-130J	VMGR-252	168071	[BH]	KC-130J	VMGR-252	
	166512	[QB]	KC-130J	VMGR-352	168072	[QB]	KC-130J	VMGR-352	
	166513	[BH]	KC-130J	VMGR-252	168073	[QH]	KC-130J	VMGR-234	
	166514		KC-130J	VX-20	168074	[QD]	KC-130J	VMGR-152	
	166762	[QB]	KC-130J	VMGR-352	168075	[QD]	KC-130J	VMGR-152	
	166763	[QD]	KC-130J	VMGR-152					

US-BASED US COAST GUARD AIRCRAFT

Notes	Type		Type		Notes
	Gulfstream Aerospace		1706	HC-130H Clearwater	
	C-37A Gulfstream V		1707	HC-130H Barbers Point	
	USCG, Washington DC		1709	HC-130H Kodiak	
	01		1711	HC-130H Clearwater	
	02		1712	HC-130H Kodiak	
			1713	HC-130H Kodiak	
	Lockheed C-130 Hercules		1714	HC-130H Kodiak	
	USCGS Barbers Point, Hawaii;		1715	HC-130H Kodiak	
	USCGS Clearwater, Florida;		1716	HC-130H Clearwater	
	USCGS Elizabeth City, North Carolina;		1718	HC-130H Clearwater	
	USCGS Kodiak, Alaska;		1719	HC-130H Clearwater	
	USCGS Sacramento, California		1720	HC-130H Barbers Point	
	1502	HC-130H Clearwater	1790	HC-130H Kodiak	
	1503	HC-130H Clearwater	2001	HC-130J Elizabeth City	
	1700	HC-130H Clearwater	2002	HC-130J Elizabeth City	
	1701	HC-130H Sacramento	2003	HC-130J Elizabeth City	
	1702	HC-130H Sacramento	2004	HC-130J Elizabeth City	
	1703	HC-130H Sacramento	2005	HC-130J Elizabeth City	
	1704	HC-130H Kodiak	2006	HC-130J Elizabeth City	

Aircraft in US Government or Military Service with Civil Registrations

Notes	Type	Type	Notes
	Canadair CL.601/CL.604*		
	Challenger		
	Federal Aviation Administration,		
	Oklahoma		
	N85		
	N86		
	N87		
	N88*		

The list below is not intended to be a complete list of military aviation sites on the Internet. The sites listed cover Museums, Locations, Air Forces, Companies and Organisations that are mentioned elsewhere in 'Military Aircraft Markings'. Sites listed are in English or contain sufficient English to be reasonably easily understood. Each site address was correct at the time of going to press. Additions are welcome, via the usual address found at the front of the book, or via e-mail to admin@aviation-links.co.uk. An up to date copy of this list is to be found at The 'Military Aircraft Markings' Web Site, http://www.militaryaircraftmarkings.co.uk/.

Name of Site	Web Address All prefixed 'http://')
MILITARY SITES-UK	
No 1 Sqn	www.raf.mod.uk/organisation/1squadron.cfm
No 2 Sqn	www.raf.mod.uk/organisation/2squadron.cfm
No 3 Sqn	www.raf.mod.uk/organisation/3squadron.cfm
No 4(R) Sqn	www.raf.mod.uk/organisation/4squadron.cfm
No 5 Sqn	www.raf.mod.uk/organisation/5squadron.cfm
No 6 Sqn	www.raf.mod.uk/organisation/6squadron.cfm
No 7 Sqn	www.raf.mod.uk/organisation/7squadron.cfm
No 8 Sqn	8squadron.co.uk/
No 9 Sqn	www.raf.mod.uk/organisation/9squadron.cfm
No 11 Sqn	www.raf.mod.uk/organisation/11squadron.cfm
No 14 Sqn	www.raf.mod.uk/organisation/14squadron.cfm
No 15(R) Sqn	www.raf.mod.uk/organisation/15squadron.cfm
No 17(R) Sqn	www.raf.mod.uk/organisation/17squadron.cfm
No 18 Sqn	www.raf.mod.uk/organisation/18squadron.cfm
No 22 Sqn	www.raf.mod.uk/organisation/22squadron.cfm
No 24 Sqn	www.raf.mod.uk/organisation/24squadron.cfm
No 27 Sqn	www.raf.mod.uk/organisation/27squadron.cfm
No 28 Sqn	www.raf.mod.uk/organisation/28squadron.cfm
No 29(R) Sqn	www.raf.mod.uk/organisation/29squadron.cfm
No 30 Sqn	www.raf.mod.uk/organisation/30squadron.cfm
No 31 Sqn	www.raf.mod.uk/organisation/31squadron.cfm
No 32(The Royal) Sqn	www.raf.mod.uk/organisation/32squadron.cfm
No 33 Sqn	www.raf.mod.uk/organisation/33squadron.cfm
No 39 Sqn	www.raf.mod.uk/organisation/39squadron.cfm
No 41(R) Sqn	www.raf.mod.uk/organisation/41squadron.cfm
No 45(R) Sqn	www.raf.mod.uk/organisation/45squadron.cfm
No 47 Sqn	www.raf.mod.uk/organisation/47squadron.cfm
No 51 Sqn	www.raf.mod.uk/organisation/51squadron.cfm
No 55(R) Sqn	www.raf.mod.uk/organisation/55squadron.cfm
No 60(R) Sqn	www.raf.mod.uk/organisation/60squadron.cfm
No 72(R) Sqn	www.raf.mod.uk/organisation/72squadron.cfm
No 84 Sqn	www.raf.mod.uk/organisation/84squadron.cfm
No 99 Sqn	www.raf.mod.uk/organisation/99squadron.cfm
No 100 Sqn	www.raf.mod.uk/organisation/100squadron.cfm
No 101 Sqn	www.raf.mod.uk/organisation/101squadron.cfm
No 202 Sqn	www.raf.mod.uk/organisation/202squadron.cfm
No 208(R) Sqn	www.raf.mod.uk/organisation/208squadron.cfm
No 230 Sqn	www.raf.mod.uk/organisation/230squadron.cfm
No 617 Sqn	www.raf.mod.uk/organisation/617squadron.cfm
No 703 NAS	www.royalnavy.mod.uk/our-organisation/the-fighting-arms/fleet-air-arm/support-and-training/703-naval-air-squadronng
No 727 NAS	www.royalnavy.mod.uk/our-organisation/the-fighting-arms/fleet-air-arm/support-and-training/727-naval-air-squadron
No 736 NAS	www.royalnavy.mod.uk/our-organisation/the-fighting-arms/fleet-air-arm/hawk-jets/736-naval-air-squadron
No 750 NAS	www.royalnavy.mod.uk/our-organisation/the-fighting-arms/fleet-air-arm/support-and-training/750-naval-air-squadron
No 771 NAS	www.royalnavy.mod.uk/our-organisation/the-fighting-arms/fleet-air-arm/helicopter-squadrons/sea-king-mk5/771-naval-air-squadron
No 809 NAS	www.royalnavy.mod.uk/our-organisation/the-fighting-arms/fleet-air-arm/future-aircraft/809-naval-air-squadron
No 814 NAS	www.royalnavy.mod.uk/our-organisation/the-fighting-arms/fleet-air-arm/helicopter-squadrons/merlin-mk1/814-naval-air-squadron

Name of Site	Web Address All prefixed 'http://')
No 815 NAS	www.royalnavy.mod.uk/our-organisation/the-fighting-arms/fleet-air-arm/helicopter-squadrons/lynx-mk8/815-naval-air-squadron
No 820 NAS	www.royalnavy.mod.uk/our-organisation/the-fighting-arms/fleet-air-arm/helicopter-squadrons/merlin-mk2/820-naval-air-squadron
No 824 NAS	www.royalnavy.mod.uk/our-organisation/the-fighting-arms/fleet-air-arm/helicopter-squadrons/merlin-mk2/824-naval-air-squadron
No 825 NAS	www.royalnavy.mod.uk/our-organisation/the-fighting-arms/fleet-air-arm/helicopter-squadrons/wildcat/825-naval-air-squadron
No 829 NAS	www.royalnavy.mod.uk/our-organisation/the-fighting-arms/fleet-air-arm/helicopter-squadrons/merlin-mk1/829-naval-air-squadron
No 845 NAS	www.royalnavy.mod.uk/our-organisation/the-fighting-arms/fleet-air-arm/helicopter-squadrons/sea-king-mk4/845-naval-air-squadron
No 846 NAS	www.royalnavy.mod.uk/our-organisation/the-fighting-arms/fleet-air-arm/helicopter-squadrons/sea-king-mk4/846-naval-air-squadron
No 847 NAS	www.royalnavy.mod.uk/our-organisation/the-fighting-arms/fleet-air-arm/helicopter-squadrons/wildcat/847-naval-air-squadron
No 849 NAS	www.royalnavy.mod.uk/our-organisation/the-fighting-arms/fleet-air-arm/helicopter-squadrons/sea-king-asac-mk7/849-naval-air-squadron
No 854 NAS	www.royalnavy.mod.uk/our-organisation/the-fighting-arms/fleet-air-arm/helicopter-squadrons/sea-king-asac-mk7/854-naval-air-squadron
No 857 NAS	www.royalnavy.mod.uk/our-organisation/the-fighting-arms/fleet-air-arm/helicopter-squadrons/sea-king-asac-mk7/857-naval-air-squadron
Aberdeen, Dundee and St Andrews UAS	dialspace.dial.pipex.com/town/way/gba87/adstauas/
The Army Air Corps	www.army.mod.uk/aviation/air.aspx
Cambridge University Air Squadron	www.raf.mod.uk/cambridgeuas/
Fleet Air Arm	www.royalnavy.mod.uk/our-organisation/the-fighting-arms/fleet-air-arm
Liverpool University Air Squadron	www.raf.mod.uk/liverpooluas/
Ministry of Defence	www.mod.uk/
Oxford University Air Sqn	www.raf.mod.uk/universityairsquadrons/findasquadron/ ouas.cfm
QinetiQ	www.qinetiq.com/
RAF Benson	www.raf.mod.uk/rafbenson/
RAF Brize Norton	www.raf.mod.uk/rafbrizenorton/
RAF Coningsby	www.raf.mod.uk/rafconingsby/
RAF Cosford	www.raf.mod.uk/rafcosford/
RAF Linton-on-Ouse	www.raf.mod.uk/raflintononouse/
RAF Lossiemouth	www.raf.mod.uk/raflossiemouth/
RAF Marham	www.raf.mod.uk/rafmarham/
RAF Northolt	www.raf.mod.uk/rafnortholt/
RAF Odiham	www.raf.mod.uk/rafodiham/
RAF Shawbury	www.raf.mod.uk/rafshawbury/
RAF Valley	www.raf.mod.uk/rafvalley/
RAF Waddington	www.raf.mod.uk/rafwaddington/
RAF Wittering	www.raf.mod.uk/rafwittering/
Red Arrows	www.raf.mod.uk/reds/
Royal Air Force	www.raf.mod.uk/
Royal Air Force Reserves	www.raf.mod.uk/rafreserves/
University of London Air Sqn	www.ulas.org.uk/
Yorkshire UAS	www.raf.mod.uk/yorkshireuas/
MILITARY SITES-US	
Air Combat Command	www.acc.af.mil/
Air Force Reserve Command	www.afrc.af.mil/
Air National Guard	www.ang.af.mil/
Aviano Air Base	www.aviano.af.mil/
Liberty Wing Home Page (48th FW)	www.lakenheath.af.mil/
NASA	www.nasa.gov/
Mildenhall	www.mildenhall.af.mil/
Ramstein Air Base	www.ramstein.af.mil/
Spangdahlem Air Base	www.spangdahlem.af.mil/
USAF	www.af.mil/
USAF Europe	www.usafe.af.mil/
USAF World Wide Web Sites	www.af.mil/publicwebsites/index.asp

Name of Site	Web Address All prefixed 'http://')
US Army	www.army.mil/
US Marine Corps	www.marines.mil/
US Navy	www.navy.mil/
US Navy Patrol Squadrons (unofficial)	www.vpnavy.com/

MILITARY SITES-ELSEWHERE

Armée de l'Air	www.defense.gouv.fr/air/
Aeronautica Militare	www.aeronautica.difesa.it
Austrian Armed Forces (in German)	www.bmlv.gv.at/
Belgian Air Component	www.mil.be/nl/luchtcomponent/
Finnish Defence Force	www.puolustusvoimat.fi/en/
Forca Aerea Portuguesa	www.emfa.pt/
Frecce Tricolori	www.aeronautica.difesa.it/PAN/PAN_ENG/Pagine/default.aspx
German Marine	www.deutschemarine.de/
Greek Air Force	www.haf.gr/en/
Irish Air Corps	www.military.ie/air-corps
Israeli Defence Force/Air Force	www.idf.il/english/
Luftforsvaret	www.mil.no/
Luftwaffe	www.luftwaffe.de/
NATO	www.nato.int/
Royal Australian Air Force	www.airforce.gov.au/
Royal Canadian Air Force	www.airforce.forces.gc.ca/
Royal Danish Air Force (in Danish)	forsvaret.dk/
Royal Netherlands AF	www.defensie.nl/luchtmacht
Royal New Zealand AF	www.airforce.mil.nz/
Singapore Air Force	www.mindef.gov.sg/rsaf/
South African AF Site (unofficial)	www.saairforce.co.za/
Swedish Air Force	www.forsvarsmakten.se/sv/Forband-och-formagor/Flygvapnet/
Turkish Air Force	www.hvkk.tsk.tr/EN/Index.aspx

AIRCRAFT & AERO ENGINE MANUFACTURERS

AgustaWestland	www.agustawestland.com/
Airbus Defence & Space	militaryaircraft-airbusds.com/
Airbus Group	www.airbusgroup.com/
BAE Systems	www.baesystems.com/
Beechcraft	www.beechcraft.com/
Bell Helicopter Textron	www.bellhelicopter.textron.com/
Boeing	www.boeing.com/
Bombardier	www.bombardier.com/
Britten-Norman	www.britten-norman.com/
CFM International	www.cfm56.com/
Dassault	www.dassault-aviation.com/
Embraer	www.embraer.com/
General Electric	www.ge.com/
Gulfstream Aerospace	www.gulfstream.com/
Kaman Aerospace	www.kaman.com/
Lockheed Martin	www.lockheedmartin.com/
Rolls-Royce	www.rolls-royce.com/
Sikorsky	www.sikorsky.com/

UK AVIATION MUSEUMS

Boscombe Down Aviation Collection, Old Sarum	www.boscombedownaviationcollection.co.uk/
Bournemouth Aviation Museum	www.aviation-museum.co.uk/
Brooklands Museum	www.brooklandsmuseum.com/
City of Norwich Aviation Museum	www.cnam.co.uk/
de Havilland Aircraft Heritage Centre	www.dehavillandmuseum.co.uk/
Dumfries & Galloway Aviation Museum	www.dumfriesaviationmuseum.com/
Fleet Air Arm Museum	www.fleetairarm.com/
Gatwick Aviation Museum	www.gatwick-aviation-museum.co.uk/
Imperial War Museum, Duxford	www.iwm.org.uk/duxford/
Imperial War Museum, Duxford (unofficial)	www.axtd59.dsl.pipex.com/
The Jet Age Museum	www.jetagemuseum.btck.co.uk/

Name of Site	Web Address All prefixed 'http://')
Lincs Aviation Heritage Centre	www.lincsaviation.co.uk/
Midland Air Museum	www.midlandairmuseum.co.uk/
Museum of Army Flying	www.flying-museum.org.uk/
Museum of Berkshire Aviation	www.museumofberkshireaviation.co.uk/
Museum of Science & Industry, Manchester	www.mosi.org.uk/
National Museum of Flight, East Fortune	www.nms.ac.uk/our_museums/museum_of_flight.aspx
Newark Air Museum	www.newarkairmuseum.org/
North East Aircraft Museum	www.nelsam.org.uk/NEAM/NEAM.htm
RAF Museum, Cosford & Hendon	www.rafmuseum.org.uk/
Science Museum, South Kensington	www.sciencemuseum.org.uk/
South Yorkshire Aircraft Museum	www.southyorkshireaircraftmuseum.org.uk/
Yorkshire Air Museum, Elvington	www.yorkshireairmuseum.org/

AVIATION SOCIETIES

Air Britain	www.air-britain.com/
British Aircraft Preservation Council	www.bapc.org.uk/
Cleveland Aviation Society	homepage.ntlworld.com/phillip.charlton/cashome.html
LAAS International	www.laasdata.com/
Royal Aeronautical Society	aerosociety.com/
Scottish Air News	www.scottishairnews.co.uk/
Scramble (Dutch Aviation Society)	www.scramble.nl/
Solent Aviation Society	www.solent-aviation-society.co.uk/
Spitfire Society	www.spitfiresociety.com/
The Aviation Society Manchester	www.tasmanchester.com/
Ulster Aviation Society	www.ulsteraviationsociety.org/
Wolverhampton Aviation Group	www.wolverhamptonaviationgroup.co.uk/

OPERATORS OF HISTORIC AIRCRAFT

The Aircraft Restoration Company	www.arc-duxford.co.uk/
Battle of Britain Memorial Flight	www.raf.mod.uk/bbmf/
The Catalina Society	www.catalina.org.uk/
Hangar 11 Collection	www.hangar11.co.uk/
Horizon Aircraft Services	www.horizonaircraft.co.uk/
Old Flying Machine Company	www.ofmc.co.uk/
Royal Navy Historic Flight	www.royalnavyhistoricflight.org.uk/
The Fighter Collection	www.fighter-collection.com/
The Real Aeroplane Company	www.realaero.com/
The Shuttleworth Collection	www.shuttleworth.org/
The Vulcan to the Sky Trust	www.tvoc.co.uk/

SITES RELATING TO SPECIFIC TYPES OF MILITARY AIRCRAFT

The 655 Maintenance & Preservation Society	www.xm655.com/
B-24 Liberator	www.b24bestweb.com/
C-130 Hercules	www.c-130hercules.net/
EE Canberra	www.bywat.co.uk/
English Electric Lightning - Vertical Reality	www.aviation-picture-hangar.co.uk/Lightning.html
The Eurofighter site	www.eurofighter.com/
The ex FRADU Canberra Site	www.fradu-canberras.co.uk/
The ex FRADU Hunter Site	www.fradu-hunters.co.uk/
F-4 Phantom II Society	www.f4phantom.com/
F-15E.info: Strike Eagle	www.f-15e.info/
F-16: The Complete Reference	www.f-16.net/
The Gripen	www.gripen.com/
Jet Provost Heaven	www.jetprovosts.com
K5083 - Home Page (Hawker Hurricane)	www.k5083.mistral.co.uk/
Lockheed SR-71 Blackbird	www.wvi.com/~lelandh/sr-71~1.htm
The MiG-21 Page	www.topedge.com/panels/aircraft/sites/kraft/mig.htm
P-3 Orion Research Group	www.p3orion.nl/
Thunder & Lightnings (Postwar British Aircraft)	www.thunder-and-lightnings.co.uk/
UK Apache Resource Centre	www.ukapache.com/

Name of Site	Web Address All prefixed 'http://')
MISCELLANEOUS	
Aerodata Software Ltd.	www.aerodata.org/
AeroResource	www.aeroresource.co.uk/
The AirNet Web Site	www.aviation-links.co.uk/
Delta Reflex	www.deltareflex.com/forum
Demobbed - Out of Service British Military Aircraft	demobbed.org.uk/
Euro Demobbed	www.eurodemobbed.org.uk/
Fighter Control	fightercontrol.co.uk/
Iconic Aircraft'	www.iconicaircraft.co.uk/
Joseph F. Baugher's US Military Serials Site	www.joebaugher.com/
The 'Military Aircraft Markings' Web Site	www.militaryaircraftmarkings.co.uk/
Military Aviation	www.crakehal.demon.co.uk/aviation/aviation.htm
Military Aviation Review/MAP	www.mar.co.uk/
Pacific Aviation Database Organisation	www.gfiapac.com/
PlaneBaseNG	www.planebase.biz/
Target Lock Military Aviation E-zine	www.targetlock.org.uk/
Thunder & Lightnings: Airfield Viewing Guides	www.thunder-and-lightnings.co.uk/spotting/
UK Airshow Review	www.airshows.co.uk/
UK Military Aircraft Serials Resource Centre	www.ukserials.com/

ZJ801 is a Typhoon T3, coded BJ and wearing the markings of 29(R) Squadron based at RAF Coningsby. Still being built and delivered to the RAF, this Typhoon is already more than ten years old.

Serial	Type	Operator

Serial	Type	Operator